Luminescent Silicon Nanostructures

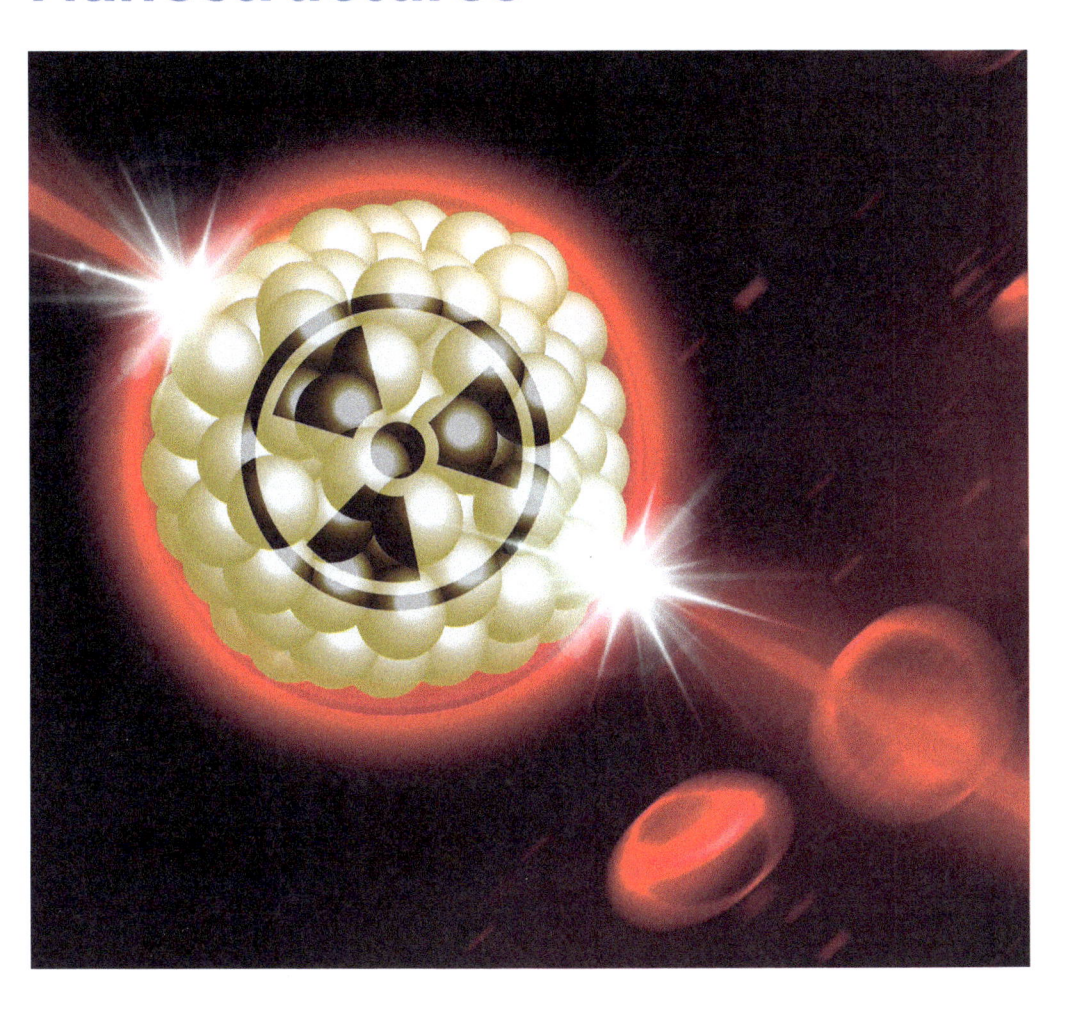

FARADAY DISCUSSIONS
Volume 222, 2020

ROYAL SOCIETY OF CHEMISTRY

The **Faraday Division** of the Royal Society of Chemistry, previously the Faraday Society, was founded in 1903 to promote the study of sciences lying between chemistry, physics and biology.

Editorial Staff

Executive Editor
Anna Simpson

Deputy Editor
Heather Montgomery

Development Editor
Kathryn Gempf

Editorial Production Manager
Gisela Scott

Senior Publishing Editor
William Bergius

Publishing Editors
Emma Lockyer, Ellis Wilde

Editorial Assistant
Daphne Houston

Publishing Assistants
David Bishop, Huw Hedges

Publisher
Jeanne Andres

Faraday Discussions (Print ISSN 1359-6640, Electronic ISSN 1364-5498) is published 8 times a year by the Royal Society of Chemistry, Thomas Graham House, Science Park, Milton Road, Cambridge, UK CB4 0WF.

Volume 222 ISBN 978-1-78801-908-8

2020 annual subscription price: print+electronic £1220 US $2148; electronic only £1162, US $2046.
Customers in Canada will be subject to a surcharge to cover GST. Customers in the EU subscribing to the electronic version only will be charged VAT.

All orders, with cheques made payable to the Royal Society of Chemistry, should be sent to the Royal Society of Chemistry Order Department, Royal Society of Chemistry, Thomas Graham House, Science Park, Milton Road, Cambridge, CB4 0WF, UK
Tel +44 (0)1223 432398; E-mail orders@rsc.org

If you take an institutional subscription to any Royal Society of Chemistry journal you are entitled to free, site-wide web access to that journal. You can arrange access via Internet Protocol (IP) address at www.rsc.org/ip

Customers should make payments by cheque in sterling payable on a UK clearing bank or in US dollars payable on a US clearing bank.

Whilst this material has been produced with all due care, the Royal Society of Chemistry cannot be held responsible or liable for its accuracy and completeness, nor for any consequences arising from any errors or the use of the information contained in this publication. The publication of advertisements does not constitute any endorsement by the Royal Society of Chemistry or Authors of any products advertised. The views and opinions advanced by contributors do not necessarily reflect those of the Royal Society of Chemistry which shall not be liable for any resulting loss or damage arising as a result of reliance upon this material. The Royal Society of Chemistry is a charity, registered in England and Wales, Number 207890, and a company incorporated in England by Royal Charter (Registered No. RC000524), registered office: Burlington House, Piccadilly, London W1J 0BA, UK, Telephone: +44 (0) 207 4378 6556.

Printed in the UK

Faraday Discussions

Faraday Discussions are unique international discussion meetings that focus on rapidly developing areas of chemistry and its interfaces with other scientific disciplines.

Scientific Committee volume 222

Chair
Paola Ceroni, University of Bologna, Italy
Brian Korgel, The University of Texas at Austin, USA
Katerina Dohnalova, University of Amsterdam, Netherlands
Lorenzo Pavesi, University of Trento, Italy
Michael Sailor, University of California, USA
Jonathan Veinot, University of Alberta, Canada

Faraday Standing Committee on Conferences

Chair
John M. Seddon, Imperial College London, UK

Secretary
Susan Weatherby, Royal Society of Chemistry, UK
Graeme M. Day, University of Southampton, UK
David Lennon, University of Glasgow, UK

Angelos Michaelides, University College London, UK
Jenny Nelson, Imperial College London, UK
Susan Perkin, University of Oxford
Ivana Radosavljević Evans, Durham University, UK
Paul Raithby, University of Bath, UK
Pat Unwin, University of Warwick, UK
Claire Vallance, University of Oxford, UK

Advisory Board

Vic Arcus, The University of Waikato, New Zealand
Timothy Easun, Cardiff University, UK
Dirk Guldi, University of Erlangen-Nuremberg, Germany
Luiz Liz-Marzán, CIC biomaGUNE, Spain
Marina Kuimova, Imperial College London, UK
Andrew Mount, University of Edinburgh, UK

Frank Neese, Max Planck Institute for Chemical Energy Conversion, Germany
Michel Orrit, Leiden University, The Netherlands
Zhong-Qun Tian, Xiamen University, China
Siva Umapathy, Indian Institute of Science, Bangalore, India
Bert Weckhuysen, Utrecht University, The Netherlands
Julia Weinstein, University of Sheffield, UK
Sihai Yang, University of Manchester, UK

Information for Authors

This journal is © the Royal Society of Chemistry 2020 Apart from fair dealing for the purposes of research or private study for non-commercial purposes, or criticism or review, as permitted under the Copyright, Designs and Patents Act 1988 and the Copyright and Related Rights Regulation 2003, this publication may only be reproduced, stored or transmitted, in any form or by any means, with the prior permission in writing of the Publishers or in the case of reprographic reproduction in accordance with the terms of licences issued by the Copyright Licensing Agency in the UK. US copyright law is applicable to users in the USA.

∞ The paper used in this publication meets the requirements of ANSI/NISO Z39.48–1992
(Permanence of Paper).

Registered charity number: 207890

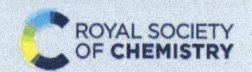

Luminescent Silicon Nanostructures

Faraday Discussions
www.rsc.org/faraday_d

A General Discussion was held on Luminescent Silicon Nanostructures.

The Royal Society of Chemistry is the world's leading chemistry community. Through our high impact journals and publications we connect the world with the chemical sciences and invest the profits back into the chemistry community.

CONTENTS

ISSN 1359-6640; ISBN 978-1-78801-908-8

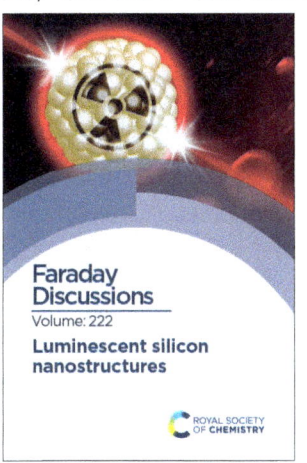

Cover
See Stephan et al., Faraday Discuss., 2020, **222**, 362–383.

Dual-labelled silicon nanoparticles allow the tracking of in vivo distribution.

Image reproduced by permission of Dr Holger Stephan from Faraday Discuss., 2020, **222**, 362.

EDITORIAL

Luminescent silicon nanostructures and COVID-19
Susan Weatherby, John M. Seddon and Paola Ceroni

INTRODUCTORY LECTURE

Introductory lecture: origins and applications of efficient visible photoluminescence from silicon-based nanostructures
Leigh Canham

PAPERS AND DISCUSSIONS

Fast room-temperature functionalization of silicon nanoparticles using alkyl silanols
Alyssa F. J. van den Boom, Sidharam P. Pujari, Fatma Bannani, Hafedh Driss and Han Zuilhof

CORRECTION

95 **Fast room-temperature functionalization of silicon nanoparticles using alkyl silanols**
Alyssa F. J. van den Boom, Sidharam P. Pujari, Fatma Bannani, Hafedh Driss and Han Zuilhof

PAPERS AND DISCUSSIONS

96 **Photophysical properties of ball milled silicon nanostructures**
Ankit Goyal, Menno Demmenie, Chia-Ching Huang, Peter Schall and Katerina Dohnalova

108 **Amine functionalised silicon nanocrystals with bright red and long-lived emission**
Giacomo Morselli, Francesco Romano and Paola Ceroni

122 **Dual-emission fluorescent silicon nanoparticle-based nanothermometer for ratiometric detection of intracellular temperature in living cells**
Jinhua Wang, Airui Jiang, Jingyang Wang, Bin Song and Yao He

135 **The shell matters: one step synthesis of core–shell silicon nanoparticles with room temperature ultranarrow emission linewidth**
Anna Fucikova, Ilya Sychugov and Jan Linnros

149 **Critical assessment of wet-chemical oxidation synthesis of silicon quantum dots**
Jonathan L. Wilbrink, Chia-Ching Huang, Katerina Dohnalova and Jos M. J. Paulusse

166 **Synthesis and functionalisation of silicon nanostructures: general discussion**

176 **The influence of hydrofluoric acid etching processes on the photocatalytic hydrogen evolution reaction using mesoporous silicon nanoparticles**
Sarah A. Martell, Ulrike Werner-Zwanziger and Mita Dasog

190 **Low temperature radical initiated hydrosilylation of silicon quantum dots**
Timothy T. Koh, Tingting Huang, Joseph Schwan, Pan Xia, Sean T. Roberts, Lorenzo Mangolini and Ming L. Tang

201 **Modulating donor–acceptor transition energies in phosphorus–boron co-doped silicon nanocrystals via X- and L-type ligands**
Gregory F. Pach, Gerard M. Carroll, Hanyu Zhang and Nathan R. Neale

217 ***Ab initio* studies of the optoelectronic structure of undoped and doped silicon nanocrystals and nanowires: the role of size, passivation, symmetry and phase**
Stefano Ossicini, Ivan Marri, Michele Amato, Maurizia Palummo, Enric Canadell and Riccardo Rurali

240 **The red and blue luminescence in silicon nanocrystals with an oxidized, nitrogen-containing shell**
Pavel Galář, Tomáš Popelář, Josef Khun, Irena Matulková, Ivan Němec, Kateřina Dohnalova Newell, Alena Michalcová, Vladimír Scholtz and Kateřina Kůsová

258 **Tight-binding calculations of the optical properties of Si nanocrystals in a SiO_2 matrix**
Mikhail O. Nestoklon, Ivan D. Avdeev, Alexey V. Belolipetskiy, Ilya Sychugov, Federico Pevere, Jan Linnros and Irina N. Yassievich

Power-dependent photoluminescence decay kinetics of silicon nanocrystals under continuous and pulsed excitation
Michael Greben and Jan Valenta

Optical and electronic properties: from theory to experiments: general discussion

Luminescent silicon nanoparticles for distinctive tracking of cellular targeting and trafficking
Gi-Heon Kim, Goun Lee, Myoung-Hee Kang, Minjong Kim, Yusung Jin, Sungjun Beck, Jihyun Cheon, Junyeong Sung and Jinmyoung Joo

The effects of drying technique and surface pre-treatment on the cytotoxicity and dissolution rate of luminescent porous silicon quantum dots in model fluids and living cells
Maxim B. Gongalsky, Uliana A. Tsurikova, Catherine J. Storey, Yana V. Evstratova, Andrew A. Kudryavtsev, Leigh T. Canham and Liubov A. Osminkina

Synthesis and characterisation of isothiocyanate functionalised silicon nanoparticles and their uptake in cultured colonic cells
Yimin Chao, Ashley I. Marsh, Mehrnaz Behray, Feng Guan, Anders Engdahl, Yueyang Chao, Qi Wang and Yongping Bao

Shedding light on the aqueous synthesis of silicon nanoparticles by reduction of silanes with citrates
John L. Z. Ddungu, Simone Silvestrini, Alessandra Tassoni and Luisa De Cola

Ultrasmall silicon nanoparticles as a promising platform for multimodal imaging
Garima Singh, John L. Z. Ddungu, Nadia Licciardello, Ralf Bergmann, Luisa De Cola and Holger Stephan

Silicon nanostructures for sensing and bioimaging: general discussion

Bridging energy bands to the crystalline and amorphous states of Si QDs
Bruno Alessi, Manuel Macias-Montero, Chiranjeevi Maddi, Paul Maguire, Vladimir Svrcek and Davide Mariotti

Silicon photosensitisation using molecular layers
Lefteris Danos, Nathan R. Halcovitch, Ben Wood, Henry Banks, Michael P. Coogan, Nicholas Alderman, Liping Fang, Branislav Dzurnak and Tom Markvart

The next big thing for silicon nanostructures – CO_2 photocatalysis
Wei Sun, Xiaoliang Yan, Chenxi Qian, Paul N. Duchesne, Sai Govind Hari Kumar and Geoffrey A. Ozin

Silicon nanostructures for energy conversion and devices: general discussion

Faraday Discussions

EDITORIAL

Luminescent silicon nanostructures and COVID-19

Susan Weatherby, [*a] John M. Seddon [b] and Paola Ceroni [c]

DOI: 10.1039/d0fd90009e

This Faraday Discussion volume is unique in the hundred plus year history of the Faraday Discussion series, being produced at a time of unprecedented circumstances worldwide and without the preceding Faraday Discussion conference having taken place.

The conference, due to take place on 12–14 February 2020 in York, UK, was cancelled at the beginning of February due to health and travel uncertainties created by the news of the emerging COVID-19 virus. At the beginning of 2020 the world first began to hear about the COVID-19 virus, and by early February 2020 we were learning more about the dangers of the virus and what this would mean for a conference such as the Faraday Discussion. In early February, given that the conference was due to take place imminently, it was necessary to take fast and clear action to remove any uncertainty regarding delegates' travel plans. Cancelling the conference was a very difficult decision, however, the wellbeing, health and safety of the conference delegates were the primary concern, and cancelling the conference was the only realistic decision that could have been made at the time.

Cancelling the conference left us with a big question over the accompanying and subsequent Faraday Discussion volume. The Faraday Discussion papers had already been submitted and published online, but of course, the unique aspect of the published Discussion volume is a record of the questions and comments made at the Discussion conference and which are published alongside the papers. How would we go about recreating this when the Discussion conference did not take place? The answer came quickly – we were very fortunate to receive the good will, support and commitment of the scientific committee, authors and researchers of this community who came forward and submitted questions and comments in writing. The Discussion comments you see in the forthcoming pages are an important record of the questions the community had about the

[a] Royal Society of Chemistry, Thomas Graham House, Science Park, Milton Road, Cambridge CB4 0WF, UK. E-mail: weatherbys@rsc.org
[b] Chemistry Department, Molecular Sciences Research Hub, Imperial College London, Wood Lane, London W12 0BZ, UK
[c] Department of Chemistry "Giacomo Ciamician", University of Bologna, via Selmi 2, Bologna, Italy

papers and topic in 2020. Uniquely for a Faraday Discussion these questions were asked and collected in writing rather than being asked in person at a conference.

The Discussion volume sections focus on:

• the synthesis and functionalisation of silicon nanocrystals and porous silicon and how the synthetic procedure impacts on the optical and electronic properties of the material;

• the luminescence and electronic properties of silicon nanostructures as a function of dimension, size, crystallinity and their surface chemistry, from both theoretical and experimental perspectives;

• the use of porous silicon and silicon nanocrystals for sensing and bioimaging applications;

• the application of silicon nanostructures for energy conversion devices and photocatalysis.

Another unique aspect of the Discussion conference would have been the presentation of the Spiers Memorial Award and it is appropriate for us to mention this here. Up to two Spiers Awards are made each year, in recognition of an individual who has made an outstanding contribution to the field of a Faraday Discussion. The Award commemorates Frederick S. Spiers, who is best known for his work as Secretary of the Faraday Society, which he helped to found in 1902. The Introductory Lecturer of this Discussion, Professor Leigh Canham, was selected to receive the Spiers Memorial Award for his pioneering work in silicon quantum dots and contributions to practical applications of silicon nanostructures in the electronics, photonics and biomedical fields. Professor Canham is Professor of Nanomaterials for Biomedicine and Photonics at the University of Birmingham, and we congratulate him on his Award.

It remains for us to say that in these very difficult and unprecedented times we are very grateful for the commitment of the scientific committee and the community of researchers who have made such a positive contribution to this Faraday Discussion. We are hopeful that this Discussion volume reflects their time and commitment, and importantly that it will become a valuable record for the scientific community going forward into the future.

Susan Weatherby, Programme Manager, Royal Society of Chemistry

John M. Seddon, Chair, Faraday Standing Committee on Conferences and Imperial College London

Paola Ceroni, Chair, Faraday Discussion Scientific Committee and University of Bologna

Faraday Discussions

PAPER

Introductory lecture: origins and applications of efficient visible photoluminescence from silicon-based nanostructures

Leigh Canham

Received 31st January 2020, Accepted 20th April 2020
DOI: 10.1039/d0fd00018c

A variety of silicon-based nanostructures with dimensions in the 1–5 nm range now emit tunable photoluminescence (PL) spanning the visible range. Achievement of high photoluminescence quantum efficiency (PLQY) relies critically on their surface chemistry passivation and an impressive "tool box" of options have been developed. Two distinct PL bands are dominant. The "S-Band" (red–green emission with *Slow* microsecond decay) has PLQY that has steadily improved from ∼3% in 1990 to 65 ± 5% by 2017. The "F-Band" (blue–yellow with *Fast* nanosecond decay) has reported PLQY values that have improved from ∼0.1% in 1994 to as high as ∼90% by 2016. The vast literature on both bands is surveyed and for the S-band, size-structure-PL correlations and selective photo-excitation studies are highlighted. Resonant photoexcitation and single quantum dot studies have revealed the key role of quantum confinement and the excitonic phonon-assisted nature of the radiative transitions. For the F-band, in contrast, specific phenomenological studies are highlighted that demonstrate similar emission without the presence of silicon nanostructures. Low PLQY F-band emission from pure silicon–silica core shell systems is probably associated with oxide-related defects, but ultrahigh PLQY from many lower temperature synthesis routes is likely to be from carbon-based nanostructures or chromophores, not silicon nanostructures. Potential applications for both PL bands include sensing, medical imaging, theranostics, photovoltaics, LED colour converters and nano-thermometry. Emerging "green" synthesis routes are mentioned. If scalability and cost are significantly improved then a number of other proposed uses of ultra-efficient PL from "nano-Si" could become viable in cosmetics, catalysis, security and forensics.

1. Introduction

1.1. Scope of review

Nanostructuring has introduced an exciting new capability to semiconducting silicon: the ability to emit visible light very efficiently under photoexcitation

School of Physics and Astronomy, University of Birmingham, Edgbaston, Birmingham, B15 2TT, UK. E-mail: l.t.canham@bham.ac.uk

(photoluminescence (PL)). This property was quickly recognized in the early 1990s as a route to extend the functionality of an already powerful silicon technology. The topic has thus received intense study over the last 30 years, with now thousands of publications. This review focuses on the visible photoluminescence of silicon nanostructures; collating both experimental and theoretical work on most types of nanostructure; providing quantitative performance metrics, particularly on photoluminescence quantum yield (PLQY); photoluminescence wavelength tunability (PLWT) and photoluminescence decay times (PLDT).

There are two dominant PL emissions in the visible (see Fig. 1): the "Slow" (S) band in the red–yellow spectral range with long microsecond decay times (see Section 2) and the "Fast" (F) band in the blue–green with faster nanosecond decay times (see Section 3). In a given nanostructure, one or the other, or both can be present. Often, as shown in Fig. 1, the intensity of the blue emission will often increase with storage time. The origins of such behaviour have received intense study.

This review highlights many spectroscopy-based studies and selected phenomenological studies that provide insight into their likely PL mechanisms. The review also covers six application areas of efficient visible PL: sensing, photovoltaics, medical imaging, theranostics, white LEDs and nanothermometry. It also proposes a number of ways in which the field might further progress, and other application areas that might emerge if cost and scalability issues are surmounted. There is no detailed discussion of

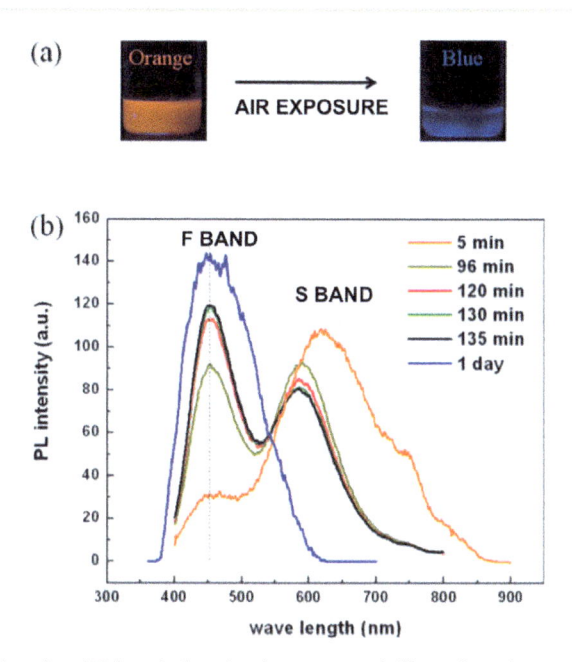

Fig. 1 The S-band and F-band photoluminescence of silicon-based nanostructures.[1] (a) Photographs of nanocrystal suspensions under UV illumination. Freshly etched nanocrystals: orange emission. After one day of air exposure: blue emission. (b) Evolution of S-band and F-band spectra with varying air exposure of the times indicated. Adapted with permission from A. Gupta and H. Wiggers, *Nanotechnology*, 2011, **22**, 055707. Copyright IOP Publishing, 2011. All rights reserved.

nanostructure synthesis techniques and surface chemistry manipulation, as these have already been the focus of many prior reviews that are referenced.

Data on electrical transport, electroluminescence, optical gain, and thereby optoelectronics, are also omitted in order to entirely focus here on the potential of photoluminescent properties. Such a platform property is however often combined with other crucial ones for specific uses, such as biodegradability and biocompatibility for medical applications. These additional properties are covered briefly in the relevant application sections. In summary, this review collates and discusses some of the very large amount of data now available on many types of photoluminescent silicon nanostructures and many applications thereof. It is hoped to be of use both as a broad introduction to the vast literature of this maturing field, but should also stimulate specific areas of discussion and perhaps further research in highlighted topics.

1.2. Electron–hole recombination processes in bulk silicon

When a semiconductor such as silicon is stimulated by light (photoexcitation) it may return to equilibrium by a variety of processes.[2] Some involve the re-emission of light (photoluminescence or fluorescence – these terms are used interchangeably in the literature on luminescent silicon nanostructures) and are so called *radiative* transitions. Others redistribute the absorbed energy amongst vibrational modes of the lattice or electronic excitations and ultimately generate heat – so-called *non-radiative* transitions.

Radiative processes in silicon include recombination from electrons and holes[3] free excitons,[4] biexcitons or excitonic molecules,[5-7] polyexcitons,[8] electron–hole liquids,[9,10] bound excitons[11] and donor–acceptor pair luminescence.[12-14] Free excitons will diffuse through the silicon lattice[15] and can be trapped and emit light at isolated dopants,[16] clusters of dopants,[17] isoelectronic traps[18] and sometimes near extended defects[19-22] such as dislocations and precipitates. Such trapping can occur in the bulk and near surfaces. The majority of these radiative excitonic processes however are only observed at very low cryogenic temperatures, due to the small binding energy of the exciton. Not only that, but their efficiency is also normally low. The efficiency of a radiative process generally depends on the electronic bandstructure of the solid and for crystalline semiconductors the perfection of its lattice and surfaces, since most defects and surfaces promote non-radiative processes. Non-radiative processes[23] in silicon are dominated by Auger recombination[24] and recombination at "deep level" defects such as transition metal impurities in the bulk[25] or defects at surfaces and interfaces such as that created by ambient or thermal oxidation of silicon.[26]

1.3. Photoluminescence of crystalline bulk silicon

It is instructive to first survey the PL of bulk Si because this reveals certain "spectroscopic fingerprints" of radiative excitonic recombination processes in this crystalline material. The fundamental indirect energy gap in crystalline bulk silicon, between the top of the valence band and lowest conduction band is 1.170 eV at 0 K, decreasing monotonically to 1.125 eV at 300 K.[27] Interband near infrared photoluminescence from silicon was first observed by Haynes and co-workers in the 1950s.[3] At room temperature its near bandgap emission consists of a single weak band at ∼1.09 eV (1134 nm) with a full width at half maximum

(FWHM) that is about 100 meV.[3,28,29] At cryogenic temperatures recombination becomes excitonic. The exciton in bulk silicon has a Bohr radius of 4.2 nm and a binding energy of only 14 meV.[30] Fig. 2 shows how the spectra broaden considerably above room temperature and sharpen considerably at cryogenic temperatures, with a spectral position that follows the temperature dependence of the silicon bandgap.

At low temperatures (<40 K) and in pure silicon, free exciton (FE) recombination totally dominates over bound exciton (BE) recombination (Fig. 3). Transition linewidths are now narrow enough to resolve fine structures corresponding to the TA and TO phonons as shown in Fig. 2 and 3. Two phonon replicas are very weak. Note the extremely weak "zero-phonon" peak I^0 near 1.16 eV in Fig. 3 where electron and hole recombine without a momentum conserving phonon. The spacings of the features I^{TA} and I^{TO} from I^0 in the spectrum are determined by the corresponding phonon energies. Currently accepted values for bulk silicon are: TA phonon = 18.4 ± 0.2 meV; LO phonon 56.2 ± 1 meV; TO phonon = 58.0 ± 1 meV. The relative ratios of the phonon-assisted peaks are temperature dependent but the TO mode is dominant with approximate values in Fig. 3 being 1 [TO] : 0.1 [LO] : 0.03 [TA].[31]

Concerning defect-related emission let us first consider the simplest point defects: substitutional donors, like P, As and Sb; interstitial donors like Li; and substitutional acceptors like B, Al, Ga and In. Free excitons migrating through the lattice can bind to the neutral donor or acceptor and binding energies lie in the range 3–50 meV.[31] When the exciton is localised on the impurity, there is a relaxation of the wave vector selection rule so the zero phonon line is thus much stronger for bound excitons (e.g. In^0 in Fig. 4) than for free excitons (e.g. I^0 in Fig. 3).

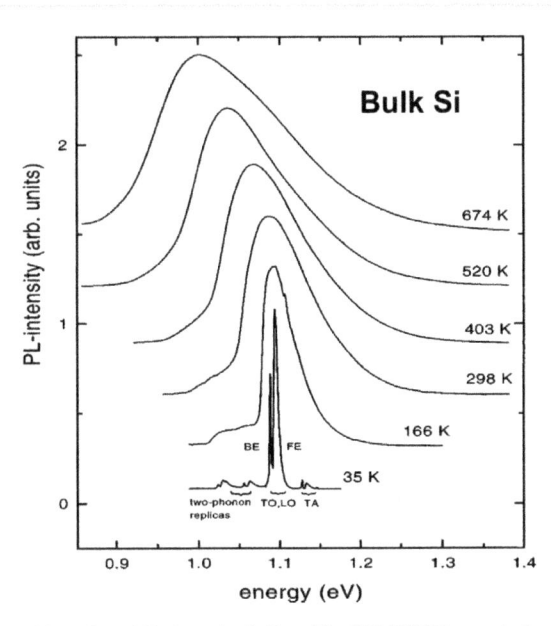

Fig. 2 Normalized interband PL from bulk Si at 35–674 K.[29] BE stands for bound exciton, FE for free exciton (see text). Reprinted with permission from V. Alex, S. Finkbeiner and J. Weber, *J. Appl. Phys.*, 1996, **79**, 6943. Copyright AIP Publishing.

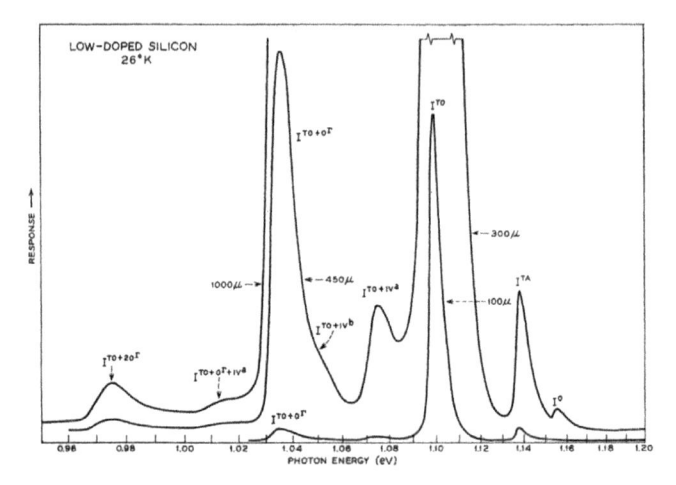

Fig. 3 Intrinsic free exciton PL from high purity silicon at 26 K.[16] The LO and TO phonon components are not resolved. Reprinted with permission from P. J. Dean, J. R. Haynes and W. F. Flood, *Phys. Rev.*, **161**, 711–729. Copyright the American Physical Society, 1967.

Rapid progress in identifying many radiative defects began in the late 1970s when Ge photodetectors became available and were orders of magnitude more sensitive than PbS detectors in the near infrared.[31] The huge increase in resolution enabled isotope shifts in sharp zero phonon line transitions to be studied, thereby unambiguously identifying which impurities were involved in specific vibronic band emission. When combined with uniaxial stress measurements to elucidate the defect symmetry, and correlations with other spectroscopic techniques, detailed models of radiative point defects emerged.[31]

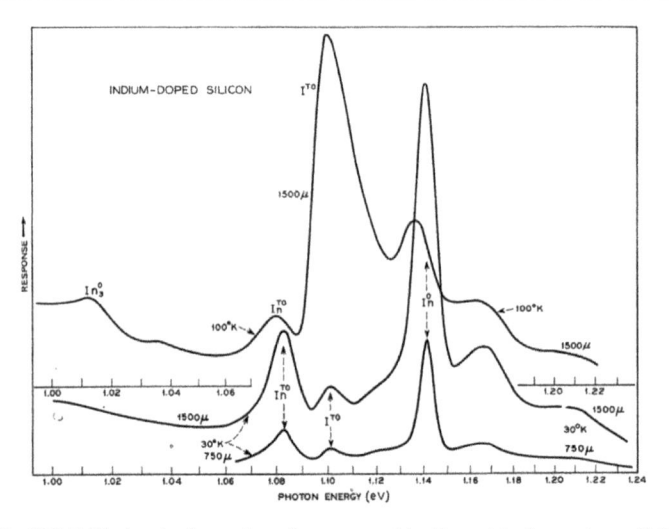

Fig. 4 30–100 K PL due to bound exciton recombination at indium atoms.[16] Reprinted with permission from P. J. Dean, J. R. Haynes and W. F. Flood, *Phys. Rev.*, **161**, 711–729, 1967. Copyright the American Physical Society, 1967.

Unfortunately, some luminescent bulk silicon crystals also display broad featureless bands at sub-bandgap wavelengths[19,21,32] providing a challenge for identification that researchers of today's silicon nanostructures are more than well aware. Fig. 5 shows an example of a broad and featureless band at 0.94 eV attributed to extended defects (hydrogen platelets) in bulk silicon.[21]

1.4. Surface passivation chemistries for bulk silicon

Surface or interface "passivation" refers to the removal of electrically or optically active defects and has become increasingly important in silicon solar cell technology due to the cost-driven reduction of cell thickness and corresponding increase in surface to volume ratio. There has been an associated significant effort in understanding the origin of, and minimizing, such non-radiative recombination. That literature is utilized here to identify some of the most useful passivation chemistries that could be applied to luminescent silicon nanostructures, and indeed many, but not all, already been used. Given that nanostructures have huge surface areas, the importance of minimizing non-radiative surface recombination becomes of paramount importance if high PLQY is to be realized.

Surfaces disrupt the band structure, creating energy states in the bandgap via strained or un-terminated ("dangling") bonds. These states can capture electrons or holes with capture velocities S_{no} and S_{po} that are governed by the product of carrier diffusion, capture cross-section and defect density, D_{it}. A surface will have different steady-state carrier concentrations (n_s, p_s) of electrons and holes to those in the bulk, and different to the intrinsic carrier concentration (n_i).

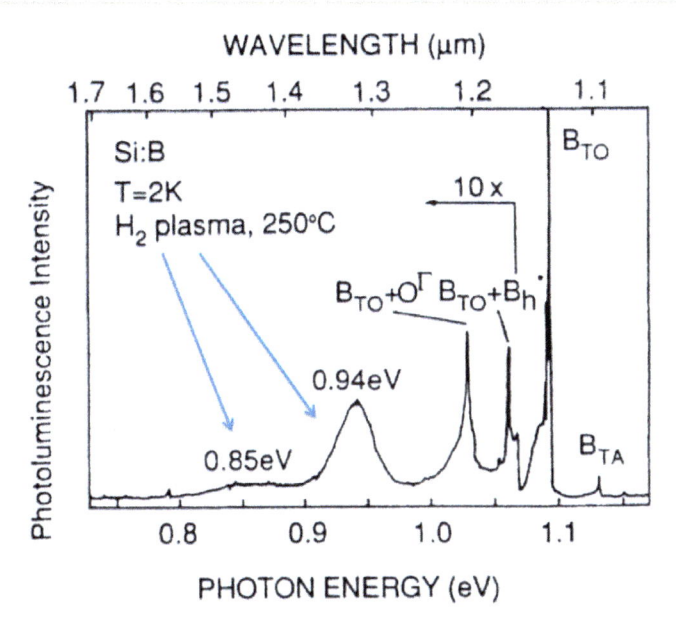

Fig. 5 Low temperature PL from plasma etched bulk silicon with high levels of hydrogen.[21] Reprinted and adapted with permission from H. Weman, B. Monemar, G. S. Oehrlein and S. J. Jeng, *Phys. Rev. B*, **42**, 3109–3112, 1990. Copyright the American Physical Society, 1990.

The total recombination at a surface can be characterized by an effective "surface recombination velocity (SRV)" that *quantitatively* assesses surface or interface passivation:

$$S_{eff} = (1/\Delta n_d) \times (n_s p_s - n_i^2)/((n_s + n_i)/S_{p0} + (p_s + n_i)/S_{n0}) \quad (1)$$

where Δn_d is the excess minority carrier concentration that dominates overall recombination rates.[33] What eqn (1) nicely illustrates are the two complementary techniques that have been investigated for reducing surface recombination in bulk silicon. One is *"chemical"* in the sense of new bonding arrangements of surface dielectric films or organic monolayers and chemical species like hydrogen. These can lower capture rates by having fewer bandgap states (defects) or lowering the capture cross sections of existing defects. They therefore lower the carrier capture velocities S_{p0} and S_{n0}. The other is *"physical"* in the sense of utilizing surface charge and thereby electric fields to lower the concentration of minority carriers at the surface or interface. This can be achieved by having a raised charge density at the surface or within the interfacial layer that acts to reduce the concentration of one type of carrier at the surface, where most defects reside. They therefore lower n_s or p_s or change their ratio. The lowest values of SRV for bulk silicon have been achieved by utilizing *both* techniques.[33] Table 1 shows SRV values for a range of bulk Si surface and interface chemistries.

Some observations from surveying this field and the data in Table 1:

Table 1 SRV values for a range of bulk Si surface and interface chemistries

Si structure	Surface chemistry	Passivation method	SRV (cm s^{-1})	References
Bulk Si (111)	Silicon hydride	*In situ* in HF and other acids	<1 (0.25)	(Yablonovitch 1986)[34]
Bulk Si	Silicon hydride	HF-based cleans	<0.05 → 0.1	(Sun 2018)[35]
Bulk Si (111)	Native oxide	HF and air exposure (24 days)	>1000	(Nemanik 2006)[36]
Bulk Si (100)	Thermal oxide	Annealed silica	<1	(Kerr 2002)[37]
Bulk Si	Thermal oxide	Charged annealed silica	<1 (0.44)	(Collet 2017)[38]
Bulk Si	Silicon nitride	PECVD	<1 (0.67)	(Wan 2013)[39]
Bulk Si	Alumina	Plasma ALD	2–6	(Hoex 2008)[40]
Bulk Si	Amorphous Si	PECVD a-Si/SiO$_x$/SiN$_x$	<1 (0.06)	(Bonilla 2016)[41]
Bulk Si (111)	Alkyl groups	Chloro-alkylation	<25	(Royea 2000)[42]
Bulk Si (111)	Methyl, ethyl, propyl, butyl, phenyl	Chloro-alkylation and 24 days in air	<100 (44–80)	(Nemanik 2006)[36]
Bulk Si	Superacid	HF-dip coating	3–10	(Bullock 2016)[43]
Bulk Si (100)	Superacid	Pre-clean/etches-SA Soln. Coating	0.3–3	(Grant 2017)[44]
Bulk Si (100)	Nafion	Spin coating	1.5	(Chen 2018)[45]

• Hydride passivation is effective but susceptible to native oxide growth which can seriously raise non-radiative rates.[36]

• Organic monolayers with Si–C bonding are also effective, but passivation also deteriorates with extended air exposure.[36]

• SRV values below 1 cm s^{-1} are achievable at bulk Si interfaces with optimized a-Si, thermal oxides and nitride coatings. Alumina, titania and silicon carbide films can also passivate silicon surfaces but SRV values are generally a little higher.

• Interface densities are much lower for atomically smooth interfaces.[46]

• Surface/interface charging has not been explored thoroughly *versus* PLQY of nanostructures.

• Superacids and Nafion can provide very low SRV, at least with bulk Si of moderate surface areas.[43–45] They are yet to be explored with silicon nanostructures.

• SRV values are generally much higher for nanostructured silicon (see Section 5.1).

1.5. Luminescent Si nanostructure diversification

The easily reproduced demonstration of efficient[47] and wavelength-tunable[47,48] visible emission from porous silicon and silicon nanocrystals around 1990 led to a wide range of silicon nanostructures being explored, together with a variety of synthesis techniques. Fig. 6 illustrates many of these schematically. Each class of nanostructure is at least mentioned in this review, but with emphasis that very much reflects their respective levels of study and development, and also their success in achieving high photoluminescence efficiency.

A variety of different terms for luminescent silicon nanostructures are used in the literature, so we will start by mentioning them here. The terms silicon "quantum" wells, wires, and dots are the smallest 1D, 2D and 3D confined crystalline silicon structures, respectively, with diameters below that of the Bohr exciton radius (∼4 nm) in bulk silicon (see Section 1.3). Such structures are expected to have bandstructures that are size-dependent due to quantum confinement effects as discussed in Section 1.6. Silicon "nanoparticles", nanowires, nanopillars or nanorods can often, but not always, have larger diameters in the range 5–100 nm. These can exhibit carrier confinement effects on PL but their bandstructure is similar to that of bulk Si. In the literature both terms are used interchangeably. "Nanoclusters", for example, often denote the tiniest structures

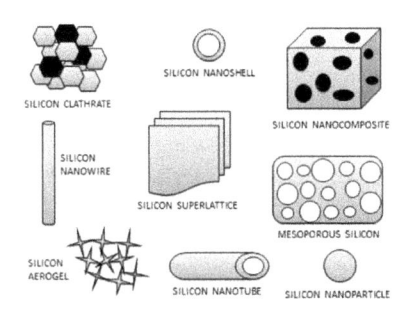

Fig. 6 Example classes of silicon nanostructures.

consisting of under 50 atoms. Likewise, "nanosheets" usually refer to free-standing layers of, at most, a few atoms thickness.

The experimental literature is currently dominated by luminescent silicon nanocrystals, porous silicon, silicon nanowires and silicon superlattices, in that order. The theoretical literature is dominated by calculations of silicon quantum wires and silicon quantum dots. Other silicon nanostructures have received some, but not substantial development. Examples of silicon nanostructures where efficient visible luminescence at 300 K is either not reported or not quantified to date include silicon nanoribbons,[49] nanoshells and nanotubes,[50] nanosheets,[51–54] molecular clusters,[55] allotropes[56] and metastable polymorphs.[57]

1.6. Theory of quantum confinement effects in Si nanostructures

Theoretical studies of how the silicon bandstructure evolves with size and shape in the critical 1–10 nm size range began in the early 1990s and there is now a substantial body of work. I am not aware of a prior review that tries to comprehensively analyze the considerable theoretical literature on silicon nanostructures and will not attempt to here. Numerous publications are provided here and a few concepts and trends are selected where there is a consensus with respect to issues that affect PLQY, PLWT and PLDT of silicon nanostructures.

Starting with the earliest calculations in 1992–1993 (ref. 58–60) and many others, theoretical work has continuously grown over the last 30 years, due to a myriad of potential applications for *nano-Si*, many of which do not utilize photoluminescence. Studies are grouped in Table 2, not by calculation technique, but by nanostructure morphology and surface/interface chemistry. The aim is to help readers access the available theory most relevant to their specific luminescent nanostructures. A large number of theoretical studies have focused on silicon quantum wires, but the literature on quantum dots is growing. Luminescent porous silicon has been modelled as idealized quantum wires and quantum dots or by introducing an artificial periodicity of holes *via* supercell techniques.[109] In reality, morphology is variable but often of mixed 2–3D confinement due to undulating wire thickness (see Section 2.1) combined with other geometrical shape variations.[110]

Surveying the theoretical data of Table 2 there would appear to be a consensus that:

Table 2 Theoretical literature on different classes of model Si nanostructures

Model nanostructure morphology	Surface chemistry	Size range studied	References
Quantum slabs (free-standing)	Si–H$_x$	0.3–6.0 nm	61–64
Quantum wells (embedded)	Si–SiO$_x$, Si–CaF$_2$	0.3–6.0 nm	65–69
Quantum wires (free-standing)	Si–H$_x$	0.8–3.0 nm	70–85
Quantum wires (core–shell)	Si–SiO$_x$	1.0–3.0 nm core	86–88
Quantum wires	Si–CH$_3$	1.0–3.0 nm	89
Quantum dots	Si–H$_x$	1.0–5.0 nm	74 and 90–94
Quantum dots	Si–SiO$_x$	1.0–5.0 nm	95–99
Quantum dots	Si–organic ligands	1.0–5.0 nm	100–106
Ultrasmall Si clusters	Si–H$_x$	1–50 atoms	107 and 108

- The silicon bandgap starts to widen significantly for quantum wells (slabs/nanosheets) with widths under 2 nm, quantum wires with diameters under 3 nm and quantum dots with diameters under 5 nm.
- In principle the entire visible range is accessible for quantum dots and wires, with bandgap tunability extending from 1 eV to 3 eV.
- Exciton binding energy is significantly enhanced due to quantum confinement reaching values as high as a few hundred meV for 1.5–3 nm diameter quantum dots or 1–2 nm diameter quantum wires.
- Surface chemistry significantly affects the bandstructure of smaller nanostructures, *e.g.* oxidation narrows the bandgap of hydride passivated nanostructures.

Table 3 provides some theoretical data for hydride passivated silicon quantum dots, quantifying how the bandgap and strength of exciton binding increase dramatically with decreasing size for 1–5 nm nanocrystals. Note how the predicted radiative recombination rates however, only become comparable with those of direct bandgap semiconductors for the smallest silicon nanocrystals around 1 nm diameter.

Aside from size, shape and surface chemistry, there are other factors that affect bandstructure and therefore PL. There have, for example, been a series of theoretical effects investigating the roles of strain.[116–119] For quantum wires, crystallographic orientation also has a significant effect on bandstructure.[75,77,83,120]

1.7. PL quantum yield (PLQY)

For any photoluminescent material or molecular species the efficiency of its radiative processes or "photoluminescence quantum yield" (PLQY) of its luminescence is an important quantitative property that will impact on its likely use. The PLQY depends on direct competition between the specific radiative process and all non-radiative processes:

Table 3 Theoretical estimates of bandgaps, exciton binding energies and radiative recombination rates for 1–5 nm diameter silicon nanocrystals (quantum dots) with hydride passivation

Size	1 nm	1.5 nm	2.0 nm	2.5 nm	3.0 nm	3.5 nm	4.0 nm	5.0 nm	References
Number of silicon atoms	29	87	191	389	705	1087	>1500	>3000	111
Number of hydrogen surface atoms	36	76	148	254	300	436	>500	>750	112
Excitonic bandgap (eV) (nm) associated spectral range	>3.5, <354, UV	>2.5, <496, blue	2.2, 563, green	1.9, 653, red	1.7, 729, red	1.6, 775, near IR	1.5, 827, near IR	1.35, 918, near IR	112
Exciton binding energy (meV)			~400	~300	200	160	125	95	113 and 114
Radiative recombination rate (s^{-1})	>10^7	~10^6	~10^5	~10^4	<10^4				115

$$\text{PLQY (\%)} = \text{photons emitted/photons absorbed} = k_r/(k_r + \Sigma k_{nr}) \quad (2)$$

where k_r is the radiative recombination rate and k_{nr} the non-radiative rate.

Measurement of both PLQY η and luminescence decay time τ_{PL} enables one to separate the radiative and non-radiative lifetimes of excited states using the simple relation

$$\text{PLQY} = \eta = \tau_{PL}/\tau_r = 1/\tau_r/(1/\tau_r + 1/\tau_{nr}) \quad (3)$$

The intrinsic radiative recombination coefficient k_r of bulk c-Si has been estimated to be 2×10^{-15} cm^3 s^{-1} which is about 100 000 times lower than direct bandgap semiconductors like GaAs. Absolute quantum efficiency estimates of interband recombination (see Fig. 2) in standard silicon crystals (wafers) with native oxide surfaces (see Table 1) typically range from 10^{-6} to 10^{-4}. However, eqn (3) emphasizes that if *all* non-radiative processes can be virtually removed, then PLQY can be quite high. This has been elegantly demonstrated by solar cell specialists (Green and co-workers). In 2003 they showed that in the highest purity float zone silicon with optimized surface passivation, PLQY of near infrared emission could be as high as 6.1 ± 0.9% at 300 K.[121]

Accurate measurement of PLQY is not as straightforward as measuring PL decay times and a number of different protocols exist, but they can be divided into *comparative* ones that utilize photoluminescence standards of known PLQY and *absolute* ones that directly measure the fraction of absorbed photons that are emitted as photoluminescence or the fraction of photons lost by non-radiative recombination. The latter use integrating spheres to collect all emitted light, or use photoacoustic/calorimetric methods to quantify heat production. Methodology can depend on the physical form of the photoluminescent material (*e.g.*, film *versus* dry powder *versus* colloidal solution) and the spectral positions of both excitation and emission. A number of important reviews are available that detail the theory, methodologies, standards, and sources of error involved.[122–127] Of particular relevance here is the review by Valenta on PLQY measurements of nanomaterials.[127]

For comparative methods, the most established standards are quinine sulphate in 0.5 M sulphuric acid, fluorescein in 0.1 M NaOH and rhodamine 6G in ethanol with peak emissions at 451, 515 and 552 nm, respectively. Their PLQY accuracy is better than 4% under well-defined conditions.[124] These standards are well suited to the spectral positions of the F-band (see Section 3).

The physical form of silicon nanostructures can have a pronounced effect on PLQY. For example, for thin solid films light entrapment effects due to total internal reflections and Fresnel transmission at the film–air interface can dramatically lower PLQY, even if the internal quantum efficiency (IQE) is very high. A light extraction efficiency Ex_{PL} for a thin film on an absorbing substrate has been estimated.[128] The escape probability of emitted light is strongly dependent on refractive index:

$$Ex_{PL} = 1/[n_c(n_c + 1)^2] \quad (4)$$

So, for example, with a solid silicon film of refractive index 3.5 only 1.4% of light escapes directly from the top surface. For silicon nanocrystals embedded in silica films with a refractive index of 1.64, only 8.7% escapes internal reflections. For porous silicon nanostructures, PLQY and IQE values start to get closer as refractive index is lowered. High PLQY pSi structures have porosities in the range 70–90% with refractive indices as low as 1.1. Now more than 20% of light emitted internally can directly escape, lowering its self-absorption. Forming luminescent nanocomposites will often normally lower the perceived brightness due to refractive indices then being higher again.

Table 4 collates the highest PLQY values reported to date for all types of silicon nanostructure. It is clear that the highest PLQY values have been obtained in colloidal quantum dots, quantum dot-based superlattices and porous silicon and so these materials are given more consideration in this review. Amorphous nanoparticles have much lower PLQY than crystalline ones and are therefore not discussed. Molecular size clusters and *size-dependent* PL from ultrathin quantum wells have not had PLQY quantified to date. Nanotube studies have not reported visible PL.

2. S-Band photoluminescence

2.1. Porous Si (2D to 3D confinement)

Historically, relatively efficient (~3%) S-band emission at room temperature was first observed in high porosity silicon[47] which is why we first consider this nanostructure, despite its complex range of morphologies which make theoretical modelling a challenge. Freshly etched red-emitting high porosity silicon was found to contain "undulating" quantum wires[147] of below 3 nm average width (Fig. 7b).

Efficient S-band emissions are also evident in material where hydride passivation is replaced by native oxides,[137,149] anodic oxides,[150] high temperature oxides[30] and organic monolayers.[151] All these surface chemistries, with the notable exception of native oxides in some cases (see Table 1) produce low levels of non-radiative surface recombination in bulk Si. In oxide-free material tuning from the near infrared to the blue spectral region (1300 to 400 nm) has been achieved, as shown in Fig. 7a.[148,152] In oxidized material, from the near infrared to the yellow (1300 to 590 nm).[148] Typical PL lifetimes (see Table 5) are in the microsecond range but depend on wavelength and surface passivation quality (SRV) (see eqn (3)).

The data in Table 5 shows variability in PLDT at given wavelengths, due to differing silicon skeleton morphology and interconnectivity, levels of polycrystallinity and surface passivation. Nonetheless, the longest PLDT values of Gelloz 2005 probably reflect primarily the lowest SRV values obtained to date by their wet oxide passivation technique.

2.2. Single quantum wells and nanosheets (1D confinement)

Takahashi *et al.*[157] and Saeta *et al.*[158] were the first to report visible emission from single 2D crystalline silicon structures. Both groups used SIMOX wafers, thinning the silicon layer with high precision *via* thermal oxidation. A red and near infrared emission band (1.65 eV) was observed with an intensity dependent on the well thickness but spectral position varying little. No luminescence was observed unless the quantum well thickness was less than 5–8 nm. PLQY peaked at ~2 nm

Table 4 Si nanostructure classes with the highest PL efficiencies (PLQY). For data on the F-band there is considerable uncertainty over the origin of the emission and specifically whether it arises from silicon nanostructures and question marks are included in the table to emphasize this (see Section 3)

Class of nanostructure	Structure details	PL peak position	PL efficiency (PLQY at 300 K) and PL band (F or S)	References
Si QDs (?)	4 nm NPs and fluorescein ligands in colloidal solution	522 nm, 2.38 eV	90% (F)	(Zhong 2019)[129]
Si QDs (?)	5 nm NPs and arylamine ligands in colloidal solution	555 nm, 2.23 eV	90% (F)	(Li 2016)[130]
Porous silicon nanoparticles	3 nm alkyl-Si NCs in colloidal solution	740 nm, 1.87 eV	68 ± 5% (S)	(Yuan 2017)[131]
Si QDs	3 nm alkyl-Si NCs in colloidal solution	638 nm, 1.94 eV	66 ± 8% (S)	(Yang 2015)[132]
Si QD-based nanocomposites	4.5 nm alkyl-Si NC in polymer matrix	775 nm, 1.6 eV	65 ± 5% (S)	(Marinins 2017)[133]
Si QDs	5 nm methyl-Si NC colloidal solution	775 nm, 1.6 eV	65 ± 6% (S)	(Sangghaleh 2015)[134]
Si QDs	4 nm octadecyne-Si NC colloidal solution	789 nm, 1.57 eV	62 ± 11% (S)	(Jurbergs 2006)[135]
Si-QD based superlattices	Embedded 4.5 nm NC with 3 nm silica barriers	992 nm, 1.25 eV	50% (S)	(Valenta 2019)[136]
Porous Si microparticles	SCD pSi with <3 nm interlinked NCs and 1150 m² g⁻¹	685 nm, 1.81 eV	32% (S)	(Joo 2016)[137]
Si nanorods (?)	~20 nm wide NWs	500 nm, 2.48 eV	15% (F)	(Song 2016)[138]
Si nanosheets (?)	Buckled silicon monolayers	500 nm, 2.48 eV	9% (F)	(Ryan 2019)[139]
Porous Si nanowires	20–200 nm wide NWs with interlinked 3 nm NCs	740 nm, 1.68 eV	5% (S)	(Gonchar 2012)[140]
Solid Si quantum wires	3–4 nm wide NWs	730 nm, 1.70 eV	4.3% (S)	(Lu 2013)[141]
a-Si:O$_x$ QDs	Embedded 2 nm NPs in oxide	600 nm, 2.07 eV	1–4% (S)	(Goni 2014)[142]
a-Si:H QDs	Embedded 3–5 nm NPs	850 nm, 1.46 eV	<2% (S)	(Anthony 2009)[143]
Solid Si quantum wires	5 nm Si NWs	750 nm, 1.65 eV	0.5% (S)	(Irrera 2012)[144]
Si quantum wells	1.1–2.7 nm layers with silica barriers	700 nm, 1.77 eV, 912 nm, 1.36 eV	—	(Cho 2007)[145]
Si molecular clusters	Octasilacubane (Si$_8$)–butyldimethylsilyl crystal	751 nm, 1.65 eV	—	(Kanemitsu 1995)[55]
Si nanosheet nanocomposite	Si$_6$ H$_6$ sheets in polystyrene	510 nm, 2.43 eV	—	(Helbich 2016)[146]
Solid Si nanotubes	3–7 nm c-Si tubular walls	—	—	(Tang 2005)[50]

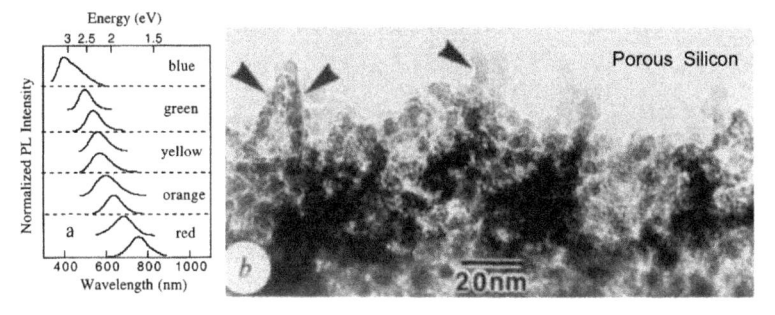

Fig. 7 (a) S-band PL tuning *via* porosity of hydride passivated porous silicon.[148] Reprinted and adapted with permission from M. V. Wolkin, J. Jorne, P. M. Fauchet, G. Allan and C. Delerue, *Phys. Rev. Lett.*, **82**(1), 197–200, 1999. Copyright 1999, the American Physical Society. (b) Complex morphology of photoluminescent porous silicon of 78% porosity.[147] Adapted with permission from A. G. Cullis and L. T. Canham, *Nature*, **353**, 335–337, 1991. Copyright 1991, Nature Publishing Group

Table 5 Experimental S-band decay times (PLDT) in porous silicon *versus* wavelength/energy for hydride or oxide passivation. SCD: SuperCritically Dried, HP: High Pressure

Wavelength (nm)	Energy (eV)	Decay time (μs)	Decay time (μs)	Decay time (μs)	Decay time (μs)	Decay time (μs)	Decay time (μs)
Surface passivation		Native oxide	Native oxide	Hydride	Native oxide	SCD native oxide	HP wet oxide
References		153	154	152	155	137	156
800	1.55		10	20	50	65	100
700	1.77	9	10	12	30	42	50
600	2.07	2.5	5	9	12	20	20
550	2.25	1.5	1	3			10
500	2.48			0.7			2.5
450	2.76			0.1			0.2
400	3.10			0.02			

thickness but was very low at 10^{-4}.[158,159] Okamoto *et al.*[160] reported 2 K PL from SIMOX-derived quantum wells of only 0.6 and 1.6 nm. Emission was again at 1.66 eV. Pauc *et al.*[161] studied PL at 6 K from SIMOX derived wells of 3.9 down to 0.6 nm. A 1.6 eV band appeared for the narrowest wells with an estimated PLQY of only 0.1% at 6 K.

The first really size-tunable emission was achieved by Green and co-workers.[162] They used ELTRAN structures with superior defect densities (<50 cm^{-2}), atomically flat interfaces and lower SRVs. Fig. 8 shows TEM data and the 650–950 nm PL spectra as a function of well thickness over the range 1.1–2.7 nm. Their data on processed SIMOX wafers also reported an emission whose peak wavelength was insensitive to QW thickness but correlated instead with oxide thickness. Zhu *et al.*[163] studied (110) SIMOX wafers with ∼1–4.4 nm thick wells. Both QC and interface effects were discussed in relation to their 750–800 nm emission. Wagner

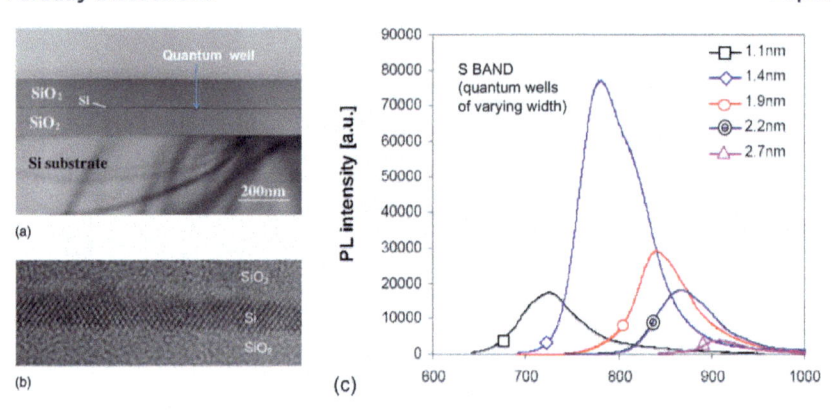

Fig. 8 Single silicon quantum well and associated S-band PL for varying well thickness.[162] Reprinted and adapted with permission from E. C. Cho, M. A. Green, J. Xia, R. Corkish, P. Reece, M. Gal and S. H. Lee. *J. Appl. Phys.*, 2007, **101**, 024321. Copyright AIP Publishing.

et al.[68] also achieved size dependent 75 K emissions over 1.6 to 1.2 eV for 1–4 nm well widths.

2.3. Superlattices and multilayers (1D to 3D confinement)

Multilayer and superlattice structures have primarily been based on silica passivation,[164–171] but photoluminescent lattice matched Si/CaF_2 superlattices[172–174] have also been achieved. Lockwood and co-workers[164,165] first reported a photoluminescent MBE grown Si/SiO_2 superlattice. The 6 period structure had 1.0–5.4 nm thick Si quantum wells with 1 nm thick oxide barriers. Visible emission was only observed for wells below 3 nm. PL peak energies were tunable over the 1.7–2.3 eV range but PLQY values were not given. A major advance was made by the Zacharias group in a series of

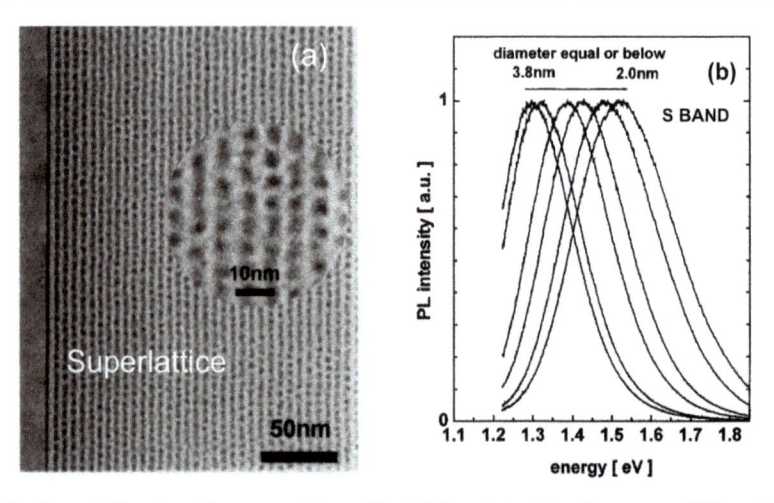

Fig. 9 Superlattice-based nanocrystals and their S-band photoluminescence.[171] Reprinted and adapted with permission from M. Zacharias, J. Heitmann, R. Scholz, U. Kahler, M. Schmidt and J. Blasing, *Appl. Phys. Lett.*, 2002, **80**(4), 661–663. Copyright AIP Publishing.

papers[171,175,176] using superlattices to exploit 3D rather than 1D confinement. They also dramatically raised PLQY *via* improved interface passivation, size control and nanocrystal density. Fig. 9 shows a typical structure and the tunability of the PL.[171] Valenta *et al.*[176] provided detailed information on PLQY values, lying in the range 10–20% and doubling when the silica barrier layer thickness was increased from 1 to 3 nm. PLQY variation with temperature was also studied, peaking at 30% at 100 K. Very recently,[136] PLQY values reported for near infrared emission (1.35 eV/0.97 micron) corresponding to 4.5 nm nanocrystals were as high as 50%.

2.4. Nanowires and quantum wires (2D confinement)

The first free-standing isolated NWs of widths under 10 nm realized by lithographic etching techniques showed very weak photoluminescence[177-179] or no visible luminescence at all, ascribed to insufficient surface passivation.[180] Using laser ablation to create 13 nm wide ultralong nanowires, and then oxidation to thin the silicon core to 3.5 nm, Feng and co-workers reported red emission of unknown PLQY.[181] Complete oxidation of the wires removed the red emission but a distinct green emission band persisted (see Section 3.2). Brongersma and co-workers used catalysed CVD growth to create 20 nm wide nanowires and then oxidation to decrease core widths to below 5 nm. Tunable 800 to 720 nm emission with microsecond decay times was achieved.[182]

Examination of Table 4 reveals that nanowire fabrication, excluding porous silicon, has been less successful to date with regards to achieving the highest PLQY values for S-band emission. Korgel's group reported a PLQY of 4.3% for 3–4 nm wide solid quantum wires.[141] For nanowires we need to distinguish those that are solid silicon from those that have larger diameters but are themselves porous. The latter "porous silicon nanowire arrays" will behave in a similar manner to porous silicon films but with lowered nanocrystal densities and, to date, lower PLQY (see Table 4). Some much larger solid nanowires also have oxidized rough sidewalls containing nanocrystals that generate the visible photoluminescence.[183]

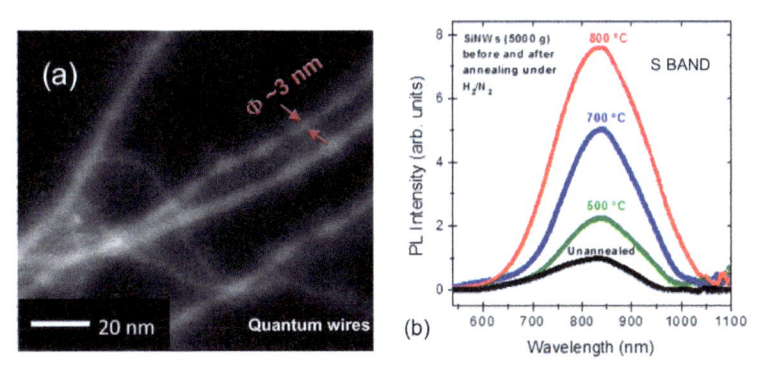

Fig. 10 Luminescent ultrathin silicon nanowires from plasma synthesis and their S-band 300 K PL at varying levels of surface passivation.[184] Adapted with permission from V. Le Borgne, M. Agati, S. Boninelli, P. Castrucci, M. de Crescenzi, R. Dolbec and M. A. E. Khakani, *Nanotechnology*, 2017, **28**, 285702. Copyright IOP Publishing. All rights reserved.

Fig. 10 shows an energy filtered TEM image of ultrathin luminescent solid quantum wires reported recently.[184] These were extracted by-products of a plasma torch spherodization of micro-particulate powder. Once extracted *via* centrifugation, the various fractions showed PL that shifted from 950 to 680 nm as the average Si core diameter evolved from ∼5 to ∼3 nm.

2.5. Nanoparticles and quantum dots (3D confinement)

S-band PL from visibly luminescent silicon nanocrystals (nanoparticles) synthesized by a broad range of techniques have now received a vast amount of study. Fabrication techniques include plasma-assisted decomposition of silane or silicon tetrachloride; laser pyrolysis of silane; liquid phase synthesis; laser ablation; high energy milling and microemulsion growth. The reviews of Mangolini,[185] Huan *et al.*[186] and Cheng *et al.*[187] provide details on the majority of these synthesis techniques. Highlighted here is the recent study of the Ceroni and Korgel Groups using dodecene passivated nanocrystals in colloidal suspension.[188] Fig. 11 illustrates the excellent control over size. Fig. 12 shows how spectral position smoothly "blueshifts" with decreasing nanocrystal size right from the 1.1 eV bandgap of bulk silicon (for 9.1 nm average size) to 2.1 eV (1.8 nm average size).

Table 6 shows the decay times from a number of nanocrystal studies and it is informative to compare values therein with Table 5 (data from porous silicon). For example, at 800 nm, pSi decay times (Table 5) lie in the range 10–100 microseconds; isolated nanocrystals emitting at 800 nm (Table 6) have much longer decay

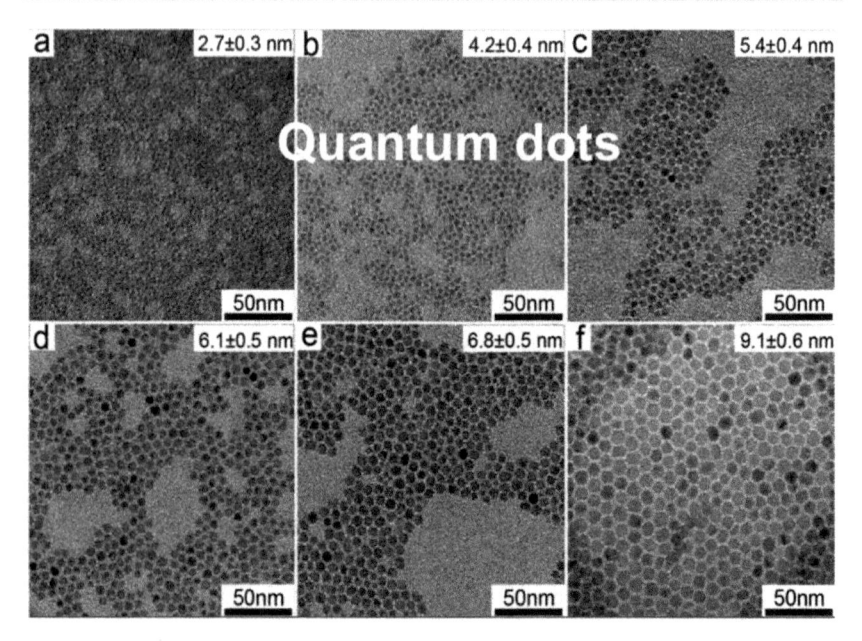

Fig. 11 S-band photoluminescent silicon nanocrystals of controllable size.[188] Adapted with permission from Y. Yu, G. Fan, A. Fermi, R. Mazzaro, V. Morandi, P. Ceroni, D. M. Smilgies and B. A. Korgel, *J. Phys. Chem. C*, 2017, **121**, 23240–23248. Copyright 2007 American Chemical Society.

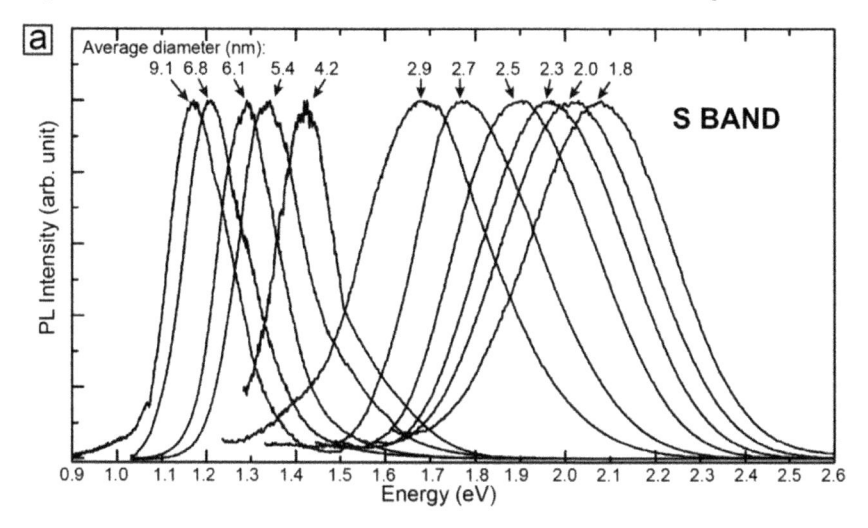

Fig. 12 S-band 300 K PL from the various-sized nanocrystals of Fig. 11.[188] Adapted with permission from Y. Yu, G. Fan, A. Fermi, R. Mazzaro, V. Morandi, P. Ceroni, D. M. Smilgies and B. A. Korgel, J. Phys. Chem. C, 2017, 121, 23240–23248. Copyright 2007 American Chemical Society.

Table 6 Experimental S-band decay times (PLDT) in Si nanocrystals. Decay time units are in microseconds, as in Table 5

Wavelength (nm)	Energy (eV)	Ref. 154	Ref. 189	Ref. 190	Ref. 188	Ref. 191	Ref. 192
1000	1.24				600	800	
900	1.38		120		300	420	
800	1.55	100	80	175	100	200	247
700	1.77	35	40	90	30	100	134
600	2.07	8		40	6	40	69
550	2.25			25			36

times in the range 80–247 microseconds. It is likely that the internal surfaces of porous silicon nanostructures are more challenging to passivate than isolated or embedded nanocrystals.

2.6. PLQY of S-band

Table 7 summarizes the remarkable improvements made in raising the PLQY of S-band luminescence alone since 1990. Early values in the range 3–5% probably reflected the lower levels of size control, combined with higher SRV levels. The highly interconnected silicon skeleton of porous silicon facilitates rapid exciton migration to lower bandgap parts of the network that have slower radiative rates.[137,193] Size control is thus key. This, in combination with differing light extraction efficiencies, explains why colloidal pSi nanoparticles typically have higher PLQY than microparticles and films (see Table 7).

Table 7 Improvements in the PLQY of the S-band over time (1990–2020)

Maximum PLQY (%)	Nanostructure	Peak PL wavelength (nm)	Mean size/size range (nm)	Surface chemistry	Fabrication technique	References
68	pSi NPs	740	3	Dodecyl	pSi ablation	(Yuan 2017)[131]
66	NP	700	3	Dodecyl	HSQ decomposition	(Yang 2015)[132]
65	NP	690	4	Native oxide	Silane pyrolysis	(Potrick 2011)[194]
62	NP	789	4	Octadecene	Plasma synthesis	(Jurbergs 2006)[135]
62	NP	825	4	High pressure wet oxide	Plasma synthesis	(Gelloz 2019)[192]
50	SL	954	4.5	Thermal oxide	PECVD	(Valenta 2019)[136]
60	NP	775	3	Dodecyl	HSQ decomposition	(Sanggahelah 2015)[134]
59	NP	750	2 to 3	Thermal oxide	Ion implantation and anneal	(Walters 2006)[195]
45	NP	827	3.5	Dodecene	HSQ decomposition	(Yu 2017)[188]
43	NP	760	1.9 to 2.5	Allylbenzene	HSQ decomposition	(Mastronardi 2011)[196]
32	pSi particles	685	<2.5	Native oxide	Anodisation and supercritical drying	(Joo 2016)[137]
31	pSi films	735	<2.7	High pressure wet oxide	Anodisation	(Gelloz 2019)[192]
30	NP	600–850	2 to 7	Native oxide	Laser pyrolysis	(Ledoux 2002)[200]
25	NP	650	1.6	Allylbenzene	HSQ decomposition	(Mastronardi 2011)[196]
19	SL	950	4.5	Thermal oxide	PECVD	(Valenta 2014)[176]
0.3	SL	1170	12	Thermal oxide	Sputtering	(Zacharias 1998)[197]
5	NP	600–800	1 to 2	Thermal oxide	Disilane pyrolysis	(Brus 1995)[198]
3	pSi films	700–850	2 to 10	Anodic oxide	Electrochemical etch and oxidation	(Vial 1992)[150]
3	pSi films	750–850	2 to 10	Hydride	Electrochemical etching	(Canham 1990–1991)[47]

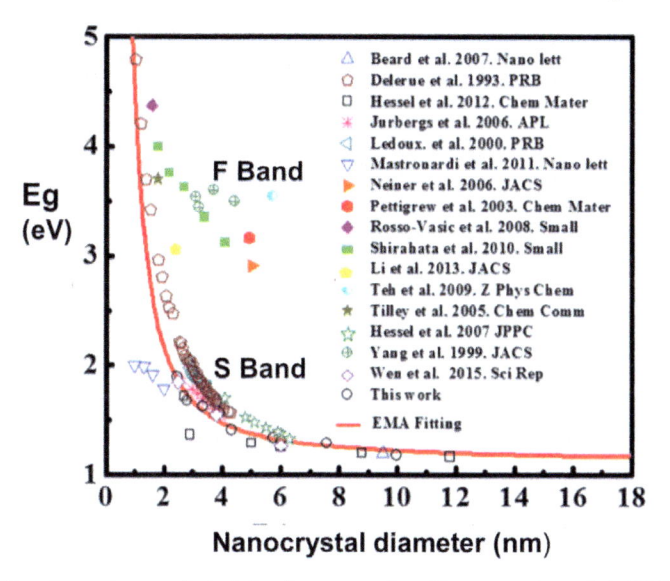

Fig. 13 Size dependence of photoluminescence for silicon quantum dots.[208] Reprinted with permission from X. Liu, Y. Zhang, T. Yu, X. Qiao, R. Gresback, X. Pi and D. Yang. *Part. Part. Syst. Charact.*, **33**(1), 44–52, 2016. Copyright 2015, WILEY-VCH Verlag GmbH & Co. KGaA, Weinheim. Some of the featured studies reported F-band data rather than S-band data, as highlighted here.

2.7. Wavelength tuning of S-band *via* nanostructure size and shape

A wealth of data now exists confirming that S-band spectral position depends on size distribution for a variety of nanostructures: 1.1–2.7 nm quantum wells within oxide;[162] 3–5 nm quantum wires from plasma synthesis;[184] 3–8 nm nanocrystals from disilane pyrolysis;[199] 3.4–4.8 nm nanocrystals from silane pyrolysis;[200] 3–8 nm and 4–9 nm nanocrystals embedded in silica *via* sputtering;[200–202] 2–5 nm nanocrystals embedded in silica *via* ion implantation;[203] 1.5–4 nm nanocrystals embedded in silica *via* PECVD;[204] and hydride passivated porous silicon of varying porosity.[47,148,152,205–207]

Fig. 13 presents a published and very useful compilation of PL data from 17 studies on nanocrystals made by a variety of techniques[208] and effective mass approximation (EMA) fitting of nanocrystal bandgap *versus* size. There is clearly considerable scatter of quantitative data, partly because both S-band (this section) and F-band (see Section 3) data were used. For the S-band data, another complicating factor are the size-dependent optical absorption coefficients: smaller silicon nanocrystals have lower optical absorption which biases the PL spectra towards larger nanocrystals.[188] Fig. 14 illustrates how one parameter (nanocrystal size dispersion) can therefore strongly influence the theoretical EMA modelling of PL tunability *versus* average size. Good agreement between such theory and experimental data is achieved once this is taken into account.

2.8. Wavelength tuning of S-band *via* surface chemistry

The ability to tune PL emissions by surface chemistry passivation, as well as by nanostructure size, would be desirable, particularly for fabrication techniques

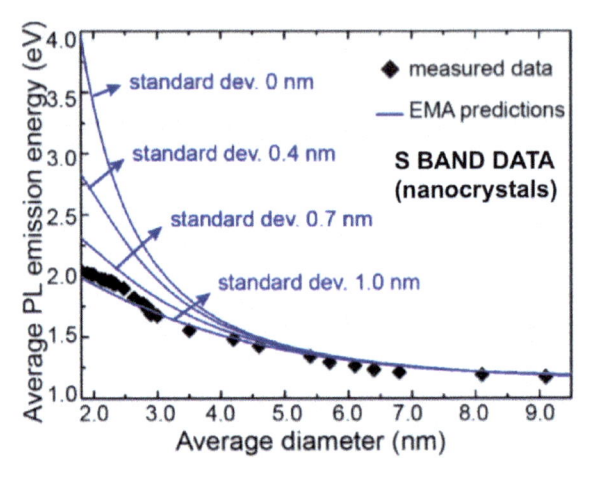

Fig. 14 Effective mass approximation (EMA) modeling of S-band spectral position *versus* nanocrystal size for various levels of size polydispersity (standard deviations).[188] Adapted with permission from Y. Yu, G. Fan, A. Fermi, R. Mazzaro, V. Morandi, P. Ceroni, D. M. Smilgies and B. A. Korgel, *J. Phys. Chem. C*, 2017, **121**, 23240–23248. Copyright 2007, American Chemical Society.

which offered high volume throughput, but lacked fine tunability of size. Theoretical calculations have shown that the bandstructure of ultrasmall nanostructures such as 1–1.5 nm diameter quantum dots and quantum wires are strongly affected by surface chemistry.[86,95,96] Spectral position of the S-band might then also be tunable by surface chemistry. However direct experimental evidence for extensive S-band tuning is sparse. One study reported that for larger nanocrystals (1.64 eV PL corresponding to ~3.5 nm nanocrystals) changing from hydride to alkyl (dodecene) passivation had little effect on spectral position.[209] Not surprisingly however, PLQY and PLDT were affected, as the SRV will differ. In contrast, a study of amide and alkoxy functionalization relative to alkyl did find a small but measurable effect on spectral position for 3.5–6.4 nm nanocrystals.[210] In addition, another study using alkynyl(aryl) groups also observed PL shifts (in the range 685–800 nm). They attributed the effect to *"surface-state induced bandgap changes"*.[211] In instances where chemical tuning *"right across the visible range"* is claimed for silicon nanocrystals,[212,213] it would be important to investigate whether the blue emission is in fact F-band emission by measuring PLDT as a function of tuning (see Section 3) and therefore of quite different origin to the yellow or red S-band emission. The extent to which the S-band can be tuned over a broad wavelength range *via* surface chemistry alone needs further investigation.

2.9. Selective photoexcitation of S-band emission

As we saw with many radiative transitions in bulk silicon (Section 1.3), sharp features in PL spectra can allow the determination of precise energies of electronic states, the symmetry of luminescent defects and the phonon energies involved. The large linewidth of the S-band PL obscures the spectroscopic information it contains. Photoexcitation at resonant wavelengths *via* tunable lasers can selectively excite subsets of the nanocrystal ensemble and lower the effects of

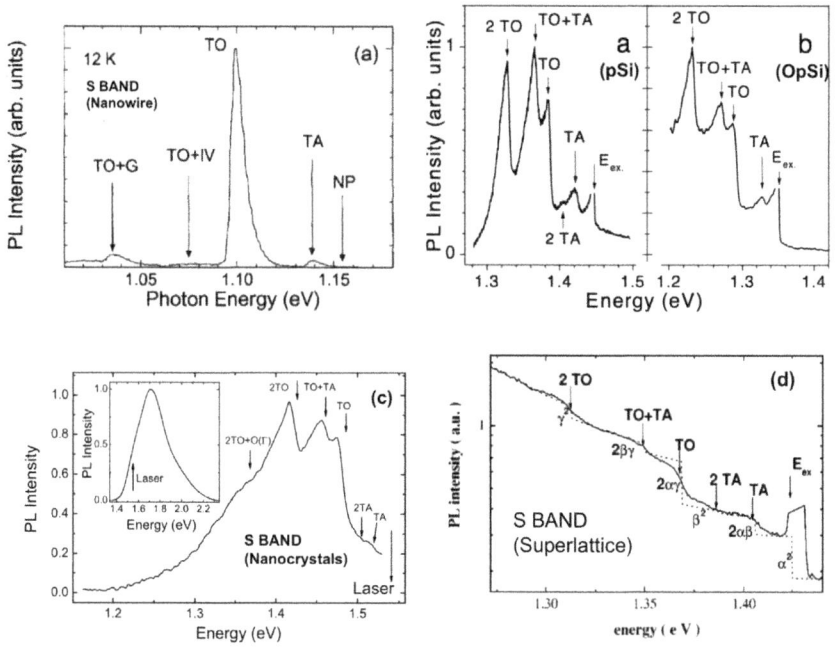

Fig. 15 Resonant photoexcitation of silicon nanostructures. (a) Large silicon nanowires,[225] (b) freshly etched porous silicon (pSi) and oxidized porous silicon (OpSi),[228] (c) free standing silicon nanocrystals,[230] (d) silicon nanocrystals embedded in oxide-based superlattice.[224] Adapted with permission from Y. Kanemitsu, H. Sato, S. Nihonyanagi and Y. Hirai, *Phys. Stat. Solidi*, **190**, 755–758, 2002; D. Kovalev, H. Heckler, M. Ben-Chorin, G. Polisski, M. Schwartzkopf and F. Koch, *Phys. Rev. Lett.*, 1998, **81**(13), 2803–2806; B. Goller, S. Polisski, H. Wiggers and D. Kovalev, *Appl. Phys. Lett.*, 2010, **97**, 041110 and J. Heitman, F. Muller, L. Yi, M. Zacharias, D. Kovalev and F. Eichhorn, *Phys. Rev. B*, 2004, **69**, 195309, respectively.

inhomogeneous broadening, thereby revealing otherwise hidden spectral features. Starting with the studies of Calcott, Nash and co-workers it has now been investigated extensively with luminescent porous silicon[198,214–222] but also with silicon superlattice nanocrystals,[223,224] large silicon nanowires,[225] quantum wells[226] and nanocrystals made by various other techniques.[198,227–230] Initial concerns raised about its applicability to high PLQY porous silicon emission at room temperature and the precise number of phonon features were resolved by 1997.[30] Highlighted here in Fig. 15 are S-band resonant photoexcitation data from a range of different silicon nanostructures. All show the TO and TA phonon-related features so characteristic of exciton recombination in crystalline silicon. For the S-band PL from a given nanostructure, at increasing levels of confinement (shorter wavelength emission), the exciton binding energy increases (Table 3) and the no-phonon (NP) to phonon (TO, TA) intensity ratio increases, just as for localized bound excitons in bulk silicon (Fig. 3 and 4).

Even more revealing than resonant photoexcitation can be the photoexcitation of isolated individual nanostructures. Single nanostructure spectroscopy can reveal homogeneous linewidths and most previously hidden spectral features, but is experimentally challenging. The first such spectroscopy studies were performed on porous silicon particles in 1998 by Buratto and co-workers.[231] Progress has been recently thoroughly reviewed by the Linnros group[232] so only a few comments will be made here that relate to PLQY of nanostructure ensembles and the origin of S-band PL:

• Individual nanocrystals "blink" (periods of no emission under constant photoexcitation) and some are completely "dark"; others are "bright" with near unity IQE. The ensemble PLQY can thus be raised significantly by lowering the density of dark silicon nanocrystals that have a very efficient non-radiative channel(s), by decreasing blinking rates, or by both routes.

• PL linewidths for single nanocrystals are substantially narrower than for the ensemble (typically 100–160 meV at 300 K but values as low as 17 meV are now being reported[233]). Both temperature and the medium surrounding the core have a strong influence.

• Substantial TO phonon involvement was again invoked to explain the emission spectral data.

2.10. Origin of S-band PL

The wavelength tuning *via* size (Section 2.7), right from the bulk silicon bandgap[201,205] to at least the yellow spectral region, convinced many that quantum confinement was indeed responsible for S-band emission. Resonant photoexcitation experiments (Fig. 15), providing a spectroscopic signature of established silicon phonon-assisted transitions (Section 1.3), convinced many more. Single nanocrystal spectroscopy (Section 2.9) holds the most potential for unravelling the complexity of ensemble nanosystems where size and shape control is still improving, but not yet sufficient to avoid significant inhomogeneous broadening and rapid carrier migration effects.[234] Nonetheless, after many hundreds of studies, there is now substantial evidence, both experimental and theoretical, supporting a quantum confinement model for S-band emission, at least for near infrared to green emission. Agreement between theory and experiment exists for PL position (excitonic bandgap) *versus* size and shape (Fig. 14), together with strongly wavelength-dependent measured decay times (Tables 5 and 6) and extracted radiative decay times. The reasons for rapid decline in PLQY for yellow to blue S-band emission; the ultimate PL linewidths achievable; the avoidance of "blinking"; improvements in SRV for smaller nanostructures – these are all topics requiring further study.

3. F-band photoluminescence

3.1. Porous Si

Reports of blue–green emission ("fast"-nanosecond decay time and hence "F-band emission") from porous silicon started to appear[235,236] in 1992. Kovalev and co-workers[237] showed that its intensity could be significantly raised by rapid thermal oxidation and since then it has received a series of studies. It is noteworthy that even moderate PLQY requires oxidation[238] and/or heat treatment of the porous silicon. Freshly oxidized porous silicon often does not display F-band emission and it only emerges after storage in ambient air.[239]

3.2. Nanowires

Growth of nanowires *via* catalyzed laser ablation VLS growth or thermal evaporation often results in short wavelength emission alone.[240–244] A silicon oxide phase is invariably present and often silicon core width greatly exceeds that required for silicon bandgap enlargement (<3 nm). Sometimes there is no silicon phase at all.[240] PLQY values have not generally been provided.

3.3. Nanoparticles and quantum dots

For silicon quantum dots prepared by high temperature synthesis, such as silane decomposition, F-band emission also appears after air storage (see Fig. 1).

A very large number of reports of F-band emission of high PLQY (only some are in Table 8) have appeared since the 2013 report of "one-pot aqueous synthesis"[245] by He and co-workers. Here, efficient blue–green emission is achieved immediately after reduction synthesis using inexpensive precursors, low temperatures (e.g. <200 °C) and short reaction times. Highlighted here though instead, is the recent reappraisal study of Oliinyk and co-workers.[246] They cast doubt on the likelihood of such conditions generating crystalline silicon nanostructures.

Table 8 Improvements in the PLQY of the F-band over time (1994–2020)

Maximum PLQY (%)	Peak PL wavelength (nm)	Mean decay time (ns)	Mean size of NP component (nm)	Fabrication technique	References
90	522	—	4	Solution synthesis	(Zhong 2019)[129]
85	425	—	3.5	Solution synthesis	(Zheng 2019)[249]
76	440	7	2.8	Laser ablation in organic solvents	(Wang 2019)[250]
62	440	8.3	2	Solution synthesis	(Geng 2018)[251]
90	550	3	5.2	Solution synthesis	(Li 2016)[130]
50	600	1.0	3	Solution synthesis	(Zhong 2016)[252]
82	450	14.3	3	Solution synthesis	(Wu 2015)[253]
75	480	31	2.5	Solution synthesis	(Wang 2015)[254]
85	520	3.9	3.5	Solution synthesis	(Cho 2014)[255]
25	375	—	3	Solution synthesis	(Wang 2011)[256]
23	425	—	3.4	Solution synthesis	(Atkins 2011)[257]
60	425	—	<4	Ball milling in organic solvents	(Heintz 2009)[258]
44	450	—	3	Electrochemical etching and sonication in organic solvents	(Portoles 2009)[259]
18	480	<100	1.5	Silane decomposition and etching	(Fojtik 2006)[260]
<1	500	—	3.7	Gas evaporation or vacuum sublimation	(Kimura 1999)[261]
0.1	470	1	<5	Electrochemical etching and oxidation	(Tsybeskov 1994)[238]

Importantly, they also report that repeating the synthesis with a *control* (reducing agent and a silicon-free organic amine) produced remarkably similar photoluminescence. The de Cola group have also repeated such a synthesis under both short (15 minute) microwave irradiation and lengthy (300 minute) hydrothermal conditions.[247] After thorough characterization of repeat and *control* formulations (lacking reducing agent or silicon source) they attribute the blue-emitting product to the formation of silica and carbon dots rather than silicon nanocrystals.

A further reappraisal study[248] has now also been conducted on a different wet chemical synthesis route that uses a magnesium silicide precursor developed by the Kauzlarich group. Once again the key control run, where the source of silicon was omitted, led to similar F-band emission to where the silicon precursor was included. These three studies[246–248] highlight the current uncertainty that surrounds the true origin(s) of F-band emission (see Section 3.5).

3.4. PLQY of F-band

Table 7 illustrates the dramatic rise in PLQY for the F-band over the last 5 years. In the first reports of emission from oxidized porous silicon films and microparticles, quantum yield was probably at best 0.1%.[238] Colloidal NP emission prior to 2010 was much better, but in the range 1–60%. Since 2015 however there have been a number of reports of PLQY above 75%, even as high as 90%. The table also emphasizes the dominance of bottom-up solution synthesis techniques in achieving the highest PLQY values. The question is, as discussed in the previous section, to what extent does such high PLQY have anything to do with silicon nanocrystals?

3.5. Selective photoexcitation and "fine structure" of F-band

As with the S-band, selective photoexcitation studies of F-band PL might provide some insight by revealing hidden spectral features. Currently, very few studies have been conducted[262,263] but these did reveal potentially useful fine structures that need further study. However, even under non-resonant excitation, some interesting structures of the emission band have been reported in some instances,[264] especially for NPs made by laser ablation in solvents.[265–267] By conducting *control* experiments where the solvent (*e.g.* toluene or octene) alone was UV irradiated, there appeared to be a good match of F-band spectra with those where silicon nanoparticles were created. Fluorescent solvent by-products need to be considered in such laser ablation studies.

The study of Kobayashi and co-workers[268] is also particularly insightful with regards to the potential influence of nanostructure-adsorbed impurities at low concentrations. They utilized high energy milling to generate 5 nm NPs that were then suspended in hexane and generated a blue–green PL band with numerous sub-peaks. Control spectra of hexane alone, without NPs, yielded very much weaker spectra in the same location (Fig. 16). They attributed the PL to adsorbed dimethylanthracene (DMA), a highly fluorescent impurity in hexane. DMA has been used as a standard to estimate comparative PLQY for blue-emitting silicon nanoparticles.[258] Importantly, intentional addition of only a 20 nM solution of DMA in hexane to another DMA-free solvent containing silicon NPs was sufficient for the vibronic structure to appear in spectra. Exciting light absorption *via* energy transfer from the silicon NP was estimated to raise PL output of DMA in hexane by

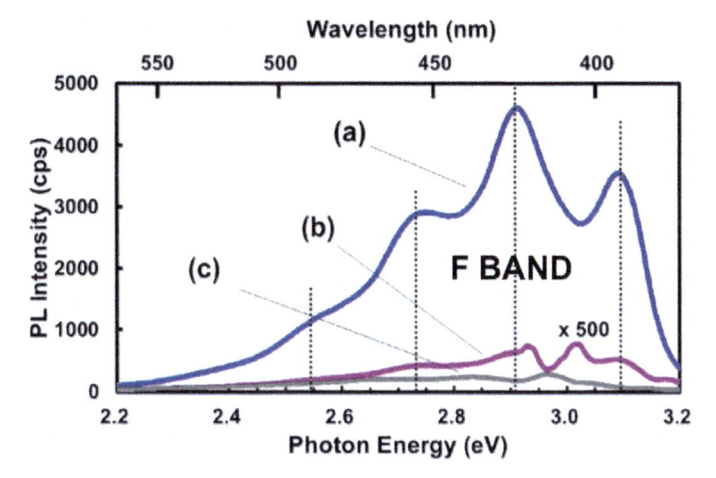

Fig. 16 Photoluminescence of adsorbed fluorescent organic contaminants on silicon nanoparticles.[268] Adapted with permission from T. Matsumoto, M. Maeda and H. Kobayashi, *Nanoscale Res. Lett.*, 2016, **11**(7), available via an open access CC-BY license.

a factor of about 3000. This study highlights the need for extensive characterization of the purity of potentially luminescent silicon nanostructures.

3.6. Origin(s) of F-band PL

Studies focused on elucidating the origin of the F-band emission have broadly drawn four proposals to date: that the emission is connected to the core of small silicon NPs;[269–273] radiative silicon–oxide interface states;[1,274,275] defect states in pure oxide;[1,276,277] or contamination of that oxide.[239,278,279]

Valenta et al.[270], for example, studied F-band emissions from small nanoparticles obtained by electrochemical etching, hydrogen peroxide treatment and filtration. They interpreted the very large spectral shifts of the band under varying photoexcitation wavelengths (270–420 nm) to the behaviour of Si QD core emission, rather than defective oxide states. Data was fitted to theoretical predictions of bandgap for small hydride-terminated nanoparticles. No FTIR data or chemical composition data was presented however, so the level of hydride remaining vs. oxide passivation (known to generate different bandgaps) is unknown.

F-band emission assignment to the oxide phase of oxidized silicon nanostructures is based on a number of observations, not least that it bears similarities to emissions from pure defective silica structures[280] and can still be observed when the nanocrystalline silicon phase has been removed.[240] It is interesting that high surface area silica, for example, shows similar "aging" and "regeneration" effects for its blue photoluminescence to oxidized porous silicon and nanocrystals from high temperature synthesis, attributed to interactions with water vapour and organic molecules in ambient air.[281–283]

Are there specific contaminants that might be responsible? A very recent study[284] used pulverized pSi particles to investigate the effects of nitrogen contamination on F-band emission, following the work of Dasog et al.[285] reporting significant influence of nitrogen incorporation in nanocrystals made by other routes. The porous Si structures were heavily oxidized (FTIR spectra) and intentionally subjected to environments rich in nitrate and nitrates. They found increased nitrogen levels diminished, rather than induced, blue emission.

The study of Chen et al.[286] in 2013 instead provides strong support for a link to carbon contamination as one potential origin of the F-band in silicon nanostructures containing organic molecules that are subjected to oxidation, heat treatments, microwave treatments or UV laser treatments. They calcined mesoporous silica NPs containing organic molecules at 400 °C and observed a 450–475 nm emission band. The fluorescent particles had retained an ordered silica framework with 3 nm pores. When these ~80 nm mesoporous silica spheres were subsequently soaked in 37% HF to dissolve the silica phase, the blue output was however completely preserved. Under subsequent TEM examination, the residue (see Fig. 17) was found to contain a high density of carbon nanoparticles (~3 nm diameter). The presence of carbon dots does not show that they are necessarily the origin of the luminescence.[287,288] It could be from molecular-sized organic chromophore by-products.

What this does demonstrate however, is that porous silica and oxidized porous silicon nanostructures can act as unintentional "nanoreactors" for carbon dot synthesis. Careful experimentation is required to check that F-band-like emission is not a result of carbon contamination of an initially hydrophobic and organophilic nanostructure. As pointed out many years ago, there is direct evidence of carbon contamination of hydride-terminated porous silicon from a variety of sources: extensive water rinsing used to remove HF residues;[289] atmospheric and vacuum storage;[149] plastic wafer and sample container outgassing.[279] Low levels of carbon contamination are likely to generate F-band emission that is detectable, but of low PLQY. Intentional carbon impregnation of mesoporous silica (e.g. by glycerol) is now used by the carbon dot scientific community to generate high PLQY[290] in the blue–green spectral range. It would be informative to search for carbon dots as well as silicon quantum dots in the now varied protocols for realizing ultrahigh F-band PLQY (see Table 7). It is also noteworthy that the carbon dot research community has similar issues with regards to unintentional fluorescent impurities.[287,288] Without sufficient purification steps, molecular carbon-based chromophores can contribute the majority of the emission from bottom-up carbon dot formulations. Intentional introduction of both silicon and

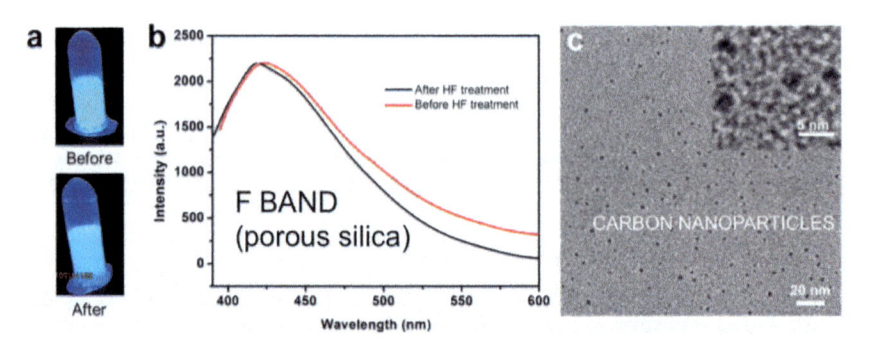

Fig. 17 F-band like photoluminescence and carbon dot contamination of porous silica.[286] Adapted with permission from H. Chen, Z. Zhen, W. Tang, T. Todd, Y. Chuang, L. Wang, Z. Pan and J. Xie, Theranostics, 2013, 3(9), 650–657. Copyright IvySpring.

carbon dots into silica has led to white light emission (see Section 4.5), so optimizing PLQY for both S and F-bands can be of use, even if their origins are completely different.

4. PL applications

4.1. Sensing

Very soon after the demonstration of efficient visible photoluminescence from porous silicon[47] it became apparent that PL output could be sensitive to a range of adsorbates.[291] This has been the basis of many studies exploiting the photochemical properties of nanoscale silicon to attain sensors for gas and liquid analytes. Prior reviews include those of Sailor et al.,[292] Gonzalez et al.[293] and Ji et al.[294] Chemical toxicants, pollutants and biomedical analytes have been the primary targets. Both PL quenching[295–299] and enhancements ("photobrightening") have been reported.[300,301] The vast majority of sensors have utilized PL quenching, which in many cases is reversible, but irreversible quenching has also been considered for single-use sensors.

Fig. 18 illustrates S-band sensing of various explosives with silicon nanocrystals embedded within a convenient paper substrate.

Quenching can be static (binding), dynamic (collisional), oxidation-induced or deposition-induced.[302,303] Table 9 and 10 collate the majority of sensing studies performed together with Limit of Detection (LOD) values and separate them into those that use the S-band (red–yellow PL) from the many recent studies exploiting the F-band (blue–green PL).

General advantages of fluorimetric (PL) sensing include low background signals and thereby high sensitivity, fast data acquisition and low cost. General challenges include sensor stability, specificity, repeatability, linearity, response and recovery times, and calibration. For Si-based sensors (Tables 9 and 10), approaches to improve specificity have included the immobilization of enzymes;[319] ratiometric PL spectroscopy;[302] combining PL quenching at a single wavelength with peak shifts;[303] or decay time changes.[307] With porous Si, multilayer microcavities offer multiparametric sensing.[317] Recently, pSi-based biosensors have focussed on reflectance-based interferometry rather than photoluminescence-based analysis[324] but as PLQY and more importantly, its stability, continues to improve, there could be a resurgence of interest in exploiting the S-band.

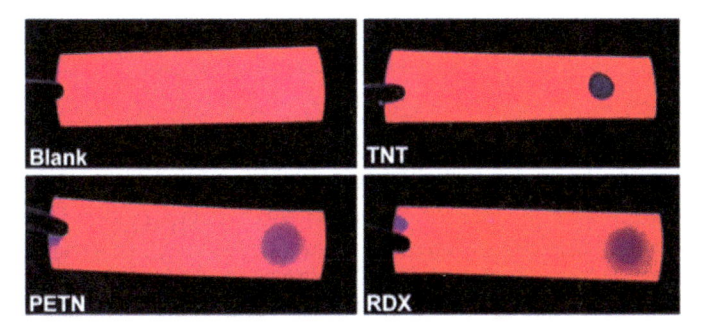

Fig. 18 S-band sensing of explosives.[314] Reproduced with permission from C. M. Gonzalez, M. Iqbal, M. Dasog, D. G. Piercey, R. Lockwood, T. M. Kiapotke and J. G. C. Veinot, *Nanoscale*, 2014. Copyright Royal Society of Chemistry.

Table 9 Sensing with S-band (red–green) photoluminescence

Analyte class	Analyte	Sensing mode	LOD	Silicon nanostructure	References
Environmental contaminants and toxins	Aromatic hydrocarbons	PL quenching	0.5 mM	pSi-hydride	(Fisher 1995)[295]
	Sulphur dioxide	PL quenching	440 ppb	pSi-oxide	(Kelly 1996)[304]
	Nitrogen dioxide	PL quenching and conductance	5 ppm	pSi multilayer	(Baratto 2002)[305]
	Alcohols	PL quenching	40 ppm	pSi-undecanoate	(Holec 2002)[306]
	Ethanol	PL quenching and PL decay rates	<620 ppm (vapour)	pSi- native oxide	(Dian 2007)[307]
	Alcohols	PL quenching	0.2 µM (vapour)	pSi-undecanoate	(Dian 2010)[308]
	Ethanol	PL quenching	380 ppm (vapour)	NP-hydride	(Zhang 2013)[309]
	Heavy metals	Immunoassay	10 nM	pSi-enzyme	(Syshchyk 2014)[310]
	Copper	PL quenching	1 µM	pSi-alkyl	(Hwang 2016)[311]
	Aflatoxin	PL quenching	2.5 pg mL^{-1}	pSi-Au-protein	(Myndrul 2017)[312]
	Metal ion discrimination	PL quenching and peak shift	10 µM	pSi-native oxide	(Jin 2019)[303]
Explosives	TNT, DNT	PL quenching	1–2 ppb	pSi-hydride	(Content 2000)[313]
	DNB, DNT	PL quenching and PL decay rates	<50 µM	NP-oxide	(Germanenko 2001)[314]
	DNT	PL quenching	18.2 ng (solid)	NP-alkyl	(Gonzalez 2014)[315]
	DNT	PL quenching	6 ppb (vapour)	NP-alkyl	(Nguyen 2016)[316]
Medical analytes	Nitric oxide	PL quenching	2 ppb	pSi-hydride	(Harper 1996)[296]
	Viral DNA	PL shift	194 fM	pSi multilayer	(Chan 2000)[317]
	Gram negative bacteria	PL shift	1.7 µg	pSi multilayer	(Chan 2001)[318]
	Urea	PL quenching	30 µM	pSi and enzyme	(Chaudhari 2005)[319]
	Ascorbic acid	PL quenching	2 mM	pSi-oxide	(La Ferrara 2012)[320]
	Intracellular Cu	PL quenching	2.5 µM	pSi-carboxyl	(Xia 2013)[321]
	AChE inhibitor	PL enhancement	1.25 µg mL^{-1}	pSi-enzyme	(Saleem 2014)[322]
	Human IgG	PL quenching	500 fM	pSi-protein	(Cho 2015)[323]
	Dopamine	Ratiometric PL quenching	4 µM	pSi-silica	(Hollett 2019)[302]

Table 10 Sensing with F-band (blue–yellow) photoluminescence

Analyte class	Analyte	Sensing mode	LOD	Silicon nanostructure	References
Environmental contaminants and toxins	Pesticides	PL enhancement	7–200 ng L^{-1}	NP-enzyme	(Yi 2013)[325]
	Mercury	PL quenching	50 nM	NP-amine	(Zhang 2014)[326]
	Copper	PL quenching	8 nM	NP-hydride	(Zhao 2014)[327]
	Tetracyclines		20-28 nM	NP-amine	(Lin 2015)[328]
	Chromium	PL quenching	2.7 µM	NP-dendrimer	(Campos 2015)[329]
	Copper		0.5 µM	NP-oxide	(Liao 2016)[330]
	Sudan I	PL quenching	39 nM	NP-hexadecylamine	(Jose 2016)[331]
	Hypochlorite	PL quenching	10 nM	NP-amine	(Guo 2016)[332]
	Chromium, hydrogen, sulphide	PL quenching and restoration	28 nM, 22 nM	NP-amine	(Zhu 2017)[333]
	Crystal violet	PL quenching	25 ng mL^{-1}	NP-amine	(Han 2018)[334]
	Ethyl carbamate	RF-ELISA	2.6 µg L^{-1}	NP-enzyme	(Luo 2018)[335]
	Nitrophenol	PL quenching	29 nM	NP-amine	(Han 2019)[336]
Explosives	TNT	PL quenching	1 nM	NP-amine	(Ban 2015)[337]
Medical analytes	Extracellular pH	PL quenching	pH 4–10	NP-amine	(Feng 2014)[338]
	Dopamine	PL quenching	0.3 nM	NP-amine	(Zhang 2015)[339]
	Intracellular pH	Ratiometric	pH 4–10	NP-amine	(Chu 2016)[340]
	ALP	PL quenching	1 U L^{-1}	NP-carboxyl	(Li 2018)[341]
	Creatinine	PL quenching	0.14 µM	NP-amine	(Meng 2019)[342]
	Hemoglobin	PL quenching	40 nM	NP-amine	(Li 2019)[343]

4.2. *In vitro* and *in vivo* imaging

Cellular and biomolecular imaging *in vitro* enables contactless studies of numerous physiological processes. Medical imaging facilitates noninvasive imaging of disease progression *in vivo*. These are critical tools in clinical diagnostics, physiology and cell biology.

A significant volume of literature now exists on using photoluminescent silicon nanostructures for both *in vitro* and *in vivo* imaging. It has been the focus of a few prior reviews[344–348] including a very recent one,[349] so only a short overview will be given here.

The S-band emission has been used extensively *in vitro* as a "fluorescent label" in cell cultures.[349–355] It is also increasingly being evaluated for *in vivo* use.[357–366]

The facile synthesis for achieving very high PLQY of F-band emission has led to a recent surge in its use as a label *in vitro* and there is now also a significant body of work on this.[367–380] In contrast, *in vivo* usage of F-band emission is very rare.[381]

S-band emission can be used *in vitro* and *in vivo*, whereas F-band emission is generally restricted to *in vitro* use, because of high tissue absorption of both UV excitation wavelengths and blue–green emission wavelengths. Tissue has two "transmission windows" in the near infrared – 700–900 nm ("*NIR 1*") and 1000–1700 nm ("*NIR II*"), where scattering and absorption of light is lower than adjacent wavelengths.[382] The S-band emission of Si nanostructures normally falls within the NIR1 window. Interestingly, with degenerate doping of nanocrystals to shrink their bandgap, room temperature PL output in NIR II has also been achieved, but the PLQY was relatively low at about 2%.[383] An ideal luminescent *in vivo* probe would have high PLQY with both excitation and emission occurring within these windows. Unfortunately, efficient PLQY of S-band emission necessitates short wavelength 300–450 nm excitation which limits the depth that can be imaged *in vivo*. For this reason there have been recent studies exploring two-photon excitation of the S-band[384,385] where both excitation and emission target the NIR I window. Sailor and co-workers have recently shown the feasibility of this with 850 nm excitation under two photon microscopy (TPM) and 560–740 nm emission. Although the efficiency of this process with 60 nm pSiNPs was significantly lower than an established TPM molecular probe like rhodamine 6G, the porous silicon probe showed much better photostability, the ability to target tumours and carry drug payloads.[366]

Gooding and co-workers have also utilized two-photon excitation, in combination with *in vitro* fluorescence lifetime imaging, to excite F-band emission.[386]

A wide range of luminescent nanostructured materials are being assessed for *in vivo* medical imaging[387] so I will summarize the perceived merits and challenges of using either the S-band or the F-band PL for biomedical imaging. There are 3 primary drivers for using S-band luminescence in Si nanostructures for *in vivo* imaging:

(a) Wavelength tunable and now very efficient luminescence output (see Table 7) within tissue window(s);

(b) Time-gated imaging that exploits the microsecond decay times to improve contrast;[357,388,389]

(c) Biodegradability[356,390] combined with good *in vivo* biocompatibility in different body sites.[391–393]

Challenges for *in vivo* use of the S-band include low optical absorption coefficients, the relatively broad FWHM of PL, colloidal stability of the PL in body fluids, and one-photon excitation wavelengths lying outside the tissue windows.

When using the S-band for *in vitro* cell imaging the primary advantages are tunable emission, the low cytotoxicity of silicon nanoparticles[394–400] and the flexible surface chemistry "toolbox" that has been realized (see Section 5.3).

For *in vitro* imaging exploiting the F-band emission, the primary advantages are very high PLQY (see Table 8), PL stability (less photobleaching and stable over a large pH range[401]) and narrower FWHM of PL than the S-band. Challenges for the F-band include the short excitation and emission wavelengths which overlap with tissue scattering and biological autofluorescence, and current uncertainties with regards to the origin of PL.

It should also be noted that whereas fluorescence labelling is a core technique used in *in vitro* cell biology, using luminescent agents *in vivo* is currently very much a developing research tool, rather than a mainstream clinical imaging modality. The latter point has relevance with regards to the likelihood of fast clinical translation of Si NP-based luminescence imaging and hence short-term impact. In this regard, the development of *biomodal imaging* probes, where photoluminescence imaging is combined with a more clinically established technique, has more potential for clinical translation. Unfortunately, silicon nanoparticles have low X-ray contrast compared to metals. Therefore more complex "nanoconstructs" are needed. There could be, for example, an opportunity for developing porous silicon with entrapped gas for a photoluminescent and biodegradable ultrasonic contrast agent. Table 11 provides examples from the literature – attention to date has been on accessing MRI rather than ultrasonic imaging.

4.3. Theranostics

Theranostic nanomedicine, where nanostructures simultaneously facilitate disease therapy and diagnosis, is currently dominated by drug delivery based therapies.[407] In particular, the pre-clinical evaluation of pharmacokinetics and the target site and healthy organ accumulation of drugs.[408] The potential of theranostics with nano-Si was initiated by the pioneering studies of Sailor and co-workers [357] who demonstrated *in vivo* imaging of S-band and organ targeting. The field has been very recently reviewed[349] so only a short perspective is given here.

Whilst for many therapies the luminescence provides only a means to track the nanostructures, the photosensitization of oxygen scales with PLQY.[409] In contrast to photodynamic therapy, for photothermal therapy one wants to maximize non-radiative processes, so PL is unwanted. Thus luminescent Si nanostructures can

Table 11 Dual imaging with photoluminescent Si nanostructures

Photoluminescence emission	2nd imaging mode	Silicon nanostructure	References
F-band (540 nm)	MRI	3 nm NPs with iron oxides	(Sato 2011)[402]
F-band (430 nm)	MRI	3 nm NPs doped with iron	(Singh 2012)[405]
S-band (750 nm)	MRI	170 nm complex with 3 nm NPs and chelated Gd^{3+}	(Erogbogbo 2012)[361]
S-band (500–800 nm)	MRI	pSi and iron oxide NPs	(Xia 2017)[403]
F-band (460 nm)	MRI	3.85 nm NPs with ^{19}F ligands	(Li 2018)[404]
F-band (400–510 nm)	MRI	6.6 nm NPs doped with gadolinium	(Li 2018)[381]
F-band	PET	2.4 nm NPs with NIR dye and ^{64}Cu radiolabel	(Singh 2020)[406]

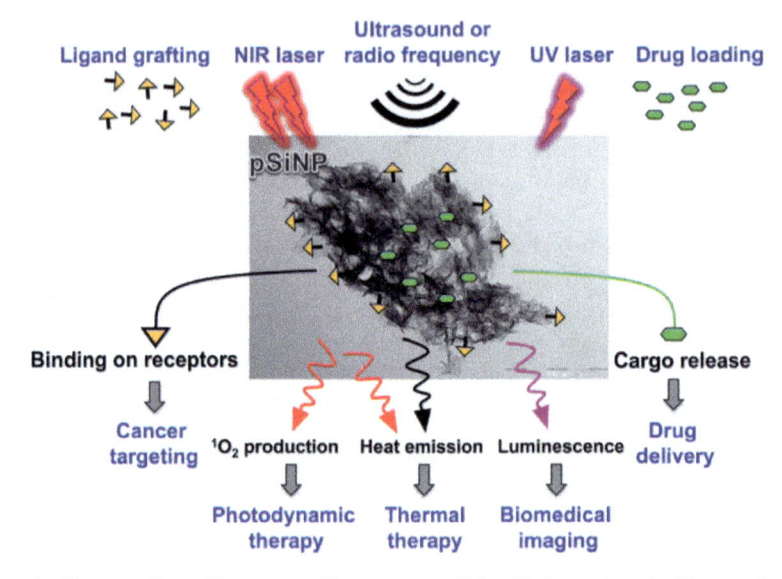

Fig. 19 Theranostics with porous silicon nanoparticles.[412] Reproduced with permission from V. Stojanovic, F. Cunin, J. O. Durand, M. Garcia and M. Gary-Bobo, *J. Mater. Chem. B*, 2016, 4, 7050–7059. Copyright Royal Society of Chemistry.

themselves provide therapy (photosensitization[409,410]) or carry payloads that perform therapy (drugs, radioisotopes, metals[349]). Luminescent pSi is ideally suited to theranostics[411,412] because of the much higher drug payloads achievable *via* filling the pore volume, compared to solid nanoparticles where the drug is attached only to the outer surfaces. At 80% porosity for example, drug payloads in pSi can exceed 50 wt% by melt loading techniques and the drug is itself nanostructured by entrapment within 5–50 nm mesopores. Control over pharmacokinetics is achieved *via* a combination of restricted diffusion and carrier biodegradability. Fig. 19 illustrates the multifunctionality of porous silicon nanoparticles (pSiNP).[412]

4.4. Photovoltaics

A major limiting factor in the ultimate efficiency of silicon solar cells is the spectral mismatch between the energy distribution of the incident solar spectrum and the bandgap. An efficient photoluminescent material can act as a spectral converter to raise the efficiency of a solar cell by down-shifting (DS), quantum cutting (QC) or "upconversion" (UC) as shown schematically in Fig. 20.[413]

Visibly photoluminescent silicon can provide "down-shifters"[414–420] or quantum cutters[421–425] for traditional solar cells.

A second, recently emerged potential use is in "luminescent solar concentrators" (LSCs) for building integrated photovoltaics.[426–428] Here, the large Stokes shift between strong UV absorption and S-band emission is of great benefit in minimizing light re-absorption over long distances. Solar concentrators are large plastic plates filled with photoluminescent material (commonly organic dyes) that redirect and concentrate solar energy to small photovoltaic cells at their perimeter (see Fig. 21). A large fraction of the light can be effectively waveguided in the

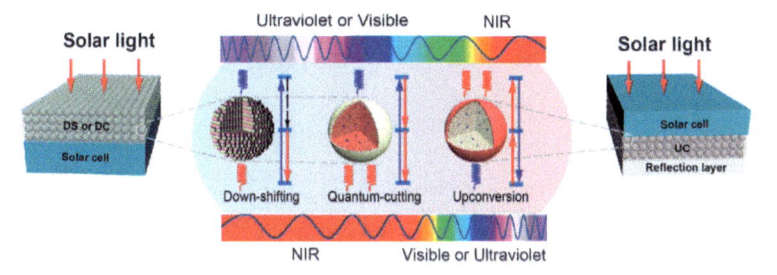

Fig. 20 Classes of spectral converters.[413] Reproduced with permission from C. X. Huang, S. Han, W. Huang and X. Liu, *Chem. Soc. Rev.*, 2013, **42**, 173–201. Copyright Royal Society of Chemistry.

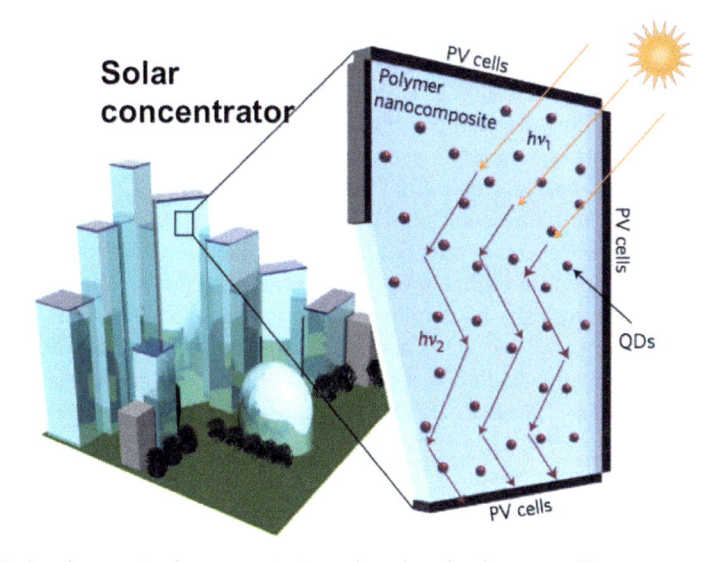

Fig. 21 Luminescent solar concentrator using photoluminescent silicon nanocrystals.[426] Reprinted with permission from F. Meinardi, S. Ehrenberg, L. Dhamo, F. Carulli, M. Mauri, F. Bruni, R. Simonutti, U. Kortshagen and S. Brovelli, *Nature Photonics.*, 2017, **11**, 177–186. Copyright 2017, Springer Nature.

plastic sheet due to total internal reflections. The tunable bandgaps of silicon nanostructures offer design freedom with regards to aesthetics and optical performance. Challenges for silicon solar concentrator use include low absorptivity, scattering due to nanocrystal agglomeration, scalability of production and above all, cost.

4.5. Colour converters for white LEDs

White light emitting LEDs of 5 lm W^{-1} brightness first became commercially available in 1996, and by 2010, brightness had already improved to the 150–250 lm W^{-1} range. The performance and cost of white LEDs are now competitive for general lighting, automotive, communication and medical uses.[429] There are 3 methods for generating white light *via* an LED: (a) combining blue, green and red

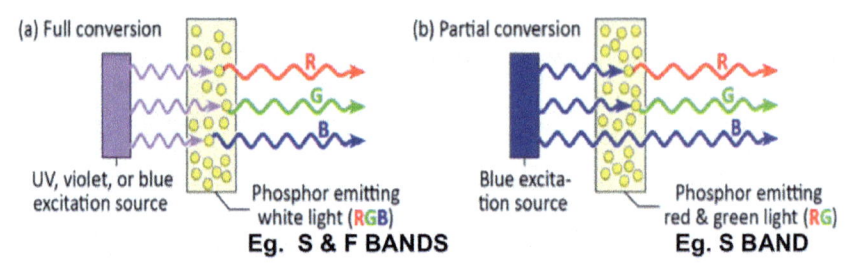

Fig. 22 White light generation using red, green or blue photoluminescent materials.[429] Adapted with permission from J. Cho, J. H. Park, J. K. Kim and E. F. Schubert, *Laser Photonics. Rev.*, 2017, **11**(2), 1600147. Copyright John Wiley & Sons.

LEDs; (b) combing a UV LED with blue, green and red phosphors; and (c) using a blue LED to excite green or yellow and red phosphors. Fig. 22 illustrates methods (b) and (c). The latter approach (c) has received increasing use following the development of ultra-bright blue GaN LEDs in the 1990s.

For general lighting the total power emitted, power efficiency and "whiteness" (colour temperature and rendering) are critical technical parameters. A 60 W incandescent tungsten light bulb, for example, currently has a luminous flux of 1000 lumins (lm) and a luminous efficiency of 17 lm W^{-1}. The colour rendering index (CRI) is a quantitative measure of a light source's ability to reproduce the colours of objects against natural or reference colours. Perfect performance is 100. Fluorescent lights typically have values in the range 50–98. LEDs are typically around 80. Colour correlated temperature (CCT) quantifies the temperature of an ideal blackbody radiating light of a colour comparable to the light source. Daylight ("cool") white has a CCT values above 5000 K, neutral white 3000–5000 K, whilst incandescent ("warm") white corresponds to a CCT value in the range 2500–3000 K.

In 2014 the start-up company Lumisands Inc, USA, reported the first use of visibly photoluminescent Si to generate "warm" white LEDs by method (b), and combined the use with other green and blue phosphors.[430] The Si NPs were derived from porous silicon. Surprisingly high PLQY values for the S-band were estimated under blue (405 nm) excitation, especially since the authors reported their particles had "*micron-size silicon cores with nanosize Si QD attached*". One would expect significant scattering and re-absorption of red emission with this type of structure. Colour rendering close to incandescent bulbs with a CCT ~2800 K and CRI ~95 were reported, without a value for luminous efficiency.

Broad band white PL has been reported from a number of silicon-containing nanostructures: heavily oxidized porous silicon,[431] oxidized and carbon containing porous silicon,[432,433] carbon doped silicon-rich silicon oxide[434] and silicon and carbon co-implanted silica films.[435] Ghosh *et al.*[436] have also used method (b) but with white-light emitting Si nanocrystals containing material derived from rice husks (see Section 5.4) whose broad PL (230 nm FWHM) was possibly a combination of the S-band and F-band. In this regard, of particular relevance is the study where laser ablation of organosilica also produced white-emitting nanoparticles.[437] Both silicon and carbon nanoparticles were discovered by TEM and are jointly attributed with white light generation. The carbon dots were revealed after HF removal of the silica phase, as used by Chen *et al.*[286] and discussed in Section 3.6.

A study with commercial blue LEDs and porous silicon NPs embedded in an overlying polymer film demonstrated the flexibility of colour tuning by varying Si NP concentrations.[438] Luminous efficiency was again not reported. For maximizing brightness, thin down-converting films should have low refractive indices to maximize light extraction efficiencies.[128]

The F-band has also been investigated for this application.[439–441] Solution synthesis of nanoparticles was adapted to achieve broad yellow-green emissions centred at 540 nm under 410 nm excitation from a blue LED.[439] The 43.5% EQE of the blue LED was reported to be lowered to 8% for the resulting white LED. A luminance efficiency value of 13 lm W^{-1} was provided together with CCT of 7982 K and a CRI of 82. In a second study by He's group, luminescent nanorods achieved a similar luminance efficiency.[440] In a third study, green emitting NPs with PLQY of 60%, derived from rice husks, were combined with blue and red emitting materials. Under 3 W UV LED excitation the composite film generated white light with a luminance efficiency of 75 lm W^{-1}. Photostability of the green-emitting material was monitored over 1 month (720 hours) and PLQY dropped by 12%.

However, high reliability and a very long operating lifetime (50 000–100 000 hours) is one of the key features of LEDs that have led to their remarkable uptake in recent years. Successful inorganic color-converting phosphors like YAG:Ce have ultrahigh chemical stability that can withstand continuous ~100 mW mm^{-2} radiance from a blue LED. Long-term photostability testing under intense blue/UV excitation is needed for Si nanostructures, as this might be a daunting challenge for this application area.

4.6. Nanothermometry

Measuring temperature with nanometer scale spatial resolution has potential widespread use in electronics, photonics and biomedicine but also a range of other niche areas.[442] There are diverse methods but using luminescent probes of temperature are one of the most attractive, as they are non-invasive, contactless and easy to use.[443] Most PL parameters are dependent to some extent on temperature: intensity (efficiency), spectral position, spectral shape, bandwidth, polarization and decay time. Which PL parameter(s) and which applications have the most potential for Si nanostructures? This topic is only starting to receive attention so it is difficult to gauge. Suggested specific applications to date range from tribology[446] to cryosurgery.[444]

Researchers at INSA, Lyon, were the first to realize the potential of S-band PL in this regard and they chose tribological studies of mechanical contacts as the target application area.[445–450] Functionalized 4–8 nm porous Si NPs with S-band emission were studied in alkanes and alkenes from 293–393 K. As expected, with increasing temperature, red shifts of PL peak position were universally observed but data analysis revealed that shift magnitude also depended on solvent viscosity. They attributed this to Brownian motion causing energy transfer between nanoparticles of different size.[450] A consequence is that temperature sensitivity will have a complex dependence on liquid viscosity, NP concentration, NP size distribution and NP passivation. A significant advance was therefore shown by Ferreira and Kortshagen's groups[451] who demonstrated that with better size control (2.0 ± 0.2 nm) of surface oxidized NPs, exciton migration effects were

Table 12 Thermal sensitivities of S and F-band PL for potential nanothermometry applications

PL band and peak emission (nm)	PL parameter	Temperature range (K)	Thermal sensitivity	References
S (1.65 eV)	Decay time, peak position, intensity	293–393, 293–373	0.1–0.3 µs K^{-1} (<0.75%/K), 0.18 meV K^{-1}, −0.8%/K	(Ryabchikov 2012, 2013)[445–447]
F (480) F (2.35 eV)	Decay time Peak position, bandwidth	273–333 293–393	−0.24 ns K^{-1} +0.97 meV K^{-1}, +1.2 meV K^{-1}	(Li 2013)[452] (Ryabchikov 2014)[448]
S (630)	Peak position	303–383	−0.9 meV K^{-1}	(Hajjaji 2014)[449]
S (1.7 eV)	Peak position	303–390	<0.052%/K	(Hamza 2015)[450]
S (650)	Peak position	13–480	−0.04%/K	(Botas 2016)[451]

greatly reduced and the emission thermal dependence mainly follows the shrinkage of the bandgap. A single calibration then suffices for different media.

Use of the blue–green F-band in thermometry has also been reported.[448,452] The low temperature solution synthesis route[452] yielded very stable 480 nm emission with also very high 75% PLQY and monoexponential decay, but the PL origin is not established, and hence neither is the biodegradability of the luminescent entity (see Section 3.6). They chose to utilize decay times which linearly decreased from 23 ns at 0.5 °C to 8 ns at 60 °C, yielding a sensitivity of 0.24 ns °C^{-1} with a thermal resolution under 1 °C. Interestingly, when Ryabchikov and co-workers[448] compared the thermal sensitivity of Si NCs with carbon NPs (produced by anodization of Si and SiC wafers, respectively) they found spectral shifts to be 3-4 times higher for the Si nanomaterial and of opposite sign to those of carbon dots (−0.18 vs. +0.05 meV K^{-1}, respectively). Future comparative studies of the temperature responses of high PLQY F-band emissions from Si-based formulations and those of established "carbon dots" may thus provide further insight into the origin of the former.

The quantitative thermal sensitivity values reported to date for both S and F-bands are shown in Table 12. It is clear that, at least for the S-band for biomedical uses, the standard figures of merit are currently significantly lower than the highest values (1–5%/K) of other material systems.[442] Nonetheless, with further improvements of surface passivation, size control and PLQY, the thermal sensitivity should improve, and this combined with having a thermometer system of tunable biodegradability, makes this an interesting area for further development.

5. Summary and future perspectives

5.1. PL efficiency and passivation (stability)

Progress in improving PLQY of both the S-band (Table 7) and F-band (Table 8) over the last 30 years has been spectacular. Photostability of the S-band from derivatized silicon quantum dots is now competitive against many organic dyes with regards to biotagging. For applications like LED's and photovoltaics where very long term stability is essential, further improvements in passivation seem

Table 13 Surface passivation (SRV values) for large silicon nanolayers and nanowires

Silicon nanostructure	Surface chemistry	Technique	SRV (cm s^{-1})	References
2D Si nanolayer (90–200 nm)	Hydride	1% HF treatment	22.5 → 399 (10 K) air storage	(Coustel 2011)[453]
2D Si nanolayer (90–200 nm)	Alkyl groups	Hydrosilylation	62 → 209 (10 K) air storage	(Coustel 2011)[453]
Si nanowire (114, 224 nm diameter)	Thermal oxide (7–21 nm thick)	Diluted oxygen and forming gas at 960 °C	20	(Demichel 2010)[454]
Si nanowire (25–125 nm diameter)	Thermal oxide (5–10 nm thick)	Diluted oxygen at 1000 °C	10 000–30 000	(Grumstrup 2014)[455]
Si nanowire (50–90 nm diameter)	a-Si (10 nm thick)	VLS deposition	4500	(Dan 2011)[456]

mandatory. How do the effectiveness of surface passivation chemistries developed by the PV community for bulk Si compare with those used on silicon nanostructures? There are only a few studies that have estimated SRV values for 1 and 2D confinement in relatively large nanostructures that hence showed no S-band emission, as shown in Table 13.

As for bulk silicon, hydride and organic monolayer passivation is metastable with SRV values rising with air storage. SRV values for nanowires can be remarkably high, reflecting perhaps challenges in avoiding defects along elongated nanostructures with highly curved interfaces.

5.2. Size control and wavelength tunability

If properties are size-dependent then size control and also metrology of nano-size distribution becomes critical.[457,458] Control over size can be strived for during synthesis, or after synthesis, or a combination of the two. Some of the narrowest size distributions for as-synthesized silicon nanocrystals have been demonstrated with superlattice manipulation,[459] gas phase laser ablation,[460] plasma synthesis[461] and micro-emulsion techniques.[462,463] Some data are shown in Table 14. Various methods have also been applied to narrow size distributions of already fabricated luminescent silicon nanostructures (see Table 15). These include gas phase selection techniques,[460–462,465,466] size selective precipitation,[463] size exclusion chromatography[199] and density gradient ultracentrifugation.[196,468,469]

The S-band has been tuned over a wide spectral range but currently very high PLQY (>50%) is restricted to the 700–1000 nm red-near infrared region (see Table 7). The F-band has been tuned over a complementary spectral region with very high PLQY (>50%) in the 440–600 nm blue to yellow region (see Table 8).

5.3. Surface chemistry toolbox

The well-developed surface chemistry of bulk silicon[470] was the platform from which many functionalization routes have been successfully applied to luminescent silicon nanostructures.[471] The availability of metastable hydride surfaces has been both a help and a hindrance in this regard: it has facilitated the grafting of a wide range of organic capping ligands but 100% substitution seems virtually impossible due to steric hindrance issues. The resulting derivatized surfaces then

Table 14 Size control via synthesis techniques

References	(Tilley 2005)[462], (Rosso-Vassic 2008)[463]	(Patrone 2000)[460]	(Tan 2011)[265]	(Zacharias 2002)[171]	(Kothermann 2019)[458]	(Gresback 2011)[461]
Method	Reverse micelles	Laser ablation (gas phase)	Laser ablation (liquid phase)	Superlattice (reactive evaporation)	Superlattice (PECVD)	Plasma synthesis
Mean nanocrystal size (nm)	1.6–1.8	1.8	2.37	3.4	3.7	5.0
Size variation (SD nm)	0.2	0.2	0.56	0.5	0.7	1.0
PL emission band	F	S	F	S	S	S

Table 15 Size control via post-synthesis techniques

References	(Littau 1993)[199]	(Camata 1996)[464]	(Mastronardi 2011)[467]	(Miller 2012)[468]	(Liu 2016)[208]
Method	Size exclusion chromatography	Gas phase differential mobility	Size selective precipitation	Density gradient ultracentrifugation	Density gradient ultracentrifugation
Mean size (nm)	7.5	2.8	1.61	3.60	2.81
Size variation (SD nm)	2.5	0.6	0.26	0.35	0.25
PL emission band	S	—	S	S	S

have a mixed chemical nature with the desired Si–C bonding plus residual Si–H bonding. Oxidation is suppressed but not eliminated. Complex surface chemistries are more exploitable in *ex vivo* uses like sensing than *in vivo* imaging where clinical translation is needed for success. For applications where long term PL stability is of paramount importance, attention should focus on inorganic coatings that have a proven track record with regards to surface passivation of silicon (*cf.* Table 1). The challenge is to find scalable techniques that realize full surface coverage of nanostructures with low SRV values (*cf.* Table 13).

5.4. Scalability and cost of luminescent silicon

Nanostructures are typically more much expensive than bulk materials and more difficult to manufacture at scale. Table 16 summarizes some reported throughputs for silicon nanocrystals. It is evident that the majority of these for *luminescent* nanocrystals lie in the range 0.1–10 g h^{-1} (*i.e.* less than 0.01 kg h^{-1}). Whilst pilot plant production rates of about 1 kg h^{-1} have been reported, the nanocrystals were not small enough to be photoluminescent.

Exciting progress has been made with silicon nanostructure synthesis by bottom-up high temperature techniques with molecular precursors and conversion synthesis with biogenic silica structures. Continuous belt-furnace production using "silicon accumulator plant" nanostructures[478] is one very scalable technique for producing kilogrammes of biodegradable porous silicon. However, no efficient S-band emission has been reported to date from biogenic porous silicon. Where luminescent nanostructures have been achieved from plant material it has primarily been restricted to F-band emission[441,479,480] and thus might be from carbonaceous material rather than any silicon nanoparticles also present. A significant goal is thus to achieve tunable S-band output of high PLQY from, for example, bamboo-derived silicon nanostructures. This would open a suite of new high volume and low cost uses (see next section). Dasog's group

Table 16 Throughputs of various techniques for fabricating Si nanocrystals. As in Table 4, question marks signify considerable uncertainty regarding origin of F-band PL

Silicon nanostructure	Luminescent?	Fabrication technique	Stated throughput (g h^{-1})	References
Nanocrystals, <5 nm	Yes (S-band)	Silane pyrolysis and stain etching	0.02–0.2	(Li 2003)[472]
Nanocrystals, 4–9 nm	Yes (S-band)	Silane reduction	0.1–1	(Lacour 2007)[473]
Nanocrystals, 100–500 nm	No	Silane reduction in pilot plant	750	(Hulser 2011)[474]
Nanocrystals, (?) 2 nm	Yes (F-band)	Microwave assisted chemical synthesis	0.6	(Zhong 2013)[245]
Nanocrystals, 2–10 and 50–120 nm	Yes (S-band)	Plasma decomposition of silane	6	(Dogan 2013)[475]
Nanocrystals, (?) 4–9 nm	Yes (F-band)	Triethoxysilane reduction	4–12	(Pujari 2018)[476]
Nanocrystals, 10–40 nm	No	Silane reduction	22–200	(Kunze 2019)[477]

have recently demonstrated that high S-band PLQY (∼50%) can be achieved *via* magnesiothermic reduction techniques.[481] Synthetic sol–gel silica precursors were used. If this can be replicated with inexpensive biogenic silica feedstocks then the cost of visibly luminescent silicon should be significantly lowered in the future.

5.5. Novel applications

If the cost and scalability of luminescent nanosilicon can be lowered substantially then a range of further applications could emerge as commercially viable. In most cases, ultrahigh PLQY and much lower cost would need to be supplemented with another key attribute for that application. I mention a few examples here from the diverse fields of dentistry, cosmetics, environmental remediation, forensics and security:

- Sunlight-driven photocatalyst (*e.g.* adsorbed toxicant degradation).
- Anti-ageing sunscreens (blue emitters and UV blockers).
- Dental "optical brightener" (tooth whitening formulations use blue emitters).
- Cosmetic skin hue converter (green-emitters used to mask red blemishes).
- Luminescent inks for banknotes (security features).
- Contrast agent in forensics (*e.g.* luminescent dusting powder).

6. Concluding comments

Remarkable progress has been made in raising the efficiency of visible emission from silicon-based quantum dots, superlattices and porous silicon over the last 30 years, with efficiencies of S-band red emission around 60–70% and green to blue F-band emission around 80–90% being the highest reported values. Surface chemistry plays a pivotal role in determining the efficiency and stability of emission. The widely tunable S-band arises from phonon-assisted exciton recombination within the silicon nanostructure. Emission in the F-band spectral range can have more than one origin, but carbon-based nanostructures and molecular chromophores are looking like the most likely origin of very efficient output. Some low temperature synthesis routes need to be revisited in this regard with more attention given to luminescent nanoparticle characterization and purification. Given the abundance of silica, low toxicity and biodegradability of luminescent silicon nanostructures compared to many other luminescent material nanosystems, their further development for a myriad of applications looks inevitable.

Conflicts of interest

There are no conflicts of interest to declare.

References

1 A. Gupta and H. Wiggers, Free-standing silicon quantum dots: origin of red and blue emission, *Nanotechnology*, 2011, **22**, 055707.
2 J. I. Pankove, *Optical processes in semiconductors*, Dover Press, 1975.

3 J. R. Haynes and W. C. Westphal, Radiation resulting from recombination of holes and electrons in silicon, *Phys. Rev.*, 1956, **101**, 1676–1678.
4 J. D. Cuthbert, Recombination kinetics of excitonic molecules and free excitons in intrinsic silicon, *Phys. Rev. B: Solid State*, 1970, **1**, 1552–1557.
5 J. R. Haynes, Experimental observation of the excitonic molecule, *Phys. Rev. Lett.*, 1966, **17**(16), 860–862.
6 P. L. Gourley and J. P. Wolfe, Thermodynamics of excitonic molecules in silicon, *Phys. Rev. B: Condens. Matter Mater. Phys.*, 1979, **20**(8), 3319–3327.
7 M. L. W. Thewalt and W. G. McMullen, Green and near-infrared luminescence due to the biexcitons in unperturbed silicon, *Phys. Rev. B: Condens. Matter Mater. Phys.*, 1984, **30**(10), 6232–6234.
8 A. G. Steele, W. G. McMullan and M. L. W. Thewalt, Discovery of polyexcitons, *Phys. Rev. Lett.*, 1987, **59**(25), 2899–2902.
9 R. B. Hammond, T. C. McGill and J. W. Mayer, Temperature dependence of the electron-hole liquid luminescence in silicon, *Phys. Rev. B: Solid State*, 1976, **13**(8), 3566–3575.
10 P. L. Gourley and J. P. Wolfe, Spatial condensation of strained confined excitons and excitonic molecules into an electron-hole liquid in silicon, *Phys. Rev. Lett.*, 1978, **40**(8), 526–530.
11 J. R. Haynes, Experimental proof of a new electronic complex in silicon, *Phys. Rev. Lett.*, 1960, **4**(7), 361–363.
12 R. C. Enck and A. Honig, Radiative spectra from shallow donor-acceptor electron transfer in silicon, *Phys. Rev.*, 1969, **177**(3), 1182–1193.
13 M. Tajima, T. Iwai, H. Toyota, S. Binetti and D. Macdonald, Fine structure due to donor-acceptor pair luminescence in compensated Si, *Appl. Phys. Express*, 2010, **3**, 071301.
14 M. Tajima, T. Iwai, H. Toyota, S. Binetti and D. Macdonald, Donor-acceptor pair luminescence in compensated Si for solar cells, *J. Appl. Phys.*, 2011, **110**, 043506.
15 M. A. Tamor and J. P. Wolfe, Drift and diffusion of free excitons in Si, *Phys. Rev. Lett.*, 1980, **44**(25), 1703–1706.
16 P. J. Dean, J. R. Haynes and W. F. Flood, New radiative recombination processes involving neutral donors and acceptors in silicon and germanium, *Phys. Rev.*, 1967, **161**, 711–729.
17 M. Tajima, K. Tanaka, S. Dubois, J. Veirman, K. Nakagawa and A. Ogura, Photoluminescence due to impurity cluster bound exciton in highly doped and highly compensated silicon, *Jpn. J. Appl. Phys.*, 2015, **54**, 111304.
18 J. Weber, W. Schmidt and R. Sauer, Localized exciton bound to an isoelectronic trap in silicon, *Phys. Rev. B: Condens. Matter Mater. Phys.*, 1980, **21**(6), 2401–2414.
19 N. A. Drozdov, A. A. Patrin and V. D. Tkatchev, Recombination radiation on dislocations in silicon, *JETP Lett.*, 1976, **23**(11), 597–599.
20 R. Sauer, J. Weber and J. Stolz, Dislocation-related photoluminescence in silicon, *Appl. Phys. A: Solids Surf.*, 1985, **36**, 1–13.
21 H. Weman, B. Monemar, G. S. Oehrlein and S. J. Jeng, Strain-induced quantum confinement of carriers due to extended defects in silicon, *Phys. Rev. B: Condens. Matter Mater. Phys.*, 1990, **42**(5), 3109–3112.

22 K. Bothe, R. J. Falster and J. D. Murphy, Room temperature sub-bandgap photoluminescence from silicon containing oxide precipitates, *Appl. Phys. Lett.*, 2012, **101**, 032107.
23 A. M. Stoneham, Non radiative transitions in semiconductors, *Rep. Prog. Phys.*, 1981, **44**, 1251–1295.
24 A. Richter, S. W. Glunz, F. Werner, J. Schmidt and A. Cuevas, Improved quantitative description of Auger recombination in crystalline silicon, *Phys. Rev. B: Condens. Matter Mater. Phys.*, 2012, **86**, 165202.
25 A. Hangleiter, Nonradiative recombination at deep defect levels in silicon: experiment, *Phys. Rev. B: Condens. Matter Mater. Phys.*, 1987, **35**(17), 9149–9161.
26 C. R. Helms and E. H. Poindexter, The silicon- silicon dioxide system: its microstructure and imperfections, *Rep. Prog. Phys.*, 1994, **57**(8), 791.
27 W. Bludau, A. Onton and W. Heinke, Temperature dependence of the bandgap of silicon, *J. Appl. Phys.*, 1974, **45**(4), 1846.
28 L. T. Canham, Room temperature photoluminescence from etched silicon surfaces: the effects of chemical pretreatments and gaseous ambients, *J. Phys. Chem. Solids*, 1986, **47**(4), 363–373.
29 V. Alex, S. Finkbeiner and J. Weber, Temperature dependence of the indirect energy gap in crystalline silicon, *J. Appl. Phys.*, 1996, **79**, 6943.
30 A. G. Cullis, L. T. Canham and P. D. J. Calcott, The structural and luminescence properties of porous silicon, *J. Appl. Phys.*, 1997, **82**(3), 909–956.
31 G. Davies, The optical properties of luminescent centres in silicon, *Phys. Rep.*, 1989, **176**(3–4), 83–188.
32 L. T. Canham, M. R. Dyball, W. Y. Leong, M. R. Houlton, A. G. Cullis and P. W. Smith, Radiative recombination channels due to hydrogen in crystalline silicon, *Mater. Sci. Eng.*, 1989, **B4**, 41–45.
33 R. S. Bonilla, B. Hoex, P. Hamer and P. R. Wilshaw, Dielectric surface passivation for silicon solar cells: a review, *Phys. Status Solidi A*, 2017, **214**(7), 1700293.
34 E. Yablonovitch, D. L. Allara, C. C. Chang, T. Gmitter and T. B. Bright, Unusually low surface recombination velocity on silicon and germanium surfaces, *Phys. Rev. Lett.*, 1986, **57**(2), 249–252.
35 Q. Sun, L. Melnikov, A. Mandelis and R. H. Pagliaro, Surface recombination velocity imaging of wet-cleaned silicon wafers using quantitative heterodyne lock-in carrierography, *Appl. Phys. Lett.*, 2018, **112**, 012105.
36 E. J. Nemanik, P. T. Hurley, B. S. Brunschwig and N. S. Lewis, Chemical and electrical passivation of silicon (111) surfaces through functionalization with sterically hindered alkyl groups, *J. Phys. Chem. B*, 2006, **110**, 41800–14808.
37 M. J. Kerr and A. Cuevas, Very low bulk and surface recombination in oxidiized silicon wafers, *Semicond. Sci. Technol.*, 2002, **17**, 35–38.
38 K. A. Collett, R. S. Bonilla, P. Hamer, G. Bourret-Sicotte, R. Lobo, T. Kho and P. R. Wilshaw, An enhanced alneal process to produce SRV <1 cm s^{-1} in 1 ohm cm n-type Si, *Sol. Energy Mater. Sol. Cells*, 2017, **173**, 50–58.
39 Y. Wan, K. R. McIntosh, A. F. Thomson and A. Cuevas, Low surface recombination velocity by low absorption silicon nitride on c-Si, *IEEE J. Photovolt.*, 2013, **3**, 554–559.
40 B. Hoex, J. Schmidt, P. Pohl, M. C. M. van de Sanden and W. M. Kessels, Silicon surface passivation by atomic layer deposited Al$_2$O$_3$, *J. Appl. Phys.*, 2008, **104**, 044903.

41 R. S. Bonilla, C. Reichel, M. Hermle and P. R. Wilshaw, Extremely low surface recombination in 1 ohm cm n-type monocrystalline silicon, *Phys. Status Solidi RRL*, 2017, **11**, 1600307.

42 W. J. Royea, A. Juang and N. S. Lewis, Preparation of air-stable low recombination velocity Si (111) surfaces through alkyl termination, *Appl. Phys. Lett.*, 2000, **77**(13), 1988–1990.

43 J. Bullock, D. Kiriya, N. Grant, A. Azcati, M. Hettick, T. Kho, P. Phang, H. C. Sio, D. Yan, D. Macdonald, M. A. Quevedo-Lopez, R. M. Wallace, A. Cuevas and A. Javey, Superacid passivation of crystalline Si surfaces, *ACS Appl. Mater. Interfaces*, 2016, **8**, 24205–24211.

44 N. E. Grant, T. Niewelt, N. R. Wilson, E. C. Wheeler-Jones, J. Bullock, M. Al-Amin, M. C. Schubert, A. C. van Veen, A. Javey and A. D. Murphy, Superacid treated Si surfaces: extending the limit of carrier lifetime for photovoltaic applications, *IEEE J. Photovolt. Appl.*, 2017, **7**(6), 1574–1583.

45 J. Chen, K. Ge, C. Zhang, J. Guo, L. Wang, D. Song, F. Li, Z. Xu, Y. Xu and Y. Mai, Vacuum-free room temperature organic passivation of silicon: towards very low recombination of micro/nanotextured surface structures, *ACS Appl. Mater. Interfaces*, 2018, **10**, 44890–44896.

46 Y. Yamashita, A. Asano, Y. Nishioka and H. Kobayashi, Dependence of interface states in the Si band gap on oxide atomic density and interfacial roughness, *Phys. Rev. B: Condens. Matter Mater. Phys.*, 1999, **59**(24), 15872–15881.

47 L. T. Canham, Silicon quantum wire array fabrication by electrochemical and chemical dissolution of wafers, *Appl. Phys. Lett.*, 1990, **57**(10), 1046–1048.

48 H. Takagi, H. Ogawa, Y. Yamazaki, A. Ishikazi and T. Nakagiri, Quantum size effects on photoluminescence in ultrafine Si particles, *Appl. Phys. Lett.*, 1990, **56**, 2379.

49 Y. Ding and J. Ni, Electronic structures of silicon nanoribbons, *Appl. Phys. Lett.*, 2009, **95**, 083115.

50 Y. H. Tang, L. Z. Pei, Y. W. Chen and C. Guo, Self-assembled silicon nanotubes under supercritically hydrothermal conditions, *Phys. Rev. Lett.*, 2005, **95**, 116102.

51 H. Nakano, T. Mitsuoka, M. Harada, K. Horibuchi, H. Nozaki, N. Takahashi, T. Nonaka, Y. Seno and H. Nakamura, Soft synthesis of single-crystal silicon monolayer sheets, *Angew. Chem., Int. Ed.*, 2006, **45**, 6303–6306.

52 U. Kim, I. Kim, Y. Park, K. Y. Lee, S. Y. Yim, J. G. Park, H. G. Ahn, S. H. Park and H. J. Choi, Synthesis of Si Nanosheets by a Chemical Vapor Deposition Process and Their Blue Emissions, *ACS Nano*, 2011, **5**(3), 2176–2181.

53 S. W. Kim, J. Lee, J. H. Sung, D. Seo, I. Kim, M. H. Jo, B. W. Kwon, W. K. Choi and H. J. Choi, Two-Dimensionally Grown Single-Crystal Silicon Nanosheets with Tunable Visible-Light Emissions, *ACS Nano*, 2014, **8**(7), 6556–6562.

54 D. Karar, N. R. Banddyopadhyay, A. K. Pramanick, D. Achharya, G. Conibeer, N. Banerjee, O. E. Kusmartseva and M. Ray, Quasi-two-dimensional luminescent silicon nanosheets, *J. Phys. Chem. C*, 2018, **122**, 18912–18921.

55 Y. Kanemitsu, K. Suzuki, M. Kondo and S. Kyushin, Luminescence properties of a cubic silicon cluster: octacubasilane, *Phys. Rev. B: Condens. Matter Mater. Phys.*, 1995, **51**, 10666.

56 Q. Wang, B. Xu, J. Sun, H. Liu, Z. Zhao, D. Yu, C. Fan and J. He, Direct band gap silicon allotropes, *J. Am. Chem. Soc.*, 2014, **136**, 9826–9829.

57 A. Kailer, Y. G. Gogotsi and K. G. Nickel, Phase transformations of silicon caused by contact loading, *J. Appl. Phys.*, 1997, **81**(7), 3057–3063.

58 A. J. Read, R. J. Needs, K. J. Nash, L. T. Canham, P. D. J. Calcott and A. Qteish, First principles calculations of the electronic properties of silicon quantum wires, *Phys. Rev. Lett.*, 1992, **69**(8), 1232–1235.

59 C. Delerue, G. Allan and M. Lannoo, Theoretical aspects of the luminescence of porous silicon, *Phys. Rev. B: Condens. Matter Mater. Phys.*, 1993, **48**(15), 11024.

60 M. S. Hybertsen and M. Needels, First-principles analysis of electronic states in silicon nanoscale quantum wires, *Phys. Rev. B: Condens. Matter Mater. Phys.*, 1993, **48**(7), 4608–4611.

61 B. Delley and E. F. Steigmeier, Size dependence of band gaps in silicon nanostructures, *Appl. Phys. Lett.*, 1995, **67**(16), 2370–2372.

62 S. Ke, K. Zhang and X. Xie, Electronic structures and optical absorption spectra of hydrogen-terminated Si quantum slabs, *Phys. Rev. B: Condens. Matter Mater. Phys.*, 1997, **55**(8), 5124–5128.

63 B. K. Agrawal and S. Agrawal, First principles study of one-dimensional quantum confined H-passivated ultrathin Si films, *Appl. Phys. Lett.*, 2000, **77**(19), 3039–3041.

64 L. Lin, Z. Li, J. Feng and Z. Zhang, Indirect to direct band gap transition in ultra-thin silicon films, *Phys. Chem. Chem. Phys.*, 2013, **15**, 6063–6067.

65 J. B. Xia and K. W. Cheah, Quantum confinement effect in silicon quantum well layers, *Phys. Rev. B: Condens. Matter Mater. Phys.*, 1997, **56**(23), 14925–14928.

66 M. Nishida, Theoretical study of luminescence enhancement in oxidized Si (001) ultrathin films, *Phys. Rev. B: Condens. Matter Mater. Phys.*, 1998, **58**(11), 7103–7112.

67 M. Nishida, Theoretical study of luminescence degradation by oxidation in Si (001) and Si (111) ultrathin films: gap states induced by oxidation, *Phys. Rev. B: Condens. Matter Mater. Phys.*, 1999, **66**(12), 8902–8908.

68 J. M. Wagner, K. Seino, F. Bechstedt, A. Dymiati, J. Mayer, R. Rolver, M. Forst, B. Berghoff, B. Spangenberg and H. Kurz, Electronic band gap of Si/SiO$_2$ quantum wells: comparison of ab initio calculations and photoluminescence measurements, *J. Vac. Sci. Technol., A*, 2007, **25**(6), 1500–1504.

69 S. Ossicini, A. Fasolino and F. Bernardini, Gap opening in ultrathin Si layers: role of confined and interface states, *Phys. Rev. Lett.*, 1994, **72**(7), 1044–1047.

70 A. J. Read, R. J. Needs, K. J. Nash, L. T. Canham, P. D. J. Calcott and A. Qteish, First principles calculations of the electronic properties of silicon quantum wires, *Phys. Rev. Lett.*, 1992, **69**(8), 1232–1235.

71 F. Buda, J. Kohanoff and M. Parrinello, Optical properties of porous silicon: a first principles study, *Phys. Rev. Lett.*, 1992, **69**(8), 1272–1275.

72 G. D. Sanders and Y. C. Chang, Optical properties of free-standing silicon quantum wires, *Appl. Phys. Lett.*, 1992, **60**(20), 2525–2527.

73 C. Y. Yeh, S. B. Zhang and A. Zunger, Identity of the light emitting states in porous silicon wires, *Appl. Phys. Lett.*, 1993, **63**(25), 3455–3457.

74 B. Delley and E. F. Steigmeier, Quantum confinement in Si nanocrystals, *Phys. Rev. B: Condens. Matter Mater. Phys.*, 1993, **47**(3), 1397–1400.

75 A. B. FilonovG, V. Petrov, V. A. Novikov and V. E. Borisenko, Orientation effect in electronic properties of silicon wires, *Appl. Phys. Lett.*, 1995, **67**(8), 1090–1091.

76 S. Ossicini, C. M. Bertoni, M. Biagini, A. Lugli, G. Roma and O. Bisi, Optical properties of isolated and interacting silicon quantum wires, *Thin Solid Films*, 1997, **297**, 154–162.
77 A. M. Saitta, F. Buda, G. Fiumara and P. V. Giaquinta, Ab initio molecular dynamics studies of electronic and optical properties of silicon quantum wires: orientational effects, *Phys. Rev. B: Condens. Matter Mater. Phys.*, 1996, **53**(3), 1446–1451.
78 X. Zianni and A. G. Nassiopoulu, Photoluminescence lifetimes of Si quantum wires, *Phys. Rev. B: Condens. Matter Mater. Phys.*, 2002, **66**, 205323.
79 R. Q. Zhang, Y. Lifshitz, D. D. D. Ma, Y. L. Zhao, T. Frauenheim, S. T. Lee and S. Y. Tong, Structures and energetics of hydrogen terminated Si nanowire surfaces, *J. Chem. Phys.*, 2005, **123**, 144703.
80 T. Vo, A. J. Williamson and G. Galli, First principles simulations of the structural and electronic properties of silicon nanowires, *Phys. Rev. B: Condens. Matter Mater. Phys.*, 2006, **74**, 045116.
81 D. Yao, G. Zhang and B. Li, A universal expression of band gap for silicon nanowires of different cross-section geometries, *Nano Lett.*, 2008, **8**(12), 4557–4561.
82 Z. Wu, J. B. Neaton and J. C. Grossman, Quantum confinement and electronic properties of tapered silicon nanowires, *Phys. Rev. Lett.*, 2008, **100**, 246804.
83 K. Zhuo and M. Y. Chou, Surface passivation and orientation dependence in the electronic properties of silicon nanowires, *J. Phys.: Condens. Matter*, 2013, **25**, 145501.
84 N. Zonias, P. Lagoudakis and C. K. Skylaris, Large scale first principles and tight binding density functional calcuations on hydrogen passivated silicon nanorods, *J. Phys.: Condens. Matter*, 2013, **22**, 025303.
85 I. J. T. Jensen, A. G. Ulyashin and O. M. Lovvik, Direct to indirect bandgap transitions in ⟨110⟩ silicon nanowires, *J. Appl. Phys.*, 2016, **119**, 015702.
86 M. Nolan, S. O. Callaghan, G. Fagas, J. C. Greer and T. Frauenheim, Silicon nanowire bandgap modification, *Nano Lett.*, 2007, **7**(1), 34–38.
87 M. Gao, S. You and Y. Wang, First principles study of silicon nanowires with different surfaces, *Jpn. J. Appl. Phys.*, 2008, **47**(4S), 3303–3309.
88 D. B. Migas and V. E. Borisenko, Effects of oxygen, fluorine and hydroxyl passivation on electronic properties of (001) oriented silicon nanowires, *J. Appl. Phys.*, 2008, **104**, 024314.
89 K. Zhuo and M. Y. Chu, Surface passivation and orientation dependence in the electronic properties of silicon nanowires, *J. Phys.: Condens. Matter*, 2013, **25**, 145501.
90 I. Vasiliev, J. R. Chelikowsky and R. M. Martin, Surface oxidation effects on the optical properties of silicon nanocrystals, *Phys. Rev. B: Condens. Matter Mater. Phys.*, 2002, **65**, 121302.
91 C. S. Garoufalis, A. D. Zdetsis and S. Grimme, High Level A*b Initio* Calculations of the Optical Gap of Small Silicon Quantum Dots, *Phys. Rev. Lett.*, 2001, **87**(27), 276402.
92 J. See, P. Dolfus and S. Galdin, Comparison between a sp3d5 tight binding and an effective mass description of Si quantum dots, *Phys. Rev. B: Condens. Matter Mater. Phys.*, 2002, **66**, 193307.
93 N. Tit, Z. H. Yamani, J. Graham and A. Ayesh, Effects of the passivating coating on the properties of silicon nanocrystals, *Mater. Chem. Phys.*, 2010, **124**, 927–935.

94 P. Hapala, K. Kusova, I. Pelant and P. Jelinek, Theoretical analysis of electronic band structure of 2-3 nm Si nanocrystals, *Phys. Rev. B: Condens. Matter Mater. Phys.*, 2013, **87**, 195420.

95 A. Puzder, A. J. Williamson, J. C. Grossman and G. Galli, Surface control of optical properties in silicon nanoclusters, *J. Chem. Phys.*, 2002, **117**(14), 6721–6729.

96 Z. Zhou, L. Brus and R. Friesner, Electronic Structure and Luminescence of 1.1 and 1.4 nm Silicon Nanocrystals: Oxide Shell versus hydrogen Passivation, *Nano Lett.*, 2003, **3**(2), 163–167.

97 M. Luppi and S. Ossicini, Multiple Si=O bonds at the silicon cluster surface, *J. Appl. Phys.*, 2003, **94**, 2130.

98 G. Hadjisavvas and P. C. Kelires, Structure and energetics of Si nanocrystals embedded in a-SiO$_2$, *Phys. Rev. Lett.*, 2004, **93**, 226104.

99 X. Chen, X. Pi and D. Yang, Bonding of Oxygen at the Oxide/Nanocrystal Interface of Oxidized Silicon Nanocrystals: An *Ab Initio* Study, *J. Phys. Chem. C*, 2010, **114**, 8774–8781.

100 R. Wang, X. Pi and D. Yang, First-Principles Study on the Surface Chemistry of 1.4 nm Silicon Nanocrystals: Case of Hydrosilylation, *J. Phys. Chem. C*, 2012, **116**, 19434–19443.

101 Z. Ni, X. Pi and D. Yang, Density functional theory study on a 1.4 nm silicon nanocrystal coated with carbon, *RSC Adv.*, 2012, **2**, 11227–11230.

102 V. Kocevski, O. Eriksson and J. Rusz, Transition between direct and indirect band gap in silicon nanocrystals, *Phys. Rev. B: Condens. Matter Mater. Phys.*, 2013, **87**, 245401.

103 P. Hapala, K. Kusova, I. Pelant and P. Jelinek, Theoretical analysis of electronic band structure of 2-3 nm Si nanocrystals, *Phys. Rev. B: Condens. Matter Mater. Phys.*, 2013, **87**, 195420.

104 H. Li, Z. Wu, T. Zhou, A. Sellinger and M. T. Lusk, Tailoring the optical gap of silicon quantum dots without changing their size, *Phys. Chem. Chem. Phys.*, 2014, **16**, 19275–19281.

105 T. Zhou, R. T. Anderson, H. Li, J. Bell, Y. Yang, B. P. Gorman, S. Pylypenko, M. T. Lusk and A. Sellinger, Bandgap Tuning of Silicon Quantum Dots by Surface Functionalization with Conjugated Organic Groups, *Nano Lett.*, 2015, **15**(6), 3657–3663.

106 V. Kocevski, Temperature dependence of radiative lifetime, optical and electronic properties of silicon nanocrystals capped with various organic ligands, *J. Chem. Phys.*, 2018, **149**, 054301.

107 M. Rohlfing and S. G. Loiue, Excitonic effects and the optical absorption spectrum of hydrogenated Si clusters, *Phys. Rev. Lett.*, 1998, **80**(15), 3320–3323.

108 O. Lehtonen and D. Sundholm, Density functional studies of excited states of silicon nanoclusters, *Phys. Rev. B: Condens. Matter Mater. Phys.*, 2005, **72**, 085424.

109 M. Cruz, M. R. Beltran, C. Wang, J. Taguena-Martinez and Y. G. Rubo, Supercell approach to the properties of porous silicon, *Phys. Rev. B: Condens. Matter Mater. Phys.*, 1999, **59**(23), 15381–15387.

110 B. Sapoval, S. Russ and J. N. Chazalviel, Eigenstates in irregular quantum wells: application to porous silicon, *J. Phys.: Condens. Matter*, 1996, **8**, 6235–6249.

111 F. A. Reboredo, A. Franceschetti and A. Zunger, Dark excitons due to direct Coulomb interactions in silicon quantum dots, *Phys. Rev. B: Condens. Matter Mater. Phys.*, 2000, **61**(19), 13073–13087.

112 D. Gabay, X. Wang, V. Lomakin, A. Boag, M. Jain and A. Natan, Size dependent electronic properties of silicon quantum dots – an analysis with hybrid, screened hybrid and local density functional theory, *Comput. Phys. Commun.*, 2017, **221**, 95–10.

113 T. Takagahara and T. Takeda, Theory of the quantum confinement effect on excitons in quantum dots of indirect gap materials, *Phys. Rev. B: Condens. Matter Mater. Phys.*, 1992, **46**(23), 15578–15581.

114 D. H. Feng, Z. Z. Xu, T. Q. Jia, X. X. Li and S. Q. Gong, Quantum size effects on exciton states in indirect gap quantum dots, *Phys. Rev. B: Condens. Matter Mater. Phys.*, 2003, **68**, 035334.

115 C. Delerue, G. Allan and M. Lannoo, Electron-phonon coupling and optical transitions for indirect semiconductor nanocrystals, *Phys. Rev. B: Condens. Matter Mater. Phys.*, 2001, **64**, 193402.

116 X. H. Peng, S. Ganti, A. Alizadeh, N. Bhate, K. K. Varanasi, S. K. Kumar and S. K. Nayak, Strain-engineered photoluminescence of silicon nanoclusters, *Phys. Rev. B: Condens. Matter Mater. Phys.*, 2006, **74**, 035339.

117 X. H. Peng, A. Alizadeh, P. Sharma, S. K. Kumar and S. K. Nayak, First principles investigation of strain effects on the energy gaps in silicon nanoclusters, *J. Phys.: Condens. Matter*, 2007, **19**, 266212.

118 S. Z. Karazhanov, A. Davletova and A. Ulyashin, Strain induced modulation of band structure of silicon, *J. Appl. Phys.*, 2008, **104**, 024501.

119 K. Kusova, P. Hapala, J. Valenta, P. Jelinek, O. Cibulka, L. Ondic and I. Pelant, Direct bandgap silicon: tensile strained silicon, *Adv. Mater. Interfaces*, 2014, **1**, 1300042.

120 S. Horiguchi, Conditions for a direct band gap in Si quantum wires, *Superlattices Microstruct.*, 1998, **23**(2), 355–364.

121 T. Trupke, J. Zhao, A. Wang, R. Corkish and M. A. Green, Very efficient light emission from bulk crystalline silicon, *Appl. Phys. Lett.*, 2003, **82**(18), 2996–2998.

122 G. A. Crosby and J. N. Denas, Measurement of photoluminescence quantum yields. Review, *J. Phys. Chem.*, 1971, **75**(8), 991–1024.

123 T. S. Ahn, R. O. Al-Kasyi, A. M. Muller, K. M. Wentz and C. J. Bardeen, Self-absorption correction for solid-state photoluminescence quantum yields obtained from integrating sphere measurements, *Rev. Sci. Instrum.*, 2007, **78**, 086105.

124 A. M. Brouwer, Standards for photoluminescence quantum yield measurements in solution (IUPAC Technical Report), *Pure Appl. Chem.*, 2011, **83**(12), 2213–2228.

125 C. Wurth, M. Grabolle, J. Pauli, M. Spieles and U. Resh-Genger, Comparison of methods and achievable uncertainties for the relative and absolute measurements of photoluminescence quantum yields, *Anal. Chem.*, 2011, **83**(9), 3431–3439.

126 S. Leyre, E. Coutino-Gonzalez, J. J. Joos, J. Ryckaert, Y. Meuret, D. Poelman, P. F. Smet, G. Durinck, J. Hofkens, G. Deconinck and P. Hanselaer, Absolute determination of photoluminescence quantum efficiency using an integrating sphere setup, *Rev. Sci. Instrum.*, 2014, **85**, 123115.

127 J. Valenta, Determination of absolute quantum yields of luminescing nanomaterials over a broad spectral range: from the integrating sphere theory to the correct methodology, *Nanosci. Methods*, 2014, **3**(1), 11–27.

128 Z. T. Kang, B. Arnold, C. J. Summers and B. K. Wagner, Synthesis of silicon quantum dot buried SiO_x films with controlled luminescent properties for solid state lighting, *Nanotechnology*, 2006, **17**, 4477–4482.

129 Y. Zhong, B. Song, X. Shen, D. Guo and Y. He, Fluorescein sodium ligand modified silicon nanoparticles produce ultrahigh fluorescence with robust pH- and photo-stability, *Chem. Commun.*, 2019, **55**, 365–368.

130 Q. Li, T. Y. Luo, M. Zhou, H. Abroshan, J. Huang, H. J. Kim, N. L. Rosi, Z. Shao and R. Jin, Silicon nanoparticles with surface nitrogen: 90% quantum yield with narrow luminescence bandwidth and the ligand structure based energy law, *ACS Nano*, 2016, **10**, 8385–8393.

131 Z. Yuan, T. Nakamura, S. Adachi and K. Matsuishi, Improvement of laser processing for colloidal silicon nanocrystal formation in a reactive solvent, *J. Phys. Chem. C*, 2017, **121**, 8623–8629.

132 Z. Yang, G. B. de los Reyes, L. V. Titova, I. Sychugov, M. Dasog, J. Linnros, F. A. Hegmann and J. G. C. Veinot, Evolution of the Ultrafast Photoluminescence of Colloidal Silicon Nanocrystals with Changing Surface Chemistry, *ACS Photonics*, 2015, **2**(5), 595–604.

133 A. Marinins, R. Z. Shafagh, W. van der Wijngaart, T. Haraldsson, J. Linnros and J. G. C. Veinot, Light-converting polymer/Si nanocrystal composites with stable 60–70% quantum efficiency and their glass laminates, *ACS Appl. Mater. Interfaces*, 2017, **9**, 30267–30272.

134 F. Sangghaleh, I. Sychugov, Z. Yang, J. G. C. Veinot and J. Linnros, Near-Unity Internal Quantum Efficiency of Luminescent Silicon Nanocrystals with Ligand Passivation, *ACS Nano*, 2015, **9**(7), 7097–7104.

135 D. Jurbergs, E. Rogojina, L. Mangoloni and U. Kortshagen, Silicon nanocrystals with ensemble quantum yields exceeding 60%, *Appl. Phys. Lett.*, 2006, **88**, 233116.

136 J. Valenta, M. Greben, S. A. Dyakov, N. A. Gippius, D. Hiller, S. Gutsch and M. Zacharias, Nearly perfect near infrared luminescence efficiency of Si nanocrystals: a comprehensive quantum yield study employing the Purcell effect, *Sci. Rep.*, 2019, **9**, 11214.

137 J. Joo, T. Defforge, A. Loni, D. Kim, Z. Y. Li, M. J. Sailor, G. Gautier and L. T. Canham, Enhanced quantum yield of photoluminescent porous silicon prepared by supercritical drying, *Appl. Phys. Lett.*, 2016, **108**, 153111.

138 B. Song, Y. Zhong, S. Wu, B. Chu, Y. Su and Y. He, One dimensional fluorescent silicon nanorods featuring ultrahigh photostability, favourable biocompatibility and excitation wavelength dependent emission spectra, *J. Am. Chem. Soc.*, 2016, **138**, 4824–4831.

139 B. J. Ryan, M. P. Hanrahan, Y. Wang, U. Ramesh, C. K. A. Nyamekye, R. D. Nelson, Z. Liu, C. Huang, B. Whitehead, J. Wang, L. T. Roling, E. A. Smith, A. J. Rosini and M. G. Panthani, Silicene, siloxene or silicane? Revealing the structural and optical properties of silicon nanosheets derived from calcium disilicide, *Chem. Mater.*, 2020, **32**(2), 795–804.

140 K. A. Gonchar, L. A. Osminkina, R. A. Galkin, M. B. Gongalsky, V. S. Marshov, V. Yu. Timoshenko, M. N. Kulmas, V. V. Solovyev, A. A. Kudryavtsev and V. A. Sivakov, Growth, Structure and Optical properties of Silicon

Nanowires formed by Metal-Assisted Chemical Etching, *J. Nanoelectron. Optoelectron.*, 2012, **7**(6), 602–606.
141 X. Lu, C. M. Hessel, Y. Yu, T. D. Bogart and B. A. Korgel, Colloidal Luminescent Silicon Nanorods, *Nano Lett.*, 2013, **13**, 3101–3105.
142 A. R. Goni, L. R. Muniz, J. S. Reparaz, M. I. Alonso, M. Garriga, A. F. Lopeandia, J. Rodriguez-Viejo, J. Arbitol and R. Rurali, Using high pressure to unravel the mechanism of visible emission in amorphous Si/SiO$_x$ nanoparticles, *Phys. Rev. B: Condens. Matter Mater. Phys.*, 2014, **89**, 045428.
143 R. Anthony and U. Kortshagen, Photoluminescence quantum yields of amorphous and crystalline silicon nanoparticles, *Phys. Rev. B: Condens. Matter Mater. Phys.*, 2009, **80**, 115407.
144 A. Irrera, M. J. L. Faro, C. D'Andrea, A. A. Leonardi, P. Artoni, B. Fazio, R. A. Picca, N. Cioffi, S. Trusso, G. Franzo, P. Musumeci, F. Priolo and F. Iacona, Light emitting silicon nanowires obtained by metal-assisted chemical etching, *Semicond. Sci. Technol.*, 2017, **32**, 043004.
145 E. C. Cho, M. A. Green, J. Xia, R. Corkish, P. Reece and M. Gal, Clear quantum-confined luminescence from crystalline silicon/SiO$_2$ single quantum wells, *Appl. Phys. Lett.*, 2004, **84**(13), 2286–2288.
146 T. Helbich, A. Lyuleeva, T. Ludwig, L. M. Schwerf, T. F. Fassier, P. Lugli and B. Rieger, One-step synthesis of photoluminescent covalent polymeric nanocomposites from 2D silicon nanosheets, *Adv. Funct. Mater.*, 2006, **26**, 6711–6718.
147 A. G. Cullis and L. T. Canham, Visible light emission due to quantum size effects in highly porous silicon, *Nature*, 1991, **353**, 335–337.
148 M. V. Wolkin, J. Jorne, P. M. Fauchet, G. Allan and C. Delerue, Electronic States and Luminescence in Porous Silicon Quantum Dots: The Role of Oxygen, *Phys. Rev. Lett.*, 1999, **82**(1), 197–200.
149 L. T. Canham, M. R. Houlton, W. Y. Leong, C. Pickering and J. M. Keen, Atmospheric impregnation of porous silicon at room temperature, *J. Appl. Phys.*, 1991, **70**, 422–431.
150 J. C. Vial, A. Bsiesy, F. Gaspard, R. Herino, M. Ligeon, F. Muller, R. Romestain and R. F. Macfarlane, Mechanisms of visible-light emission from electrooxidized porous silicon, *Phys. Rev. B: Condens. Matter Mater. Phys.*, 1992, **45**(24), 14171–14176.
151 R. Boukeherroub, D. D. M. Wayner, D. J. Lockwood and L. T. Canham, Passivated luminescent porous silicon, *J. Electrochem. Soc.*, 2001, **148**(9), H91–H97.
152 H. Mizuno, H. Koyama and N. Koshida, Oxide-free blue photoluminescence form photochemically etched porous silicon, *Appl. Phys. Lett.*, 1996, **69**(25), 3779–3781.
153 Y. H. Xie, W. L. Wilson, F. M. Ross, J. A. Mucha, E. A. Fitzgerald, J. M. Macauley and T. D. Harris, Luminescence and structural study of porous silicon films, *J. Appl. Phys.*, 1992, **71**, 2403.
154 J. Linnros, N. Lalic, A. Galeckas and V. Grivackas, Analysis of the stretched exponential photoluminescence decay from nanometer-sized silicon crystals in SiO$_2$, *J. Appl. Phys.*, 1999, **86**(11), 6128–6134.
155 L. Pavesi and M. Ceschini, Stretched exponential decay of the luminescence in porous silicon, *Phys. Rev. B: Condens. Matter Mater. Phys.*, 1993, **48**(23), 17625–17628.

156 B. Gelloz and N. Koshida, Mechanism of a remarkable enhancement in the light emission from nanocrystalline porous silicon annealed in high pressure water vapor, *J. Appl. Phys.*, 2005, **98**, 123509.
157 Y. Takahashi, T. Furuta, Y. Ono, T. Ishiyama and M. Tabe, Photoluminescence from a silicon quantum well formed on separation by implanted oxygen substrate, *Jpn. J. Appl. Phys.*, 1995, **34**, 950–954.
158 P. N. Saeta and A. C. Gallagher, Visible luminescence from single crystal silicon quantum wells, *Appl. Phys. Lett.*, 1995, **77**(9), 4639–4641.
159 P. N. Saeta and A. C. Gallagher, Photoluminescence properties of silicon quantum well layers, *Phys. Rev. B: Condens. Matter Mater. Phys.*, 1997, **55**(7), 4563–4574.
160 S. Okamoto and Y. Kanemitsu, Quantum confinement and interface effects on photoluminescence from silicon quantum wells, *Solid State Commun.*, 1997, **103**(10), 573–576.
161 N. Pauc, V. Calvo, J. Eymery, F. Fournel and N. Magnea, Photoluminescence of nanometric single silicon quantum wells, *Opt. Mater.*, 2005, **27**, 1000–1003.
162 E. C. Cho, M. A. Green, J. Xia, R. Corkish, P. Reece, M. Gal and S. H. Lee, Photoluminescence in crystalline silicon quantum wells, *J. Appl. Phys.*, 2007, **101**, 024321.
163 X. Zhu, J. Lu, Y. Gao, S. Yuan, N. Zhou, M. Xie, Z. Zheng and Y. Zhao, Strong photoluminescence from ultrathin silicon (110) quantum wells at room temperature, *Jpn. J. Appl. Phys.*, 2017, **56**, 020305.
164 Z. H. Lu, D. J. Lockwood and J. M. Baribeau, Quantum confinement and light emission in SiO_2/Si superlattices, *Nature*, 1995, **378**, 258–260.
165 D. J. Lockwood, Z. H. Lu and J. M. Baribeau, Quantum confined luminescence in Si/SiO_2 superlattices, *Phys. Rev. Lett.*, 1996, **76**(3), 539–541.
166 S. V. Novikov, J. Sinnkonen, O. Kilpela and S. V. Gastev, Visible luminescence from Si/SiO_2 superlattices, *J. Vac. Sci. Technol., B: Microelectron. Nanometer Struct.–Process., Meas., Phenom.*, 1997, **15**(4), 1471–1473.
167 P. Photopoulos, A. G. Nassiopoulou, D. N. Kouvatsos and A. Travios, Photoluminescence from nanocrystalline silicon in Si/SiO_2 superlattices, *Appl. Phys. Lett.*, 2000, **76**(24), 3588–3590.
168 V. Vinciguerra, G. Franzo, F. Priolo, F. Iacona and C. Spinella, Quantum confinement and recombination dynamics in silicon nanocrystals embedded in Si/SiO_2 superlattices, *J. Appl. Phys.*, 2000, **87**(11), 8165–8173.
169 Y. Kanemitsu, Y. Fukenishi, M. Iiboshi and S. Okamoto, Visible luminescence from Si/SiO_2 quantum wells and dots: confinement and localization of excitons, *Physica E: Low-dimensional Systems and Nanostructures*, 2000, **7**, 456–460.
170 F. Gourbilleau, X. Portier, C. Ternon, P. Voivenei, R. Madelon and R. Rizk, Si-rich/SiO_2 nanostructured multilayer's by reactive magnetron sputtering, *Appl. Phys. Lett.*, 2001, **78**(20), 3058–3060.
171 M. Zacharias, J. Heitmann, R. Scholz, U. Kahler, M. Schmidt and J. Blasing, Size-controlled highly luminescent silicon nanocrystals: A SiO/SiO_2 Superlattice approach, *Appl. Phys. Lett.*, 2002, **80**(4), 661–663.
172 F. A. D'Avitaya, L. Vervoort, F. Bassani, S. Ossicini, A. Fasolino and F. Bernardini, Light emission at room temperature from Si/CaF_2 multilayers, *Europhys. Lett.*, 1995, **31**(1), 25–30.

173 L. Vervoort, F. Bassani, I. Mihalcescu, J. C. Vial and F. Arnaud D'Avitaya, Efficient visible light emission from Si/CaF$_2$(111) heterostructures grown by molecular beam epitaxy, *Phys. Status Solidi B*, 1995, **190**, 123–127.

174 M. Watanabe, F. IIzuka and M. Asada, Visible light emission from nanocrystalline silicon embedded in CaF$_2$ layers on Si (111), *IEEE Trans. Electron Devices*, 1996, **E79-C**, 1562–1567.

175 J. Heitmann, F. Muller, L. Yi, M. Zacharias, D. Kovalev and F. Eichhorn, Excitons in Si nanocrystals: Confinement and migration effects, *Phys. Rev. B: Condens. Matter Mater. Phys.*, 2004, **69**, 195309.

176 J. Valenta, M. Greben, S. A. Dyakov, N. A. Gippius, S. Gutsch, D. Hiller and M. Zacharias, Effects of Inter-nanocrystal distance on luminescence quantum yield in ensembles of Si nanocrystals, *Appl. Phys. Lett.*, 2014, **105**, 243107.

177 H. Namatsu, M. Nagasse, K. Kurihara, K. Iwadate, T. Furuta and K. Murase, Fabrication of sub-10 nm silicon lines with minimum fluctuation, *J. Vac. Sci. Technol., B: Microelectron. Nanometer Struct.–Process., Meas., Phenom.*, 1995, **13**(4), 1473–1476.

178 A. G. Nassiopoulos, S. Grigoropoulos and E. Gogolides, Visible luminescence from 1-dimensional and 2-dimensional silicon structures produced by conventional lithographic and reactive ion etching techniques, *Appl. Phys. Lett.*, 1995, **66**, 1114.

179 S. H. Zaidi and S. R. J. Brueck, Optical properties of nanoscale, one-dimensional silicon grating structures, *J. Appl. Phys.*, 1996, **80**(12), 6997–7007.

180 M. Gotza, M. Dutoit and M. Ilegems, Fabrication and photoluminescence investigation of silicon nanowires on silicon-on-insulator material, *J. Vac. Sci. Technol., B: Microelectron. Nanometer Struct.–Process., Meas., Phenom.*, 1998, **16**(2), 582.

181 D. P. Yu, Z. G. Bai, J. J. Wang, Y. H. Zou, W. Qian, J. S. Fu, H. Z. Zhang, Y. Ding, G. C. Xiong, L. P. You, J. Xu and S. Q. Feng, Direct evidence of quantum confinement from the size dependence of the photoluminescence of silicon quantum wires, *Phys. Rev. B: Condens. Matter Mater. Phys.*, 1999, **59**(4), R2898–R2501.

182 A. R. Guichard, D. N. Barsic, S. Sharma, T. I. Kamins and M. L. Brongersma, Tunable light emission from quantum-confined excitons in TiSi2- catalyzed silicon nanowires, *Nano Lett.*, 2006, **6**(9), 2140–2144.

183 V. A. Sivakov, F. Voigt, A. Berger, G. Bauer and S. H. Christiansen, Roughness of silicon nanowire sidewalls and room temperature photoluminescence, *Phys. Rev. B: Condens. Matter Mater. Phys.*, 2010, **82**, 125446.

184 V. Le Borgne, M. Agati, S. Boninelli, P. Castrucci, M. de Crescenzi, R. Dolbec and M. Y. A. E. Khakani, Structural and photoluminescence properties of silicon nanowires extracted by means of a centrifugation process from plasma torch synthesized silicon nanopowder, *Nanotechnology*, 2017, **28**, 285702.

185 L. Mangolini, Synthesis, properties and applications of silicon nanocrystals, *J. Vac. Sci. Technol., B: Nanotechnol. Microelectron.: Mater., Process., Meas., Phenom.*, 2013, **31**(2), 020801.

186 C. Huan and S. S. Qing, Silicon nanoparticles: Preparation, properties, and applications, *Chin. Phys. B*, 2014, **23**(8), 088102.

187 X. Cheng, S. B. Lowe, P. J. Reece and J. J. Gooding, Colloidal silicon quantum dots: from preparation to the modification of self-assembled monolayers (SAMs) for bio-applications, *Chem. Soc. Rev.*, 2014, **43**, 2680–2700.
188 Y. Yu, G. Fan, A. Fermi, R. Mazzaro, V. Morandi, P. Ceroni, D. M. Smilgies and B. A. Korgel, Size-dependent photoluminescence efficiency of silicon nanocrystal quantum dots, *J. Phys. Chem. C*, 2017, **121**, 23240–23248.
189 C. Garcia, B. Garrido, P. Pellegrino, R. Ferre, J. A. Moreno, L. Pavesi, M. Cazanelli and J. R. Morante, Size dependence of lifetime and absorption cross section of Si nanocrystals embedded in SiO_2, *Appl. Phys. Lett.*, 2003, **82**, 1595.
190 C. Delerue, G. Allan, C. Reynaud, O. Guillard, G. Ledoux and F. Huisken, Multiexponential photoluminescence decay in indirect gap semiconductor nanocrystals, *Phys. Rev. B: Condens. Matter Mater. Phys.*, 2006, **73**, 235318.
191 M. Greben, P. Khoroshyy, X. Liu, X. Pi and J. Valenta, Fully radiative relaxation of silicon nanocrystals in colloidal ensemble revealed by advance treatment of decay kinetics, *J. Appl. Phys.*, 2017, **122**, 034304.
192 B. Gelloz, F. B. Juangsa, T. Nozaki, K. Asaka, N. Koshida and L. Jin, Si/SiO_2 core/shell photoluminescent silicon nanocrystals and porous silicon powders with high quantum yield, long lifetime and good stability, *Front. Phys.*, 2019, **7**, 47.
193 K. Furuta, M. Fujii, H. Sugimoto and K. Imakita, Energy Transfer in Silicon Nanocrystal Solids Made from All-Inorganic Colloidal Silicon Nanocrystals, *J. Phys. Chem. Lett.*, 2015, **6**, 2761–2766.
194 K. Potrick, T. Schmidt, S. Bublitz, Chr. Muhlig, W. Paa and F. Huisken, Determination of the photoluminescence quantum efficiency of silicon nanocrystals by laser-induced deflection, *Appl. Phys. Lett.*, 2011, **98**, 083111.
195 R. J. Walters, J. Kalkman, A. Polman, H. A. Atwater and M. J. A. de Dood, Photoluminescence quantum efficiency of dense silicon nanocrystals ensembles in SiO_2, *Phys. Rev. B: Condens. Matter Mater. Phys.*, 2006, **73**, 132302.
196 M. L. Mastronardi, F. Maier-Flag, D. Faulkner, E. J. Henderson, C. Kubel, U. Lemmer and G. A. Ozin, Size-dependent absolute quantum yields for size-separated colloidally-stable silicon nanocrystals, *Nano Lett.*, 2011, **12**, 337–342.
197 M. Zacharias, L. Tsybeskov, K. D. Hirschman, P. M. Fauchet, J. Blasing, P. Kohlert and P. Veit, Nanocrystalline silicon superlattices: fabrication and characterization, *J. Non-Cryst. Solids*, 1998, **227–230**, 1132–1136.
198 L. E. Brus, P. F. Szajowski, W. L. Wilson, T. D. Harris, S. Schuppler and P. H. Citrin, Electronic Spectroscopy and Photophysics of Si Nanocrystals: Relationship to Bulk c-Si and Porous Si, *J. Am. Chem. Soc.*, 1995, **117**, 2915–2922.
199 K. A. Littau, A. J. Muller, P. J. Szawoski, A. R. Kortan and L. Brus, A luminescent silicon nanocrystal colloid via a high temperature aerosol reaction, *J. Phys. Chem.*, 1993, **97**, 1224–1230.
200 G. Ledoux, O. Guillois, D. Porterat, C. Reynaud, F. Huisken, B. Kohn and V. Paillard, Photoluminescence properties of silicon nanocrystals as a function of their size, *Phys. Rev. B: Condens. Matter Mater. Phys.*, 2000, **62**(23), 15942–15951.

201 S. Takeoka, M. Fujii and S. Hayashi, Size-dependent photoluminescence from surface-oxidized Si nanocrystals in a weak confinement regime, *Phys. Rev. B: Condens. Matter Mater. Phys.*, 2000, **62**, 16820–16825.

202 M. I. Alonso, I. C. Marcus, M. Garriga, A. R. Goni, J. Jedrzejewski and I. Balberg, Evidence of quantum confinement effects on interband optical transitions in Si nanocrystals, *Phys. Rev. B: Condens. Matter Mater. Phys.*, 2010, **82**, 045302.

203 M. L. Brongersma, A. Polman, K. S. Min, E. Boer, T. Tambo and H. A. Atwater, Tuning the emission wavelength of Si nanocrystals in SiO_2 by oxidation, *Appl. Phys. Lett.*, 1998, **72**(20), 2577–2579.

204 F. Iacona, G. Franzo and C. Spinella, Correlation between luminescence and structural properties of Si nanocrystals, *J. Appl. Phys.*, 2000, **87**(5), 1295–1303.

205 G. Polisski, H. Heckler, D. Kovalev, M. Schwartzkopff and F. Koch, Luminescence of porous silicon in a weak confinement regime, *Appl. Phys. Lett.*, 1998, **73**(8), 1107–1119.

206 J. Von Behren, T. van Buuren, M. Zacharias, E. H. Chimowitz and P. M. Fauchet, Quantum confinement in nanoscale silicon: the correlation of size with bandgap and luminescence, *Solid State Commun.*, 1998, **105**(5), 317–322.

207 J. Choi, N. S. Wang and V. Reipa, Photoassisted Tuning of Silicon Nanocrystal Photoluminescence, *Langmuir*, 2007, **23**, 3388–3394.

208 X. Liu, Y. Zhang, T. Yu, X. Qiao, R. Gresback, X. Pi and D. Yang, Optimum quantum yield of the light emission from 2 to 10 nm hydrosilylated silicon quantum dots, *Part. Part. Syst. Charact.*, 2016, **33**(1), 44–52.

209 A. M. P. Botas, R. J. Anthony, J. Wu, D. J. Rowe, N. J. O. Silva, U. Kortshagen, R. N. Pereira and R. A. S. Ferreira, Influence of the surface termination on the light emission of crystalline silicon nanoparticles, *Nanotechnology*, 2016, **27**, 325703.

210 G. M. Carroll, R. Limpens and N. R. Neale, Tuning confinement in colloidal silicon nanocrystals with saturated surface ligands, *Nano.Lett.*, 2018, **18**(5), 3118–3124.

211 A. Angi, R. Sinelnikov, H. H. Heenen, A. Meldrum, J. G. C. Veinot, D. Azulay, I. Balberg, O. Millo and B. Rieger, The influence of conjugated alkylnyl (aryl) surface groups on the optical properties of silicon nanocrystals: photoluminescence through in-gap states, *Nanotechnology*, 2018, **29**, 355705.

212 M. Dasog, G. B. De los Reyes, L. V. Titova, F. A. Hegmann and J. G. C. Veinot, Size vs. Surface: Tuning the photoluminescence for Freestanding Silicon Nanocrystals Across the Visible Spectrum via Surface Groups, *ACS Nano*, 2014, **8**(9), 9636–9648.

213 M. Dasog, K. Bader and J. G. C. Veinot, Influence of halides on the optical properties of silicon quantum dots, *Chem. Mater.*, 2015, **27**, 1153–1156.

214 P. D. J. Calcott, K. J. Nash, L. T. Canham, M. J. Kane and D. Brumhead, Identification of radiative transitions in highly porous silicon, *J. Phys.: Condens. Matter*, 1993, **5**, L91–L98.

215 P. D. J. Calcott, K. J. Nash, L. T. Canham, M. J. Kane and D. Brumhead, Spectroscopic identification of the luminescence mechanism of highly porous silicon, *J. Lumin.*, 1995, **57**, 257–269.

216 M. Rosenbauer, M. Stutzmann, S. Finkbeiner, J. Weber and E. Bustaret, Resonantly excited photoluminescence spectra of porous silicon, *Phys. Rev. B: Condens. Matter Mater. Phys.*, 1995, **51**(16), 10539–10547.

217 M. Rosenbauer, M. Stutzmann, S. Finkbeiner, J. Weber and E. Bustaret, Reply to "Comment on Resonantly excited photoluminescence spectra of porous silicon", *Phys. Rev. B: Condens. Matter Mater. Phys.*, 1997, **55**(15), 10117–10118.
218 T. Suemoto and A. Saito, Comment on "Resonantly excited photoluminescence spectra of porous silicon", *Phys. Rev. B: Condens. Matter Mater. Phys.*, 1997, **55**(15), 10115–10116.
219 Y. Kanemitsu and S. Okamato, Resonantly excited photoluminescence from porous silicon: effects of surface oxidation on resonant luminescence spectra, *Phys. Rev. B: Condens. Matter Mater. Phys.*, 1997, **56**(4), R1696–R1699.
220 Y. Kanemitsu and S. Okamato, Phonon structures and Stokes shift in resonantly excited luminescence of silicon nanocrystals, *Phys. Rev. B: Condens. Matter Mater. Phys.*, 1998, **58**, 9652.
221 J. L. Gole and S. M. Prokes, Resonantly excited photoluminescence from porous silicon and the question of bulk phonon replicates, *Phys. Rev. B: Condens. Matter Mater. Phys.*, 1998, **58**(8), 4761–4770.
222 H. Heckler, D. Kovalev, G. Polisski, N. N. Zinovev and F. Koch, Magneto-optical effects in photoluminescence of Si nanocrystals, *Phys. Rev. B: Condens. Matter Mater. Phys.*, 1999, **60**(11), 7718–7721.
223 G. F. Grom, D. J. Lockwood, J. P. McCaffrey, H. J. Labbe, P. M. Fauchet, B. White Jr, J. Diener, D. Kovalev, F. Koch and L. Tsybeskov, Ordering and self-organization in nanocrystalline silicon, *Nature*, 2000, **407**, 358–361.
224 J. Heitman, F. Muller, L. Yi, M. Zacharias, D. Kovalev and F. Eichhorn, Excitons in silicon nanocrystals: confinement and migration effects, *Phys. Rev. B: Condens. Matter Mater. Phys.*, 2004, **69**, 195309.
225 Y. Kanemitsu, H. Sato, S. Nihonyanage and Y. Hirai, Efficient radiative recombination of indirect excitons in silicon nanowires, *Phys. Status Solidi A*, 2002, **190**, 755–758.
226 Y. Kanemitsu and S. Okamoto, Photoluminescence from Si/SiO$_2$ single quantum wells by selective excitation, *Phys. Rev. B: Condens. Matter Mater. Phys.*, 1997, **56**, 15561.
227 U. Kahler and H. Hofmeister, Size evolution and photoluminescence of silicon nanocrystallites in evaporated SiO$_x$ thin films upon thermal processing, *Appl. Phys. A: Mater. Sci. Process.*, 2002, **74**, 13–17.
228 D. Kovalev, H. Heckler, M. Ben-Chorin, G. Polisski, M. Schwartzkopf and F. Koch, Breakdown of the k-conservation rule in Si nanocrystals, *Phys. Rev. Lett.*, 1998, **81**(13), 2803–2806.
229 Y. Kanemitsu and S. Okamoto, Optical properties of hydrogen terminated silicon nanocrystals, *Solid-State Electron.*, 1998, **42**(7–8), 1315–1318.
230 B. Goller, S. Polisski, H. Wiggers and D. Kovalev, Freestanding spherical silicon nanocrystals: A model system for studying confined excitons, *Appl. Phys. Lett.*, 2010, **97**, 041110.
231 M. D. Mason, G. M. Credo, K. D. Weston and S. K. Buratto, Luminescence of individual porous Si chromophores, *Phys. Rev. Lett.*, 1998, **80**, 5405–5408.
232 I. Sychugov, J. Valenta and J. Linnros, Probing silicon quantum dots by single-dot techniques, *Nanotechnology*, 2017, **28**, 072002.
233 A. Fucikova, I. Sychugov and J. Linnros, The shell matters: one step synthesis of core-shell silicon nanoparticles with room temperature ultranarrow emission linewidth, *Faraday Discuss.*, 2020, DOI: 10.1039/C9FD00093C.

234 L. V. Titova, T. L. Cocker, D. G. Cooke, X. Wang, A. Meldrum and F. A. Hegmann, Ultrafast percolative transport in silicon nanocrystal films, *Phys. Rev. B: Condens. Matter Mater. Phys.*, 2011, **83**, 085403.

235 A. V. Andrianov, D. I. Kovalev, V. B. Shuman and I. D. Yaroshetski, Short-lived green band and time evolution of the photoluminescence spectrum of porous silicon, *JETP Lett.*, 1992, **56**(5), 236–239.

236 T. Ito, T. Ohta and A. Hiraki, Light emission from microcrystalline Si confined in SiO_2 matrix through partial oxidation of porous silicon, *Jpn. J. Appl. Phys.*, 1992, **31**, L1–L3.

237 D. Kovalev, I. D. Yaroshetzkii, T. Muschik, V. Petrova-Koch and F. Koch, Fast and slow visible luminescence bands of oxidized porous Si, *Appl. Phys. Lett.*, 1994, **64**(2), 214–216.

238 L. Tsybeskov, J. V. Vandyshev and P. M. Fauchet, Blue emission in porous silicon : oxygen related photoluminescence, *Phys. Rev. B: Condens. Matter Mater. Phys.*, 1994, **49**(11), 7821–7824.

239 A. Loni, A. J. Simons, P. D. J. Calcott and L. T. Canham, Blue photoluminescence from rapidly thermally oxidized porous silicon following storage in ambient air, *J. Appl. Phys.*, 1995, **77**(7), 3557–3559.

240 D. P. Yu, Q. L. Hang, Y. Ding, H. Z. Zhang, Z. G. Bai, J. J. Wang, Y. H. Zou, W. Qian, G. C. Xiong and S. Q. Feng, Amorphous silica nanowires: intensive blue light emitters, *Appl. Phys. Lett.*, 1998, **73**(21), 3076–3078.

241 J. Qi, A. M. Belcher and J. M. White, Spectroscopy of individual silicon nanowires, *Appl. Phys. Lett.*, 2003, **82**(16), 2616–2619.

242 P. Noe, J. Guignard, P. Gentile, E. Demeladeine and V. Calvo, Enhancement of the photoluminescence of silicon oxide defect states by combining silicon oxide with siicon nanowires, *J. Appl. Phys.*, 2007, **102**, 016103.

243 M. Shao, L. Cheng, M. Zhang, D. D. D. Ma, J. A. Zapien, S. T. Lee and X. Zhang, Nitrogen doped silicon nanowires:synthesis and their blue cathodoluminescence and photoluminescence, *Appl. Phys. Lett.*, 2009, **95**, 143110.

244 R. Li, X. Zhang, D. Zhang, Y. Zhang and G. Xiang, Optical properties and quantum confinement in ultrafine single crystal silicon nanowires synthesized by thermal evaporation without catalyst, *RSC Adv.*, 2020, DOI: 10.1039/C9FD00093C.

245 Y. Zhong, F. Peng, F. Bao, S. Wang, X. Ji, L. Yang, Y. Su, S. T. Lee and Y. He, Large-scale aqueous synthesis of fluorescent and biocompatible silicon nanoparticles and their use as highly photostable biological probes, *J. Am. Chem. Soc.*, 2013, **135**(22), 8350–8356.

246 B. V. Oliinyk, D. Korytko, V. Lysenko and S. Alekseev, Are fluorescent silicon nanoparticles formed in a one-pot aqueous synthesis?, *Chem. Mater.*, 2019, **31**, 7167–7172.

247 J. L. Z. Ddungu, S. Silvestrini, A. Tassoni and L. de Cola, Shedding light on the aqueous synthesis of silicon nanoparticles by reduction of silanes with citrates, *Faraday Discuss.*, 2020, DOI: 10.1039/C9FD00127A.

248 J. L. Wilbrink, C. C. Huang, K. Dohnalova and J. M. J. Paulusse, Critical assessment of wet-chemical oxidation synthesis of silicon quantum dots, *Faraday Discuss.*, 2020, DOI: 10.1039/C9FD00099B.

249 X. Zheng, D. Zhang, Z. Fan, Z. Huang, H. Mao and Y. Ma, One-step hydrothermal synthesis of ultrabright water-soluble silicon nanoparticles for folate-receptor-mediated bioimaging, *J. Mater. Sci.*, 2019, **54**, 9707–9717.

250 J. Wang, Y. Zhang, H. Hao and W. Shen, Structural evolution and effective improvement of emission quantum yields for silicon nanocrystals synthesized by femtosecond laser ablation in HF-containing solution, *Nanotechnology*, 2019, **30**, 015705.

251 X. Geng, Z. Li, Y. Hu, H. Liu, Y. Sun, H. Meng, Y. Wang, L. Qu and Y. Lin, One pot green synthesis of ultrabright N-doped fluorescent silicon nanoparticles for cellular imaging by using ethylenediaminetetracetic acid disodium salt as an effective reductant, *ACS Appl. Mater. Interfaces*, 2018, **10**, 27979–27986.

252 Y. Zhong, B. Song, F. Peng, Y. Wu, S. Wu, Y. Su and Y. He, In situ rapid growth of fluorescent silicon nanoparticles at room temperature and under atmospheric pressure, *Chem. Commun.*, 2016, **52**, 13444–13447.

253 F. Wu, X. Zhang, S. Kai, M. Zhang, H. Wang, J. N. Myers, Y. Weng, P. Liu, N. Gu and Z. Chen, One-Step Synthesis of Superbright Water-Soluble Silicon Nanoparticles with Photoluminescence Quantum Yield Exceeding 80%, *Adv. Mater. Interfaces*, 2015, 1500360.

254 L. Wang, Q. Li, H. Wang, J. Huang, R. Zhang, Q. Chen, H. Xu, W. H. Z. Shao and H. Sun, Ultrafast optical spectroscopy of surface-modified silicon quantum dots: unraveling the underlying mechanism of the Ultrabright and color-tunable photoluminescence, *Light: Sci. Appl.*, 2015, **4**, e245.

255 B. Cho, S. Baek, H. G. Woo and H. Sohn, Synthesis of silicon quantum dots showing high quantum efficiency, *J. Nanosci. Nanotechnol.*, 2014, **14**, 5868–5872.

256 J. Wang, S. Sun, F. Peng, L. Cao and L. Sun, Efficient one-pot synthesis of highly photoluminescent alkyl-functionalized silicon nanocrystals, *Chem. Commun.*, 2011, **47**, 4941–4943.

257 T. M. Atkins, A. Thibert, D. S. Larsen, S. Dey, N. D. Browning and S. M. Kaulrich, Femtosecond Ligand/Core Dynamics of Microwave Assisted Synthesized Silicon Quantum Dots in Aqueous Solution, *J. Am. Chem. Soc.*, 2011, **133**(51), 20664–20667.

258 A. S. Heintz, M. J. Fink and B. S. Mitchell, Mechanochemical synthesis of blue luminescent alkyl/alkenyl passivated silicon nanoparticles, *Adv. Mater.*, 2007, **19**(22), 3984–3988.

259 M. J. L. Portoles, F. R. Nieto, D. B. Soria, J. I. Amalvy, P. J. Peruzzo, D. O. Martire, M. Kotler, O. Holub and M. C. Gonzalez, Photophysical properties of blue – emitting silicon nanoparticles, *J. Phys. Chem. C*, 2009, **113**(31), 13694–13702.

260 A. Fojtik and A. Henglein, Surface Chemistry of Luminescent Colloidal Silicon Nanoparticles, *J. Phys. Chem. B*, 2006, **110**, 1994–1998.

261 K. Kimura, Blue luminescence from silicon nanoparticles suspended in organic liquids, *J. Cluster Sci.*, 1999, **10**(2), 359–380.

262 D. Dobrovolskas, J. Mickevicius, G. Tamulaitis and V. Reipa, Photoluminescence of Si nanocrystals under selective excitation, *J. Phys. Chem. Solids*, 2009, **70**, 439–443.

263 T. Schmidt, A. L. Chizhik, A. M. Chizhik, K. Potrick, A. J. Meixner and F. Huisken, Radiative exciton recombination and defect luminescence observed in single silicon nanocrystals, *Phys. Rev. B: Condens. Matter Mater. Phys.*, 2012, **86**, 125302.

264 K. Kimura and S. Iwasaki, Vibronic fine structure found in the blue luminescence from silicon nanocolloids, *Jpn. J. Appl. Phys.*, 1999, **38**, 609–612.

265 D. Tan, Z. Ma, B. Xu, Y. Dai, G. Ma, M. He, Z. Jin and J. Qiu, Surface passivated silicon nanocrystals with stable luminescence synthesized by femtosecond laser ablation in solution, *Phys. Chem. Chem. Phys.*, 2011, **13**, 20255–20261.
266 R. Intartaglia, K. Bagga, A. Genovese, A. Athanassiou, R. Cingolani, A. Diaspro and F. Brandi, Influence of organic solvent on optical and structural properties of ultra-small silicon dots synthesized by UV laser ablation in liquid, *Phys. Chem. Chem. Phys.*, 2012, **14**, 15406–15411.
267 S. Dewan, J. H. Odhner, K. M. Tibbetts, S. Afsari, R. J. Levis and E. Borguet, Resolving the source of blue luminescence from alkyl—capped silicon nanoparticles synthesized by laser pulse ablation, *J. Mater. Chem. C*, 2016, **4**, 6894–6899.
268 T. Matsumoto, M. Maeda and H. Kobayashi, Photoluminescence enhancement of adsorbed species on Si nanoparticles, *Nanoscale Res. Lett.*, 2016, **11**, 7.
269 Y. Kanemitsu, T. Futagi, T. Matsumoto and H. Mimura, Origin of the blue and red luminescence from oxidized porous silicon, *Phys. Rev. B: Condens. Matter Mater. Phys.*, 1994, **49**(20), 14732–14735.
270 J. Valenta, A. Fucikova, I. Pelant, K. Kusova, K. Dohnalova, A. Aleknvicius, O. Cibulka, A. Fojtik and G. Kada, On the origin of the fast photoluminescence band in small silicon nanoparticles, *New J. Phys.*, 2008, **10**, 073022.
271 A. A. Prokofiev, A. S. Moskalenko and I. N. Yassievich, Direct bandgap optical transitions in Si nanocrystals, *JETP Lett.*, 2009, **90**(12), 758–762.
272 K. Dohnalova, A. Fucikova, C. P. Umesh, J. Humpolickova, J. M. J. Paulusse, J. Valenta, J. Huilhof, M. Hof and T. Gregorkiewicz, Microscopic origin of the fast blue-green luminescence of chemically synthesized non-oxidized silicon quantum dots, *Small*, 2012, **8**(20), 3185–3191.
273 L. Ondic, K. Kusova, M. Ziegler, L. Fekete, V. Gartnerova, V. Chab, V. Holy, O. Cibulka, K. Herynkova, M. Gallart, P. Gilliot, B. Honerlage and I. Pelant, A complex study of the fast blue luminescence of oxidized silicon nanocrystals: the role of the core, *Nanoscale*, 2014, **6**, 3837–3845.
274 G. G. Qin, X. S. Liu, S. Y. Ma, J. Lin, G. Q. Yao, X. Y. Lin and K. X. Lin, Photoluminescence mechanism for blue light emitting porous silicon, *Phys. Rev. B: Condens. Matter Mater. Phys.*, 1997, **55**(19), 12876–12879.
275 S. Yang, W. Li, B. Cao, H. Zeng and W. Cai, Origin of Blue Emission from Silicon Nanoparticles: Direct Transition and Interface Recombination, *J. Phys. Chem. C*, 2011, **115**, 21056–21062.
276 T. Uchino, N. Kurumoto and N. Sagawa, Structure and formation of blue light emitting centres in silicon and silica-based nanostructured materials, *Phys. Rev. B: Condens. Matter Mater. Phys.*, 2006, **73**, 233203.
277 B. Gelloz, R. Mentek and N. Koshida, Specific blue light emission from nanocrystalline porous Si treated by high-pressure water vapor annealing, *Jpn. J. Appl. Phys.*, 2009, **48**(4), 04C119.
278 L. T. Canham, A. Loni, P. D. J. Calcott, A. J. Simons, C. J. Reeves, M. R. Houlton, J. P. Newey, K. J. Nash and T. I. Cox, On the origin of blue luminescence arising from atmospheric impregnation of oxidized porous silicon, *Thin Solid Films*, 1996, **276**, 112–115.
279 A. Loni, A. J. Simons, P. D. J. Calcott, J. P. Newey, T. I. Cox and L. T. Canham, Relationship between storage media and blue photoluminescence for oxidized porous silicon, *Appl. Phys. Lett.*, 1997, **71**, 107–109.

280 Y. D. Glinka, A. S. Zyubin, A. M. Mebel, S. H. Lin, L. P. Hwang and Y. T. Chen, Photoluminescence from mesoporous silica akin to that from nanoscale silicon: the nature of the light-emitters, *Chem. Phys. Lett.*, 2002, **358**, 180–186.
281 J. Ewles, Luminescence of silica, *Nature*, 1950, **165**, 812–813.
282 J. Ewles and R. F. Youell, Luminescence effects associated with the production of silicon monoxide and with oxygen deficit in silica, *Trans. Faraday Soc.*, 1951, **47**, 1060–1064.
283 C. M. Carbonaro, P. C. Rici, R. Corpino, M. Marceddu and A. Anedda, Photoluminescence of aged and regenerated mesoporous silica, *J. Non-Cryst. Solids*, 2011, **357**, 1904–1907.
284 P. Galar, T. Popelar, J. Khun, I. Matulkova, I. Nemec, K. Dohnalova Newell, A. Michalcova, V. Scholtz and K. Kusova, The red and blue luminescence in silicon nanocrystals with an oxidized, nitrogen-containing shell, *Faraday Discuss.*, 2020, DOI: 10.1039/C9FD00092E.
285 M. Dasog, Z. Yang, S. Regli, T. M. Atkins, A. Faramus, M. P. Singh, E. Muthuswamy, S. M. Kauzlarich, R. D. Tilley and J. G. C. Veinot, Chemical Insight into the Origin of Red and Blue Photoluminescence Arising from Freestanding Silicon Nanocrystals, *ACS Nano*, 2013, **7**(3), 2676–2685.
286 H. Chen, Z. Zhen, W. Tang, T. Todd, Y. Chuang, L. Wang, Z. Pan and J. Xie, Label-Free Luminescent Mesoporous Silica Nanoparticles for Imaging and Drug Delivery, *Theranostics*, 2013, **3**(9), 650–657.
287 J. B. Essner, J. A. Kist, L. Pola-Parada and G. A. Baker, Artifacts and errors associated with the ubiquitous presence of fluorescent impurities in carbon nanodots, *Chem. Mater.*, 2018, **30**, 1878–1887.
288 N. C. Verma, A. Yadav and C. K. Nandi, Paving the path to the future of carbogenic nanodots, *Nat. Commun.*, 2019, **10**, 2391.
289 L. T. Canham and A. J. Groszek, Characterization of microporous silicon by flow calorimetry: comparison with a hydrophobic silica molecular sieve, *J. Appl. Phys.*, 1992, **72**(4), 1558–1565.
290 C. W. Lai, Y. H. Hsaio, Y. K. Peng and P. T. Chou, Facile synthesis of highly emissive carbon dots from pyrolysis of glycerol: gram scale production of carbon dots/mSiO$_2$ for cell imaging and drug release, *J.Mater.Chem.*, 2012, **22**, 14403.
291 J. M. Lauerhaas and M. J. Sailor, Chemical modification of the photoluminescence quenching of porous silicon, *Science*, 1993, **261**, 1567–1568.
292 M. J. Sailor and E. C. Wu, Photoluminescence-based sensing with porous silicon films, microparticles and nanoparticles, *Adv. Funct. Mater.*, 2009, **19**(20), 3195–3208.
293 C. M. Gonzalez and J. G. C. Veinot, Silicon nanocrystals for the development of sensing platforms, *J.Mater.Chem.*, 2016, **C4**, 4836–4846.
294 X. Ji, H. Wang, B. Song, B. Chu and Y. He, Silicon nanomaterials for biosensing and bioimaging analysis, *Front. Chem.*, 2018, **6**, 38.
295 D. L. Fisher, J. Harper and M. J. Sailor, Energy transfer quenching of porous Si photoluminescence by aromatic molecules, *J. Am. Chem. Soc.*, 1995, **117**, 7846–7847.
296 J. Harper and M. J. Sailor, Detection of nitric oxide and nitrogen dioxide using photoluminescent porous silicon, *Anal. Chem.*, 1996, **68**(21), 3713–3717.

297 M. T. Kelly and A. B. Bocarsly, Effects of SO_2 and I_2 on the photoluminescence of oxidized porous silicon, *Chem. Mater.*, 1997, **9**, 1659–1664.
298 W. J. Salcedo, F. J. R. Fernandez and J. C. Rubim, Photoluminescence quenching effect on porous silicon films for gas sensors application, *Spectrochim. Acta, Part A*, 2004, **60**, 1065–1070.
299 J. Dian, T. Chvojka, V. Vrkoslav and I. Jelinek, Photoluminescence quenching of porous silicon in gas and liquid phases – the role of dielectric quenching and capillary condensation effects, *Phys. Status Solidi C*, 2005, **2**(9), 3481–3485.
300 T. Dittrich, E. A. Konstantinova and V. Y. Timoshenko, Influence of molecule adsorption on porous silicon photoluminescence, *Thin Solid Films*, 1995, **255**, 238–240.
301 F. Yin, X. R. Xiao, X. P. Li, Z. Z. Zhang, B. W. Zhang, Y. Cao, G. H. Li and Z. P. Wang, Photoluminescence enhancement of porous silicon by organic cyano compounds, *J. Phys. Chem. B*, 1998, **102**, 7978–7982.
302 G. Hollett, D. S. Roberts, M. Sewell, E. Wensley, J. Wagner, W. Murray, A. Krotz, B. Toth, V. Vijayakumar and M. J. Sailor, Quantum ensembles of silicon nanoparticles: discrimination of static and dynamic photoluminescence quenching processes, *J. Phys. Chem. C*, 2019, **128**, 17976–17986.
303 Y. Jin, W. Duan, F. Wo and J. Wu, Two-dimensional fluorescent strategy based on porous silicon quantum dots for metal ion detection and recognition, *ACS Appl. Nano Mater.*, 2019, **2**(10), 6110–6115.
304 M. T. Kelly, J. K. M. Chun and A. B. Bocarsly, A silicon sensor for SO_2, *Nature*, 1996, **382**, 214–215.
305 C. Baratto, G. Faglia, G. Sberveglieri, Z. Gaburro, L. Pancheri, C. Oton and L. Pavesi, Multiparametric porous silicon sensors, *Sensors*, 2002, **2**, 121–126.
306 T. Holec, T. Chvojka, I. Jelinek, J. Jindřich, I. Nemec, I. Pelant, J. Valenta and J. Dian, Determination of sensoric parameters of porous silicon in sensing of organic vapors, *Mater. Sci. Eng., C*, 2002, **19**, 251–254.
307 J. Dian, V. Vrkoslav and I. Jelinek, Chemical sensing by simultaneous measurement of photoluminescence intensity and photoluminescence decay time of porous silicon, *Phys. Status Solidi*, 2007, **4**(6), 2078–2082.
308 J. Dian, V. Vrkoslav and I. Jelinek, Recognition enhancement of oxidized and methyl-10-undecenoate functionalized porous silicon in gas phase photoluminescence sensing, *Sens. Actuators, B*, 2010, **147**, 406–410.
309 Z. H. Zhang, R. Lockwood, J. G. C. Veinot and A. Meldrum, Detection of ethanol and water vapour with silicon quantum dots coupled to an optical fiber, *Sens. Actuators, B*, 2013, **181**, 523–528.
310 O. Syshchyk, V. A. Skyshrevsky, O. O. Soldatykin and A. P. Soldatykin, Enzyme biosensor systems based on porous silicon photoluminescence for detection of glucose, urea and heavy metals, *Biosens. Bioelectron.*, 2015, **66**, 89–94.
311 J. Hwang, M. P. Hwang, M. Choi, Y. Seo, Y. Jo, J. Son, J. Hong and J. Choi, Sensitive detection of copper ions via ion-responsive fluorescence quenching of engineered porous silicon nanoparticles, *Sci. Rep.*, 2016, **6**, 35565.
312 V. Myndrul, R. Viter, M. Savchuk, M. Koval, N. Starodub, V. Silamikelis, V. Smyntyna, A. Ramanavicius and I. Iatsunskyi, Porous silicon based

photoluminescence immunosensor for rapid and highly sensitive detection of ochratoxin A, *Biosens. Bioelectron.*, 2018, **102**, 661–667.
313 S. Content, W. C. Trogler and M. J. Sailor, Detection of nitrobenzene, DNT and TNT vapors by quenching of porous silicon photoluminescence, *Chem.–Eur. J.*, 2000, **6**(12), 2205–2213.
314 I. N. Germanenko, S. Li and M. S. El-Shall, Decay dynamics and quenching of photoluminescence from silicon nanocrystals by aromatic nitro compounds, *J. Phys. Chem.*, 2001, **B105**, 59–66.
315 C. M. Gonzalez, M. Iqbal, M. Dasog, D. G. Piercey, R. Lockwood, T. M. Kiapotke and J. G. C. Veinot, Detection of high energy compounds using photoluminescent silicon nanocrystal paper based sensors, *Nanoscale*, 2014, **6**, 2608–2612.
316 A. Nguyen, C. M. Gonzalez, R. Sinelnikov, W. Newman, S. Sun, R. Lockwood, J. G. C. Veinot and A. Meldrum, Detection of nitroaromatics in the solid, solution and vapor phases using silicon quantum dot sensors, *Naotechnology*, 2016, **27**(10), 105501.
317 S. Chan, P. M. Fauchet, Y. Li, L. J. Rothberg and B. L. Miller, Porous silicon microcavities for biosensing applications, *Phys. Status Solidi A*, 2000, **182**(1), 541–546.
318 S. Chan, S. R. Horner, P. M. Fauchet and B. L. Miller, Identification of gram-negative bacteria using nanoscale silicon microcavities, *J. Am. Chem. Soc.*, 2001, **123**, 11797–11798.
319 P. S. Chaudhuri, A. Gokarna, M. Kulkarni, M. S. Karve and S. V. Bhoraskar, Porous silicon as an entrapping matrix for the immobilization of urease, *Sens. Actuators, B*, 2005, **107**, 258–263.
320 V. La Ferrara, G. Fiorentino, G. Rametta and G. Di Francia, Luminescence quenching of porous silicon nanoparticles in presence of ascorbic acid, *Phys. Status Solidi A*, 2012, **209**(4), 736–740.
321 B. Xia, W. Zhang, J. Shi and S. Xiao, Fluorescence quenching in luminescent porous silicon nanoparticles for the detection of intracellular Cu^{2+}, *Analyst*, 2013, **138**, 3629–3632.
322 M. Saleem, L. P. Lee and K. H. Lee, Photoluminescent sensor for acetylcholinesterase inhibitor determination, *J. Mater. Chem.*, 2014, **2**, 6802–6808.
323 B. Cho, S. Kim, H. Woo, S. Kim and H. Sohn, Detection of Human IgG Using Photoluminescent Porous Silicon Interometer, *J. Nanosci. Nanotechnol.*, 2015, **15**, 1083–1087.
324 S. Arshavsky-Graham, N. Massad-Ivanir, E. Segal and S. Weiss, Porous silicon based photonic biosensors: current status and emerging appplications, *Anal. Chem.*, 2018, **91**(1), 441–467.
325 Y. Yi, G. Zhu, C. Liu, Y. Huang, Y. Zhang, H. Li, J. Zhao and S. Yao, A label-free silicon quantum dots-based photoluminescence sensor for ultrasensitive detection of pesticides, *Anal. Chem.*, 2013, **85**, 11464–11470.
326 J. Zhang and S. H. Yu, Highly photoluminescent silicon nanocrystals for rapid, label-free and recyclable detection of mercuric ions, *Nanoscale*, 2014, **6**(8), 4096–4101.
327 J. Zhao, J. Deng, Y. Yi, H. Li, Y. Zhang and S. Yao, Label-free silicon quantum dots as fluorescent probe for selective and sensitive detection of copper ions, *Talanta*, 2014, **125**, 372–377.

328 J. Lin and Q. Wang, Role of novel silicon nanoparticles in luminescent detection of a family of antibiotics, *RSC Adv.*, 2015, **5**, 27458–27463.
329 B. B. Campos, M. Algarra, B. Alonso, C. M. Casado, J. Jimenez-Jimenez, E. Rodriguez-Castellon and J. C. G. Esteves da Silva, Fluorescent sensor for Cr (VI) based on functionalized silicon quantum dots with dendrimers, *Talanta*, 2015, **144**, 862–867.
330 B. Liao, W. Wang, X. Deng, B. He, W. Zeng, Z. Tang and Q. Liu, A facile one-step synthesis of fluorescent silicon quantum dots and their application for detecting Cu^{2+}, *RSC Adv.*, 2016, **6**, 14465–14467.
331 A. R. Jose, U. Sivasankaran, S. Menson and K. G. Kumar, A silicon nanoparticle based turn off fluorescent sensor for sudan I, *Anal. Methods*, 2016, **8**, 5701–5706.
332 Y. Guo, L. Zhang, F. Cao, L. Mang, X. Lei, S. Cheng and J. Song, Hydrothermal synthesis of blue-emitting silicon quantum dots for fluorescent detection of hypochlorite in tap water, *Anal. Methods*, 2016, **8**, 2723–2728.
333 L. Zhu, X. Peng, H. Li, Y. Zhang and S. Yao, On-off-on fluorescent silicon nanoparticles for recognition of chromium VII and hydrogen sulphide based on the inner filter effect, *Sens. Actuators, B*, 2017, **238**, 196–203.
334 Y. Han, Y. Chen, J. Liu, X. Niu, Y. Ma, S. Ma and X. Chen, Room temperature synthesis of yellow-emitting fluorescent silicon nanoparticles for sensitive and selective determination of crystal violet in fish tissues, *Sens. Actuators, B*, 2018, **263**, 508–516.
335 L. Luo, Y. Song, C. Zhu, S. Fu, Q. Shi, Y. Sun, B. Jia, D. Du, Z. Xu and Y. Lin, Fluorescent silicon nanoparticle-based ratiometric fluorescence immunoassay for sensitive detection of ethyl carbamate in red wine, *Sens. Actuators, B*, 2018, **225**, 2742–2749.
336 Y. Han, Y. Chen, J. Feng, M. Na, J. Liu, Y. Ma, S. Ma and X. Chen, Investigation of nitrogen content effect in reducing agent to prepare wavelength controllable fluorescent silicon nanoparticles and its application in detection of 2-nitrophenol, *Talanta*, 2019, **194**, 822–829.
337 R. Ban, F. Zheng and J. Zhang, A highly sensitive fluorescent assayfor 2,4,6-trinitrotoluene using amine-capped silicon quantum dots as a probe, *Anal. Methods*, 2015, **7(5)**, 1732–1737.
338 Y. Feng, Y. Liu, C. Su, X. Ji and Z. He, New fluorescent pH sensor based on label-free silicon nanodots, *Sens. Actuators, B*, 2014, **203**, 795–801.
339 X. Zhang, X. Chen, S. Kai, H. Y. Wang, J. Yang, F. G. Wu and Z. Chen, Highly selective and sensitive detection of dopamine using one pot synthesized highly photoluminescent silicon nanoparticles, *Anal. Chem.*, 2015, **87**, 3360–3365.
340 B. Chu, H. Wang, B. Song, F. Peng, Y. Su and Y. He, Fluorescent and photostable silicon nanoparticles sensors for real time and long term intracellular pH measurements in live cells, *Anal. Chem.*, 2016, **88**, 9235–9242.
341 Z. Li, X. Ren, C. Hao, X. Meng and Z. Li, Silicon quantum dots with tunable emission synthesized via one-step hydrothermal method and their application in alkaline phosphatase detection, *Sens. Actuators, B*, 2018, **260**, 426–431.
342 L. Meng, C. Lan, Z. Liu, J. Yin and N. Xu, Nitrogen-terminated silicon nanoparticles obtained via chemical etching and passivation are specific fluorescent probes for creatinine, *Microchim. Acta*, 2019, **186**, 387.

343 Q. Li, K. Peng, Y. Yu, X. Ruan and Y. Wei, One-pot synthesis of highly fluorescent silicon nanoparticles for sensitive and selective detection of hemoglobin, *Electrophoresis*, 2019, **40**, 2129–2134.

344 N. O'Farrell, A. Houlton and B. R. Horrocks, Silicon nanoparticles: applications in cell biology and medicine, *Int. J. Nanomed.*, 2006, **1**(4), 451–472.

345 H. A. Santos, L. M. Bimbo, B. Herranz, M. A. Shahbazi, J. Hirvonen and J. Salonen, Nanostructured porous silicon in pre-clinical imaging: moving from bench to bedside, *J.Mater.Chem.*, 2013, **28**(2), 152–164.

346 B. F. P. McVey and R. D. Tilley, Solution Synthesis Optical Properties, and Bioimaging Application of Silicon Nanocrystals, *Acc. Chem. Res.*, 2014, **47**(10), 3045–3051.

347 X. Cheng, S. B. Lowe, P. J. Reece and J. J. Gooding, Colloidal silicon quantum dots: from preparation to the modification of self-assembled monolayers (SAMs) for bio-applications, *Chem. Soc. Rev.*, 2014, **43**, 2680–2700.

348 M. Montalti, A. Cantelli and G. Batistelli, Nanodiamonds and silicon quantum dots : ultrastable and biocompatible luminescent nanoprobes for long-term bioimaging, *Chem. Soc. Rev.*, 2015, **44**, 4853–4921.

349 Y. Park, J. Yoo, M. Kang, W. Kwon and J. Joo, Photoluminescent and biodegradable porous silicon nanoparticles for biomedical imaging, *J. Mater. Chem. B*, 2019, **7**, 6271–6292.

350 F. Erogbogbo, K. Yong, I. Roy, G. Xu, P. N. Prasad and M. T. Swihart, Biocompatible luminescent silicon quantum dots for imaging of cancer cells, *ACS Nano*, 2008, **2**(5), 873–878.

351 S. K. Chiu, B. A. Manhat, W. J. I. deBenedetti, A. L. Brown, K. Fichter, T. Vu, M. Eastman, J. Jiao and A. M. Gosforth, Aqueous red-emitting silicon nanoparticles for cellular imaging: Consequences of protecting against surface passivation by hydroxide and water for stable red emission, *J. Mater. Res.*, 2013, **28**(2), 216–230.

352 M. B. Gongalsky, L. A. Osminkina, A. Pereira, A. A. Manankov, A. A. Federenko, A. N. Vasiliev, V. V. Solyvev, A. A. Kudryavstev, M. Sentis, A. V. Kabashin and V. Y. Timoshenko, Laser-synthesized oxide-passivated bright Si quantum dots for bioimaging, *Sci. Rep.*, 2016, **6**, 24732.

353 E. J. Henderson, A. J. Shuhendler, P. Prasad, V. Baumann, F. Maier-Flag, D. O. Faulkner, U. Lemmer, X. Y. Lu and G. A. Ozin, Colloidally stable silicon nanocrystals with near infrared photoluminescence for biological fluorescence imaging, *Small*, 2011, **7**(17), 2507–2516.

354 Q. Wang, H. Ni, A. Pietsch, F. Hennies, Y. Bao and Y. Chao, Synthesis of water-dispersible photoluminescent silicon nanoparticles and their use in biological fluorescent imaging, *J. Nanopart. Res.*, 2011, **13**, 405–413.

355 J. Wang, Y. Liu, F. Peng, C. Chen, Y. He, H. Ma, L. Cao and S. Sun, A general route to efficient functionalization of silicon quantum dots for high performance fluorescent probes, *Small*, 2012, **8**(15), 2430–2435.

356 M. B. Gongalsky, U. A. Tsurikova, C. J. Storey, Y. V. Evstratova, A. A. Kudryavstev, L. T. Canham and L. A. Osminkina, The effects of drying technique and surface pre-treatment on the cytotoxicity and dissolution rate of luminescent porous silicon quantum dots in model fluids and living cells, *Faraday Discuss.*, 2020, DOI: 10.1039/C9FD00107G.

357 J. H. Park, L. Gu, G. von Maltzahn, E. Ruoslhati, S. N. Bhatia and M. J. Sailor, Biodegradable luminescent porous silicon nanoparticles for in vivo applications, *Nat. Mater.*, 2009, **8**(4), 331–336.

358 C. M. Hessel, M. R. Rasch, J. L. Hueso, B. W. Goodfellow, V. A. Akhavan, P. Puvanakrishnan, J. W. Tunnel and B. A. Korgel, Alkyl Passivation and Amphiphilic Polymer Coating of Silicon Nanocrystals for Diagnostic Imaging, *Small*, 2010, **6**(10), 2026–2034.

359 F. Erogbogbo, K. Yong, R. Hu, W. Law, H. Ding, C. Chang, P. N. Prasad and M. T. Swihart, Biocompatible magnetofluorescent probes : luminescent silicon quantum dots coupled with superparamagnetic iron(III) oxide, *ACS Nano*, 2010, **4**(9), 5131–5138.

360 F. Erogbogbo, K. T. Yong, I. Roy, R. Hu, W. C. Law, W. Zhao, H. Ding, F. Wu, R. Kumar, P. N. Prasad and M. T. Swihart, *In Vivo* Targeted Cancer Imaging, Sentinel Lymph Node Mapping and Multi-Channel Imaging with Biocompatible Silicon Nanocrystals, *ACS Nano*, 2011, **5**(1), 413–423.

361 F. Erogbogbo, C. Chang, J. L. May, L. Liu, R. Kumar, W. Law, H. Ding, K. T. Yong, I. Roy, M. Sheshadri, P. N. Prasad and M. T. Swihart, Bioconjugation of luminescent silicon quantum dots to gadolinium ions for bioimagingapplications, *Nanoscale*, 2012, **4**, 5483–5489.

362 B. Xia, B. Wang, J. Shi, W. Zhang and S. Xiao, Engineering near-infrared fluorescent styrene terminated porous silicon nanocomposites with bovine serum albumin encapsiulation for in vivo imaging, *J. Mater. Chem. B*, 2014, **2**, 8314–8320.

363 S. Chandra, B. Ghosh, G. Beaune, U. Nagarajan, T. Yasui, J. Nakamura, T. Tsuruoka, Y. Baba, N. Shirahata and F. M. Winnik, Functional double-shelled silicon nanocrystals for two-photon fluorescence cell imaging: spectral evolution and tuning, *Nanoscale*, 2016, **8**, 9009–9019.

364 B. J. Furey, D. A. Silbaugh, Y. Yu, A. C. Guillausier, A. D. Estrada, C. Stevens, J. A. Maynard, B. A. Korgel and M. C. Downer, Measurement of two-photon absorption of silicon nanocrystals in colloidal suspension for bio-imaging applications, *Phys. Status Solidi B*, 2018, **255**, 1700501.

365 M. Sakiyama, H. Sugimoto and M. Fujii, Long-lived luminescence of colloidal silicon quantum dots for time-gated fluorescence imaging in the second near infrared window in biological tissue, *Nanoscale*, 2018, **10**, 13902–13907.

366 D. Kim, J. Kang, T. Wang, H. G. Ryu, J. M. Zuidema, J. Joo, M. Kim, Y. Huh, J. Jung, K. H. Ahn, K. H. Kim and M. J. Sailor, Two-photon *in vivo* imaging with porous silicon nanoparticles, *Adv. Mater.*, 2019, **29**, 1703309.

367 L. Wang, V. Reipa and J. Blasic, Silicon Nanoparticles as a luminescent label to DNA, *Bioconjugate Chem.*, 2004, **15**(2), 409–412.

368 J. H. Warner, A. Hoshino, K. Yamamato and R. D. Tilley, Water-Soluble Photoluminescent Silicon Quantum Dots, *Angew. Chem., Int. Ed.*, 2005, **44**, 2–6.

369 R. D. Tilley and K. Yamamoto, The microemulsion synthesis of hydrophobic and hydrophilic silicon nanocrystals, *Adv. Mater.*, 2006, **18**, 2053–2056.

370 K. Fujioka, M. Hiruoka, K. Sato, N. Manabe, R. Miyasaka, S. Hanada, A. Hoshino, R. D. Tilley, Y. Manome, K. Hirakuri and K. Yamamoto, Luminescent passive-oxidized silicon quantum dots as biological staining labels and their cytotoxicity effects at high concentration, *Nanotechnology*, 2008, **19**, 415102.

371 Y. He, Z. Kang, Q. Li, C. H. A. Tsang, C. Fan and S. T. Lee, Ultrastable highly fluorescent and water-dispersed silicon-based nanospheres as cellular probes, *Angew. Chem., Int. Ed.*, 2009, **48**, 128–132.

372 M. Rosso-Vassic, E. Sprujit, Z. Popovic, K. Overgaag, B. Lagen, B. Gradidier, D. Vanmaekelbergh, D. Dominguez-Gutierrez, L. de Cola and H. Zuilhof, Amine-terminated silicon nanoparticles: synthesis, optical properties and their use in bioimaging, *J. Mater. Chem.*, 2009, **19**, 5926–5933.

373 B. A. Manhatu, A. L. Brown, L. A. Black, J. B. A. Ross, K. Fichter, T. Vu, E. Richman and A. M. Goforth, One step Melt Synthesis of Water Soluble, Photoluminescent, Surface-Oxidized Silicon Nanoparticles for Cellular Imaging, *Chem. Mater.*, 2011, **23**(9), 2407–2418.

374 A. Shiohara, S. Prabakar, A. Faramus, C. Hsu, P. Lai, P. T. Northcote and R. D. Tilley, Sized controlled synthesis, purification and cell studies with silicon quantum dots, *Nanoscale*, 2011, **3**, 3364.

375 P. Shen, S. Ohta, S. Inasawa and Y. Yamaguchi, Selective labelling of the endoplasmic reticulum in live cells with silicon quantumdots, *Chem. Commun.*, 2011, **47**, 8409–8411.

376 J. H. Ahire, Q. Wang, P. R. Coxon, G. Malhotra, R. Brydson, R. Chen and Y. Chao, Highly luminescent and nontoxic amine-capped nanoparticles for porous silicon: synthesis and their use in biomedical imaging, *ACS Appl. Mater. Interfaces*, 2012, **4**, 3285–3292.

377 J. H. Ahire, I. Chambrier, A. Mueller, Y. Bao and Y. Chao, Synthesis of d-mannose capped silicon nanoparticles and their interactions with MCF-7 human breast cancerous cells, *ACS Appl. Mater. Interfaces*, 2013, **5**, 7384–7391.

378 J. Wang, D. Ye, G. Liang, J. Chang, J. Kong and J. Chen, One-step synthesis of water-dispersible silicon nanoparticles and their use in fluorescence lifetime imaging of living cells, *J. Mater. Chem. B*, 2014, **2**, 4338–4345.

379 X. Zhang, X. Chen, J. Yang, H. R. Jia, Y. H. Li, Z. Chen and F. G. Wu, Quarternized silicon nanoparticles with polarity-sensitive fluorescence for selectively imaging and killing gram-positive bacteria, *Adv. Funct. Mater.*, 2016, **26**, 5958–5970.

380 R. Xing, K. Li, Y. Zhou, Y. Su, S. Yan, K. Zhang, S. Wu, Y. Sima, K. Zhang, Y. He and S. Xu, Impact of fluorescent silicon nanoparticles on circulating hemolymph and hematopoiesis in an invertebrate organism, *Chemosphere*, 2016, **159**, 628–637.

381 S. Li, F. Wang, X. He, W. Li and Y. Zhang, One-pot hydrothermal preparation of gadolinium doped silicon nanoparticles as a dual-modal probe for multicolor fluorescence and magnetic resonance imaging, *J. Mater. Chem. B*, 2018, **6**, 3358–3365.

382 G. Hong, A. L. Antaris and H. Dai, Near infrared fluorophores for biomedical imaging, *Nat. Biomed. Eng.*, 2017, **1**, 0010.

383 M. Sakiyama, H. Sugimoto and M. Fujii, Long-lived luminescence of colloidal silicon quantum dots for time-gated fluorescence imaging in the second near infrared window in biological tissue, *Nanoscale*, 2018, **10**, 13902–13907.

384 G. S. He, Q. Zheng, K. Yong, F. Erogbogbo, M. T. Swihart and P. N. Prasad, Two and three photon absorption and frequency upconverted emission of silicon quantum dots, *Nano Lett.*, 2020, DOI: 10.1039/C9FD00107G.

385 L. Ravotto, Q. Chen, Y. Ma, S. A. Vinogradov, M. Locritani, G. Bergamini, F. Negri, Y. Xu, B. A. Korgel and P. Ceroni, Bright long lived luminescence

of silicon nanocrystals sensitized by two-photon absorbing antenna, *Chem*, 2017, **2**, 550–560.
386 X. Cheng, E. Hinde, D. M. Owen, S. B. Lowe, P. J. Reece, K. Gaus and J. J. Gooding, Enhancing Quantum Dots for Bioimaging using Advanced Surface Chemistry and Advanced Optical Microscopy: Application to Silicon Quantum Dots (SiQDs), *Adv. Mater.*, 2015, **27**(40), 6144–6150.
387 E. Cassette, M. Helle, L. Bezdetnaya, F. Marchal, B. Dubertret and T. Pons, Design of new quantum dot materials for deep tissue infrared imaging, *Adv. Drug Delivery Rev.*, 2013, **65**, 719–731.
388 L. Gu, D. J. Hall, Z. Qin, E. Anglin, J. Joo, D. J. Mooney, S. B. Howell and M. J. Sailor, *In Vivo* time-gated fluorescence imaging with biodegradable luminescent porous silicon nanoparticles, *Nat. Commun.*, 2013, **4**, 2326.
389 C. Tu, K. Awasthi, K. Chen, C. Lin, M. Hamada, N. Ohta and Y. Li, Time-gated imaging on live cancer cells using silicon quantum dot nanoparticles with long lived fluorescence, *ACS Photonics*, 2017, **4**, 1306–1315.
390 L. T. Canham, Bioactive silicon structure fabrication through nanoetching techniques, *Adv. Mater.*, 1995, **7**(12), 1033–1037.
391 S. P. Low, N. H. Voelcker, L. Canham and K. A. Williams, The biocompatibility of porous silicon in tissues of the eye, *Biomaterials*, 2009, **30**(15), 2873–2880.
392 F. Johansson, L. Wallman, N. Danielson, J. Schouenberg and M. Kanje, Porous silicon as a potential electrode material in a nerve repair setting: tissue reactions, *Acta Biomater.*, 2009, **5**(6), 2230–2237.
393 M. A. Tolli, M. P. A. Ferreira, S. M. Kinnunen, J. Rysa, E. M. Makila, Z. Szabo, R. E. Serpi, P. J. Ohukaihnen, M. J. Valimaki, A. M. R. Korreira, J. J. Salonen, J. T. Hirvonen, H. J. Ruskoaho and H. A. Santos, In vivo biocompatibility of porous silicon biomaterials for drug delivery to the heart, *Biomaterials*, 2014, **35**(29), 8394–8405.
394 K. Fujioka, M. Hiruoka, K. Sato, N. Manabe, R. Miyasaka, S. Hanada, A. Hoshino, R. D. Tilley, Y. Manome, K. Hirakuri and K. Yamamoto, Luminescent passive-oxidized silicon quantum dots as biological staining labels and their cytotoxicity effects at high concentration, *Nanotechnology*, 2008, **19**, 415102.
395 H. A. Santos, J. Rikonen, J. Salonen, E. Makila, T. Heikkila, T. Laaksonen, L. Peltonen, V. P. Lehto and V. Hirvonen, In vitro cytotoxicity of porous silicon microparticles: effect of the particle concentration, surface chemistry and size, *Acta Biomater.*, 2010, **6**(7), 2721–2731.
396 L. Ostrovska, A. Broz, A. Fucikova, T. Belinova, H. Sugimoto, T. Kanno, M. Fujii, J. Vaslenta and M. H. Kalbacova, The impact of doped silicon quantum dots on human osteoblasts, *RSC Adv.*, 2016, **6**, 63403.
397 Z. Cao, F. Peng, Z. Hu, B. Chu, Y. Zhong, Y. Su, S. He and Y. He, In vitro cellular behaviours and toxicity assays of small-sized fluorescent silicon nanoparticles, *Nanoscale*, 2017, **9**, 7602–7611.
398 A. E. Kusi-Appiah, M. L. Mastronardi, C. Qian, K. K. Chan, L. Ghazanfari, P. Prommapan, C. Kubel, G. A. Ozin and S. Lenhert, Enhanced cellular uptake of size-separated lipophilic silicon nanoparticles, *Sci. Rep.*, 2017, **7**, 43731.
399 T. Belinova, L. Vrabcova, I. Machova, A. Fucikova, J. Valenta, H. Suginmoto, M. Fujii and M. H. Kalbacova, Silicon quantum dots and their impact on different human cells, *Phys. Status Solidi B*, 2018, **255**, 1700597.

400 W. Phatvej, H. K. Datta, S. C. Wilkinson, E. Mutch, A. K. Daly and B. R. Horrocks, Endocytosis and lack of cytotoxicity of alkyl-capped silicon quantum dots prepared from porous silicon, *Materials*, 2019, **12**, 1702.
401 Y. He, Y. Su, X. Yang, Z. Kang, T. Xu, R. Zhang, C. Fan and S. T. Lee, Photo and pH Stable, High-Luminescent Silicon Nanospheres and Their Bioconjugates for Immunofluorescent Cell Imaging, *J. Am. Chem. Soc.*, 2009, **131**(12), 4434–4438.
402 K. Sato, S. Yokosuka, Y. Takigami, K. Hirakuri, K. Fujioka, Y. Manome, H. Sukegawa, H. Iwai and N. Fukata, Size-tunable silicon/iron oxide hybrid nanoparticles with fluorescence, superparamagnetism, and biocompatibility, *J. Am. Chem. Soc.*, 2011, **133**, 18626–18633.
403 B. Xia, J. Li, J. Shi, Y. Zhang, Q. Zhang, Z. Chen and B. Wang, Biodegradable and magnetic-fluorescent porous silicon @ iron oxide nanocomposites for fluorescence/magnetic resonance bimodal imaging of tumor in vivo, *ACS Biomater. Sci. Eng.*, 2017, **3**(10), 2579–2587.
404 S. Li, Y. Yuan, Y. Yang, C. Li, M. C. McMahon, M. Liu, S. Chen and X. Zhou, Potential detection of cancer with fluorinated silicon nanoparticles in 19F MRI and fluorescence imaging, *J.Mater.Chem.*, 2018, **B6**, 4293–4300.
405 M. P. Singh, T. M. Atkins, E. Mathuswamy, S. Kamali, C. Tu, A. Y. Louie and S. M. Kauzlarich, Development of Iron-Doped Silicon Nanoparticles As Bimodal Imaging Agents, *ACS Nano*, 2012, **6**(6), 5596–5604.
406 G. Singh, J. L. Z. Ddughu, N. Licciardello, R. Bergmann, L. de Cola and H. Stephan, Ultrasmall silicon nanoparticles as a promising platform for multimodal imaging, *Faraday Discuss.*, 2020, DOI: 10.1039/C9FD00091G.
407 J. Xie, S. Lee and X. Chen, Nanoparticle-based theranostic agents, *Adv. Drug Delivery Rev.*, 2010, **62**(11), 1064–1079.
408 T. Lammers, S. Aime, W. E. Hennick, G. Storm and F. Kiesling, Theranostic nanomedicine, *Acc. Chem. Res.*, 2011, **44**(10), 1029–1038.
409 D. Kovalev and M. Fujii, Silicon nanocrystals: photosensitizers for oxygen molecules, *Adv. Mater.*, 2005, **17**, 2531–2544.
410 L. A. Osminkina and V. Y. Timoshenko, Porous silicon as a sensitizer for biomedical applications, *Mesoporous Biomater.*, 2016, **3**, 39–48.
411 L. A. Osminkina, K. P. Tamarov, A. P. Sviridov, R. A. Galkin, M. B. Gongalsky, V. V. Solovyev, A. A. Kudryavstev and V. Y. Timoshenko, Photoluminescent biocompatible silicon nanoparticles for cancer theranostic applications, *J. Biophotonics*, 2012, **5**(7), 529–535.
412 V. Stojanovic, F. Cunin, J. O. Durand, M. Garcia and M. Gary-Bobo, Potential of porous silicon nanoparticles as an emerging platform for cancer theranostics, *J. Mater. Chem. B*, 2016, **4**, 7050–7059.
413 C. X. Huang, S. Han, W. Huang and X. Liu, Enhancing solar cell efficiency: the search for luminescent materials as spectral converters, *Chem. Soc. Rev.*, 2013, **42**, 173–201.
414 V. Svrcek, A. Slaoui and J. C. Muller, Silicon nanocrystals as light converter for solar cells, *Thin Solid Films*, 2004, **451–452**, 384–388.
415 M. Stupca, M. Alsalhi, T. A. Saud, A. Almuhanna and M. H. Nayfeh, Enhancement of polycrystalline silicon solar cells using ultrathin films of silicon nanoparticles, *Appl. Phys. Lett.*, 2007, **91**, 063107.
416 X. Pi, Q. Di, L. Li and D. Yang, Spin-coating silicon quantum dot ink to improve solar cell efficiency, *Sol. Energy Mater. Sol. Cells*, 2011, **95**, 2941–2945.

417 X. Pi, L. Zhang and D. Yang, Enhancing the efficiency of multicrystalline silicon solar cells by the inkjet printing of silicon-quantum-dot ink, *J. Phys. Chem. C*, 2012, **116**, 21240–21243.

418 Z. Yuan, G. Pucker, A. Marconi, F. Sgrignuoli, A. Anopchenko, Y. Jestin, L. Ferrario, P. Bellutti and L. Pavesi, Silicon nanocrystals as a photoluminescence down shifter of solar cells, *Sol. Energy Mater. Sol. Cells*, 2011, **95**, 1224–1227.

419 F. Sgrignuoli, P. Ingenhoven, G. Pucker, V. D. Mihailetchi, E. Froner, Y. Jestin, E. Moser, G. Sanchez and L. Pavesi, Purcell effect and luminescent downshifting in silicon nanocrystals coated back-contact solar cells, *Sol. Energy Mater. Sol. Cells*, 2012, **132**, 267–274.

420 H. C. Hao, W. Shi, J. R. Chen and M. Lu, Mass production of Si quantum dots for commercial c-Si solar cell efficiency improvement, *Mater. Lett.*, 2014, **133**, 80–82.

421 M. C. Beard, K. P. Knutsen, P. Yu, J. M. Luther, Q. Song, W. K. Metzger, R. J. Ellingson and A. J. Nozik, Multiple Exciton Generation in Colloidal Silicon Nanocrystals, *Nano Lett.*, 2007, **7**(8), 2506–2512.

422 D. Timmerman, I. Izeddin, P. Stallinga, I. N. Yassievich and T. Gregorkiewicz, Space-separated quantum cutting with silicon nanocrystals for photovoltaic applications, *Nat. Photonics*, 2008, **2**, 105–109.

423 M. T. Trinh, R. Limpens, W. D. A. M. de Boer, J. M. Shins, L. D. A. Siebbeles and T. Gregorkiewicz, Direct generation of multiple excitons in adjacent silicon nanocrystals revealed by induced absorption, *Nat. Photonics*, 2012, **36**, 1–6.

424 M. Govoni, I. Marri and S. Ossicini, Carrier multiplication between interacting crystals for fostering silicon-based photovoltaics, *Nat. Photonics*, 2012, **6**, 672–679.

425 P. Zhang, Y. Feng, X. Wen, W. Cao, R. Anthony, U. Kortshagen, G. Conibeer and S. Huang, Generation of hot carrier population in colloidal silicon quantum dots for high efficiency photovoltaics, *Sol. Energy Mater. Sol. Cells*, 2016, **145**, 391–396.

426 F. Meinardi, S. Ehrenberg, L. Dhamo, F. Carulli, M. Mauri, F. Bruni, R. Simonutti, U. Kortshagen and S. Brovelli, Highly efficient luminescent solar concentrators based on earth abundant indirect bandgap silicon quantum dots, *Nat. Photonics*, 2017, **11**, 177–186.

427 S. K. E. Hill, R. Connell, C. Peterson, J. Hollinger, M. A. Hillmyer, U. Kortshagen and V. E. Ferry, Silicon quantum dot - poly(methyl methacrylate) nanocomposites with reduced light scattering for luminescent solar concentrators, *ACS Photonics*, 2019, **6**, 170–180.

428 R. Mazarro, A. Gradone, S. Angeloni, G. Morselli, P. G. Cozzi, F. Romano, A. Vomiero and P. Ceroni, Hybrid silicon nanocrystals for color-neutral and transparent luminescent solar concentrators, *ACS Photonics*, 2019, **6**, 2303–2311.

429 J. Cho, J. H. Park, J. K. Kim and E. F. Schubert, White light emitting diodes: history, progress and future, *Laser Photonics Rev.*, 2017, **11**(2), 1600147.

430 C. C. Tu, J. H. Hoo, K. F. Bohringer, L. Y. Lin and G. Cao, Red-emitting silicon quantum dot phosphors in warm white LEDs with excellent color rendering, *Opt. Express*, 2014, **22**(S2), A276–A28.

431 K. Dohnalova, L. Ondic, K. Kusova, I. Pelant, J. L. Rehspringer and R. R. Mafouna, White-emitting oxidized silicon nanocrystals: Discontinuity in spectral development with reducing size, *J. Appl. Phys.*, 2010, **107**, 053102.

432 A. V. Vasin, Y. Ishikawa, N. Shibata, J. Salonen and V. P. Lehto, Strong White Photoluminescence for Carbon Incorporated Silicon Oxide Fabrication by Preferential Oxidation of Silicon in Nano-Structured Si: C Layer, *Jpn. J. Appl. Phys.*, 2007, **46**(19), L465–L467.

433 Y. Ishikawa, A. V. Vasin, J. Salonen, S. Muto, V. S. Lysenko, A. N. Nazarov, N. Shibata and V. P. Lehto, Color control of white photoluminescence from carbon-incorporated silicon oxide, *J. Appl. Phys.*, 2008, **104**, 083522.

434 S. Y. Seo, K. S. Cho and J. H. Shin, Intense blue-white luminescence from carbon-doped silicon-rich silicon oxide, *Appl. Phys. Lett.*, 2004, **84**(5), 717–719.

435 A. Peres-Rodriguez, O. Gonzalez-Verona, B. Garrido, P. Pellegrino, J. R. Morante, C. Bonatos, M. Carrada and A. Claverie, White luminescence from Si^+ and C^+ ion implanted SiO_2 films, *J. Appl. Phys.*, 2003, **94**(1), 254–262.

436 B. Ghosh, M. Ogawara, Y. Sakka and N. Shirahata, White-light-emitting Liquefiable Silicon Nanocrystals, *Chem. Lett.*, 2012, **41**, 1157–1159.

437 C. Liu, Z. Zhao, R. Zhang, L. Yang, Z. Wang, J. Yang, H. Jiang, M. Y. Han, B. Liu and Z. Zhang, Strong infrared laser ablation produces white light emitting materials via the formation of silicon and carbon quantum dots in silica nanoparticles, *J. Phys. Chem. C*, 2015, **119**, 8266–8272.

438 G. Barillaro and L. M. Strambini, Color tuning of light-emitting-diodes by modulating the concentration of red-emitting silicon nanocrystal phosphors, *Appl. Phys. Lett.*, 2014, **104**, 091102.

439 L. Yang, Y. Liu, Y. L. Zhong, X. X. Jiang, B. Song, X. Y. Ji, Y. Y. Su, L. S. Liao and Y. He, Fluorescent silicon nanoparticles utilized as stable color converters for white light-emitting diodes, *Appl. Phys. Lett.*, 2015, **106**, 173–109.

440 B. Song, Y. Zhong, S. Wu, B. Chu and Y. He, One dimensional fluorescent silicon nanorods featuring ultrahigh photostability, favourable biocompatibility and excitation wavelength dependent emission spectra, *J. Am. Chem. Soc.*, 2016, **138**, 4824–4831.

441 S. Bose, M. A. Ganayee, B. Mondal, A. Baidya, S. Chennu, J. S. Mohanty and T. Pradeep, Synthesis of silicon nanoparticles from rice husk and their use as sustainable fluorophores for white light emission, *ACS Sustainable Chem. Eng.*, 2018, **6**(5), 6203–6210.

442 C. D. Brites, P. P. Lima, N. J. O. Silva, A. Millan, V. S. Amaral, F. Palacio and L. D. Carlos, Thermometry at the nanoscale, *Nanoscale*, 2012, **4**, 4799.

443 D. Jaque and F. Vetrone, Luminescence nanothermometry, *Nanoscale*, 2012, **4**, 4301–4326.

444 M. A. Cardona-Castro, A. Morales-Sanchez, L. Licea-Jimenez and J. Alvarez-Quintana, Si-nanocrystal-based nanofluids for nanothermometry, *Nanotechnology*, 2016, **27**, 235502.

445 Y. B. Ryabchikov, S. Alekseev, V. Lysenko, G. Bremond and J. M. Bluet, Luminescence behavior of silicon and carbon nanoparticles dispersed in low polar liquids, *Nanoscale Res. Lett.*, 2012, **7**, 365.

446 Y. B. Ryabchikov, S. Alekseev, V. Lysenko, G. Bremond and J. M. Bluet, Photoluminescence of silicon nanoparticles chemically modified by alkyl groups and dispersed in low polar liquids, *J. Nanopart. Res.*, 2013, **15**, 1535.

447 Y. B. Ryabchikov, S. Alekseev, V. Lysenko, G. Bremond and J. M. Bluet, Photoluminescence thermometry with alkyl-terminated silicon nanoparticles dispersed in low-polar liquids, *Phys. Status Solidi RRL*, 2013, **7**(6), 414–417.

448 Y. B. Ryabchikov, V. Lysenko and T. Nychporuk, Enhanced Thermal Sensitivity of Silicon Nanoparticles Embedded in (Nano-Ag/)SiN$_x$ for Luminescent Thermometry, *J. Phys. Chem. C*, 2014, **118**, 12515–12519.

449 H. Hajjaji, S. Alekseev, G. Giullot, N. P. Blanchard, V. Monnier, Y. Chevolet, G. Bremond, M. Querry, D. Philippon, P. Vergne and J. M. Bluet, Luminescence nanothermometry with alkyl-cappedsilicon nanoparticles dispersed in nonpolar liquids, *Nanoscale Res. Lett.*, 2014, **9**, 94.

450 H. Hamza, S. M. B. Albahrani, G. Guillot, M. Maillard, D. Philippon, P. Vergne and J. M. Bluet, Temperature and Viscosity Effects on the Photoluminescence Properties of Alkyl-Capped Silicon Nanoparticles Dispersed in Nonpolar Liquids, *J. Phys. Chem. C*, 2015, **119**(29), 16897–16904.

451 A. M. P. Botas, C. D. S. Brites, J. Wu, U. Kortshagen, R. N. Pereira, L. D. Carlos and R. A. S. Ferreira, A new generation of primary luminescent thermometers based on silicon nanoparticles and operating in different media, *Part. Part. Syst. Charact.*, 2016, **33**, 740–774.

452 Q. Li, Y. He, J. Chang, L. Wang, H. Chen, Y. Tan, H. Wang and Z. Shao, Surface-Modified Silicon Nanoparticles with Ultrabright Photoluminescence and single-exponential decay for nanoscale fluorescence lifetime imaging of temperature, *J. Am. Chem. Soc.*, 2013, **135**, 14924–14927.

453 R. Coustel, Q. B. Guillaume, V. Calvo, O. Renoit, L. Dubois, F. Duclairoir and N. Pauc, Measurement of the surface recombination velocity in organically passivated silicon nanostructures: the case of silicon on insulator, *J. Phys. Chem. C*, 2011, **115**(45), 22265–22270.

454 O. Demichel, V. Calvo, A. Besson, P. Noe, B. Salem, N. Pauc, F. Oehler, P. Gentile and N. Magnea, Surface recombination velocity measurements of efficiently passivated gold-catalyszed silicon nanowires by a new optical method, *Nano Lett.*, 2010, **10**, 2323–2329.

455 E. M. Grumstrup, M. M. Gabriel, E. M. Cating, C. W. Pinion, J. D. Christesen, J. R. Kirschbrown, E. L. Vallorz, J. F. Cahoon and J. M. Papanikolas, Ultrafast carrier dynamics in individual silicon nanowires; characterization of diameter-dependent carrier lifetime and surface recombination with pump-probe microscopy, *J. Phys. Chem. C*, 2014, **118**(16), 8634–8640.

456 Y. Dan, K. Seo, K. Takei, J. H. Meza, A. Javey and K. B. Crozier, Dramatic reduction of surface recombination by in situ surface passivation of silicon nanowires, *Nano Lett.*, 2011, **11**, 2527–2532.

457 M. L. Mastronardi, E. J. Henderson, D. P. Puzzo and G. A. Ozin, Small Silicon, Big Opportunities: The Development and Future of Colloidally-Stable Monodisperse Silicon Nanocrystals, *Adv. Mater.*, 2012, **24**(43), 5890–5898.

458 R. Kothemann, N. Weber, J. K. N. Lindner and C. Meier, High precision determination of silicon nanocrystals: optical spectroscopy versus electron microscopy, *Semicond. Sci. Technol.*, 2019, **34**, 095009.

459 J. Heitmann, F. Muller, M. Zacharias and U. Gosele, Silicon nanocrystals: size matters, *Adv. Mater.*, 2005, **17**(7), 795–803.

460 L. Patrone, D. Nelson, V. I. Safarov, M. Sentis, W. Marine and S. Giorgio, Photoluminescence of silicon nanoclusters with reduced size dispersion produced by laser ablation, *J. Appl. Phys.*, 2000, **87**, 3829.

461 R. Gresback, T. Nozaki and K. Okazaki, Synthesis and oxidation of luminescent silicon nanocrystals from silicon tetrachloride by very high frequency nonthermal plasma, *Nanotechnology*, 2011, **22**, 305605.

462 R. D. Tilley, J. H. Warner, K. Yamamoto, I. Matsui and H. Fujimori, Microemulsion synthesis of monodisperse surface stabilized silicon nanocrystals, *Chem. Commun.*, 2005, 1833–1835.

463 M. Rosso-Vassic, E. Spruijt, B. van Lagen, L. de Cola and H. Zuilhof, Alkyl-functionalized oxide-free silicon nanoparticles: synthesis and optical properties, *Small*, 2008, **4**(10), 1835–1841.

464 R. P. Camata, H. A. Atwater, K. H. Vahala and R. C. Flagan, Size classification of silicon nanocrystals, *Appl. Phys. Lett.*, 1996, **68**(22), 3162–3164.

465 N. Suzuki, T. Makino, Y. Yamada, T. Yoshida and T. Seto, Monodispersed, nonagglomerated silicon nanocrystallites, *Appl. Phys. Lett.*, 2001, **78**(14), 2043–2045.

466 G. Ledoux, J. Gong, F. Huisken, O. Guillois and C. Reynaud, Photoluminescence of size-separated silicon nanocrystals: confirmation of quantum confinement, *Appl. Phys. Lett.*, 2002, **80**(25), 4834–4836.

467 M. L. Mastronardi, F. Hennrich, E. J. Henderson, F. Maier-Flag, C. Blurn, J. Reichenbach, U. Lemmer, C. Kubel, D. Wang, M. M. Kappes and G. A. Ozin, Preparation of monodisperse silicon nanocrystals using density gradient ultracentrifugation, *J. Am. Chem. Soc.*, 2011, **133**, 11928–11931.

468 J. B. Miller, A. R. van Sickle, R. J. Anthony, D. J. Kroll, U. R. Kortshagen and E. K. Hobbie, Ensemble Brightening and Enhanced Quantum Yield in Size-Purified Silicon Nanocrystals, *ACS Nano*, 2012, **6**(8), 7389–7396.

469 S. L. Brown, J. B. Miller, R. J. Anthony, U. R. Kortshagen, A. Kryjevski and E. K. Hobbie, Abrupt size partitioning of multimodal photoluminescence relaxation in monodisperse silicon nanocrystals, *ACS Nano*, 2017, **11**, 1597–1603.

470 J. M. Buriak, Organometallic chemistry on silicon and germanium surfaces, *Chem. Rev.*, 2002, **102**(5), 1271–1308.

471 J. G. C. Veinot, Synthesis surface functionalization and properties of freestanding silicon nanocrystals, *Chem. Commun.*, 2006, **40**, 4160–4168.

472 X. Li, Y. He, S. S. Talukdar and M. T. Swihart, Process for Preparing Macroscopic Quantities of Brightly Photoluminescent Silicon Nanoparticles with Emission Spanning the Visible Spectrum, *Langmuir*, 2003, **19**, 8490–8496.

473 F. Lacour, O. Guillois, X. Portier, H. Perez, N. Herlin and C. Reynaud, Laser pyrolysis synthesis and characterization of luminescent silicon nanocrystals, *Physica E: Low-dimensional Systems and Nanostructures*, 2007, **38**, 11–15.

474 T. Hulser, S. M. Schnurre and J. Wiggers, Gas phase synthesis of nanoscale silicon as an economical route towards sustainable energy technology, *KONA Powder Part. J.*, 2011, **29**, 191–207.

475 I. Dogan, N. J. Kramer, R. H. J. Westermann, K. Dohnalova, A. H. M. Smets, M. A. Verheijen, T. Gregorkiewicz and M. C. M. van der Sanden, Ultrahigh

throughput plasma processing of free-standing silicon nanocrystals with lognormal size distribution, *J. Appl. Phys.*, 2013, **113**, 134306.
476 S. P. Pujari, H. Driss, F. Bannani, B. van Lagen and H. Zuilhof, One-pot gram scale synthesis of hydrogen-terminated silicon nanoparticles, *Chem. Mater.*, 2018, **30**(18), 6503–6512.
477 F. Kunze, S. Kuns, M. Spree, T. Hulser, C. Schulz, H. Wiggers and S. M. Schnurre, Synthesis of silicon nanoparticles in a pilot-scale microwave plasma reactor: impact of flow rates and precursor concentration on the nanoparticle size and aggregation, *Powder Technol.*, 2019, **342**, 880–886.
478 L. Batchelor, A. Loni, L. T. Canham, M. Hasan and J. L. Coffer, Manufacture of mesoporous silicon from living plants and agricultural waste: an environmentally friendly and scalable process, *Silicon*, 2012, **4**, 259–266.
479 Y. Wu, Y. Zhong, B. Chu, B. Sun, B. Song, S. Wu, Y. Su and Y. He, Plant derived fluorescent silicon nanoparticles featuring excitation wavelength dependent fluorescence spectra for anti-counterfeiting applications, *Chem. Commun.*, 2016, **52**, 7047–7050.
480 Z. Wang, S. Zeng, Y. Li, W. Wang, Z. Zhang, H. Zeng, W. Wang and L. Sun, Luminescence mechanism of carbon-incorporated silica nanoparticles derived from rice husk biomass, *Ind. Eng. Chem. Res.*, 2017, **56**, 5906–5912.
481 M. J. Kirschenbaum, M. G. Boebinger, M. J. Katz, M. T. McDowell and M. Dasog, Solid state synthesis route for scalable luminescent silicon and germanium nanocrystals, *Chem. Nanomater.*, 2018, **4**(4), 423–429.

Faraday Discussions

PAPER

Fast room-temperature functionalization of silicon nanoparticles using alkyl silanols†

Alyssa F. J. van den Boom,[a] Sidharam P. Pujari,[a] Fatma Bannani,[b] Hafedh Driss[c] and Han Zuilhof[*acd]

Received 3rd October 2019, Accepted 2nd January 2020
DOI: 10.1039/c9fd00102f

Silicon nanoparticles (Si NPs) are a good alternative to conventional heavy metal-containing quantum dots in many applications, due to their low toxicity, low cost, and the high natural abundance of the starting material. Recently, much synthetic progress has been made, and crystalline Si NPs can now be prepared in a matter of hours. However, the passivation of these particles is still a time-consuming and difficult process, usually requiring high temperatures and/or harsh reaction conditions. In this paper, we report an easy method for the room-temperature functionalization of hydrogen-terminated Si NPs. Using silanol compounds, a range of functionalized Si NPs could be produced in only 1 h reaction time at room temperature. The coated NPs were fully characterized to determine the efficiency of binding and the effects of coating on the optical properties of the NPs. It was found that Si NPs were effectively functionalized, and that coated NPs could be extracted from the reaction mixture in a straightforward manner. The silanol coating increases the quantum yield of fluorescence, decreases the spectral width and causes a small (~50 nm) blue-shift in both the excitation and emission spectra of the Si NPs, compared to unfunctionalized particles.

Introduction

Semiconductor nanoparticles have proven to be useful for many different applications, including bio-imaging,[1–3] LEDs,[4–6] and solar cells.[7–12] Silicon nanoparticles (Si NPs) are, in this regard, particularly interesting, due to their low toxicity and the high natural abundance of the starting material.[13,14] However, several issues have so far stood in the way of their widespread use. First, synthesis

[a]Laboratory of Organic Chemistry, Wageningen University, Stippeneng 4, 6708 WE Wageningen, The Netherlands. E-mail: han.zuilhof@wur.nl
[b]Department of Chemistry, King Abdulaziz University, Jeddah, Saudi Arabia
[c]Chemical and Materials Engineering, Faculty of Science, King Abdulaziz University, Jeddah, Saudi Arabia
[d]School of Pharmaceutical Science and Technology, Tianjin University, 92 Weijin Road, Tianjin, P. R. China
† Electronic supplementary information (ESI) available. See DOI: 10.1039/c9fd00102f

of the NPs is time-consuming and difficult, often requiring high temperatures or harsh chemicals.[15-25] Second, once the Si NPs are formed, they have to be rapidly passivated to prevent the loss of their original properties.[26] This is especially the case for hydrogen-terminated Si NPs (H-Si NPs). Such particles are of significant interest due to their near-ideal electronic and electrical properties, yet their susceptibility to oxidation requires extensive protection. In the literature, passivation of H-Si NPs has been achieved by heating to 190 °C in the presence of a suitable ligand,[27] etching using phosphorous pentachloride followed by modification with an amine,[20,28] thermally-promoted thiolation reactions,[29] reacting with bifunctional alkenes,[26] halogenation followed by a Grignard reaction,[24,30] platinum-catalyzed reactions,[23,25,31] or reactions in the presence of radical initiators.[32,33] These procedures generally suffer from long reaction times and/or the use of non-trivial reaction conditions, making them less attractive for application on a larger scale. Furthermore, the type of ligand that can be attached in such procedures is usually limited by the reaction conditions used. In most cases, the ligand may also contain only one functional group, as having multiple functional groups interferes with the reaction. This limits the possibility for further functionalization. Clearly, an easy, fast, and versatile passivation method for H-Si NPs is still needed.

In this paper, we report the use of silanol compounds with a range of functional groups to passivate H-Si NPs (obtained *via* a microwave-based procedure developed in our group,[34] building upon earlier protocols from He's group for other types of Si NPs).[35,36] This silanol reaction showed promise due to the discovery that flat Si surfaces could readily be functionalized in this manner,[37] yielding densely packed, hydrolytically stable monolayers in <1 h at room temperature. As this reaction proceeded without the need for additional solvents or catalysts, this constitutes a significant advantage over previously mentioned routes. Here, only mono-hydroxylated silanols were considered, as silanols with two or three hydroxyl groups might yield multilayers on the surface of the Si NPs, as reported for flat surfaces.[37] We show that passivation with silanols is an easy and fast method to obtain functionalized Si NPs. The silanol coating can easily be tuned by selecting silanols with different side groups, and due to the specificity of the reaction, even silanols with multiple functional groups can be used. Finally, we show it is possible to obtain pure, passivated Si NPs with good optical properties in less than 4 h of combined reaction (1 h) and purification (3 h) time.

Experimental

Synthetic procedures

Synthesis of silanol compounds. Silanol compounds were synthesized from their corresponding chlorosilanes using an adapted literature procedure (Scheme 1).[37] In brief, an alkyldimethylchlorosilane (1 eq.) was added to an ice-cold mixture of triethylamine (1.5 eq.), water (4 eq.), and diethyl ether. The mixture was stirred for 1 h at room temperature, after which the liquid phase was isolated. Hexane was added, and the diethyl ether was evaporated *in vacuo*. The reaction mixture was cooled on an ice bath, and washed with ice-cold 0.5 M aqueous HCl. The organic phase was dried using magnesium sulfate, and the solvent was evaporated *in vacuo* to obtain the silanol. Full characterization of the silanols used in this study can be found in the ESI.†

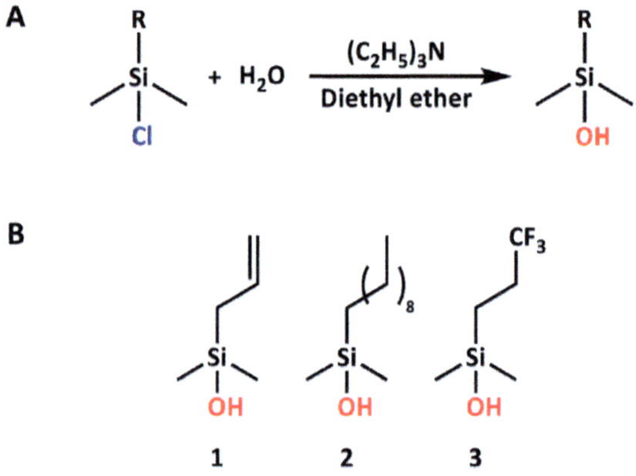

Scheme 1 (A) Reaction scheme for silanol synthesis. (B) Overview of silanols used in this study.

Synthesis of hydrogen-terminated silicon nanoparticles. H-Si NPs were prepared according to the protocol developed by Pujari et al.[34] In brief: inside a glovebox (O_2 < 0.1 ppm, H_2O < 0.1 ppm), 20 mg (0.10 mmol) of sodium ascorbate and 0.8 ml of DMSO were added to a microwave tube equipped with a stirrer bar. 0.2 ml (1.08 mmol) of triethoxysilane was added, after which the tube was quickly sealed and transported to the microwave. The mixture was allowed to react for 10 min at 170 °C while stirring at 600 rpm. After the reaction, the tube was allowed to cool down and was transported back into the glovebox. There, it was opened, and the mixture was filtered with a 0.2 μm PTFE syringe filter. The filtered brown solution was directly used for passivation.

Functionalization of silicon nanoparticles. Inside a glovebox, 0.2 ml of the selected silanol is added to the filtered nanoparticle dispersion, and the resulting mixture is stirred for 1 h at room temperature. After this passivation step, the reaction mixture is taken out of the glovebox, and the DMSO is evaporated *in vacuo* (70 °C, 1 mbar); excess silanol is subsequently extracted using diethyl ether. After drying, the resulting dry brown flakes are finely ground with 15 ml of toluene, added in several small fractions, to extract the functionalized NPs; large particles are allowed to sediment after the addition of each fraction, and the top layer of toluene is pipetted off. The combined toluene phases are dried prior to analysis. In a typical experiment, 20–35 mg of functionalized Si NPs were obtained in this way, depending on the ligand and the number of toluene fractions used.

Preparation of water-soluble particles. To prepare water-soluble Si NPs, a thiol sugar was attached to the allyldimethylsilanol-functionalized Si NPs *via* a thiol-ene click reaction. To a purified sample of 30 mg of the functionalized Si NPs, a solution of 0.36 g (1.0 mmol) 1-thio-β-D-glucose tetraacetate and 0.05 g (0.2 mmol) 2,2-dimethoxy-2-phenylacetophenone (DMPA) in 1 : 1 chlorobenzene and DMF was added inside a glovebox.[38,39] The reaction mixture was stirred at room temperature for 3.5 h with 365 nm light irradiation. The distance between the lamp and the solution was ~4 cm, so as not to heat the sample. The lamp was removed, and the solvents were evaporated *in vacuo*. A light brown gel formed, from which Si NPs and DMPA were extracted using chlorobenzene. The solvent was again removed *in vacuo*, and the functionalized Si NPs were extracted using diethyl ether.

To de-protect the acetate-capped sugar, the Si NPs were dispersed in a 0.015 M solution of sodium methoxide in methanol.[40] The reaction mixture was stirred for 45 min, and sodium methoxide was precipitated out of the solution by the addition of anhydrous pentane. The solvents in the clear dispersion were removed *in vacuo* to obtain 27 mg of dry thioglucose-coated Si NPs for analysis.

Materials and methods

NMR measurements. NMR measurements were conducted on a 400 MHz Bruker Avance III, and the resulting data were analyzed using MestReNova software, version 10.0.2-15465.

IR measurements. IR measurements were performed on a Bruker Tensor 27 spectrometer equipped with a diamond ATR accessory (32 scans; 4 cm^{-1} resolution). For the Si NP samples, the dried Si NP powder was firmly pressed onto the ATR crystal; for the pure silanols, a drop was placed on the crystal.

XPS measurements. XPS measurements were performed on a JEOL JPS-9200 photoelectron spectrometer, using a 12 kV and 20 mA monochromatic Al Kα source. The analyzer pass energy was 10 eV, and the take-off angle between sample and detector was set at 80°. Data was analyzed using the CasaXPS program, version 2.3.18PR1.0. Samples were prepared by drop coating a Si NP solution onto a plasma-cleaned gold surface.

TEM measurements. TEM measurements were performed on a JEOL JEM-1400 plus (WEMC – TEM 120 kV). Samples were prepared by evaporating a drop of dilute Si NP solution in chloroform onto an ultrathin Carbon Type-A support film with a 400 mesh on top of a copper grid with an approximate hole size of 42 μm (TED PELLA, Inc.). Samples were allowed to dry for 30 min prior to imaging.

High-resolution TEM measurements. High-resolution TEM measurements were performed on a Tecnai G2 F20 Super Twin TEM microscope at 200 kV with a LaB6 emitter. The microscope was equipped with a high-angle annular dark-field (HAADF) detector for scanning transmission electron microscopy (STEM) imaging. STEM imaging and all analytical work was performed with a probe size of 1 nm, resulting in a beam current of about 0.5 nA. TEM images were collected using a GATAN US1000 2K HR 200 kV CCD camera. The CCD line traces were collected fully automatically using the Tecnai G2 user interface and processed with both Digital Micrograph software Version 2.3 and the Tecnai Imaging and Analysis (TIA) software Version 1.9.162.

Optical properties. Optical properties (in DMSO, unless noted otherwise) were studied using an Edinburgh Instruments FLS900 fluorescence spectrometer (continuous: 450 W xenon lamp as the excitation source; for lifetime: 444 nm LED laser) for the fluorescence properties, and a Varian Cary 50 UV-vis spectrophotometer for the absorption measurements (scan rate 600 nm min^{-1}; spectra corrected for solvent absorption). Results were fitted with a 3-exponential decay. For the quantum yield, fluorescein in 0.1 M aqueous NaOH was used as a reference, with an absolute quantum yield of 0.88 ± 0.03.[41]

Results and discussion

Silanol-based functionalization of silicon nanoparticles

The passivation of H-Si NPs with silanol compounds (Fig. 1) proceeded readily at room temperature with a straightforward procedure: silanol was simply added to a H-Si NP suspension in DMSO (as obtained without further purification from Si NP synthesis), and the mixture was stirred for 1 h at room temperature. The ease of this reaction is in line with the calculated exothermicity, namely $\Delta H = -16.9$ kcal mol^{-1} in DMSO, as obtained from quantum-chemical wB97XD/6-311+G(d,p) calculations using a dielectric SMD model mimicking DMSO. The reason for the extra 30 min of stirring time, compared to the analogous reaction on flat surfaces,[37] was to ensure a high packing density on the Si NP surface, as this would minimize future oxidation. Escorihuela *et al.* further reported that for flat surfaces, the reaction speed could be increased by irradiation with UV light. However, this was not tested here, as the subsequent (relatively slow) evaporation of DMSO *in vacuo* meant the Si NPs would be exposed to silanol for at least 2 h anyway.

For both silanols **1** and **3**, excess silanol was removed simultaneously with DMSO, simplifying the purification procedure; only one extraction step with diethyl ether was necessary to remove the remaining traces of unbound silanol and obtain a dry powder. Meanwhile, **2** could not be removed by vacuum at the current temperature. Therefore, for this silanol, three extractions with diethyl ether were performed to obtain a dry powder. In all cases, an estimated yield of ~45% was obtained, assuming all Si NP surface atoms had reacted with the silanol, and no Si–H bonds remained.

To confirm that the passivation process did not affect the size of the Si NPs, TEM imaging was performed. For both passivated and non-passivated Si NPs, the average size of the particles was around 6 ± 1 nm (Fig. S13†), slightly (~1.5 nm) smaller than previously reported by Pujari *et al.* for H-Si NPs synthesized using similar conditions.[34]

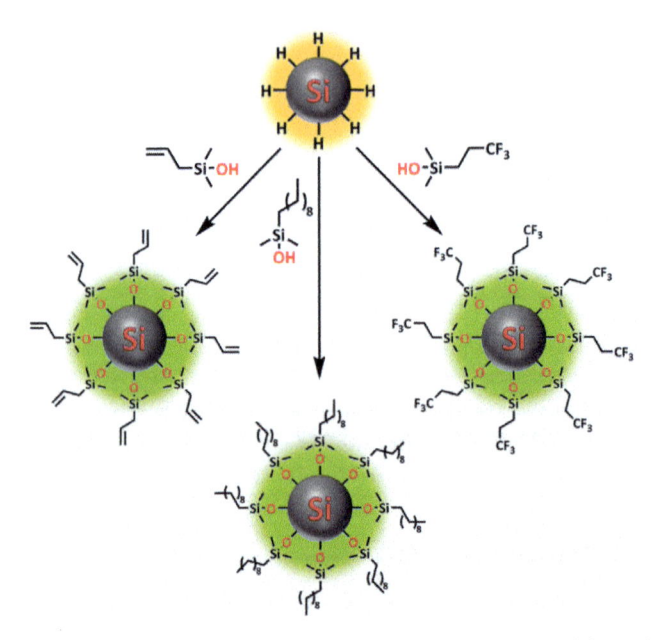

Fig. 1 Overview image showing functionalization of silicon nanoparticles using various silanol compounds.

High-resolution TEM of single Si NPs confirmed the crystallinity of the functionalized NPs, in line with data shown by Pujari et al.[34] Additionally, the interplanar spacing of the crystalline particles was measured to be 0.31 nm in the (111) direction, and 0.19 nm in the (110) direction (Fig. 2D and E). This corresponds to literature values found for crystalline silicon wafers (3.13 and 1.92 Å, respectively).[42]

Recently, a paper by Oliinyk et al. raised the question whether Si NPs could really be formed in a microwave reaction.[43] Even though they specifically doubted their formation in an aqueous solution, and suggested the formation of fluorescent NPs of a carbon or hybrid composition, their investigation is also relevant here. Based on the interplanar crystal spacing found, and the silicon signal observed using XPS (vide infra), we would argue that crystalline Si NPs were indeed formed in this reaction, rather than hybrid organo-silicon or carbon NPs. Furthermore, Oliinyk et al. attributed the source of fluorescence in the paper by Zhong et al.[36] to the formation of carbon nanoparticles and/or organic fluorophores. If this had been the case here, the organic fluorophores should have been visible in the NMR and IR spectra discussed below. However, in these spectra, only signals from the silanol coating could be seen, indicating that no other functional groups were present in the sample. Combined, these data imply the formation of well-defined, fluorescent, crystalline Si NPs, rather than non-crystalline and hybrid carbon nanodots, in the microwave reaction described above.

Characterization of silanol-functionalized particles

NMR spectroscopy. The passivated particles were first characterized using NMR to determine the purity and success of passivation. A representative spectrum of Si NPs passivated with compound **1** (**1**-Si NPs) is shown in Fig. 3 (for the remaining spectra, see ESI S2†). For clarity, the spectrum of the pure silanol is shown underneath in light grey.

All signals belonging to the alkyl groups of the silanol can clearly be seen in the spectrum of the functionalized Si NPs, while the signal from the –OH group has disappeared (5.55 ppm in DMSO-d_6, Fig. S1†). This indicates that the silanol has successfully reacted via the –OH group. In addition, the integration ratio of the peaks closely matches the ratio found for the unbound silanol, although some deviation is sometimes found due to the low signal of the functionalized NPs.

Binding is further confirmed by the broadening of the silanol peaks, caused by attachment of the organic compound to a slowly-rotating NP.[19,25] Since rotation

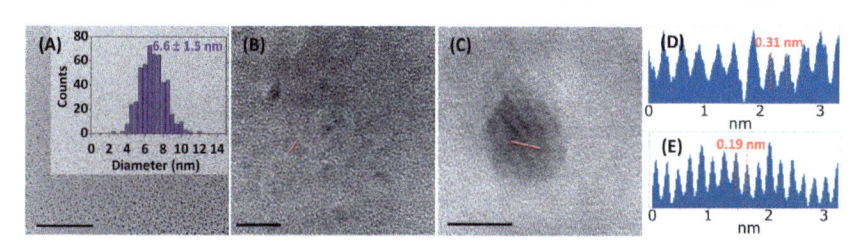

Fig. 2 High-resolution TEM images of **3**-Si NPs. Scale bars are 200 nm (A) and 10 nm (B and C). The insert in (A) shows the size distribution. (D) shows the interplanar spacing of a Si(111) NP measured along the red line in (C), and (E) shows the spacing of a Si(110) NP measured in (B).

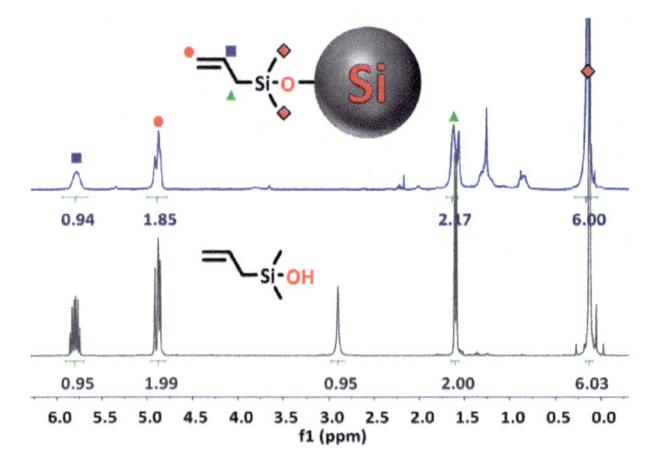

Fig. 3 ¹H NMR spectrum of **1**-Si NPs (top) and pure unbound **1** (bottom) in CDCl₃.

speed is related to size, more severe peak broadening is observed for larger particles. This is why the peak broadening here is more severe compared to the spectra of the 1–4 nm NPs reported in the literature.[19,25] Meanwhile, ¹H NMR of TiO₂ NPs of ~10.4 × 7.4 nm showed spectra with nearly unresolvable peaks.[44,45]

Final proof of successful binding is given by the slight change in chemical shift for hydrogen atoms belonging to the methyl groups of the silanol (Fig. 3, ~0.2 ppm). This shift is upfield when measured in CDCl₃, and downfield when measured in DMSO-d₆. Hydrogen atoms further up the alkyl chain of the silanol seem to be unaffected by the Si NP surface.

In addition to this, only small contaminant peaks can be found in the spectrum, indicating that the purification was successful. Since NPs, due to their slow-tumbling nature, only give very small signals compared to organic compounds, any contaminants are clearly visible in the spectrum.

IR spectroscopy. Since NMR analysis only yielded low signals and broad peaks, ATR-IR was used as extra confirmation of successful silanol attachment. Fig. 4 presents the ATR-IR spectrum of **1**-functionalized Si NPs (**1**-Si NPs), compared to that of the dimerized form of **1** (**1₂**). Both spectra contain a narrow band at 1257 cm⁻¹, indicative of the presence of Si–CH₃ groups.[24,31,46] Other characteristic peaks of silanol can also be seen in the **1**-Si NP spectrum, such as the C=C bond vibrations at 1632, 990, 932, and 893 cm⁻¹, and the =C–H vibration at 3080 cm⁻¹. Additionally, the strong band at 1059 cm⁻¹ indicates the presence of a Si–O–Si bond, as expected after passivation, while the absence of a broad peak at ~3270 cm⁻¹ reflects the disappearance of the –OH group.

For this sample in particular, the IR spectrum is interesting, as it provides further evidence that the silanol has indeed reacted *via* the –OH group, rather than the vinyl group. Not only would binding *via* the vinyl group lead to the disappearance of the C=C signals at 1632, 990, 932 and 893 cm⁻¹, it would also lead to the formation of a new signal at around 1210 cm⁻¹.[46]

The absence of this latter peak is not directly clear from the spectrum in Fig. 4. However, when the difference spectrum between passivated and non-passivated Si NPs is plotted, the lack of signal in this region becomes clear (Fig. S5†). In

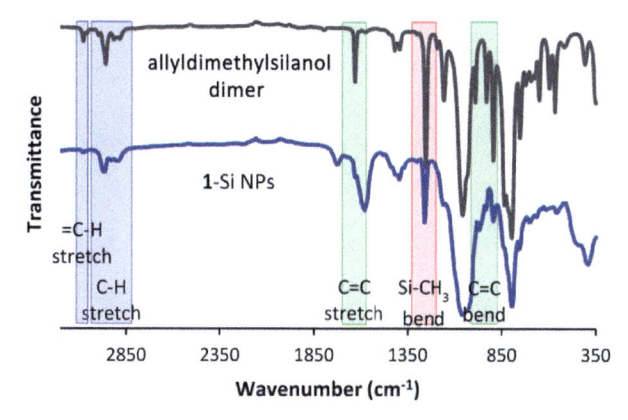

Fig. 4 ATR-IR spectra of 1_2 and 1-Si NPs. Several notable peaks are highlighted.

addition, further signals characteristic of **1** also show up in the difference spectrum, including several peaks in the fingerprint region. This is also observed for the other passivations: the difference spectra show characteristic peaks belonging to the silanol used during the reaction (Fig. S6 and S7†). Only for 2-Si NPs is the comparison less clear, due to the small number of characteristic peaks that can be uniquely attributed to **2**.

XPS analysis. To confirm that the IR signal around 1060 cm^{-1} is indeed due to the binding of the silanol, and not primarily a result of oxidation, XPS analysis was performed. For a well-passivated, non-oxidized sample, a peak <101 eV is expected in the silicon 2p narrow scan. Meanwhile, for a completely oxidized sample, the peak should be found at >103 eV.[34] For the silanol-coated particles, a clear peak was found at 101.8 eV, indicating that the Si NPs are partially oxidized (Fig. 5, wide scan in Fig. S8†). This is in line with the nature of the silanol functionalization, in which for every silanol ligand attached to the Si NPs, two silicon atoms will be directly bound to an oxygen atom. These silicon atoms need not always be located at the surface of the Si NP, as the oxygen atom can "bury" itself inside the Si NP. As a result, the binding energy of >2 silicon atoms can be affected, resulting in an average Si 2p binding energy of 102 eV. In addition, a small fraction of oxidation of the initial Si NP core is typically observed under these conditions, which can – in principle – be removed by HF etching.[34] Since the silanol functionalization itself adds oxygen to the NP, such etching was not performed in this case. This Si 2p binding energy of 101.8 eV was found for all the functionalized Si NPs, with minor (0.3 eV) differences likely due to charging effects in the XPS.[47]

As no peak at >103 eV is present in Fig. 5, we tentatively attribute the peak at 1060 cm^{-1}, as observed in the IR spectrum, to the Si–O–Si bonds from the silanol functionalization, rather than oxidation by atmospheric oxygen.

Optical properties. Next, the effect of silanol coating on the optical properties of the Si NPs was determined. To this end, the properties were compared to those of H-Si NPs. It was shown that passivation with silanol leads to a ~50 nm blue-shift in both the excitation and emission spectra (Fig. 6). This is likely due to the introduction of oxygen atoms into the Si NPs during silanol coating, which has already been shown to lead to a blue-shift in the spectra.[29] Interestingly, no shift was observed for 2-Si NPs (Table 1). It is likely that some minor oxidation of

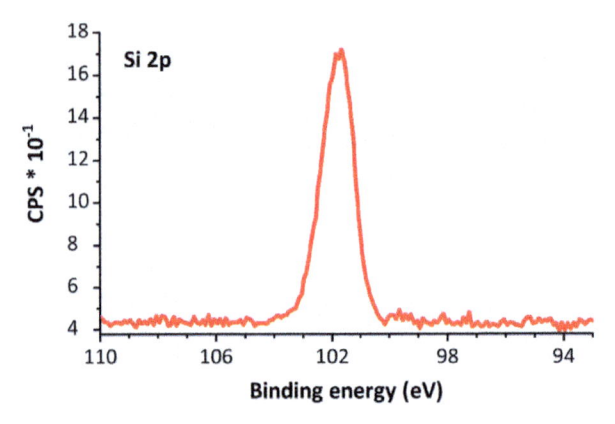

Fig. 5 XPS silicon 2p narrow scan of 2-Si NPs. The peak is centered around 101.8 eV.

1- and 3-Si NPs occurred during processing, as the smaller dimethylsilanol ligands, due to their bulkiness, cannot reach 100% passivation. Meanwhile, the large, hydrophobic alkyl chain of **2** is more effective in blocking incoming oxidants, and will therefore offer better protection against oxidation. Indeed, XPS of **3**-Si NPs after the measurement of optical properties shows the presence of a small amount of silicon oxide (Fig. S9†).

Besides shifting the positions of the excitation and emission maxima, the silanol coating seems to reduce the spectral width of the Si NPs. For non-passivated Si NPs, the spectral widths of the excitation and emission spectra were 0.68 and 0.53 eV, respectively. For **1**-passivated Si NPs the excitation width decreased to 0.57 eV, and for **3**-Si NPs the excitation and emission width decreased to 0.43 and 0.41 eV, respectively. Meanwhile, no change in spectral width was observed for **2**-Si NPs; the excitation spectrum had a spectral width of 0.69 eV, and the emission spectrum a width of 0.50 eV.

Finally, the fluorescence lifetime and quantum yield were also affected by the passivation process. The average fluorescence lifetime of **3**-Si NPs increased from 4.1 to 6.9 ns after passivation, while for **2**-Si NPs the lifetime decreased to 3.3 ns.

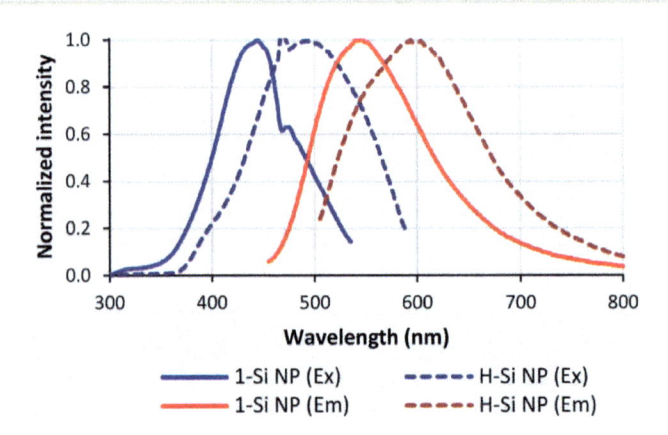

Fig. 6 Optical properties of Si NPs. Full lines: Si NPs passivated with **1**; dashed lines: non-passivated Si NPs. Excitation spectra are shown in blue, emission spectra in red.

Table 1 Optical properties of silanol-coated Si NPs

Sample	λ_{ex} max (nm)	λ_{em} max (nm)	Average lifetime (ns)	Quantum yield
H-Si NP	496	594	4.1	16%[a]
1-Si NP	445	545	4.2	30%[a]
2-Si NP	495	589	3.3	20%[a]
3-Si NP	443	539	6.9	27%[a]
Thioglucose Si NPs	370	460	—	35%[b]

[a] QY w.r.t. fluorescein in 0.1 M NaOH. [b] QY w.r.t. quinine sulfate.

These short ns lifetimes are typical for Si NPs with alkyl termination, prepared using wet-chemical methods, as shown in previous studies on free-standing Si NPs.[31,48-52] In addition, it has been shown that an electronegative cap (such as silanol) or an electronegative environment (such as DMSO) can also cause short ns lifetimes for Si NPs.[53]

The quantum yield of the Si NPs increased for all silanols, though the increase for 2-Si NPs was not as large as for the other silanols. This increase could be partially due to the purification procedure for the passivated Si NPs; as the H-Si NP sample was not purified before quantum yield determination (to minimize oxidation), this sample contains traces of non-fluorescent impurities that increase the optical absorption of the sample, and thereby slightly decrease the apparent quantum yield for the H-Si NPs compared to the purified samples. The quantum yield for these silanol-functionalized Si NPs compares favorably with other modes of functionalization.[34]

Reactivity of silanol-based Si NPs: water-soluble, glucose-coated Si NPs

As a proof of concept, to show that water-soluble, bio-functional Si NPs could be prepared in this manner, 1-thio-β-D-glucose tetraacetate was reacted with 1-Si NPs *via* a thiol-ene click reaction (Fig. 7 and 8). Previous research on thiol-ene click reactions used a reaction time of 1 h.[47] Here, due to the size and bulkiness of the sugar ligand, a longer reaction time of 3.5 h was used, to ensure a higher packing density. Due to the extensive purification needed during synthesis of these NPs, and

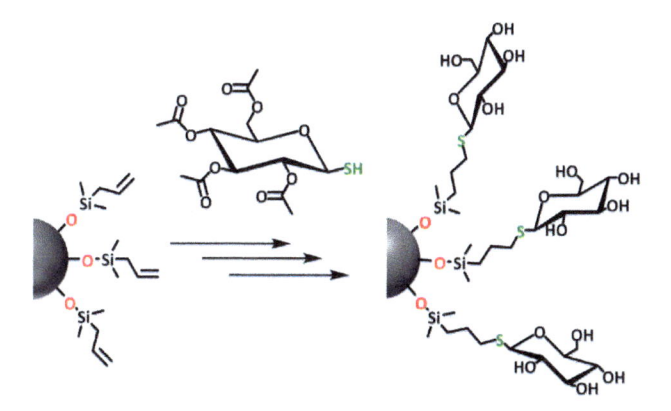

Fig. 7 Schematic overview of the synthesis of water-soluble, glucose-coated Si NPs.

Fig. 8 Glucose-coated Si NPs in water under normal light (left) and 365 nm irradiation (middle). The right picture shows the solubility of glucose-coated Si NPs in hexane (top layer) and water (bottom layer).

the small scale of the experiment (30 mg of H-Si NPs), this reaction yielded only 27 mg. Successful binding of the protected thiol-sugar was confirmed by IR, with the disappearance of the S–H stretch vibration after the click reaction (Fig. S15†).

After deprotection, the thioglucose-coated Si NPs were dispersed in water for fluorescence measurements. An excitation maximum of 370 nm and an emission maximum of 460 nm were found, with a quantum yield of 35% compared to quinine sulfate (Fig. S12†). This is a significant blue-shift compared to the Si NPs coated only with silanol, likely due to UV-induced oxidation.

Conclusions

We have demonstrated a fast, easy and efficient way to passivate hydrogen-terminated Si nanoparticles (H-Si NPs) using silanol compounds at room temperature. The reaction was specific enough to allow the presence of other functional groups in the silanol ligand, and both the passivation and purification of the Si NPs were simple procedures with opportunities for scale-up. The silanol-coated Si NPs produced here showed little oxidation and improved quantum yields. Finally, purified silanol-coated Si NPs could be produced and purified from hydrogen-terminated Si NPs in under 4 h, showing a significant improvement over previous methods for H-Si NP functionalization.

Conflicts of interest

There are no conflicts to declare.

Acknowledgements

The authors acknowledge funding from the Netherlands Organization for Scientific Research (NWO) in the framework of the Materials for Sustainability program. Low-resolution electron microscopy work was performed at the Wageningen Electron Microscopy Center (WEMC) of Wageningen University. Barend van Lagen is acknowledged for helpful instrumental assistance.

Notes and references

1 X. Cheng, S. B. Lowe, P. J. Reece and J. J. Gooding, *Chem. Soc. Rev.*, 2014, **43**, 2680.

2   Y. Su, X. Ji and Y. He, *Adv. Mater.*, 2016, **28**, 10567.
3   S. Wu, Y. Zhong, Y. Zhou, B. Song, B. Chu, X. Ji, Y. Wu, Y. Su and Y. He, *J. Am. Chem. Soc.*, 2015, **137**, 14726.
4   B. Ghosh, Y. Masuda, Y. Wakayama, Y. Imanaka, J. Inoue, K. Hashi, K. Deguchi, H. Yamada, Y. Sakka, S. Ohki, *et al.*, *Adv. Funct. Mater.*, 2014, **24**, 7151.
5   S. Coe, W.-K. Woo, M. Bawendi and V. Bulovic, *Nature*, 2002, **420**, 800.
6   L. Yang, Y. Liu, Y.-L. Zhong, X.-X. Jiang, B. Song, X.-Y. Ji, Y.-Y. Su, L.-S. Liao and Y. He, *Appl. Phys. Lett.*, 2015, **106**, 173109.
7   I. Robel, V. Subramanian, M. Kuno and P. V. Kamat, *J. Am. Chem. Soc.*, 2006, **128**, 2385.
8   G. Y. Kwak, T. G. Kim, S. Hong, A. Kim, M. H. Ha and K. J. Kim, *Sol. Energy*, 2018, **164**, 89.
9   B. Liu, J. Hu, L. Jia, J. Liu, X. Ren, X. Zhang, X. Guo and S. Liu, *Sol. Energy*, 2018, **167**, 102.
10 T. Subramani, J. Chen, Y. L. Sun, W. Jevasuwan and N. Fukata, *Nano Energy*, 2017, **35**, 154.
11 J. Duan, H. Zhang, Q. Tang, B. He and L. Yu, *J. Mater. Chem. A*, 2015, **3**, 17497.
12 B. Song and Y. He, *Nano Today*, 2019, **26**, 149.
13 S. Bhattacharjee, I. M. C. M. Rietjens, M. P. Singh, T. M. Atkins, T. K. Purkait, Z. Xu, S. Regli, A. Shukaliak, R. J. Clark, B. S. Mitchell, *et al.*, *Nanoscale*, 2013, **5**, 4870.
14 L. Ruizendaal, S. Bhattacharjee, K. Pournazari, M. Rosso-Vasic, L. H. J. De Haan, G. M. Alink, A. T. M. Marcelis and H. Zuilhof, *Nanotoxicology*, 2009, **3**, 339.
15 X. Li, Y. He, S. S. Talukdar and M. T. Swihart, *Langmuir*, 2003, **19**, 8490.
16 F. Erogbogbo, K. T. Yong, R. Hu, W. C. Law, H. Ding, C. W. Chang, P. N. Prasad and M. T. Swihart, *ACS Nano*, 2010, **4**, 5131.
17 W. Ren, Y. Wang, Q. Tan, J. Yu, U. J. Etim, Z. Zhong and F. Su, *Electrochim. Acta*, 2019, **320**, 134625.
18 F. Erogbogbo, C. A. Tien, C. W. Chang, K. T. Yong, W. C. Law, H. Ding, I. Roy, M. T. Swihart and P. N. Prasad, *Bioconjugate Chem.*, 2011, **22**, 1081.
19 A. Shiohara, S. Hanada, S. Prabakar, K. Fujioka, T. H. Lim, K. Yamamoto, P. T. Northcote and R. D. Tilley, *J. Am. Chem. Soc.*, 2010, **132**, 248.
20 M. Dasog, G. B. De los Reyes, L. V. Titova, F. A. Hegmann and J. G. C. Veinot, *ACS Nano*, 2014, **8**, 9636.
21 C. M. Hessel, E. J. Henderson and J. G. C. Veinot, *Chem. Mater.*, 2006, **18**, 6139.
22 Z. Kang, Y. Liu, C. H. A. Tsang, D. D. D. Ma, X. Fan, N.-B. Wong and S.-T. Lee, *Adv. Mater.*, 2009, **21**, 661.
23 M. Rosso-vasic, E. Spruijt, Z. Popovíc, K. Overgaag, B. van Lagen, B. Grandidier, D. Vanmaekelbergh, D. Domínguez-Gutiérrez, L. De Cola and H. Zuilhof, *J. Mater. Chem.*, 2009, **19**, 5926.
24 C.-S. Yang, R. A. Bley, S. M. Kauzlarich, H. W. H. Lee and G. R. Delgado, *J. Am. Chem. Soc.*, 1999, **121**, 5191.
25 W. Biesta, B. van Lagen, V. S. Gevaert, A. T. M. Marcelis, J. M. J. Paulusse, M. W. F. Nielen and H. Zuilhof, *Chem. Mater.*, 2012, **24**, 4311.
26 Y. Yu, C. M. Hessel, T. D. Bogart, M. G. Panthani, M. R. Rasch and B. A. Korgel, *Langmuir*, 2013, **29**, 1533.

27 J. A. Kelly, A. M. Shukaliak, M. D. Fleischauer and J. G. C. Veinot, *J. Am. Chem. Soc.*, 2011, **133**, 9564.
28 M. Dasog and J. G. C. Veinot, *Phys. Status Solidi A*, 2012, **209**, 1844.
29 Y. Yu, C. E. Rowland, R. D. Schaller and B. A. Korgel, *Langmuir*, 2015, **31**, 6886.
30 M. Dasog, K. Bader and J. G. C. Veinot, *Chem. Mater.*, 2015, **27**, 1153.
31 M. Rosso-vasic, E. Spruijt, B. Van Lagen, L. De Cola and H. Zuilhof, *Small*, 2008, **4**, 1835.
32 I. M. D. Höhlein, J. Kehrle, T. K. Purkait, J. G. C. Veinot and B. Rieger, *Nanoscale*, 2015, **7**, 914.
33 J. M. Buriak, *Chem. Mater.*, 2014, **26**, 763.
34 S. P. Pujari, H. Driss, F. Bannani, B. van Lagen and H. Zuilhof, *Chem. Mater.*, 2018, **30**, 6503.
35 Y. Zhong, X. Sun, S. Wang, F. Peng, F. Bao, Y. Su, Y. Li, S.-T. Lee and Y. He, *ACS Nano*, 2015, **9**, 5958.
36 Y. Zhong, F. Peng, F. Bao, S. Wang, X. Ji, L. Yang, Y. Su, S.-T. Lee and Y. He, *J. Am. Chem. Soc.*, 2013, **135**, 8350.
37 J. Escorihuela and H. Zuilhof, *J. Am. Chem. Soc.*, 2017, **139**, 5870.
38 N. S. Bhairamadgi, S. Gangarapu, M. A. Caipa Campos, J. M. J. Paulusse, C. J. M. Van Rijn and H. Zuilhof, *Langmuir*, 2013, **29**, 4535.
39 M. A. C. Campos, J. M. J. Paulusse and H. Zuilhof, *Chem. Commun.*, 2010, **46**, 5512.
40 K. Ágoston, A. Dobó, J. Rákó, J. Kerékgyártó and Z. Szurmai, *Carbohydr. Res.*, 2001, **330**, 183.
41 K. Suzuki, A. Kobayashi, S. Kaneko, K. Takehira, T. Yoshihara, H. Ishida, Y. Shiina, S. Oishi and S. Tobita, *Phys. Chem. Chem. Phys.*, 2009, **11**, 9850.
42 C. Xiao, J. Guo, P. Zhang, C. Chen, L. Chen and L. Qian, *Sci. Rep.*, 2017, **7**, 40750.
43 B. V. Oliinyk, D. Korytko, V. Lysenko and S. Alekseev, *Chem. Mater.*, 2019, **31**, 7167.
44 E. Schechtel, R. Dören, H. Frerichs, M. Panthöfer, M. Mondeshki and W. Tremel, *Langmuir*, 2019, **35**, 12518.
45 Z. Hens and J. C. Martins, *Chem. Mater.*, 2013, **25**, 1211.
46 B. Arkles and G. Larson, *Silicon Compounds: Silanes & Silicones*, Gelest Inc., Morrisville, 3rd edn, 2013.
47 L. Ruizendaal, S. P. Pujari, V. Gevaerts, J. M. J. Paulusse and H. Zuilhof, *Chem.–Asian J.*, 2011, **6**, 2776.
48 K. Dohnalová, A. N. Poddubny, A. A. Prokofiev, W. D. De Boer, C. P. Umesh, J. M. Paulusse, H. Zuilhof and T. Gregorkiewicz, *Light: Sci. Appl.*, 2013, **2**, e47.
49 K. Dohnalová, A. Fučíková, C. P. Umesh, J. Humpolíčková, J. M. J. Paulusse, J. Valenta, H. Zuilhof, M. Hof and T. Gregorkiewicz, *Small*, 2012, **8**, 3185.
50 D. S. English, L. E. Pell, Z. Yu, P. F. Barbara and B. A. Korgel, *Nano Lett.*, 2002, **2**, 681.
51 K. Kůsova, O. Cibulka, K. Dohnalová, I. Pelant, J. Valenta, A. Fučíková, K. Žídek, J. Lang, J. Englich, P. Matějka, *et al.*, *ACS Nano*, 2010, **4**, 4495.
52 J. H. Warner, A. Hoshino, K. Yamamoto and R. D. Tilley, *Angew. Chem., Int. Ed.*, 2005, **44**, 4550.
53 A. N. Poddubny and K. Dohnalová, *Phys. Rev. B*, 2014, **90**, 245439.

ADDITIONS AND CORRECTIONS

Fast room-temperature functionalization of silicon nanoparticles using alkyl silanols

Alyssa F. J. van den Boom, Sidharam P. Pujari, Fatma Bannani, Hafedh Driss and Han Zuilhof

Faraday Discuss., 2020, DOI: 10.1039/C9FD00102F. **Amendment published 13 May 2020.**

The published version of this manuscript included the wrong affiliation for Fatma Bannani. The correct affiliation is as follows:

Department of Chemistry, Jeddah University, Jeddah, Saudi Arabia

---

The Royal Society of Chemistry apologises for these errors and any consequent inconvenience to authors and readers.

Faraday Discussions

PAPER

Photophysical properties of ball milled silicon nanostructures†

Ankit Goyal, [iD]* Menno Demmenie, Chia-Ching Huang, Peter Schall and Katerina Dohnalova [iD]

Received 7th October 2019, Accepted 22nd October 2019
DOI: 10.1039/c9fd00105k

Luminescent silicon nanocrystals (SiNCs) have attracted scientific interest for their potential use in LEDs, displays, lasers, photovoltaic spectral-shifting filters and for biomedical applications. A lot of efforts have been made to improve the radiative emission rate in SiNCs, mostly using quantum confinement, strain and ligands. Existing methods, however, are not easily upscalable, as they do not provide the high material yield required for industrial applications. Besides, the photoluminescence (PL) efficiency of SiNCs emitting in the visible spectral range also remains very low. Hence, there is a need to develop a low-cost method for high material yield of brightly emitting SiNCs. Theoretically, strain can be used alongside quantum confinement to modify the radiative emission rates and band-gaps. In view of that, high-energy ball milling is a method that can be used to produce large quantities of highly strained SiNCs. In this technique, balls with high kinetic energy collide with the walls of a chamber and other balls, crushing the particles in between, followed by welding, fracture and re-welding phenomena, reducing the particle size and increasing strains in the samples. In this study, we have used high-energy ball milling in an inert gas atmosphere to synthesize SiNCs and study their photophysical properties. The induced accumulation of high strain, quantum confinement and possibly also impurities in the SiNCs resulted in visible light spectrum PL at room temperature. This method is low cost and easily up-scalable to industrial scale.

Introduction

Unlike bulk Si, Si nanocrystals (SiNCs) are a promising material for applications in LEDs, lasers, displays and as spectral-shifting filters in photovoltaics *etc.*, due to their increased emission rates and size-tunable band-gaps (see, *e.g.*[1–3]). Higher radiative emission rates and visible emission in the blue, green and red regions of the spectrum have been reported for various types of SiNCs.[4–10] However, production methods of SiNCs with emission in the UV and visible spectral region

---

University of Amsterdam, Institute of Physics, Science Park 904, 1098XH, Amsterdam, The Netherlands. E-mail: a.goyal@uva.nl

† Electronic supplementary information (ESI) available. See DOI: 10.1039/c9fd00105k

(and the vast majority of the NIR emitting SiNCs) have not been capable of producing the high material yields required for commercial uses so far. Production of SiNCs with diameter below <10 nm is mostly performed by bottom-up syntheses – wet-chemical or from plasma.[11–21] The important role of surface ligands often leads to limited spectral tunability.[3] Hydrogen-, oxygen- and alkyl-capped SiNCs have been studied in detail.[1,4,5,7,9,13–17,19,21,23–25] Hydrogen capped SiNCs show size-tunable emission, but are unstable and prone to oxidation upon exposure to air.[1] Resulting oxide-capped SiNCs of diameters below 5 nm show either impurity-related red emission with slow rate[23,26–28] or blue emission with fast rate[29,30] and larger sized crystals (above ∼5 nm) show size-tunable emission in the near-IR range.[22,31] Alkyl-terminated Si nanocrystals are considered very stable and resistant to oxidation, with luminescence in the visible range tunable by the ligands and size, showing direct bandgap-like radiative rates.[13,19,21,23,32,33] Further means to manipulate the optical properties can be achieved using strain.[34] Theoretical studies show that under tensile stress, it is possible to have direct bandgap-like Si nanocrystals.[35,36] Despite these advances, silicon has yet to make its way to replace toxic and less abundant options to be used in aforesaid applications because of the low-yielding synthesis methods. There is an urgent need to have a method which can produce stable, non-toxic and highly emissive Si nanocrystals at industrial scale. In view of this, high-energy ball milling which was initially developed to mix different metals and then to produce alloys, might be used for this purpose. Recently, ball milling was used for the doping of V_2O_5 into nano-Si for Li-ion battery anodes.[37] High-energy ball milling allows the synthesis of nanoparticles at kilogram scale in a short time.

Here, we present SiNCs prepared using high-energy ball mill. This method is very high yielding and leads to SiNCs that emit in the blue-green region of the visible light spectrum. The occurrence of multiple phonon replicas in the photoluminescence (PL) spectra of the milled Si particles suggests the presence of localized emission centers, possibly impurities. We interpret the phenomena based on the presence of Si–O/Si–C bonds on the surface, as well as possible doping of the SiNC particles with iron and cobalt transition metal impurities, introduced by the milling environment. The results indicate a possible energy transfer from the excited SiNCs to the metallic impurities and/or to Si–O/Si–C bonds participating in the radiative processes. We do not see evidence of quantum confinement, since no shift in the PL peak position is observed upon reducing the nanoparticle size, even down to sizes of ∼5 nm. This suggest that impurities (and maybe also strain) play larger roles than quantum confinement. We perform structural and compositional analysis to determine the structure of the milled SiNCs and correlated it with the optical properties. However, more work, especially on in-depth elemental and structural analysis, is needed to support our claims.

Materials and methods

Silicon powder (99.999% purity) with an average particle size of <20 μm was purchased from Hongwu International Group Ltd, China. An unmilled Si sample was used as reference for comparison throughout the study. To reduce the size of these Si particles to nanometer scale, a modified vertical attrition ball mill was used to perform a high-energy ball milling process. The milling was performed at room temperature using tungsten carbide (WC) balls in an inert argon gas atmosphere. The precise temperature profile during the milling process is unknown but is known to have reached 500 °C. To reduce the heating of the chamber during the milling process, continuous water cooling was provided. The Si powder was milled for 20 h in 1 hour blocks with subsequent 15 min breaks for effective cooling. The milled powder was degassed in a vacuum of 10^{-3} Torr for 2 h to remove entrapped gases and residual impurities.

Transmission electron microscopy (TEM) and high-resolution TEM (HRTEM) (Fig. S1†) were performed using a TECHNAI 20, FEI electron microscope. X-ray diffraction (XRD) studies were done using a Rigaku Miniflex II Desktop X-ray diffractometer with Cu Kα radiation ($\lambda = 1.5406$ Å). Raman spectroscopy was done using an Airix STR-500 Confocal Raman spectrophotometer with a 532 nm solid-state laser beam ∼3 mW. A FEI Nova NanoSEM 450 field emission scanning electron microscope (FESEM) equipped with a Bruker energy dispersive X-ray (EDX) analyzer was used to study the composition of the milled powder. Material analysis results are shown in ESI, Fig. S1–S4.†

For optical measurements, milled Si nanoparticles were dispersed in UV-grade ethanol and placed into quartz cuvettes (Hellma analytics). PL excitation spectra (PLE) were measured using a standard spectrofluorometer (Horiba Scientific, Fluorolog) equipped with a spectrometer (Horiba Scientific, iHR320) and a CCD detector (Horiba Scientific, Synapse). An FTIR spectrometer (PerkinElmer Spectrum Two) was used to record FTIR spectra of unmilled and milled Si powder samples. For single-nanoparticle PL emission, a highly diluted sample of milled SiNCs in UV-grade ethanol was drop-casted on a quartz cover slip (SPI supplies – 25 mm diameter; thickness: 0.15–0.18 mm) and left to dry at room temperature; for the measurement, we used an optical micro-spectroscopy setup coupled to an atomic force microscope (AFM) (Nanowizard 3, JPK Instruments). The emission was detected using a spectrometer (Princeton Instruments, Acton SP2300) with a CCD (Princeton Instruments, Pylon 400B). A diode laser (405 nm, Becker and Hickl GmbH, BDL-405-SMN) was used to excite the sample. The emitted light was collected using a 100× objective (Zeiss, Epiplan-Neofluar with NA = 0.75) and was filtered using a band pass filter (Semrock, BLP01-405R-25) to remove the scattered excitation light. The same microscopic setup and excitation (in a pulsed regime with a ∼300 ps pulse duration and a 20 MHz repetition rate) was also used to excite and collect emission for the PL lifetime measurements, with the only difference being the detection is done by TCSPC setup from Becker & Hickl GmbH (a Photon Correlator DPC-230 and a single-photon detector module ID Quantic 100). All recorded PL spectra were corrected for spectral sensitivity of the detection chain in all of the setups. To study the effect of the size of the SiNCs on the emission, we used the AFM system coupled and correlated with the optical microscopy setup used for single-dot spectroscopy. AFM was used in the tapping mode with an aluminum-coated silicon tip (Nanoworld Point probe silicon SPM sensor; resonant frequency ∼ 320 kHz). JPK software was used to capture scans and analyze the size of individual particles.

Results and discussion

The ball milled Si powder was analyzed using various methods to study the material composition and strain. First, using TEM, we found that micrometer-sized Si powder has been reduced to nanometer-sized crystallites (Fig. S1a and b†). This was further confirmed by HRTEM, where we observed SiNCs as small as ~5 nm (Fig. S1c and d†). The presence of ~5 nm particles was also confirmed later by the AFM (Fig. 3 and 4). From the HRTEM, we further see the presence of an amorphous layer on the surfaces of the nanocrystallites, which could be a result of surface oxidation of the Si nanopowder.

XRD of the unmilled and milled Si samples is shown in Fig. S2† and reveals an usual diamond cubic structure in both materials. In the milled sample we observe peak broadening, increased background and the appearance of new peaks, characteristic of β-FeSi$_2$. The peak broadening and increased background indicate the generation and accumulation of strains, dislocations and insertion of impurities from the wall and balls in the milling chamber (*e.g.* Co and Fe). Silicon is a brittle–brittle system with higher hardness than that of the transition metals. Milling with harder milling media (WC in this case) can induce wearing of soft elements, incorporating them in the Si lattice.[38] We have extracted the amount of strain, dislocation density and crystallite size using empirical formulas (Table ST1†). After milling, we observe a significant increase in the lattice strain (from 0.25% to 0.63%) and dislocation density (from 2×10^{11} to 8×10^{11}). The appearance of impurity peaks (β-FeSi$_2$) due to wearing of the WC coating from the wall suggests that most of the Fe impurities introduced into the Si material were precipitated in the form of β-FeSi$_2$ and did not have much affect on the lattice structure of the silicon nanoparticles. Insertion of Co into the SiNCs seems more likely, as evidenced by the shift in the peak shown in the inset of the Fig. S2,† which could be due to accumulation of tensile strain in the milled samples and insertion of Co into the Si lattice, leading to formation of CoSi$_2$ at the <111> position of the diamond-cubic Si.

The EDX analysis shown in Fig. S3† confirms the presence of cobalt. Furthermore, carbon and oxygen are also present in trace amounts, suggesting the possible presence of Si–O and Si–C on the surface of the SiNC particles. In general, from the above analysis we can assume the presence of Si, O, C, Fe and Co impurities in the surface shell on the surface of the SiNCs. It is possible that during the high-energy ball milling, dislocations, dangling bonds and vacancies are generated, acting as reaction sites for the elements available in the milling environment. Cobalt and iron may also been presented as an intrinsic impurity in the shell and Si core in the initial Si powder, because of their low diffusion coefficient in Si.[39]

Micro-Raman spectroscopy was performed on the unmilled and milled Si samples (Fig. S4†) to study the crystallinity and accumulation of stresses. In the unmilled sample, the bulk Si–Si phonon peak appears sharp and narrow at 520 cm^{-1}. This peak is broadened and shifted to ~500 cm^{-1} in the milled Si sample, suggesting size reduction *via* the high-energy ball milling. Interestingly, even after 20 h of milling, the Si remains crystalline, as evidenced by the absence of a peak at 480 cm^{-1}, characteristic for amorphous phase Si. Change in the position of the Raman peak from the bulk can give information about local stress. The change in frequency from 520 cm^{-1} (in the bulk) to a lower frequency in the milled SiNCs indicates the presence of tensile stress. A mechanical stress of 4250 MPa, generated by high-energy ball milling can be evaluated from the peak broadening using an empirical formula which calculates stress from the shift in frequency of the Raman peak.[40]

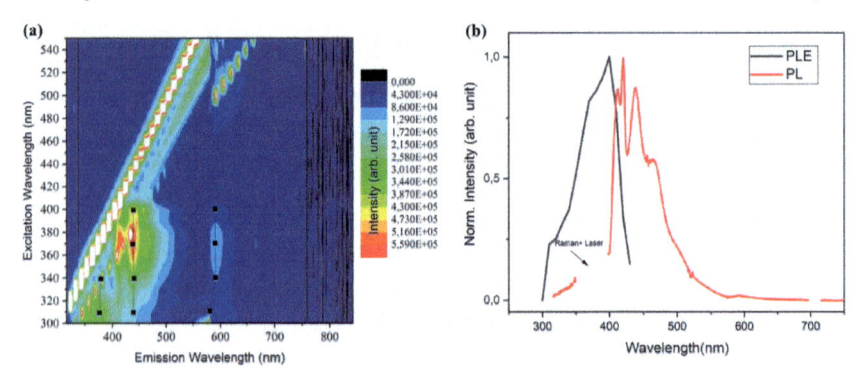

Fig. 1 (a) Contour plots of the PL spectra at varied excitation wavelengths measured from the ball-milled sample. (b) PLE (for the 450 nm PL; black line) and PL (for the 370 nm excitation; red line), for the ball-milled sample.

The optical properties of the milled SiNCs were first studied using PLE spectra in a wide range of emission and excitation wavelengths. A contour plot of the emission intensity is shown in Fig. 1a. Results for the unmilled sample are given in Fig. S5.† The most striking difference between the two samples is the occurrence of bright blue-green emission at around 450 nm in the milled sample that is completely absent in the unmilled sample, suggesting that the process of milling has caused PL in the visible range. With analysis of the peak positions of the PL for several different excitations (black dots and black lines in Fig. 1a), we find that the PL peak is independent of the excitation energy. This finding combined with the broad size distribution (not shown) of the milled SiNCs can be interpreted as a lack of size-dependence in the PL mechanism. In Fig. 1b, we show an extracted PL spectrum excited at peak excitation at 370 nm, as well as the PLE peak for the dominant emission feature at 450 nm.

Interestingly, a pronounced peak structure is observed in the PL emission of the resonantly excited PL spectrum (Fig. 1b and 2a). A similar structure is also observed in the PL excited at 300 nm (Fig. 2b), but it is less pronounced. Fitting both spectra with multiple Gaussian peaks without any constraints (not shown), we found that they can be described by individual peaks with separation energies shown by the red lines in the FTIR absorption spectrum in Fig. 2c. The peak positions coincide with several Si–C and Si–O related peaks, suggesting that the PL structure might be a result of Si–C or Si–O surface vibrations coupled to the exciton. The roughly constant energies between the peak positions (red lines) are indeed reminiscent of phonon replicas (presented as a histogram) and mostly coincide with the weak vibrations of Si–C stretching at 1265 cm^{-1} ($\sim$160 meV), similarly to observations in ref. 24 on alkyl-capped SiNCs. Strong absorption peaks at 1070 cm^{-1} can be assigned to Si–O–Si bond vibrations, confirming the presence of oxide, which is interesting and can be correlated with the experimental procedure. During the ball milling, diffusion of C from the wall and balls may have taken place to introduce Si–C bonds, but could not saturate the whole SiNC surface, allowing subsequent oxidation. Nevertheless, the Si–O bond does not seem to play an important role in the vibrational structure of the PL. This does not rule out the possible role of metallic impurities in the emission process,

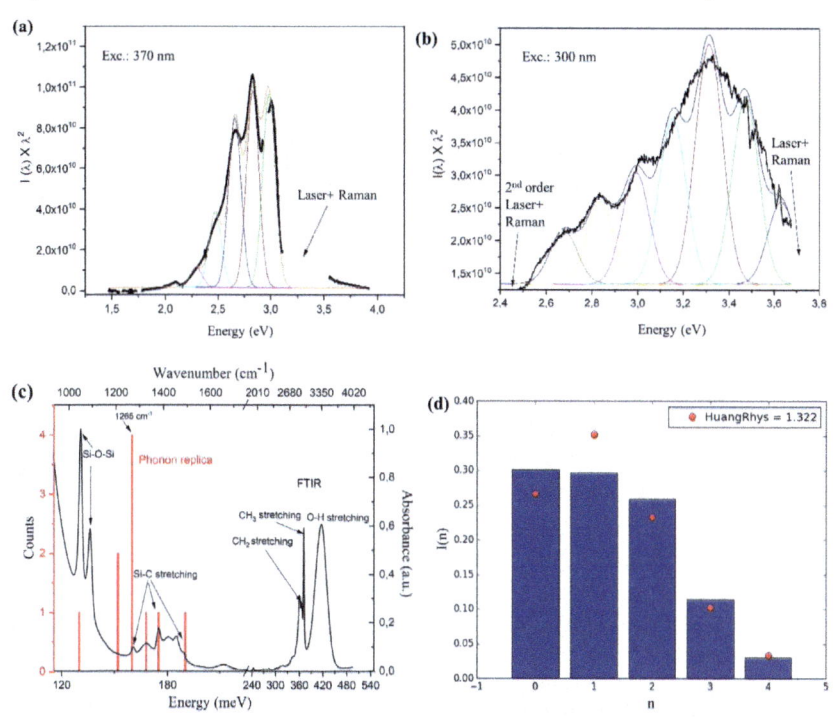

Fig. 2 PL spectra obtained at (a) excitation of 370 nm and (b) excitation at 300 nm. PL spectra were fitted with peak difference (~160 meV) and FWHM (140 meV) – parameters obtained first from fitting the most prominent three peaks in the spectrum shown in (a). The rest of the PL spectra were fitted with the fixed peak separation and FWHM and varying only the peak intensities. The colored peaks are the fits and the bold black curve is the original data. For better clarity, Raman and second order excitation peaks were removed from the data sets (labeled gaps in the data in the graphs in (a) and (b)). (c) FTIR spectrum showing infrared absorption (black) of the ball milled SiNCs dispersed in ethanol and the position of the peak separation energies obtained from the freely-fitted (i.e. without any fixed or assumed parameters) PL spectra excited at 300 and 370 nm. (d) Fitted intensities of the peaks in (a) and Huang-Rhys factor fit curve (red dots).

however, we are not aware of such vibrational states at 160 meV for Si–Fe, Co–O, Fe–C, Co–C or Si–Co bonds. The presence of O–H and C–H bond stretching in the FTIR is most likely due to the residual presence of ethanol that was used as a solvent for the measurements.

Assuming that the structure in the PL spectra is related to phonon replicas, we imposed fixed peak distances of 160 meV and a FWHM of 140 meV (parameter found from free fitting the three dominant peaks in Fig. 2a), and tried to fit the PL spectra in Fig. 2a and b, varying only the peak intensities. The resulting spectrum is shown as a thin black line in the figures and the colored lines are the fits; surprisingly, the agreement of such 'forced' fit with the whole measured PL spectral shape is very good, indicating that the PL spectrum consists of the main zero-phonon replica (at 3.0 eV for excitation at 370 nm and at 3.65 eV for excitation at 300 nm) and related series of phonon replicas. The large number of visible replicas is very surprising, since this indicates unusually strong exciton-phonon coupling.

It is known an that impurity atom can act as a localized emission center, possibly resulting in considerable exciton-phonon coupling, responsible for the occurrence of phonon replicas in PL emission (and related effective heating of the lattice of a material *via* optical excitation). The strength of the coupling can be quantified by the Huang-Rhys factor S, which can be extracted from fitting the peak intensities of the phonon replicas in the spectrum, labeled from the highest energy one (0th peak) to the lowest energy one (n^{th} peak) using the Poisson distribution

$$I(n) = e^{-S} S^n/n!, \text{ with } n = 0,1,2\ldots$$

The exciton-phonon coupling is considered weak for $S \leq 1$, medium for $S = 5$–10 and strong for $S > 20$. In the case of Si, generally, there is a very weak coupling with $S = 0.2$ to 1 where the value of S varies with the bandgap and the corresponding size of the nanocrystals.[41–43] In the SiNCs, enhanced coupling could appear due to the lowered symmetry and the increased role of the surface with flexible and vibrationally energetic ligands.[24] By fitting the structure in Fig. 2a, we found $S = 1.32$ (Fig. 2d), which is very high for a Si-based material.

Since, to the best of our knowledge, such high exciton-phonon coupling has not been reported from alkyl-capped SiNCs (Si–C) or oxidized SiNCs (Si–O), it is unlikely that the Si–O or Si–C vibrations caused such a highly structured PL, which lends credence to the possibility of the involvement of transition metal impurities in the emission process. McVey *et al.* reported a similar PL emission structure in their study of the optical properties of SiNCs doped with transition metals (Mn, Ni and Cu).[44] However, they focused more on the shift in PL peak position because of the doping of transition elements and concluded that the Si–Si bonds were the source of PL emission. In our case, the source of emission and periodic feature remains uncertain, however, the working hypothesis is that the emission is of extrinsic origin, possibly related to metallic impurities, with strong exciton-phonon coupling to the Si–C surface bonds.

To further analyze the possible size-dependence of the PL, we measured the PL from single SiNCs using an optical microspectroscopy setup, coupled to AFM, which allowed us to correlate size and PL from a single nano-object. First, we recorded a PL image in the wide-field imaging mode (Fig. 3, left top). This was then correlated with an AFM image of the same area (Fig. 3, right top). From the comparison, we selected 5 Si nanoparticles with sizes between 5 and 20 nm. For these selected Si nanoparticles (indicated by the white squares), the emission spectra were recorded by using a slit to reduce the detected area. To minimize differences due to other parameters, such as integration time and laser power gain, the spectra of all the particles were recorded with the same integration time (5 seconds) and excitation power (gain of 80%). Interestingly, we found no change in the PL peak position for the different particles. Furthermore, the PL spectral shape (width and multi-phonon peak structure) appears to be the same for the ensemble and separate nanoparticle measurements.

Analyzing the drop-casted milled SiNC sample on a quartz slide by AFM, we also found some intriguing superstructures (Fig. 4a), probably resulting from the 'coffee-stain' type drying process. The larger sized objects (~35–70 nm) showed

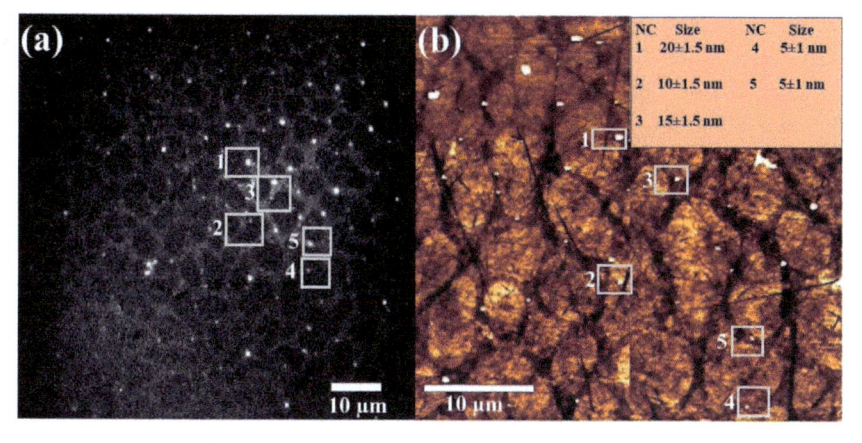

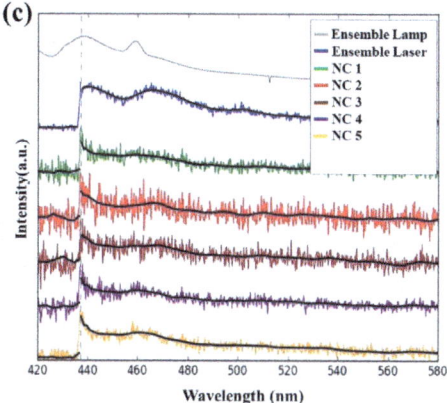

Fig. 3 (a) PL image correlated with (b) AFM scan of the ball milled sample. The 5 selected emitting dots in the PL image and AFM are presented in the white square boxes. (c) PL spectrum of the NC ensemble excited at 405 nm using a Xe lamp (in Fluorolog setup) (gray) and the NC ensemble excited at 405 nm in the microscope setup (blue), together with 5 different single NCs selected in the AFM/PL image scans. A dotted line indicates the edge of the long-pass filter at 438 nm, used in the measurement to cut out the excitation light. The black line in every spectrum is the average of the data points to reduce noise and show a clear visual of the spectrum.

high intensity emission in comparison to the smaller sized ones (∼5–20 nm), possibly due to a larger number of included SiNCs.

Fig. 4b shows PL spectra of the larger sized object (∼35–70 nm), which appears again to be very similar to the ensemble PL spectrum (Fig. 2 and 3).

All of our optical results indicate that ball-milled SiNC powders have size-independent, blue-green emission with strong phonon replica features. The observed PL has a fast mono-exponential decay of lifetime ∼6.5 ns, as measured from the ensemble of milled SiNCs using our microscope setup (at 405 nm laser excitation, integration time ∼1800 s) (Fig. S6†). Blue PL with fast decay has been previously observed from oxygen-related defect states.[45] Since the sizes of the SiNCs were found from HRTEM and AFM to be very large (>5 nm), indicating bulk-Si-like band-gaps, the emission is probably not of an intrinsic origin. Similar strong size-independent blue-green emission has been observed in the

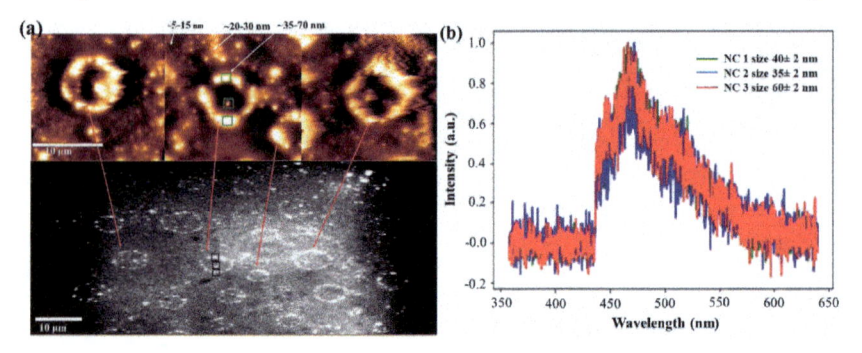

Fig. 4 (a) Formation of Si superstructures observed by AFM-spectroscopy (top) correlated with the PL imaging (bottom) in the drop-casted milled Si powder particles on a quartz substrate. (b) Normalized PL spectra recorded in three different places (shown in green (AFM) and black (PL) square boxes) with objects of sizes in the range of 35–70 nm.

past from very small oxidized SiNCs,[3,45,47] which however in larger SiNCs, such as ours, usually switches to a size-independent red emission band[46,47] or a size-dependent NIR peak in even larger SiNCs.[48] Therefore, it is unlikely that the observed emission originates from the Si–O centers. The energy of the phonon replica is observed here to be of 160 meV, which coincides with Si–C bonds. Liu et al.[49] reported similar PL spectra from SiC NCs embedded in a SiO_2 matrix with the conclusion that it could originate from either the Si–C sites or oxide defects. Another interesting work was reported by Askari et al.[50] who observed a similar PL structure from SiC NCs prepared by atmospheric pressure plasmas. They indicated an unclear origin of the multiple non-shifting spectral peaks, but ruled out defect-related emission.[50] McVay et al.[44] used UV-assisted hydrosilylation to create a strong covalent Si–C bonded passivation (hexane capped SiNCs) to protect Si from oxidation, which also suggests the possibility of such emission by Si–C related sites. They observed multiple peak emission from both doped (metals: Cu, Mn) and undoped Si which led to the belief that Si–C sites could be the possible cause of the observed emission. However, unlike us, they observed a shift in the PL spectra with change in excitation wavelength, but it was observed to be more pronounced in the undoped Si. Interestingly, they observed spectral shape and FWHM changes under the different excitation wavelengths, pointing towards the involvement of multiple active emission centers, in which metallic impurities might have played a role as well. In our case, we suggest that the most possible model that could explain the observed bright blue-green emission is emission from an extrinsic site that is excited *via* energy transfer from the Si core. The extrinsic site is either Si–C related (such as the 2.86–3.02 eV emission observed in 6H–SiC[50]) or related to metallic impurities (Co and Fe) with $E_{c,Co}$ at 2.81 eV, $E_{c,Fe}$ at 2.91 eV, $E_{v,Co}$ at 0.46 eV and $E_{v,Fe}$ at 0.09 eV in silicon.[51] Solubility of the transition metals in Si is very low because of the low diffusion coefficient, so these impurities must have been present directly on or near to the surface of the SiNCs. In such a case, it could also be possible that the emission originates from the metal impurity, which is strongly coupled to the Si–C bond vibrational states that show up in the PL spectra as phonon replicas. Nevertheless, the solubility increases at the high temperatures generated during the ball milling process. Also, the process induces several dislocations which could serve as vacancies for foreign elements, thereby causing substitutional insertion in the Si volume.

Conclusions

High-energy ball milling can be a simple, low cost and environmentally friendly way to synthesize highly emissive SiNCs. We found SiNCs as small as 5 nm in the milled sample, which makes this method very promising. The PL emission of all observed SiNCs was, however, independent of the NC size and was observed in the blue-green spectral region with a PL lifetime of a few nanoseconds. Our results suggest that the presence of metallic impurities as well as Si–C sites on the surface could play an active role. The multi-phonon structure observed in the PL emission spectra indicates that the emissive center is strongly vibrationally coupled to Si–C bonds. The effect of strain, that – beyond the mere introduction of vacancies and defects – could serve as possible sites for Fe, C, O and Co incorporation, needs to be studied in more detail in the future *e.g.* by studying samples produced with different ball milling times. The interesting superstructures observed in the drop-casted sample could be used for anti-counterfeiting or labelling applications.

Conflicts of interest

There are no conflicts to declare.

Acknowledgements

Ankit Goyal is thankful to the financial support provided by the Institute of Physics at the University of Amsterdam.

References

1. M. V. Wolkin, J. Jorne, M. Fauchet, G. Allan and C. Delerue, *Phys. Rev. Lett.*, 1999, **82**, 197, DOI: 10.1103/physrevlett.82.197.
2. N. Shirahata, T. Hasegawa, Y. Sakka and T. Tsuruoka, *Small*, 2010, **6**(8), 915–921.
3. K. Dohnalová, T. Gregorkiewicz and K. Kůsová, *J. Phys.: Condens. Matter*, 2014, **26**(17), 173201.
4. D. Tan, B. Xu, P. Chen, Y. Dai, S. Zhou, G. Ma and J. Qiu, *RSC Adv.*, 2012, **2**, 8254–8257.
5. J. H. Warner, A. Hoshino, K. Yamamoto and R. D. Tilley, *Angew. Chem.*, 2015, **117**, 4626–4630.
6. J. P. Wilcoxon, G. A. Samara and P. N. Provencio, *Phys. Rev. B: Condens. Matter Mater. Phys.*, 1999, **60**(4), 2704.
7. D. S. English, L. S. Pell, Z. Yu, P. F. Barbara and B. A. Korgel, *Nano Lett.*, 2002, **27**, 681–685.
8. J. B. Miller, A. R. Sickle, R. J. Anthony, D. M. Kroll, U. R. Kortshagen and E. K. Hobbie, *ACS Nano*, 2012, **68**, 7389–7396.

9 X. Wen, P. Zhang, T. A. Smith, R. J. Anthony, U. R. Kortshagen, P. Yu, Y. Feng, S. Shrestha, G. Coniber and S. Huang, *Sci. Rep.*, 2015, **5**, 12469.
10 Q. Li, Y. He, J. Chang, L. Wang, H. Chen, Y. W. Tan, H. Wang and Z. Shao, *J. Am. Chem. Soc.*, 2013, **135**(40), 14924–14927.
11 X. D. Pi, R. W. Liptak, J. D. Nowak, N. P. Wells, C. B. Carter, S. A. Campbell and U. Kortshagen, *Nanotechnology*, 2008, **19**(24), 245603, DOI: 10.1088/0957-4484/19/24/245603.
12 D. Jurbergs, E. Rogojina, L. Mangolini and U. Kortshagen, *Appl. Phys. Lett.*, 2006, **88**, 233116, DOI: 10.1063/1.2210788.
13 L. Mangolini, D. Jurbergs, E. Rogojina and U. Kortshagen, *J. Lumin.*, 2006, **121**(2), 327–334.
14 W. Biesta, B. Lagen, V. S. Gevaert, A. T. M. Marcelis, J. M. J. Paulusse, M. W. F. Nielen and H. Zuilhof, *Chem. Mater.*, 2012, **24**, 4311–4318.
15 J. G. C. Veinot, *Chem. Commun.*, 2006, 4160–4168, DOI: 10.1039/b607476f.
16 J. P. Wilcoxon and G. A. Samara, *Appl. Phys. Lett.*, 1999, **74**(21), 3164, DOI: 10.1063/1.124096.
17 C. S. Yang, S. M. Kauzlarich, Y. C. Wang and H. W. H. Lee, *J. Cluster Sci.*, 2000, **11**(3), 423–431.
18 M. A. Islam, R. Sinelnikov, M. A. Howlader, A. Faramus and J. G. C. Veinot, *Chem. Mater.*, 2018, **30**(24), 8925–8931.
19 J. H. Warner, H. R. Dunlop and R. D. Tilley, *J. Phys. Chem. B*, 2005, **109**(41), 19064–19067.
20 P. Mohapatra, D. M. Perez, J. M. Bobbitt, S. Shaw, B. Yuan, X. Tian, E. A. Smith and L. Cademartiri, *ACS Appl. Mater. Interfaces*, 2018, **10**(24), 20740–20747.
21 Y. Zhong, X. Sun, S. Wang, F. Peng, F. Bao, Y. Su, Y. Li, S. T. Lee and Y. He, *ACS Nano*, 2015, **9**(6), 5958–5967.
22 J. Valenta, M. Greben, S. A. Dyakov, N. A. Gippius, D. Hiller, S. Gutsch and M. Zacharias, *Sci. Rep.*, 2019, **9**, 11214.
23 K. Dohnalová, A. N. Poddubny, A. A. Prokofiev, W. D. A. M. de Boer, C. P. Umesh, J. M. J. Paulusse, H. Zuilhof and T. Gregorkiewicz, *Light: Sci. Appl.*, 2013, **2**, e47.
24 K. Dohnalová, A. Fučiková, C. P. Umesh, J. Humpoličková, J. M. J. Paulusse, J. Valenta, H. Zuilhof, M. Hof and T. Gregorkiewicz, *Small*, 2012, **8**(20), 3185–3191.
25 S. P. Pujari, H. Driss, F. Bannani, B. Lagen and H. Zuilhof, *Chem. Mater.*, 2018, **30**(18), 6503–6512.
26 M. Dasog, Z. Yang, S. Regli, T. M. Atkins, A. Faramus, M. P. Singh, E. Muthuswamy, S. M. Kauzlarich, R. D. Tilley and J. G. C. Veinot, *ACS Nano*, 2013, **73**, 2676–2685.
27 J. Fuzell, A. Thibert, T. M. Atkins, M. Dasog, E. Busby, J. G. C. Veinot, S. M. Kauzlarich and D. S. Larsen, *J. Phys. Chem. Lett.*, 2013, **4**(21), 3806–3812.
28 Y. Kanemitsu and S. Okamoto, *Phys. Rev. B: Condens. Matter Mater. Phys.*, 1997, **55**(12), 7375, DOI: 10.1103/physrevb.55.r7375.
29 G. Belomoin, J. Therrien and M. Nayfeh, *Appl. Phys. Lett.*, 2000, **77**, 779, DOI: 10.1063/1.1306659.
30 J. Valenta, A. Fucikova, I. Pelant, K. Kusova, K. Dohnalova, A. Aleknavicius, O. Cibulka, A. Fojtik and G. Kada, *New J. Phys.*, 2008, **10**, 073022.

31 Z. Kang, Y. Liu, C. H. A. Tsang, D. D. D. Ma, X. Fan, N. B. Wong and S. T. Lee, *Adv. Mater.*, 2009, **21**, 661–664.
32 F. Hua, M. T. Swihart and E. Ruckenstein, *Langmuir*, 2005, **21**(13), 6054–6062.
33 D. Beri, D. Busko, A. Mazilkin, I. A. Howard, B. S. Richards and A. Turshatov, *RSC Adv.*, 2018, **8**, 9979–9984.
34 D. M. Lyons, K. M. Ryan, M. A. Morris and J. D. Holmes, *Nano Lett.*, 2002, **28**, 811–816.
35 K. Kusova, *Phys. Status Solidi A*, 2018, **215**, 1700718.
36 K. Kůsová, P. Hapala, J. Valenta, P. Jelínek, O. Cibulka, L. Ondič and I. Pelant, *Adv. Mater. Interfaces*, 2014, **1**, 1300042.
37 G. Lv, B. Zhu, X. Li, C. Chen, J. Li, Y. Jin, X. Hu and J. Zhu, *ACS Appl. Mater. Interfaces*, 2017, **951**, 44452–44457.
38 C. Suryanarayana, *Prog. Mater. Sci.*, 2001, **46**, 1–184.
39 E. R. Weber, *Appl. Phys. A: Solids Surf.*, 1983, **30**, 1–22.
40 I. D. Wolf, *Semicond. Sci. Technol.*, 1996, **11**, 139–154.
41 T. Takagahara, *J. Lumin.*, 1996, **70**, 129–143.
42 J. Martin, F. Cichos, F. Huisken and C. Borczyskowski, *Nano Lett.*, 2008, **8**(2), 656–660.
43 I. Pelant, J. Hala, M. Ambroz, M. Vacha, J. Valenta, F. Adamec, V. Kohlova and J. Matouskova, *Impurity assessment in Si wafers by photoluminescence method V. Research report for Tesla Roznov*, Charles University in Prague, Faculty of Mathematics and Physics, Prague, 1990.
44 B. F. P. McVey, J. Butkus, J. E. Halpert, J. M. Hodgkiss and R. D. Tilley, *J. Phys. Chem. Lett.*, 2015, **6**, 1573–1576.
45 L. Tsybeskov, J. V. Vandyshev and P. M. Fauchet, *Phys. Rev. B: Condens. Matter Mater. Phys.*, 1994, **49**(11), 7821–7824.
46 K. Dohnalová, K. Kůsová and I. Pelant, *Appl. Phys. Lett.*, 2009, **94**, 211903, DOI: 10.1063/1.3141481.
47 K. Dohnalová, L. Ondič, K. Kůsová, I. Pelant, J. L. Rehspringer and R. R. Mafouana, *J. Appl. Phys.*, 2010, **107**, 053102, DOI: 10.1063/1.3289719.
48 Y. Kanzawa, T. Kageyama, S. Takeoka, M. Fujii, S. Hayashi and K. Yamamoto, *Solid State Commun.*, 1997, **102**(7), 533–537.
49 X. Liu, J. Zhang, Z. Yan, S. Ma and Y. Wang, *Mater. Phys. Mech.*, 2001, **4**, 85–88.
50 S. Askari, A. U. Haq, M. M. Montero, I. Levchenko, F. Yu, W. Zhou, K. Ostriker, P. Maguire, V. Svrcek and D. Mariotti, *Nanoscale*, 2016, **8**, 17141–17149.
51 E. B. Yakimov, Metal Impurities and Gettering in Crystalline Silicon, in *Handbook of Photovoltaic Silicon*, ed. D. Yang, Springer, Berlin, Heidelberg, 2019.

Faraday Discussions

PAPER

Amine functionalised silicon nanocrystals with bright red and long-lived emission

Giacomo Morselli, Francesco Romano and Paola Ceroni

Received 20th September 2019, Accepted 22nd October 2019
DOI: 10.1039/c9fd00089e

When functionalised with amines, silicon nanocrystals (SiNCs) are known to have surface-state emission with loss of colour tunability, low quantum yield and short nanosecond lifetimes. These changes in optical properties are produced by direct amine bonding on the silicon surface. In this article, secondary amine functionalised SiNCs with bright, red (λ_{max} = 750 nm) and long-lived emission (τ ca. 50 µs) are reported for the first time via a three-step synthetic approach. These SiNCs are colloidally stable in several polar solvents and can be further functionalised by reaction with carboxylic acid groups. We proved the feasibility of further functionalization with pyrene butyric acid: ca. 40 pyrene units per nanoparticle were attached via amide bond formation. The resulting hybrid system works as a light-harvesting antenna: excitation of pyrene units at 345 nm results in sensitised emission at 700 nm by the silicon core.

Introduction

Silicon nanocrystals (SiNCs) are luminescent semiconductor nanoparticles that are a valid biocompatible alternative to conventional quantum dots.[1–6] Quantum dots are often made of elements of the groups III–V and II–VI. The presence of elements like cadmium and lead prevents their use in consumer products or in biological applications. Silicon has advantages in these contexts since it is abundant on the Earth's crust, has a relatively low cost and most importantly, it is non-toxic.[7] Moreover, SiNCs display photophysical properties that render them interesting for several industrial and imaging applications: their emission spans the visible to near infrared spectral range according to their dimensions, and matches the optical window that most penetrates biological tissues. The emission quantum yield is reported to reach high values (up to 70%)[8] and their emission lifetimes are longer than those of direct bandgap quantum dots (tens or hundreds of microseconds). This last characteristic allows the use of time-gated detection[9] in order to obtain a clean signal which is not influenced by the noise caused by the short-lived autofluorescence of tissues or by back-scattered excitation light.[10–12]

Department of Chemistry "Giacomo Ciamician", University of Bologna, via Selmi 2, 40126, Bologna, Italy. E-mail: paola.ceroni@unibo.it

However, their wide utilization is hindered, to some extent, by the following issues: commonly used passivating ligands, *i.e.* alkyl chains, prevent suspension of SiNCs in polar solvents; the molar absorption coefficient in the visible range is inferior to that of molecules or conventional quantum dots because of the indirect bandgap nature of silicon.

Our group demonstrated that by decorating the surface of SiNCs with chromophores, their absorption coefficient is strongly enhanced in the UV-visible spectrum. Most importantly, these hybrid systems work as light-harvesting antennae: upon excitation of the organic dye, sensitised emission of the silicon core is observed by efficient energy transfer processes.[13–15] The brightness of the system (defined as the product between the molar absorption coefficient at the excitation wavelength and the emission quantum yield of the antenna) is significantly improved (up to 300% (ref. 13)). However, these antennae were obtained *via* competitive hydrosilylation reactions in which terminal olefin functionalised chromophores reacted together with a co-passivating ligand that was used also as a solvent. These reaction conditions together with high temperatures (180 °C overnight) resulted in low chromophore loading on the surface of the SiNCs, despite the large excess used during the hydrosilylation step.

To circumvent this problem, it would be desirable to passivate the surface of the SiNCs with ligands containing terminal reactive groups that can be subsequently post-functionalised under mild experimental conditions. Among the possible functional groups, amine-terminated SiNCs are ideal candidates because amide bonds can be formed by well-established amidation reactions, amide bonds are fairly stable under physiological conditions and a variety of derivatives are commercially available for amidation reactions. For example, this approach could lead to functionalisation with visible absorbing chromophores, water-solubilising groups (*e.g.*, PEG) or enzymes.

Amine functionalised SiNCs reported so far show blue PL, independent of the size of the NCs, which can be attributed to surface defects introduced with the synthetic strategies employed.[16–18] Another applicable approach would be to use hydrosilylation of hydrogen terminated SiNCs obtained with thermal disproportionation of hydrogen silsesquioxane, with amine functionalised alkenes. Unfortunately, the interaction between the nitrogen atom of a primary or secondary amine and the silicon lattice can still compromise the typical optical properties of silicon nanocrystals,[19] suppressing the bright red long-lived photoluminescence (PL) and producing a blue short-lived emission with a lower quantum yield. Recently, a strategy based on borane protection was proposed to obtain amidine functionalised SiNCs with good photophysical properties that show switchable hydrophilicity properties.[20] To the best of our knowledge, primary or secondary amine functionalised SiNCs with red emission and long lifetimes have not been reported.

In this paper, we discuss a synthetic route to obtain bright-red, long-lived emitting silicon nanocrystals functionalised with amines. This strategy consists of three steps: (i) passivation of SiNCs with chlorosilane *via* radical initiated hydrosilylation; (ii) functionalisation with amines protected by bulky groups; (iii) acid-induced deprotection of the amine functional groups. A post-functionalization through amide coupling was also successfully conducted with a very limited amount of pyrene, *ca.* 1.5% mol compared to the previously published approach by our group.[13] The resulting system contains an average of 40 pyrene units per nanocrystal and works as an efficient light-harvesting antenna.

Experimental section

Reagents and materials

All reagents were purchased from Sigma-Aldrich and used without further purification if not stated otherwise. Dry toluene was obtained *via* distillation over calcium chloride under a nitrogen atmosphere. N,N,N′,N′-Tetramethylethylenediamine was refluxed over fresh KOH and distilled under nitrogen. Nuclear magnetic resonance (^{1}H-NMR) spectra were measured using an ARX Varian INOVA 400 (400 MHz) spectrometer, chemical shifts are reported in ppm and data are reported as follows: chemical shift, multiplicity (s = singlet, d = doublet, t = triplet, q = quartet, br = broad, m = multiplet), coupling constants (Hz). GC-MS analyses were obtained using an Agilent Technologies MSD1100 equipped with an EI (70 eV) ionization system, a single quadrupole analyser, and a HP5 5% Ph-Me silicon.

Protection of *N*-methylpropargylamine with trityl chloride

In a 25 mL two-necked flask, under a nitrogen atmosphere, 920 mg of trityl chloride (3.3 mmol, 1.1 eq.), 7 mL of anhydrous dichloromethane (DCM), 450 μL of anhydrous triethylamine (TEA, 333 mg, 3.3 mmol, 1.1 eq.) and 255 μL of *N*-methylpropargylamine (207 mg, 3 mmol, 1 eq.) were added in this order. After several minutes, the formation of a precipitate occurred (the ammonium salt). The mixture was stirred at room temperature for 4 hours under an inert atmosphere. The reaction was quenched with an aqueous solution of NaHCO$_3$ and diluted with another 5 mL of DCM. The organic phase was washed twice with water, collected, dried with Na$_2$SO$_4$ and concentrated at reduced pressure. TLC analysis (1 : 1 = cyclohexane : toluene as the eluent phase) showed two spots (Rf = 0, Rf = 0.8) upon developing with KMnO$_4$. The less polar one was identified as the product and the mixture was separated with flash chromatography using cyclohexane : toluene (1 : 1) as the eluent phase. The collected fractions gave a white crystalline solid (673 mg, yield: 72%). GC-MS (*m/z*):311. ^{1}H-NMR (CDCl$_3$, 400 MHz), δ (ppm): 7.0–7.5 (15H, m); 2.93 (2H, br s); 2.25 (3H, s); 2.19 (1H, t).

Synthesis of hydride-terminated silicon nanocrystals (H-SiNCs)

Oxide embedded silicon nanocrystals were obtained *via* a literature reported procedure.[21,22] Hydride-terminated silicon nanocrystals were liberated from the silica matrix through HF etching: 300 mg of oxide-embedded silicon nanocrystals were dispersed in a mixture composed of 3 mL of ethanol, 3 mL of bidistilled water and 3 mL of a 49% solution of aqueous HF. (*Caution! HF is dangerous and must be handled with extreme care*). The mixture was stirred for 1 h and 30 minutes under ambient light at room temperature. The nanocrystals were extracted with toluene (3 × 10 mL) and then centrifuged three times in toluene (8000 rpm for 5 minutes). The supernatant was discarded and the nanocrystals were then transferred into a dry-box.

Hydrosilylation of H-SiNCs with chloro(dimethyl)vinylsilane[23]

The nanocrystals were dispersed in 2 mL of dry toluene in an 8 mL vial. Two milligrams of 4-decyldiazobenzene tetrafluoroborate (4-DDB, about 6 μmol) were added. Afterwards, 400 μL of chloro(dimethyl)vinylsilane (CDMVS, 3 mmol, 360 mg) were introduced to obtain chlorosilane-passivated silicon nanocrystals. The mixture was stirred overnight at room temperature. The mixture of chlorosilane-passivated SiNCs was then filtered, concentrated using a rotary evaporator, transferred again into the dry-box and diluted in 2 mL of dry toluene.

Functionalization of SiNCs with trityl-amines

In a two-necked 25 mL round-bottom flask, dried and filled with nitrogen, 180 mg of *N*-methyl-*N*-tritylpropargylamine (0.6 mmol) were introduced. The flask was transferred into a dry-box and 140 μL of *N,N,N′,N′*-tetramethylethylenediamine (TMEDA, 0.9 mmol, 105 mg) and 3 mL of dry toluene were added. The flask was then removed from the dry-box and was connected to a Schlenk line filled with N_2. After having cooled the reaction to −78 °C with a liquid nitrogen/acetone bath, 240 μL of *n*-butyllithium (*n*-BuLi 2.5 M in hexanes, 0.6 mmol; *Attention: pyrophoric! It must be handled under an inert atmosphere*) were added dropwise, while stirring. The acetone bath was removed after 45 minutes, and the mixture was stirred for 15 minutes at room temperature. Again at −78 °C, the suspension of silicon nanocrystals in 2 mL of toluene was slowly added to the reaction mixture. One hour later, the acetone bath was removed, and the reaction mixture was stirred for an hour at room temperature. Later, it was heated to 40 °C, and stirred for another hour. The reaction was cooled again to −78 °C and a second amount of *n*-BuLi (120 μL, 0.3 mmol) was added to complete the capping of the surface. The reaction was allowed to reach room temperature and was stirred overnight. The introduction of 7 mL of MeOH made the nanocrystals precipitate. The brownish precipitate was washed 3 times with methanol and separated from the supernatant by centrifugation (8000 rpm, 5 minutes). The obtained nanocrystals were dispersed in chloroform and the precipitate was filtered off.

Cleavage of trityl group for ammonium-terminated silicon nanocrystals

The suspension of tritylamine functionalised silicon nanocrystals was transferred into a 20 mL vial with a magnetic stir bar and was stirred slowly. Trifluoroacetic acid (TFA, 99%) was added dropwise carefully until the precipitation of silicon nanocrystals occurred (about 10 μL) due to the protonation of the amines. The suspension was left stirring for 10 minutes. Then, the suspension was centrifuged, washing three times with chloroform (3 × 8000 rpm, 5 minutes) and the precipitate was dissolved in alcoholic solvent (ethanol, methanol), acetonitrile or *N,N*-dimethylformamide (DMF). ^{1}H-NMR and a ninhydrin test before and after the deprotection confirmed the fading of the trityl group (Fig. 1 and 2). In the NMR spectrum, the signals related to the attached ligands are broadened because the motions of the molecules linked to the nanocrystal were hindered, resulting in a longer relaxation time.[23,24] The signals of the trityl group lie between 6.7 and 7.7 ppm (Fig. 1a); these signals are completely absent in the deprotected sample (Fig. 1b). A ninhydrin test is an assay used to detect the presence of amines. Primary amines react with ninhydrin, yielding a blue-coloured compound, secondary amines produce an orange compound and tertiary amines don't react.[25] The tritylamine functionalised SiNCs did not display a colour change when exposed to the ninhydrin assay (A in Fig. 2a), as expected for a tertiary amine functionalization of the surface (see *e.g.*, *N*-methyl-*N*-tritylpropargylamine, B in Fig. 2a). On the other hand, the deprotected SiNCs (E in Fig. 2c) showed

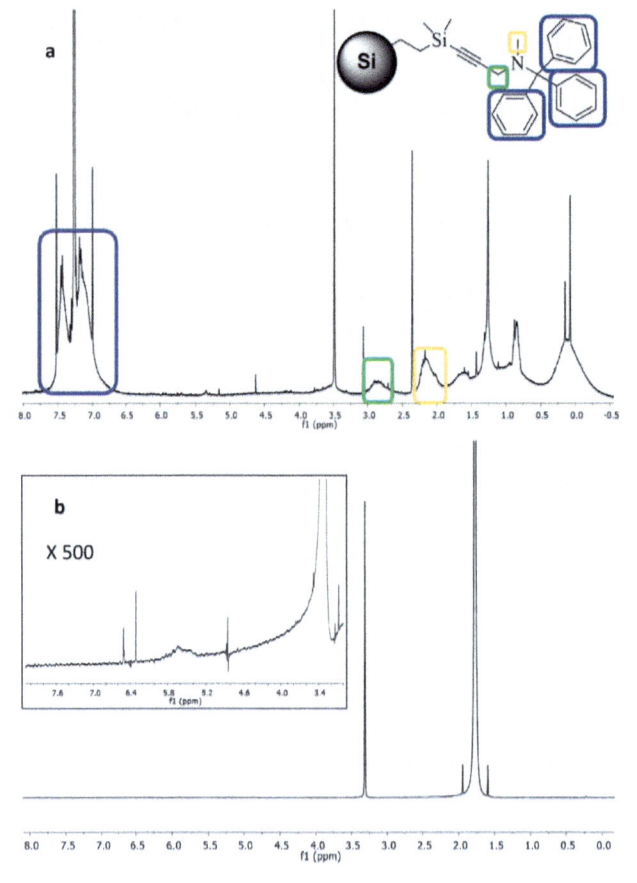

Fig. 1 NMR spectra of (a) trityl functionalised silicon nanocrystals (400 MHz, CDCl₃) and (b) ammonium functionalised silicon nanocrystals (400 MHz, CD₃OD), with a 500× magnification in the inset, showing the absence of the aromatic signals.

a yellowish color, similar to that of N-methylpropargylamine (F in Fig. 2b), suggesting the presence of a secondary amine.

Addition of triethylamine made the nanocrystals precipitate in ethanol and we could suspend them in chloroform because of the deprotonation of the ammonium groups.

Water-suspendable ammonium-terminated silicon nanocrystals

An excess of a HCl solution in methanol was added to a suspension of silicon nanocrystals in methanol. An excess of water was then introduced, and the suspension was filtered through 0.45 micrometre cut-off RC filters. The suspension was concentrated using a rotary evaporator to remove the methanol and the acid in excess.

Amide coupling with 1-pyrenebutyric acid

In a two-necked 25 mL round-bottom flask, under a nitrogen atmosphere, 7 mg of 1-pyrenebutyric acid (0.028 mmol) were dissolved in anhydrous DMF. Then, 9 μL

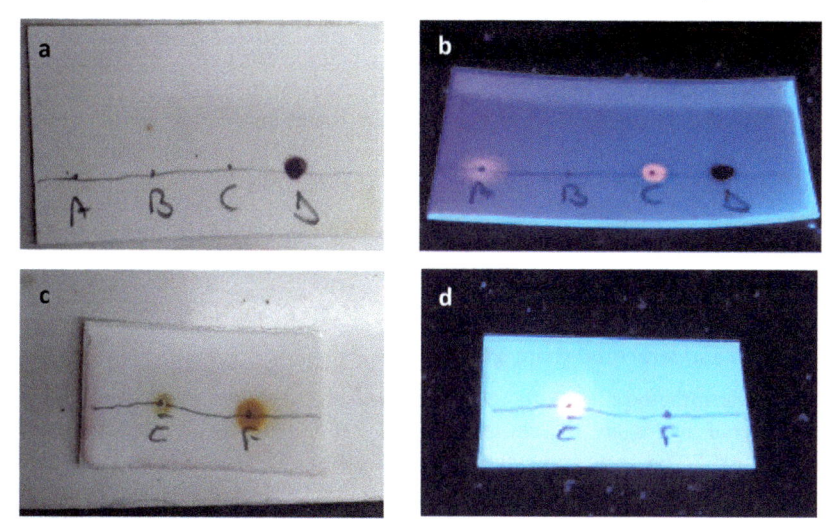

Fig. 2 Ninhydrin tests (a–c) on TLC plates for different suspensions: (A) tritylamine functionalised silicon nanocrystals, (B) a tertiary amine (here, N-methyl-N-tritylpropargylamine), (C) alkyl-passivated silicon nanocrystals (not containing amines), (D) a primary amine (in this case, bis-aminopropyl polyethylene glycol), (E) amine functionalised silicon nanocrystals (after trityl cleavage) and (F) a secondary amine (N-methylpropargylamine), under visible (a and c) or 365 nm UV (b) light (picture (d) was taken before the test, to show the luminescence of sample (E) under UV light).

of triethylamine (0.066 mmol, 6.6 mg), 5 μL of 1-ethyl-3-(3-dimethylaminopropyl) carbodiimide (EDC, 0.03 mmol, 4.4 mg) and 13 mg of (1-cyano-2-ethoxy-2-oxoethylidenaminooxy)dimethylamino – morpholino – carbenium hexafluorophosphate (COMU, 0.03 mmol) were added. The suspension was left to stir for half an hour at room temperature. Half a batch of ammonium-terminated silicon nanocrystals derived from the etching of 300 mg of silica-embedded SiNCs, dispersed in 2 mL of DMF, was added to the mixture, which was then left stirring for 1 hour and a half. Thereafter, another 5 μL of EDC and 13 mg of COMU were added, and the mixture was left stirring overnight. The addition of several mL of methanol made the nanocrystals precipitate. They were centrifuged five times with methanol (8000 rpm, 5 minutes) and the supernatant was dispersed in DMF.

Photophysical measurements

Photophysical measurements were carried out in air-equilibrated ethanol, chloroform, water or N,N-dimethylformamide at 298 K. UV-visible absorbance spectra were recorded with a PerkinElmer λ_{650} spectrophotometer, using quartz cells with 1.0 cm path length. Emission spectra were obtained using either a PerkinElmer LS-50 spectrofluorometer, equipped with a Hamamatsu R928 phototube, or an Edinburgh FLS920 spectrofluorometer equipped with a Ge-detector for emission in the NIR spectral region. Correction of the emission spectra for detector sensitivity in the 550–1000 nm spectral region was performed using a calibrated lamp.[26,27] Emission quantum yields were measured following the method of Demas and Crosby[28] (standard used: $[Ru(bpy)_3]^{2+}$ in air-equilibrated aqueous solution $\Phi = 0.0407$ and HITCI, 1,1′,3,3,3′,3′-hexamethyl-indotricarbocyanine iodide, in EtOH $\Phi = 0.308$). Emission intensity decay measurements in the range 10 μs to 1 s were performed using a home-made time-resolved phosphorimeter. The estimated experimental errors are: 2 nm on the absorption and emission band maximum, 5% on the molar absorption coefficient and luminescence lifetime, and 10% on the luminescence quantum yield.

Results and discussion

Synthesis and structural characterization

Hydride-terminated silicon nanocrystals were prepared by the thermal decomposition of hydrogen silsesquioxane (HSQ). Surface passivation, needed to prevent oxidation and aggregation of the nanocrystals, was performed by room temperature hydrosilylation with chloro(dimethyl)vinylsilane (CDMVS) in the presence of diazonium salts (4-decyl diazobenzene tetrafluoroborate, 4-DDB) as radical initiators (step (i) in Scheme 1a).[29] This reaction introduces chlorosilane electrophilic groups on the surface of the silicon nanocrystals, which can react with several nucleophilic reagents, such as Grignard reagents, alcohols, silanols and acetylides. We used a protected N-methylpropargylamine (step (ii) in Scheme 1a), that can be deprotonated by strong bases (*e.g.* LDA or *n*-BuLi) to obtain the corresponding acetylide.

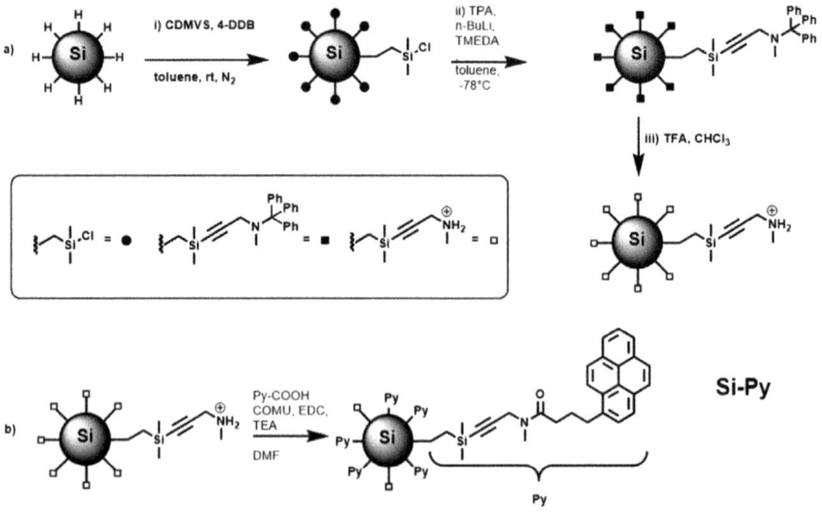

Scheme 1 (a) Three-step functionalization of SiNCs consisting of (i) a hydrosilylation step with chloro(dimethyl)vinyl silane (CDMVS) and 4-decyldiazonium tetrafluoroborate (4-DDB), (ii) functionalization with N,N-methyl-triphenylmethyl-propargyl amine (TPA), in the presence of N,N,N′,N′-tetramethylethylenediamine (TMEDA) and (iii) deprotection with trifluoroacetic acid (TFA). (b) Amidation of amine functionalised SiNCs with 1-pyrenebutyric acid (Py-COOH), in the presence of (1-cyano-2-ethoxy-2-oxoethylidenaminooxy)dimethylamino – morpholino – carbenium hexafluorophosphate (COMU), 1-ethyl-3-(3-dimethylaminopropyl)carbodiimide (EDC) and triethylamine (TEA) in dimethylformamide (DMF).

The triphenylmethyl group (better known as the trityl group) was chosen[30] because it is stable under the nucleophilic and basic reaction conditions, it can be cleaved under mild conditions (addition of trifluoroacetic acid, TFA, that does not damage silicon nanocrystals) and the protection of the amine with this group is very easy (it involves N-methylpropargylamine and trityl chloride in the presence of a non-nucleophilic base such as triethylamine, TEA).

The steric hindrance of a bulky group such as trityl has two main effects that need to be considered in optimising the reaction conditions: (i) the trityl group can substitute only one of the two hydrogen atoms of a primary amine, such as propargylamine;[30] (ii) the protected amine is too bulky to cap all of the chlorosilanes present on the surface of the silicon nanocrystals. As a consequence, unreacted chlorosilanes will be present on the surface of silicon nanocrystals and their presence is detrimental: while exposed to ambient moisture or alcohols used for the work up, they form silanols or silyl ethers that, in the presence of acidic or basic catalysis, condense with each other, leading to aggregation of the nanocrystals through siloxane bridges. Therefore, we decided to cap the chlorosilanes by the subsequent addition of an excess of n-BuLi nucleophile, as previously reported by us,[24] and using a secondary amine, namely N-methylpropargylamine, to avoid deprotonation during the reaction with excess n-BuLi.

The synthetic strategy, therefore, involves three different steps: (i) passivation of SiNCs with chlorosilane *via* radical initiated hydrosilylation; (ii) functionalisation with N,N-methyl-triphenylmethyl-propargyl amine; (iii) deprotection of the amine functional group promoted by acid (Scheme 1a). The so-obtained ammonium-terminated silicon nanocrystals were suspendable in polar solvents such as methanol (Fig. 3).

The presence of primary and secondary amines on the SiNCs surface was detected by the ninhydrin test that was positive only after cleavage of the trityl group, and by ^{1}H-NMR (see Experimental section).

We noticed that ammonium-terminated silicon nanocrystals are not dispersible in water if the counter anion is trifluoroacetate. Upon addition of HCl in methanol anion exchange takes place, replacing the trifluoroacetate with chloride. The so-obtained nanocrystals were suspendable in water. Ammonium-

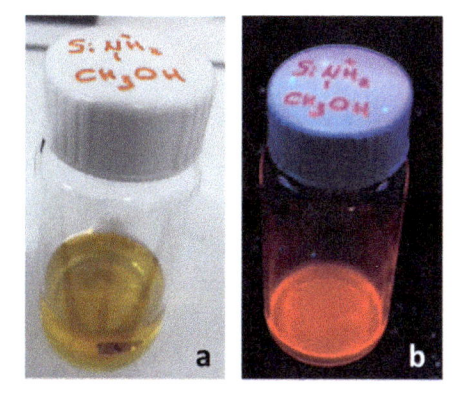

Fig. 3 Vial containing a suspension of ammonium-terminated silicon nanocrystals in methanol under ambient (a) and 365 nm UV (b) light.

terminated silicon nanocrystals are proven to be cytotoxic.[31] Therefore, considering the potential biomedical applications of those quantum dots, our research is focussed on the linkage of a most biocompatible polymer (*e.g.* PEG) through amide bonds.

The possibility to further functionalise the so-obtained SiNCs *via* amide bond formation was tested by reaction with 1-pyrenebutyric acid in the presence of amide coupling reagents (1-cyano-2-ethoxy-2-oxoethylidenaminooxy) dimethylamino – morpholino – carbenium hexafluorophosphate and 1-ethyl-3-(3-dimethylaminopropyl)carbodiimide (COMU and EDC, as shown in Scheme 1b). The resulting SiNCs functionalised with pyrene chromophores were dispersed in DMF solution and the linkage of the chromophore was confirmed by photophysical characterisations (see below).

Photophysical properties

The absorption spectrum of ammonium-terminated silicon nanocrystals in ethanol (line blue in Fig. 4) exhibits the typical trend of the absorption of alkyl-passivated SiNCs, which gradually increases at lower wavelengths.[1] The emission band (red line in Fig. 4) is centred at about 750 nm with an emission quantum yield of 24% and an emission lifetime of 75 μs.

Upon deprotonation of ammonium-terminated silicon nanocrystals with TEA in chloroform, the absorption and emission spectra maintained the same shapes, but a decrease of the emission intensity was observed (Fig. 4b), corresponding to an emission quantum yield of 4%. This quenching process can be due to a photoinduced electron transfer between the silicon nanocrystal core and the terminal amine functionalities. The emission lifetime did not change significantly. This experimental finding is likely related to the fact that we cannot detect components shorter than 5 μs with this experimental setup and the observed intensity decay is related to the emission of excitons formed in the inner part of the nanocrystal core, that are not affected by the photoinduced electron transfer process.

The absorption and emission spectra of water-suspended ammonium-terminated silicon nanocrystals are reported in Fig. 5. The slight blue-shift observed in the photoluminescence (the emission band is centred at 675 nm) is

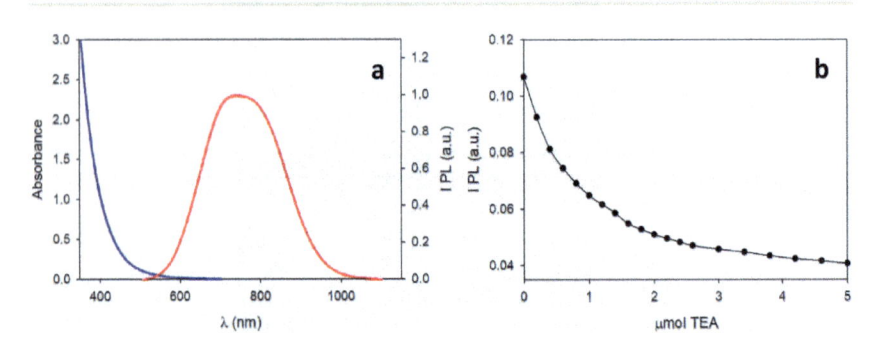

Fig. 4 (a) Absorption (blue line) and emission (red line) of ammonium functionalised silicon nanocrystals in ethanol; (b) decay of the emission intensity upon addition of triethylamine (TEA).

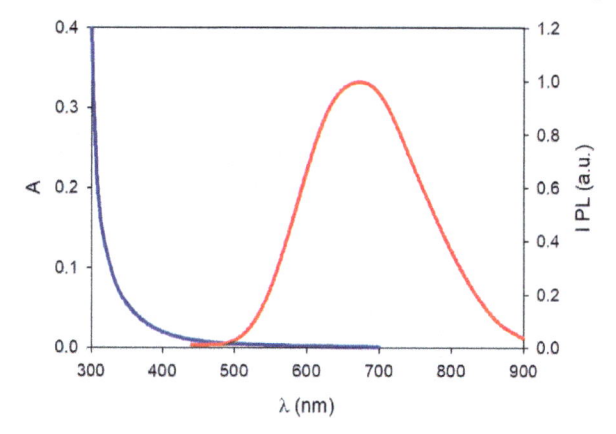

Fig. 5 Absorption (blue line) and emission (red line) spectra of ammonium-terminated silicon nanocrystals in water.

probably due to oxidation of the surface caused by a non-homogeneous coating of the surface.[32] The emission quantum yield was equal to 5% and the emission lifetime was about 50 microseconds.

Light-harvesting antenna

The absorption spectrum of **Si-Py** (blue line in Fig. 6) displays a trend associated to the sum of the absorption of the silicon core (red line) and the 1-pyrenebutyric acid one (green line), demonstrating that no significant interactions take place between the organic pyrene chromophore and the inorganic silicon core in the ground state.

From these data, knowing the molar absorption coefficient for pyrenebutyric acid (3.7×10^4 M^{-1} cm^{-1} at 344 nm, derived from spectrophotometric measurements) and that of the silicon core (1.0×10^5 M^{-1} cm^{-1} computed at

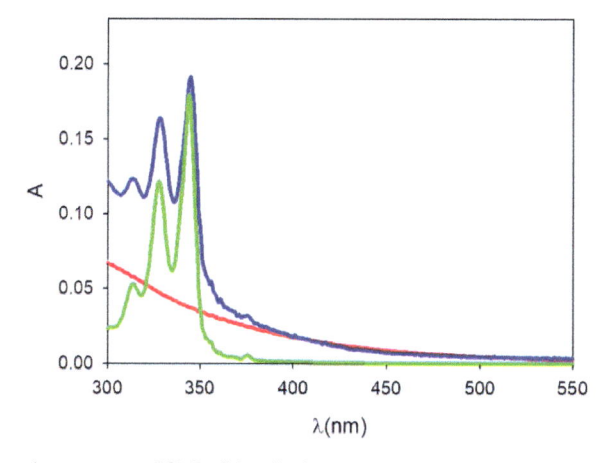

Fig. 6 Absorption spectra of **Si-Py** (blue line), alkyl passivated silicon nanocrystals (in red) and 1-pyrenebutyric acid (green).

430 nm,[22] where the absorption of the organic fluorophore does not occur) it is possible to estimate an average of 40 pyrene units per silicon nanocrystal.

Upon excitation at 450 nm where only the silicon core absorbs light, the characteristic red luminescence is present (Fig. 7a) with a photoluminescence quantum yield of 1.4%, which is increased up to 11% (τ = 20 μs) upon the addition of trifluoroacetic acid. This enhancement is compatible with the protonation of amine groups that prevent photoinduced electron transfer processes, as previously discussed for ammonium-terminated silicon nanocrystals.

Upon excitation at 340 nm, where 70% of light is absorbed by pyrene and 30% of light by the silicon core, the emission spectrum displays three main bands (Fig. 7b) attributed to: (i) the pyrene monomer emission (380–420 nm, as confirmed by the emission spectrum of the 1-pyrenebutyric acid in blue); (ii) the emission of the pyrene excimer (centred at about 480 nm); (iii) the photoluminescence of the silicon nanocrystal (from 600 nm), which is better observed using a time-gated detection (dashed line in Fig. 7b).

The intense emission of the excimer is proof of the high number of pyrene units linked to the nanocrystals, as previously observed for pyrene-functionalised nanocrystals.[33] On the other hand, the emission of the pyrene monomer is strongly quenched (10-times). Indeed, the fluorescence lifetime of pyrene at 396 nm for **Si-Py** in air-equilibrated dimethylformamide solution is 3.7 ns, compared to 37 ns for 1-pyrenebutyric acid under the same experimental conditions. The same value (3.7 ns) corresponds to the rise time in PL emission of the SiNCs at 700 nm upon excitation at 340 nm, which proves the occurrence of energy transfer from the pyrenes to the silicon core.

A further demonstration of the energy transfer between the organic chromophores and the silicon core is given by the excitation spectrum recorded at 700 nm (blue line in Fig. 8), where only the nanocrystal contributes to the emission. The excitation spectrum shows a good match with the absorption spectrum (red line in Fig. 8).

The close match of the absorption (red line in Fig. 8) and excitation spectra (blue line in Fig. 8) proves the occurrence of energy transfer with efficiency higher

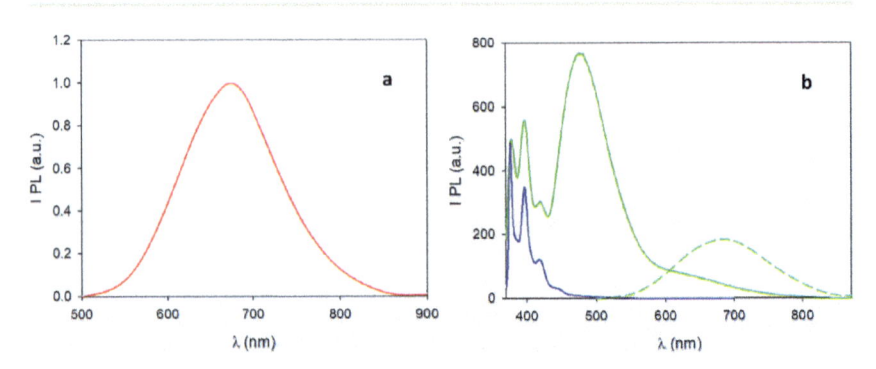

Fig. 7 (a) Emission spectrum of silicon core upon excitation at 450 nm; (b) emission spectra of **Si-Py** (green line) and 1-pyrenebutyric acid (blue line) in DMF exciting at 340 nm and emission of **Si-Py** (green dashed line) with a time-gated detection of 50 microseconds (this emission is magnified).

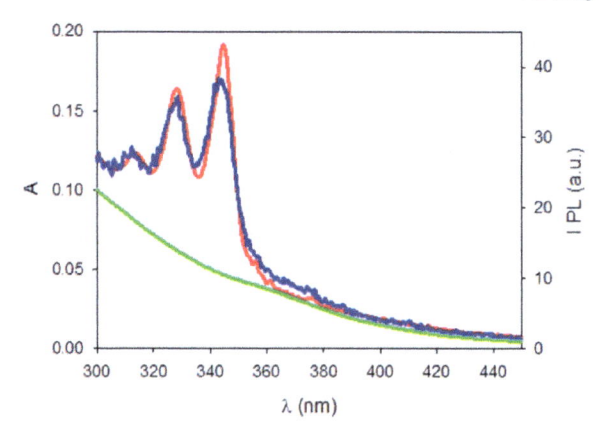

Fig. 8 Absorption (red line) and excitation spectra (blue line, λ_{em} = 700 nm, time-gated detection with delay = 50 μs and gate time = 1 ms) of **Si-Py** in DMF. For comparison purposes, the excitation spectrum (green line) of a physical mixture of 1-pyrenebutyric acid and ammonium-terminated SiNCs under the same experimental conditions is shown.

than 90%. The excitation spectrum was also measured for ammonium-terminated silicon nanocrystals mixed with free 1-pyrenebutyric acid in the appropriate ratios to match the **Si-Py** absorbance profile. In this case, the excitation spectrum (green line in Fig. 8) is superimposed onto the absorption spectrum of the ammonium-terminated SiNC sample and no contribution from the pyrene chromophores is present. As previously observed,[13,33] in the physical mixture, excitation of pyrene does not result in sensitised emission of the SiNCs: the pyrene fluorescent excited state is short lived (tens of ns) and cannot interact with non-covalently bound SiNCs present in low concentration.

Conclusions

We have successfully synthesized red/NIR emitting amine-terminated silicon nanocrystals characterized by a high emission quantum yield and long emission lifetimes, using a three-step synthetic strategy that involves a bulky protecting group cleavable in mild conditions. These SiNCs are colloidally stable in several polar solvents, such as ethanol. Upon protonation, ammonium-terminated silicon nanocrystals with chloride counter anions are obtained and they can be suspended in water.

The amine-terminated SiNCs can be post-functionalised *via* the formation of amide bonds that are fairly stable under physiological conditions. We demonstrated a successful post-functionalization with a pyrene chromophore through amidation. The resulting hybrid system works as a light-harvesting antenna that is stable in a polar environment: upon UV excitation of the pyrene chromophores, sensitised red emission of the silicon core is observed *via* an efficient energy transfer process. The amidation reaction needs much lower amount of the chromophore, compared to previously published procedures: this result is important in view of future implementation with more sophisticated and expensive chromophores. Furthermore, the same amidation approach can be used to append PEG polymers and proteins, as is currently under investigation in our laboratory.

Conflicts of interest

There are no conflicts to declare.

Acknowledgements

The University of Bologna is gratefully acknowledged.

References

1. R. Mazzaro, F. Romano and P. Ceroni, *Phys. Chem. Chem. Phys.*, 2017, **19**, 26507–26526.
2. B. F. P. Mcvey and R. D. Tilley, *Acc. Chem. Res.*, 2014, **47**, 3045–3051.
3. J. Park, L. Gu, G. Von Maltzahn, E. Ruoslahti, N. Sangeeta and M. J. Sailor, *Nat. Mater.*, 2009, **8**, 331–336.
4. S. Chinnathambi, S. Chen and S. Ganesan, *Adv. Healthcare Mater.*, 2014, **3**, 10–29.
5. M. Montalti, A. Cantelli and G. Battistelli, *Chem. Soc. Rev.*, 2015, **44**, 4853–4921.
6. X. Ji, H. Wang, B. Song, B. Chu and Y. He, *Front. Chem.*, 2018, **6**, 1–9.
7. F. Erogbogbo, K.-T. Yong, I. Roy, R. Hu, W.-C. Law, W. Zhao, H. Ding, F. Wu, R. Kumar, M. T. Swihart and P. N. Prasad, *ACS Nano*, 2011, **5**, 413–423.
8. M. A. Islam, H. Mobarok, R. Sinelnikov, T. K. Purkait and J. G. C. Veinot, *Langmuir*, 2017, **33**, 8766–8773.
9. L. Ravotto, Q. Chen, Y. Ma, S. A. Vinogradov, M. Locritani, G. Bergamini, F. Negri, Y. Yu, B. A. Korgel and P. Ceroni, *Chem*, 2017, **2**, 550–560.
10. E. J. New, D. Parker, D. G. Smith and J. W. Walton, *Curr. Opin. Chem. Biol.*, 2010, **14**, 238–246.
11. A. J. Amoroso and S. J. A. Pope, *Chem. Soc. Rev.*, 2015, **44**, 4723–4742.
12. L. Sun, R. Wei, J. Feng and H. Zhang, *Coord. Chem. Rev.*, 2018, **364**, 10–32.
13. M. Locritani, Y. Yu, G. Bergamini, M. Baroncini, J. K. Molloy, B. A. Korgel and P. Ceroni, *J. Phys. Chem. Lett.*, 2014, **5**, 3325–3329.
14. A. Fermi, M. Locritani, D. Carlo, M. Pizzotti, S. Caramori, Y. Yu, B. A. Korgel and P. Ceroni, *Faraday Discuss.*, 2015, 481–495.
15. F. Romano, Y. Yu, B. A. Korgel, G. Bergamini and P. Ceroni, *Top. Curr. Chem.*, 2016, **374**, 89–106.
16. S. Chatterjee and T. K. Mukherjee, *J. Phys. Chem. C*, 2013, **117**, 10799–10808.
17. J. H. Warner, A. Hoshino, K. Yamamoto and R. D. Tilley, *Angew. Chem., Int. Ed.*, 2005, **44**, 4550–4554.
18. G. B. De los Reyes, M. Dasog, M. Na, L. V. Titova, J. G. C. Veinot and F. A. Hegmann, *Phys. Chem. Chem. Phys.*, 2015, **17**, 30125–30133.
19. M. Dasog, Z. Yang, S. Regli, T. M. Atkins, A. Faramus, M. P. Singh, E. Muthuswamy, S. M. Kauzlarich, R. D. Tilley and J. G. C. Veinot, *ACS Nano*, 2013, **7**, 2676–2685.
20. A. N. Thiessen, T. K. Purkait, A. Faramus and J. G. C. Veinot, *Phys. Status Solidi A*, 2018, **215**, 1–5.
21. R. J. Clark, M. Aghajamali, C. M. Gonzalez, L. Hadidi, M. A. Islam, M. Javadi, H. Mobarok, T. K. Purkait, C. J. T. Robidillo, R. Sinelnikov, A. N. Thiessen, J. Washington, H. Yu and J. G. C. Veinot, *Chem. Mater.*, 2017, **29**, 80–89.

22 M. R. Hessel, C. M. Reid, D. Panthani, M. G. Rasch, B. A. Goodfellow, B. W. Wei, J. Fujii, H. Akhavan and V. Korgel, *Chem. Mater.*, 2012, **24**, 393–401.
23 I. M. D. Höhlein, J. Kehrle, T. K. Purkait, J. G. C. Veinot and B. Rieger, *Nanoscale*, 2015, **7**, 914–918.
24 R. Mazzaro, A. Gradone, S. Angeloni, G. Morselli, P. G. Cozzi, F. Romano, A. Vomiero and P. Ceroni, *ACS Photonics*, 2019, **6**, 2303–2311.
25 M. Friedman and L. D. Williams, *Bioorg. Chem.*, 1974, **3**, 267–280.
26 M. Montalti, A. Credi, L. Prodi and M. T. Gandolfi, *Handbook of photochemistry*, 2006.
27 V. Balzani, P. Ceroni and A. Juris, *Photochemistry and Photophysics - Concepts, Research, Applications*, 2015.
28 G. A. Crosby and J. N. Demas, *J. Phys. Chem.*, 1971, **75**, 991–1024.
29 I. M. D. Höhlein, J. Kehrle, T. Helbich, Z. Yang, J. G. C. Veinot and B. Rieger, *Chem.–Eur. J.*, 2014, **20**, 4212–4216.
30 P. G. M. Wuts, in *Greene's Protective Groups in Organic Synthesis*, 2014, pp. 1086–1087.
31 S. Bhattacharjee, L. H. J. De Haan, N. M. Evers, X. Jiang, A. T. M. Marcelis, H. Zuilhof, I. M. C. M. Rietjens and G. M. Alink, *Part. Fibre Toxicol.*, 2010, **7**, 1–12.
32 R. Sinelnikov, M. Dasog, J. Beamish, A. Meldrum and J. G. C. Veinot, *ACS Photonics*, 2017, **4**, 1920–1929.
33 R. Mazzaro, M. Locritani, J. K. Molloy, M. Montalti, Y. Yu, B. A. Korgel, G. Bergamini, V. Morandi and P. Ceroni, *Chem. Mater.*, 2015, **27**, 4390–4397.

Faraday Discussions

PAPER

Dual-emission fluorescent silicon nanoparticle-based nanothermometer for ratiometric detection of intracellular temperature in living cells†

Jinhua Wang,‡ Airui Jiang,‡ Jingyang Wang, Bin Song* and Yao He *

Received 20th September 2019, Accepted 24th October 2019
DOI: 10.1039/c9fd00088g

In this article, we present a kind of dual-emission fluorescent nanothermometer, which is made of europium (Eu^{3+})-doped silicon nanoparticles (Eu@SiNPs), allowing the detection of intracellular temperature in living cells with high accuracy. In particular, the presented SiNP-based thermometer features dual-emission fluorescence (blue (455 nm) and red (620 nm) emission), negligible toxicity (cell viability of treated cells remains above 90% during 24 h of treatment) and robust photostability in living cells (*i.e.*, preserving >90% of fluorescence intensity after 45 min of continuous UV irradiation). More significantly, the fluorescence intensity of the Eu@SiNPs exhibits a linear ratiometric temperature response in a broad range from 25 to 70 °C. Taking advantage of these attractive merits, the Eu@SiNP-based nanothermometer is able to accurately (~4.5% change per °C) determine dynamic changes in intracellular temperature in a quantitative and long-term (*i.e.*, 30 min) manner.

Introduction

Temperature is a fundamental parameter in physiological processes, which affects biological and chemical phenomena in a wide variety of ways. Generally, cellular processes are temperature-dependent, due to temperature influencing the reactivity and dynamics of the biochemical species inside cells.[1] Therefore, it is highly important to develop techniques for accurate sensing of intracellular temperature. For this purpose, fluorescent nanothermometers are of particular interest because they operate as "non-contact" sensors and offer the dual functions of temperature sensing and imaging on the nanoscale.[2] Recently, numerous

---

Laboratory of Nanoscale Biochemical Analysis, Institute of Functional Nano & Soft Materials (FUNSOM), Collaborative Innovation Center of Suzhou Nano Science and Technology (NANO-CIC), Soochow University, Suzhou, Jiangsu 215123, China. E-mail: yaohe@suda.edu.cn; bsong@suda.edu.cn

† Electronic supplementary information (ESI) available: Materials and devices, methods and additional data (Fig. S1–S10), and the corresponding discussion. See DOI: 10.1039/c9fd00088g

‡ These authors contributed equally.

fluorescent nanomaterials have been developed for nanothermometer applications, such as nanodiamonds, lanthanide nanoparticles, fluorescent gold nanoclusters and thermoresponsive nanogels.[3] For instance, Kucsko and co-workers have developed a smart nanowire-assisted delivery process to locate nanodiamonds and gold NPs inside cells, and then realize the measurement of sub-kelvin temperature changes inside a single cell.[3a] Rocha *et al.* reported a new lanthanide silicate orthorhombic system, exhibiting unique photoluminescence properties for self-calibration (an internal reference approach capable of eliminating environmental interferences) thermometer, which is particularly sensitive at cryogenic temperatures (<100 K).[3c-e] Despite this progress, new high-performance fluorescent nanothermometers, featuring high resolution, precise self-calibration and real-time and long-term measuring abilities, are still in high demand at present.

In recent years, fluorescent silicon (Si) nanomaterials have been intensively studied and fabricated, showing great promise for various optical applications.[4] As representative zero-dimensional Si nanostructures, fluorescent Si nanoparticles (SiNPs) with high fluorescence quantum yield (PLQY), photostability, and low-toxicity have attracted great attention for fluorescent biosensors.[4] In order to obtain a desirable sensing system, extensive efforts have been devoted to developing fluorescent SiNPs with functional dopants, in which multiple optical functionalities can be endowed by virtue of the synergistic effect between the SiNPs and the dopants.[5] Meanwhile, europium (Eu^{3+}), known as a typical element in the lanthanide group, is highly attractive for constructing high-sensitive sensors.[6] Recently, scientists have carried out elegant work on doping Eu^{3+} ions into different types of nanosystem. For example, Chi's group has developed a novel kind of dual-emission nanohybrid through combining carbon dots with blue light fluorescence and red light europium-doped metal–organic frameworks. The nanohybrids can be used as a visual colorimetric and sensitive sensor for determining the water content in organic solvents with a wide response range of 0.2–30%.[6d] In 2017, Tang *et al.* developed a microwave-assisted method to obtain a nanohybrid made of Eu complexes and carbon dots. The obtained hybrid nanosystem exhibits stimuli response on account of the energy levels of the ligands, providing a high-sensitivity sensor to realize secure fluorescent information recording.[6e]

Herein, we develop a novel kind of fluorescent nanothermometer, which is made of europium (Eu^{3+})-doped SiNPs (Eu@SiNPs). The resultant SiNP-based nanothermometer presents dual-emission fluorescence (*i.e.*, blue emission (455 nm) from the SiNPs and red (620 nm) emission from Eu) with a ratiometric temperature response from 25 to 70 °C, allowing the detection of temperature with high accuracy. In addition, the presented SiNP-based thermometer features negligible toxicity (the cell viability of treated cells remained above 90% during 24 h of treatment) and robust photostability in living cells (*i.e.*, preserving >90% of fluorescence intensity after 45 min of continuous UV irradiation). Taking advantage of these merits, the Eu@SiNP-based nanothermometer is highly accurate (~4.5% change per °C) for determining temperature in living cells for *ca.* 30 min without interruption, presenting dynamic changes in intracellular temperature in a quantitative and long-term manner.

The proposed process of fabrication and self-calibrated response to the intracellular temperature of the Eu@SiNPs are schematically illustrated in Fig. 1.

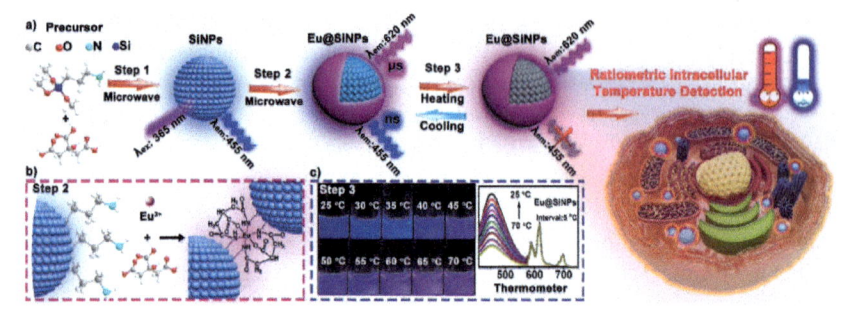

Fig. 1 (a) Schematic illustration for fabricating a Eu@SiNP-based nanothermometer for intracellular temperature detection. The process is composed of three typical steps. Step 1: synthesis of the SiNPs. Step 2: fabrication of the dual-emission Eu@SiNP-based nanothermometer. (b) Local coordination process the between SiNPs and the Eu complex. Step 3: Temperature response process of the Eu@SiNPs and their application to the detection of temperature in living cells. (c) Photographs under UV irradiation and the corresponding photoluminescence spectra of Eu@SiNPs under different temperatures.

Typically, in accordance with our previous investigation, SiNPs with blue emission (λ_{em} = ~455 nm) were obtained under microwave irradiation (step 1).[4e] In addition, the amino groups on the surface of the SiNPs and the trisodium citrate in the precursor provided plenty of coordination sites, which could directly chelate Eu^{3+} ions through the coordination process. Consequently, blue emission (~455 nm) with a short lifetime (ns) and red emission (~620 nm) with a long lifetime (μs) can be integrated into this SiNP-based nanothermometer system (step 2, Fig. 1a and b).[6e] In step 3, the intensity of the blue fluorescence is temperature-dependent in a broad range from 25 to 70 °C (the intensity of the blue emission will significantly decrease during the heating process), and the emission at 620 nm is stable with negligible change (Fig. 1a and c). By virtue of the excellent biocompatibility and photostability of the Eu@SiNPs, a Eu@SiNP-based nanothermometer was employed for the visualization of intracellular temperature in a quantitative and long-term manner, revealing important information for understanding the dynamic process of temperature fluctuations.

Fig. 2a, c and e show transmission electron microscopy (TEM) images of Eu@SiNPs of different sizes, which were prepared with different Eu^{3+} content varying from 0.25 to 2.0%. In our experiment, the sizes of the particles which are monodisperse clearly become larger with changing Eu^{3+} concentration. The size measurements based on TEM images also demonstrate the narrow size distribution of the particles (insets in Fig. 2a, c and e) and show that the sizes of the Eu@SiNPs are ~3.5, ~4.8 and ~5.4 nm. The reason is that local coordination could produce shell overgrowth on the particle surface, which leads to larger sized particles.[6c,d] Furthermore, similar results of ~3.7, ~5.5 and ~6.2 nm were obtained using dynamic light scattering (DLS) (the blue curves in the insets of Fig. 2a, c and e). The slight difference is attributed to the different characterized environments between TEM and DLS characterization, as demonstrated in our previous studies.[4e,f] When the Eu concentration increases to 2.0%, the local coordination reaches a high level; therefore, no significant change in the size of the NPs is observed beyond this concentration. According to the high-resolution

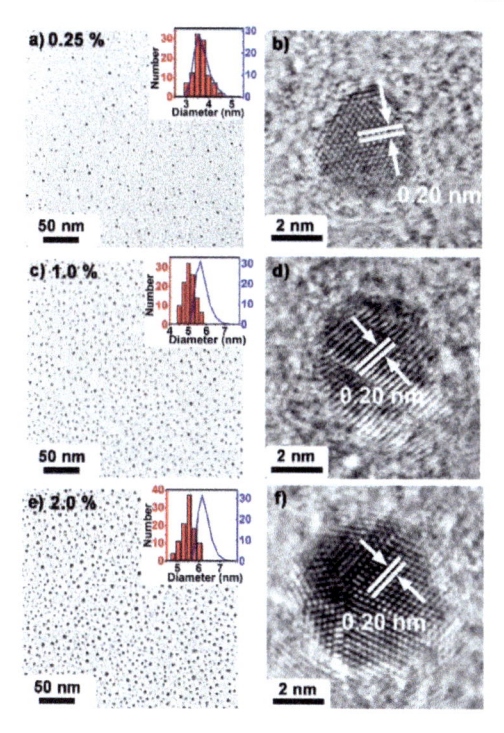

Fig. 2 (a) TEM and HRTEM images of Eu@SiNPs with differing Eu content: (a and b) 0.25%; (c and d) 1.0% and (e and f) 2.0%. The insets in (a, c and e) represent the corresponding size distribution analysis determined with TEM (histograms) and DLS spectra (blue curves).

TEM (HRTEM) images (Fig. 2b, d and f), the prepared Eu@SiNPs with different sizes show well-defined lattice fringes. The fast Fourier transform is also provided (Fig. S1†), with a set of spots with a lattice spacing of 0.20 nm, providing demonstration of the crystallinity of Eu@SiNPs.[5c]

The resultant Eu@SiNPs (the Eu@SiNPs with 1.0% Eu^{3+} content are employed as a model in the following study) were systematically characterized using Fourier transform infrared spectroscopy (FTIR), electron dispersive X-ray spectroscopy (EDX), X-ray photoelectron spectroscopy (XPS) and thermal gravimetric analysis (TGA) (Fig. 3 and S2–S4†). Fig. 3a shows the high angle annular dark field (HAADF) images (I) and EDX elemental maps (II–VI) of the Eu@SiNPs, confirming that the Eu@SiNPs contain Si, Eu, O, N and C elements, as indicated by the green (II), pink (III), yellow (IV), blue (V), and red (VI), respectively. From the XPS spectrum (Fig. 3b), five peaks at 103 eV, 286 eV, 398 eV, 533 eV and 1110 eV (ascribed to Si 2p, C 1s, N 1s, O 1s and Eu 3d) were observed, suggesting the presence of Si, C, N, O and Eu in the Eu@SiNP system. High-resolution XPS (HRXPS) of the Si 2p region (Fig. 3c) showed an intense emission at 99.3 (0), 100.3 (I), 101.3 (II), 102.3 (III), and 103.4 (IV) eV characteristic of Si surface atoms, as well as Si suboxides.[7] The spectrum of Eu 3d is displayed in Fig. 3d, in which Eu $3d_{3/2}$ and Eu $3d_{5/2}$ produce two peaks at ∼1155 and ∼1126 eV, respectively. Fig. 3e presents the typical C 1s spectrum, exhibiting four peaks at 284.8, 285.3, 286.1 and 288.5 eV, which are ascribed to C–C, C–N, C–O and C=O, respectively.[5c,6d]

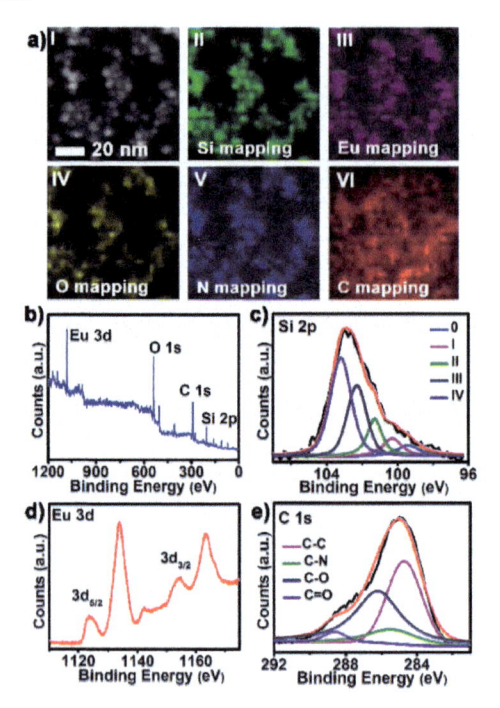

Fig. 3 (a) EDX elemental maps of the prepared Eu@SiNPs. (b) XPS spectra, (c) Si 2p, (d) Eu 3d and (e) C 1s spectra of the Eu@SiNPs.

These results demonstrate that Eu^{3+} ions are coordinated on the surface of the SiNPs.

To determine the triplet excited-state energy of the ligands, we further investigated the energy levels of the Gd^{3+} complexes on the SiNP system. The energy levels of the triplet excited states ($3\pi\pi^*$) of the ligands on the SiNPs were tested through low-temperature (77 K) phosphorescence. Based on Latva's empirical rule, an optimal energy transfer process of ligand-to-metal for Ln^{3+} (Eu^{3+}) needs $\Delta E = 3\pi\pi^*-{}^5D_J = 2500-4000$ cm^{-1}.[8] According to the low-temperature phosphorescence of the resultant Gd^{3+} and SiNP complexes (Fig. 4a), the energy level of the SNP system is more than 3200 cm^{-1} above the $^5D_1$ level of Eu^{3+} (Fig. 4b). Thus, efficient intermolecular energy transfer is possible. The optical properties of the Eu@SiNPs were thoroughly investigated. Fig. 4c shows photographs of the SiNP, EuCl$_3$ and Eu@SiNP solutions with a series of different Eu content (from 0.25% to 2.0%) under different excitation wavelengths. In contrast to the blue emission of the SiNPs and the undetectable fluorescence of EuCl$_3$, the prepared Eu@SiNPs exhibit tunable fluorescence from blue to red under 365 nm, 380 nm and 400 nm irradiation, which is attributed to the different optical transitions between the SiNP and Eu-related species. Moreover, with the increase in Eu content, red fluorescence gradually dominates the emission in the Eu@SiNP system. According to the corresponding PL spectra of Eu@SiNPs at different excitation wavelengths, the Eu@SiNPs show a blue emission centered at ~455 nm, which is ascribed to the SiNPs. On the other hand, the emission peaks at ~590, ~620, ~650 and ~690 nm are ascribed to the $^5D_0 \rightarrow {}^7F_j$ ($j = 1, 2, 3$ and 4) transitions of

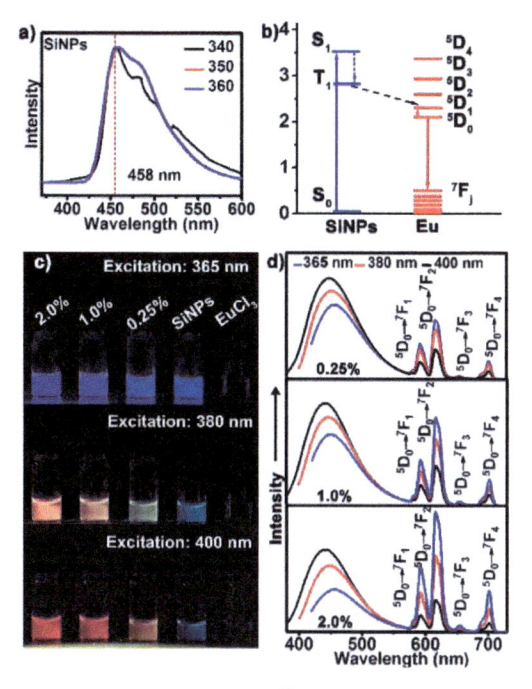

Fig. 4 (a) The phosphorescence spectra of Gd^{3+} complexes on the SiNP platform at 77 K. (b) Calculated energy levels of the SiNPs relative to Eu^{3+} with radiative and nonradiative transitions and energy-transfer pathways. (c) Photographs of the Eu@SiNPs with differing Eu content (from 0.25% to 2.0%), SiNP and EuCl$_3$ solutions under 365 nm, 380 nm and 400 nm irradiation and (d) the corresponding PL spectra.

the Eu species, respectively. This suggests that the excitation energy harvested by the linker groups can be transferred to the Eu^{3+} ions efficiently.[8b,c] Moreover, the PL intensity between the lanthanide ions and the ligands is dependent on the excitation wavelength, thus allowing the tuning of the fluorescence color. These results indicate that the fluorescence of these materials could be readily tuned through doping of Eu^{3+} with different concentrations.

Fig. 5a shows the temperature dependence of the PL spectra of the Eu@SiNPs (the sample with 1.0% Eu content is employed as a model in the following investigation) in phosphate-buffered saline (PBS). The intensity at 455 nm decreases by 71% upon increasing the temperature from 25 to 70 °C without spectral shifts in the investigated temperature window. In our case, the spatial resolution of the probe is ~1.5% change per °C and the ultimate temperature resolution is ~0.5 °C in PBS solution. Moreover, as shown in Fig. 5b, the PL intensity of the blue emission exhibits a linear temperature response in the range 25 to 70 °C, while the red emission from Eu is temperature-independent. The inset in Fig. 5b shows the PL intensity ratio ($R = I_{455}/I_{620}$) versus the temperature values ($R = 2.37 - 0.026T$, correlation coefficient: $r^2 = 0.997$). Obviously, the blue-emission provides a sensitive fluorescence response. To investigate the reusability of the Eu@SiNP-based thermometer, the blue PL intensities of the Eu@SiNPs were tested over ten cycles of cooling and heating at temperatures between 25 and 70 °C (Fig. 5c). In each cycle, the PL is measured after 5 min of equilibration. No

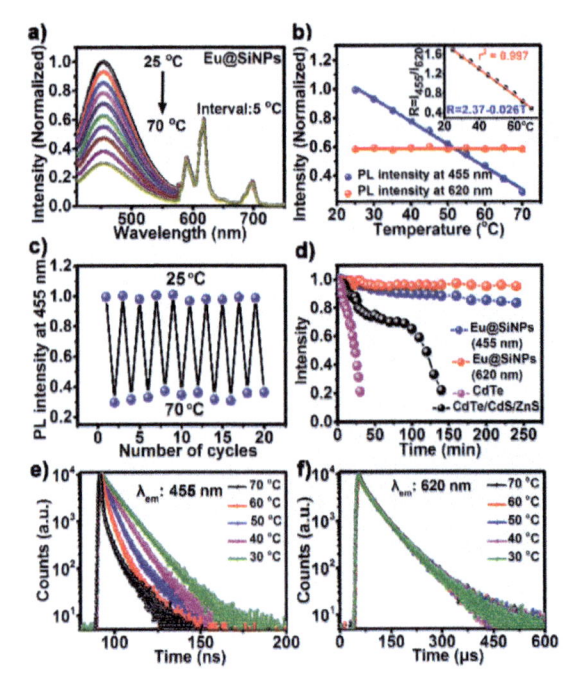

Fig. 5 (a) Normalized PL spectra of the Eu@SiNPs at different temperatures. (b) The apparent linear relationship of the PL intensity at 455 nm and 620 nm *versus* temperature. The inset presents the linear relationship between R and temperature. (c) Temperature reversibility assess of the PL intensity at 455 nm between 25 and 70 °C. (d) Photostability comparison of the II–VI QDs and Eu@SiNPs. Time-resolved fluorescence–decay curves of the Eu@SiNPs at different temperatures with the emission wavelength at 455 nm (e) and 620 nm (f).

thermal hysteresis was observed during the cooling or heating processes, demonstrating that the PL properties of the Eu@SiNPs exhibit excellent thermal stability. We also found that the Eu@SiNPs were photostable, and superior to CdTe and CdTe/CdS/ZnS quantum dots (QDs: a kind of photostable fluorescent probe). Fig. 5d shows that the optical properties of CdTe QDs vanish under high power UV irradiation, while CdSe/ZnS QDs are more stable: their fluorescence intensities decrease to <20% in 2 h (blue line) due to the surface deterioration under UV irradiation.[9] In marked contrast, both the blue emission (blue line) and the red emission (red line) of the Eu@SiNPs are highly photostable, preserving ∼93 and ∼86% of their fluorescence intensities during a 4 h UV treatment, respectively.

We next investigated the Eu@SiNP emission dynamics at different temperatures in PBS. Fig. 5e and f show the PL decay curves, at 455 and 620 nm, at different temperatures (from 30 to 70 °C), respectively. These measurements afforded luminescence time profiles, which were well fitted by triexponential functions. As shown in Fig. 5e, the PL decay curves at 455 nm originate from SiNPs. Their lifetimes (ns) decline markedly from 3.15 to 6.95 ns as the temperature increases. Meanwhile, the average triple-exponential fluorescence lifetime of Eu^{3+} at the microsecond scale is 570.22 μs (Fig. 5f), which remains stable at different temperatures. To understand the thermodynamics of the Eu@SiNP fluorescence properties (455 nm) more deeply, we also correlated the nonradiative (τ_{nr}^{-1}) and radiative (τ_r^{-1}) recombination rates of the Eu@SiNPs. The radiative rate was calculated from the time-resolved PL and PLQY measurement data. The recombination rate τ^{-1} is $\tau_r^{-1} = QY \times \tau^{-1}$ and the nonradiative relaxation rate τ_{nr}^{-1} is $\tau_{nr}^{-1} = \tau^{-1} - \tau_r^{-1}$. Both nonradiative and radiative recombination rates are shown in Fig. S9.†[10] The radiative recombination rate is lower than the nonradiative rate and remains stable at temperatures between 25 and 70 °C, remaining in the range $(1.9–2.2) \times 10^6$ s^{-1}. In sharp contrast, there is an increase (almost 3-fold, from 2.42×10^7 to 6.98×10^7 s^{-1}) in the rate of nonradiative recombination by raising the temperature. These results indicate that the activation of nonradiative relaxation channels lead to the PL quenching in the Eu@SiNPs.

In this study, taking advantage of the optical merits of the Eu@SiNP system, we further exploited the Eu@SiNPs as a novel type of intracellular nanothermometer for realizing temperature mapping in living cells. Before biological application of the Eu@SiNPs, their cytotoxicity on human U87MG glioblastoma cancer cells (U87MG cells) and human epithelial cervical cancer cells (HeLa cells) was evaluated by measuring the cellular mitochondrial activity via an MTT assay. Fig. 6a shows that the cell viability of U87MG cells and HeLa cells treated by the Eu@SiNPs maintains >90% at different concentrations (0.125–2 mg mL^{-1}) and incubation times (3–24 h). Fig. 6b shows that Eu@SiNP-treated HeLa cells preserve normal morphology over the 24 h incubation, demonstrating the low-toxicity of the Eu@SiNPs. To further investigate the intracellular distribution of the Eu@SiNPs, Eu@SiNP treated cells (U87MG and HeLa cells) were stained with green fluorescent lysosome dye (Lyso-tracker Green DND-26). As shown in Fig. 6d and e, the blue (SiNPs) and red (Eu) fluorescence of the Eu@SiNPs overlap well with the fluorescence from Green DND-26. The Pearson's colocalization coefficient (R_r) is ~0.80. This result demonstrates that Eu@SiNPs can assess in

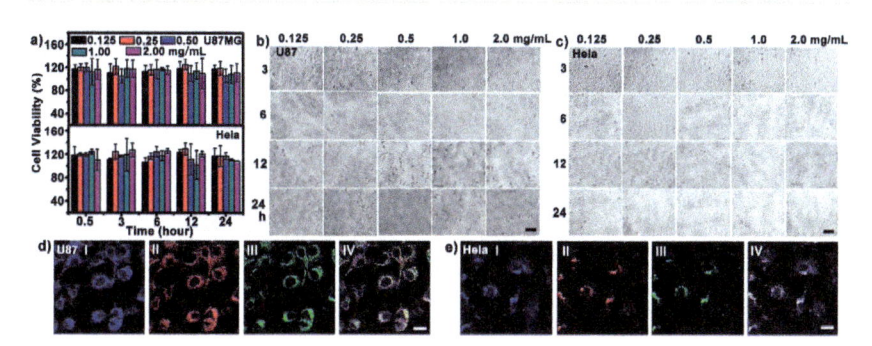

Fig. 6 (a) Cell viability of U87MG and HeLa cells treated with Eu@SiNPs for different incubation times. The viability of the control cells is considered to be 100%. (b and c) Morphology of cells incubated with Eu@SiNPs of different concentrations (0.125, 0.25, 0.5, 1.0 and 2.0 mg mL^{-1}) for 3, 6, 12 and 24 h. Scale bars = 50 μm. (d and e) Fluorescence images of the intracellular distribution of Eu@SiNPs in live U87MG and HeLa cells. Eu@SiNP fluorescence was collected at 430–480 nm and 610–650 nm with excitation at 405 nm; Lyso-tracker fluorescence was collected at 510–560 nm with excitation at 476 nm. Scale bars: 20 μm.

lysosomes specifically. These distinct features, including good reversibility, wide linear range, excellent selectivity, low-toxicity and lysosome-targeting properties, give rise to the feasibility of Eu@SiNP sensors for temperature sensing in living cells.

In particular, compared to Green DND-26 which has weak photostability, the resultant Eu@SiNPs are suitable for monitoring the dynamic fluctuation of temperature in a long-term manner, due to their strong photostability. As shown in Fig. 7(a–d), the green emission from Green DND-26 distributed in U87MG cells vanishes after 15 min of UV irradiation (i.e., the intensity decreases by ∼90%); in comparison, for the Eu@SiNPs, both the blue and red fluorescence signals maintain stability after 45 min of UV irradiation (i.e., the red and blue fluorescence intensity drops by ∼5% and ∼4% after 45 min of continual UV irradiation, respectively). The above results suggest this kind of fluorescent Eu@SiNP sensor to be a novel kind of sensing platform, capable of long-term tracking of temperature fluctuations *in vitro*.

We further explored the capability of the Eu@SiNPs for temperature sensing in a living U87MG cell *via* calibration experiments. The temperature of the environment inside a cell can be varied using a temperature-controlled sample stage using an established method. As depicted in Fig. 8a for living cell imaging, the cell PL spectra at each temperature are highly reproducible at all recorded temperatures. With increasing temperature (with a step size of 5 °C), the blue fluorescence intensity of the SiNP channel (λ_{ex} = 405 nm, λ_{em} = 430–480 nm) decreased markedly. In sharp contrast, the red fluorescence intensity of the Eu channel (λ_{ex} = 405 nm, λ_{em} = 610–650 nm) is stable in the range 25–45 °C (Fig. 8b). The fluorescence intensity of the Eu@SiNPs is much lower in cells than in PBS over the temperature range, which is possibly due to the formation of a biomolecular corona around the internalized nanomaterials that influences their optical properties.[11]

Furthermore, we carefully explored the performance of the Eu@SiNPs for long-term and real-time intracellular nanothermometry. During the heating and

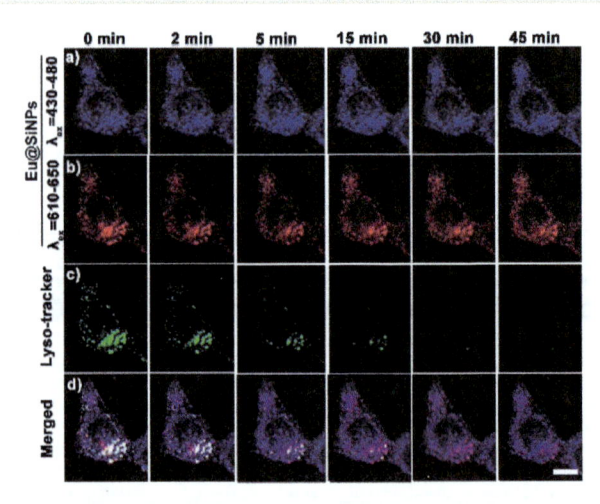

Fig. 7 (a–d) Photostabilitiy comparison of Eu@SiNPs and Lyso-tracker in live U87MG cells under continual excitation over 45 min. Scale bars: 10 μm.

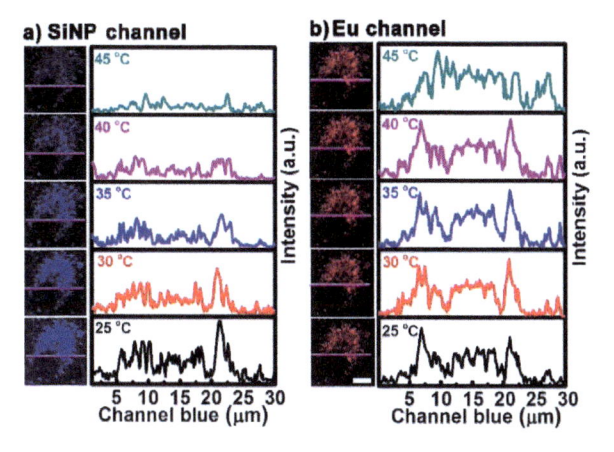

Fig. 8 Intracellular temperature response via Eu@SiNP-based temperature sensors (SiNP channel (a) and Eu channel (b)). Fluorescence images and the corresponding spectra of the fluorescence intensity of Eu@SiNPs versus a single U87MG cell at different temperatures (from 25 to 45 °C). Scale bars: 10 μm.

cooling process from 30 to 45 °C, we recorded the fluorescence intensities of U87MG cells incubated with Eu@SiNPs every 30 s for 30 min. To precisely evaluate the results of accurate intracellular temperature sensing, we plotted the ratio of the 455 nm to the 620 nm intensity as a function (10.6–5.6R) of temperature (Fig. 9a and b, blue symbols for $T_{measure}$). Both heating and cooling processes showed a fluorescence intensity change of ∼4.5% per °C. Independently, the temperature of the cell was measured through a calibrated reference thermometer (Fig. 9a and b, red curves for T_{set}). In the heating process, the average cell temperature reported by the Eu@SiNPs was 35.1 ± 0.5 °C (40 cells) for the 35.5 °C cell solution at 330 s (T_{set}, measured with a thermocouple) and it was 35.0 ± 0.9 °C (40 cells) for the 35.5 °C cell solution at 1470 s during the cooling process. The temperatures detected by the Eu@SiNPs and the calibrated reference thermometer are in good agreement, and the accuracy of the temperature sensing is further

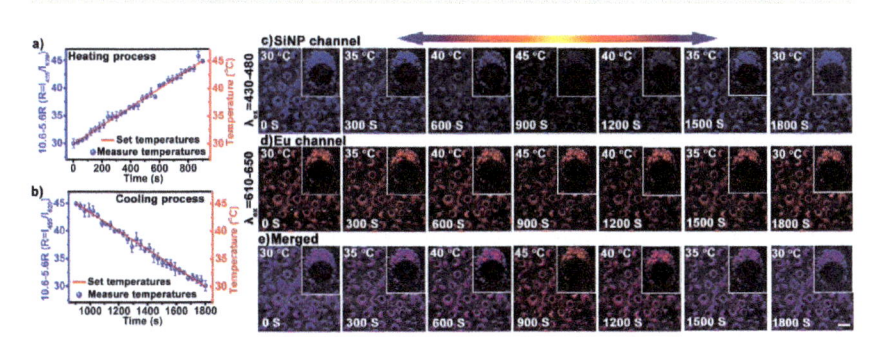

Fig. 9 (a and b) Measured temperatures determined using the scatter plots presented by the equation (10.6–5.6R) ($T_{measure}$) and set temperatures (T_{set}) against time ($R = I_{455}/I_{620}$). (c) Fluorescence images of the Eu@SiNPs in U87MG cells collected at different times. (d and e) The images of the Eu@SiNPs are, respectively, collected at 430–480 nm and at 610–650 nm with excitation at 405 nm. Scale bars: 50 μm.

demonstrated by statistical analysis of the differences between the real (reference) and measured temperatures (Fig. S10†). Moreover, the thermal response of the Eu@SiNPs is reversible upon intracellular temperature change, and there is no thermal hysteresis during the cooling and heating processes between 30 and 45 °C. These results show that the PL intensity ratio of the nanothermometer based on Eu@SiNPs can be reliably used to track intracellular temperatures in a long-term manner.

In summary, we herein introduce a kind of Eu@SiNP-based nanothermometer for the detection of intracellular temperature with high accuracy. The developed Eu@SiNP-based nanothermometer displays dual-emission fluorescence with a ratiometric temperature response in a wide range (25–70 °C). Moreover, the Eu@SiNP-based nanothermometer presents robust photostability, feeble cytotoxicity and high sensitivity towards intracellular temperature (~4.5% change per °C), enabling long-term visualization of temperature changes in living cells. Our findings suggest this fluorescent SiNP-based nanothermometer to be a novel high-performance sensing platform, providing invaluable information for understanding the dynamic process of intracellular temperature fluctuations.

Conflicts of interest

There are no conflicts to declare.

Acknowledgements

The authors appreciate the financial support provided by the National Basic Research Program of China (973 Program 2013CB934400), the National Natural Science Foundation of China (21825402, 31400860, 21575096, and 21605109), the Priority Academic Program Development of Jiangsu Higher Education Institutions (PAPD), the 111 Project, and the Collaborative Innovation Center of Suzhou Nano Science and Technology (NANO-CIC).

Notes and references

1 (*a*) P. J. Stoward, *Nature*, 1962, **194**, 977–978; (*b*) R. Sitia and I. Braakman, *Nature*, 2003, **426**, 891–894; (*c*) M. Mecklenburg, W. A. Hubbard, E. R. White, R. Dhall, S. B. Cronin, S. Aloni and B. C. Regan, *Science*, 2015, **347**, 629–632; (*d*) B. B. Lowell and B. M. Spiegelman, *Nature*, 2000, **404**, 652–660; (*e*) S. P. Ashby, J. A. Thomas, J. Garcia-Canadas, G. Min, J. Corps, A. V. Powell, H. Xu, W. Shen and Y. Chao, *Faraday Discuss.*, 2014, **176**, 349–361; (*f*) F. Peng, Y. Y. Su, Y. L. Zhong, C. H. Fan, S. T. Lee and Y. He, *Acc. Chem. Res.*, 2014, **47**, 612–623; (*g*) B. Song and Y. He, *Nano Today*, 2019, **26**, 149–163; (*h*) L. H. Lie, S. N. Patole, A. R. Pike, L. C. Ryder, B. A. Connolly, A. D. Ward, E. M. Tuite, A. Houltona and B. R. Horrocks, *Faraday Discuss.*, 2004, **125**, 235–249.

2 (*a*) L. Shang, F. Stockmar, N. Azadfar and G. U. Nienhaus, *Angew. Chem., Int. Ed.*, 2013, **52**, 11154–11157; (*b*) K. Okabe, N. Inada, C. Gota, Y. Harada, T. Funatsu and S. Uchiyama, *Nat. Commun.*, 2012, **3**, 705–714; (*c*) L. Sambe, V. R. de La Rosa, K. Belal, F. Stoffelbach, J. Lyskawa, F. Delattre, M. Bria, G. Cooke, R. Hoogenboom and P. Woisel, *Angew. Chem., Int. Ed.*, 2014, **53**,

5044–5048; (*d*) L. H. Lie, S. N. Patole, A. R. Pike, L. C. Ryder, B. A. Connolly, A. D. Ward, E. M. Tuite, A. Houltona and B. R. Horrocks, *Faraday Discuss.*, 2004, **125**, 235–249; (*e*) A. Fermi, M. Locritani, G. Di Carlo, M. Pizzotti, S. Caramori, Y. Yu, B. A. Korgel, G. Bergamini and P. Ceroni, *Faraday Discuss.*, 2015, **185**, 481–495.

3 (*a*) G. Kucsko, P. C. Maurer, N. Y. Yao, M. Kubo, H. J. Noh, P. K. Lo, H. Park and M. D. Lukin, *Nature*, 2013, **500**, 54–58; (*b*) D. M. Toyli, C. F. Casas, D. J. Christle, V. V. Dobrovitski and D. D. Awschalom, *Proc. Natl. Acad. Sci. U. S. A.*, 2013, **110**, 8417–8421; (*c*) D. Ananias, F. A. A. D. Paz, S. Yufit, L. D. Carlos and J. Rocha, *J. Am. Chem. Soc.*, 2015, **137**, 3051–3058; (*d*) F. Ye, C. Wu, Y. Jin, Y. Chan, X. Zhang and D. T. Chiu, *J. Am. Chem. Soc.*, 2011, **133**, 8146–8149; (*e*) S. Deo and H. A. Godwin, *J. Am. Chem. Soc.*, 2000, **122**, 174–175; (*f*) C. Y. Tay, M. I. Setyawati and D. T. Leong, *ACS Nano*, 2017, **11**, 5609–5617; (*g*) C. Y. Tay, L. Yuan and D. T. Leong, *ACS Nano*, 2015, **9**, 2764–2772; (*h*) M. L. Zhang, Z. Zhang, K. M. He, J. M. Wu, N. Li, R. Zhao, J. H. an, H. Xiao, Y. Y. Zhang and X. H. Fang, *Anal. Chem.*, 2018, **90**, 4282–4287; (*i*) D. Zhu, P. Song, J. W. Shen, S. Su, J. Chao, A. Aldalbahi, Z. A. Zhou, S. P. Song, C. H. Fan, X. L. Zuo, Y. Tian, L. H. Wang and H. Pei, *Anal. Chem.*, 2016, **88**, 4949–4954; (*j*) J. Su, D. F. Wang, L. Nörbel, J. L. Shen, Z. H. Zhao, Y. Z. Dou, T. H. Peng, J. Shi, S. Mathur, C. H. Fan and S. P. Song, *Anal. Chem.*, 2017, **89**, 2531–2538; (*k*) X. R. Li, H. Ma, M. Deng, A. Iqbal, X. Y. Liu, B. Li, W. S. Liu, J. P. Li and W. W. Qin, *J. Mater. Chem. C*, 2017, **5**, 2149–2152.

4 (*a*) A. V. Zhukhovitskiy, M. G. Mavros, K. T. Queeney, T. Wu, T. V. Voorhis and J. A. Johnson, *J. Am. Chem. Soc.*, 2016, **138**, 8639–8652; (*b*) A. Zhang and C. M. Lieber, *Chem. Rev.*, 2016, **116**, 215–257; (*c*) B. Song, Y. L. Zhong, S. C. Wu, B. B. Chu, Y. Y. Su and Y. He, *J. Am. Chem. Soc.*, 2016, **138**, 4824–4831; (*d*) P. D. Howes, R. Chandrawti and M. M. Stevens, *Science*, 2014, **346**, 6205; (*e*) Y. L. Zhong, F. Peng, F. Bao, S. Y. Wang, X. Y. Ji, L. Yang, Y. Y. Su, S. T. Lee and Y. He, *J. Am. Chem. Soc.*, 2013, **135**, 8350–8356; (*f*) F. Peng, Y. Y. Su, Y. L. Zhong, C. H. Fan, S. T. Lee and Y. He, *Acc. Chem. Res.*, 2014, **47**, 612–623; (*g*) S. C. Wu, Y. L. Zhong, Y. F. Zhou, B. Song, B. B. Chu, X. Y. Ji, Y. Y. Wu, Y. Y. Su and Y. He, *J. Am. Chem. Soc.*, 2015, **137**, 14726–14732; (*h*) T. K. Purkait, M. Iqbal, M. H. Wahl, K. Gottschling, C. M. Gonzalez, M. A. Islam and J. G. C. Veinot, *J. Am. Chem. Soc.*, 2014, **136**, 17914–17917; (*i*) Z. Ni, L. Ma, S. Du, Y. Xu, M. Yuan, H. Fang, Z. Wang, M. Xu, D. Li, J. Yang, W. Hu, X. Pi and D. Yang, *ACS Nano*, 2017, **11**, 9854–9862.

5 (*a*) J. Demuth, E. Fahrenkrug, L. Ma, T. Shodiya, J. I. Deitz, T. J. Grassman and S. Maldonado, *J. Am. Chem. Soc.*, 2017, **139**, 6960–6968; (*b*) S. Guo, K. Matsukawa, T. Miyata, T. Okubo, K. Kuroda and A. Shimojima, *J. Am. Chem. Soc.*, 2015, **137**, 15434–15440; (*c*) B. Song, Y. L. Zhong, S. C. Wu, B. B. Chu, Y. Y. Su and Y. He, *J. Am. Chem. Soc.*, 2016, **138**, 4824–4831; (*d*) J. Yang, R. Fainblat, S. G. Kwon, F. Muckel, J. H. Yu, H. Terlinden, B. H. Kim, D. Iavarone, M. K. Choi, I. Y. Kim, I. Park, H. K. Hong, J. Lee, J. S. Son, Z. Lee, K. Kang, S. Hwang, G. Bacher and T. Hyeon, *J. Am. Chem. Soc.*, 2015, **137**, 12776–12779; (*e*) P. Mi, D. Kokuryo, H. Cabral, H. Wu, Y. Terada, T. Saga, I. Aoki, N. Nishiyama and K. Kataoka, *Nat. Nanotechnol.*, 2016, **11**, 724–730; (*f*) J. G. Croissant, Y. Fatieiev and N. M. Khashab, *Adv. Mater.*, 2017, **29**, 1604634–1604655; (*g*) N. J. Kramer, K. S. Schramke and U. R. Kortshagen, *Nano Lett.*, 2015, **15**, 5597–5603; (*h*) S. Chandra,

Y. Masuda, N. Shirahata and F. M. Winnik, *Angew. Chem., Int. Ed.*, 2017, **56**, 6157–6160.

6 (*a*) H. Dong, S. R. Du, X. Y. Zheng, G. M. Lyu, L. D. Sun, L. D. Li, P. Z. Zhang, C. Zhang and C. H. Yan, *Chem. Rev.*, 2015, **115**, 10725–10815; (*b*) S. V. Eliseeva and J. C. G. Bunzli, *Chem. Soc. Rev.*, 2010, **39**, 189–227; (*c*) Y. S. Liu, S. Y. Zhou, D. T. Tu, Z. Chen, M. D. Huang, H. M. Zhu, E. Ma and X. Y. Chen, *J. Am. Chem. Soc.*, 2012, **134**, 15083–15090; (*d*) Y. Q. Dong, J. H. Cai, Q. Q. Fang, X. You and Y. W. Chi, *Anal. Chem.*, 2016, **88**, 1748–1752; (*e*) X. Li, Y. Xie, B. Song, H. L. Zhang, H. Chen, H. Cai, W. Liu and Y. Tang, *Angew. Chem., Int. Ed.*, 2017, **56**, 2689–2693; (*f*) G. Valenti, E. Rampazzo, S. Bonacchi, L. Petrizza, M. Marcaccio, M. Montalti, L. Prodi and F. Paolucci, *J. Am. Chem. Soc.*, 2016, **138**, 15935–15942.

7 (*a*) Z. Y. Yang, M. Iqbal, A. R. Dobbie and J. G. C. Veinot, *J. Am. Chem. Soc.*, 2013, **135**, 17595–17601; (*b*) T. K. Purkait, M. Iqbal, M. H. Wahl, K. Gottschling, C. M. Gonzalez, M. A. Islam and J. G. C. Veinot, *J. Am. Chem. Soc.*, 2014, **136**, 17914–17917.

8 (*a*) M. Latva, H. Takalo, V. M. Mukkala, C. Matachescu, J. C. Rodriguez-Ubis and J. Kankare, *J. Lumin.*, 1997, **75**, 149–169; (*b*) Y. S. Zhou, X. M. Li, L. J. Zhang, Y. Guo and Z. H. Shi, *Inorg. Chem.*, 2014, **53**, 3362–3370; (*c*) Y. Zhao, C. Shi, X. Yang, B. Shen, Y. Sun, Y. Chen, X. Xu, H. Sun, K. Yu, B. Yang and Q. Lin, *ACS Nano*, 2016, **10**, 5856–5863.

9 (*a*) C. L. Choi, H. Li, A. C. K. Olson, P. K. Jain, S. Sivasankar and A. P. Alivisatos, *Nano Lett.*, 2011, **11**, 2358–2362; (*b*) Q. Lin, N. S. Makarov, W. K. Koh, K. A. Velizhanin, C. M. Cirloganu, H. Luo, V. I. Klimov and J. M. Pietryga, *ACS Nano*, 2015, **9**, 539–547; (*c*) A. E. Albers, E. M. Chan, P. M. McBride, C. M. Ajo-Franklin, B. E. Cohen and B. A. Helms, *J. Am. Chem. Soc.*, 2012, **134**, 9565–9568.

10 Q. Li, Y. He, J. Chang, L. Wang, H. Chen, Y. W. Tan, H. Wang and Z. Shao, *J. Am. Chem. Soc.*, 2013, **135**, 14924–14927.

11 (*a*) C. Röcker, M. Pötzl, F. Zhang, W. J. Parak and G. Ulrich Nienhaus, *Nat. Nanotechnol.*, 2009, **4**, 577–580; (*b*) M. P. Monopoli, C. Aberg, A. Salvati and K. A. Dawson, *Nat. Nanotechnol.*, 2012, **7**, 779–786; (*c*) C. Li, F. Li, Y. Zhang, W. Zhang, X. Zhang and Q. Wang, *ACS Nano*, 2015, **9**, 12255–12263; (*d*) L. Sun, Y. Gao, Y. Xu, J. Chao, H. Liu, L. Wang, D. Li and C. Fan, *J. Am. Chem. Soc.*, 2017, **139**, 17525–17532.

Faraday Discussions

PAPER

The shell matters: one step synthesis of core–shell silicon nanoparticles with room temperature ultranarrow emission linewidth†

Anna Fucikova, [*a] Ilya Sychugov [b] and Jan Linnros[b]

Received 28th September 2019, Accepted 1st November 2019
DOI: 10.1039/c9fd00093c

Here we present a one-step synthesis that provides silicon nanocrystals with a thin shell composed of a ceramic-like carbonyl based compound, embedded in a porous organosilicon film. The silicon nanocrystals were synthesised from hydrogen silsesquioxane molecules, modified with organic molecules containing carbonyl groups, which were annealed at 1000 °C in a slightly reducing 5% H_2 : 95% Ar atmosphere. The organic character of the shell was preserved after annealing due to trapping of organic molecules inside the HSQ-derived oxide matrix that forms during the annealing. The individual silicon nanocrystals, studied by single dot spectroscopy, exhibited a significantly narrower emission peak at room temperature (lowest linewidth ∼ 17 meV) compared to silicon nanocrystals embedded in a silicon oxide shell (150 meV). Their emission linewidths are even significantly narrower than those of single CdSe quantum dots (>50 meV). It is hypothesized that the Si-core–thin shell structure of the nanoparticle is responsible for the unique optical properties. Its formation within one synthesis step opens new opportunities for silicon-based quantum dots. The luminescence from the produced nanocrystals covers a broad spectral range from 530–720 nm (1.7–2.3 eV) suggesting strong application potential for solar cells and LEDs, following the development of a suitable mass-fabrication protocol.

Introduction

Silicon nanocrystals (Si-NCs) have been attracting attention since the discovery of the strong photoluminescence (PL) of porous silicon[1] at room temperature by Canham in 1990, which contrasts with the practically non-existent light emission of bulk silicon. During the last three decades, the properties of Si-NCs have been extensively studied in conjunction with many research fields and possible

[a] Department of Chemical Physics and Optics, Faculty of Mathematics and Physics, Charles University, Ke Karlovu 3, 121 16 Praha 2, Czech Republic. E-mail: Anna.Fucikova@mff.cuni.cz
[b] KTH Royal Institute of Technology, Department of Applied Physics, SE-16440 Kista, Sweden
† Electronic supplementary information (ESI) available. See DOI: 10.1039/c9fd00093c

applications have been proposed. These involve quantum computing,[2] chemical sensing,[3] fluorescent markers or drug carriers in biology,[4-6] solid-state light emitting devices,[7] solar cells[8] and others. Si-NCs exhibit efficient room-temperature PL, and the quantum yield (QY) has been reported to reach up to 70%.[9-11] The mechanism of the PL of Si-NCs is still not clear today; the latest experiments are suggesting that quantum confinement and/or trap states are the origin of the luminescence.[12] However, the electronic bandgap of Si-NCs is broadened significantly due to the quantum size effect for crystals smaller than the bulk exciton Bohr radius (~4.9 nm (ref. 13)). For such nanocrystals, the luminescence can be tuned over the whole visible spectrum by changing their size. The emission of an individual Si-NC is not only influenced by the nanoparticle size, but is also affected by the shape, impurities, surface passivation and the surrounding environment.[14]

Despite these attractive applications, the production of Si-NCs with high spectral purity and narrow spectral emission is still not possible. Oxide-embedded Si-NCs show a very broad homogeneous linewidth of ~ 150 meV at room temperature.[15] Although narrowing of the full width at half maxima (FWHM) (down to 100 meV (ref. 14)) was observed for different surface passivated Si-NCs, this is still not enough, especially for experiments in biology, where narrowly emitting nanoparticles are desirable and could allow for multi-tagging experiments. The synthesis method, proposed here, could provide straightforward production of Si-NCs with a thin shell, exhibiting a narrow homogeneous luminescence linewidth, surpassing even that of CdSe quantum dots (>50 meV)[16] and also that of carbon quantum dots whose reported narrowest FWHM is around 75 meV (ref. 17) with the majority having linewidths > 100 meV.[17-19]

After the discovery of a preparation[1] for porous silicon, several methods for preparing Si-NCs have been developed *e.g.* laser assisted pyrolysis of silane[20] or low pressure nonthermal plasma synthesis.[21] Both produce larger amounts of Si-NCs, but the Si-NCs are mostly inhomogeneous in size and shape and they are usually interconnected after the synthesis. Si-NCs can be also prepared *via* solution reduction.[22] Although this method produces size defined (2–10 nm) spherical nanoparticles, its production efficiency is low and this technique is time consuming. Finally, we would like to point out the synthesis of Si-NCs based on annealing of a hydrogen silsesquioxane (HSQ) ladder polymer. This specific compound consists of Si, O, and H atoms in a specific T8 matrix.[23-25] When heated in a slightly reducing atmosphere, Si-NCs in a silicon oxide matrix are created.[23,26] They are usually released from the oxide matrix *via* alcoholic HF etching followed by surface modification and size separation.[27,28] After HF etching, the Si-NCs have their surface covered with Si–H bonds, this surface is reactive and must be passivated to prevent oxidation. Passivation can be achieved *via* various hydrosilylation reaction schemes, although other approaches including reactions with Grignard and organolithium reagents have proven useful. It is important to note that the nature of the Si-NC surface depends on the reaction employed and properties can vary dramatically.[29]

Despite impressive advances in Si-NC surface chemistry, modification of each Si–H bond is simply not possible due to the surface group packing density and steric hindrance considerations. Assuming that the presence of Si–H on the surface of the Si-NCs limits their long term stability and hinders their optimal optical performance, there is a need to explore new surface coverage. The formation of a core–shell structure has already proven very successful in optimising the optical performance of CdSe quantum dots.[30] Early reports describe materials that degraded quickly and only exhibited QY in the range of several percent. The introduction of a ZnS or CdS shell on the CdSe core resulted in dramatic improvements in the spectral tunability and gave QY values approaching 100%.[31]

In this work we describe a one step synthesis of Si-NCs with a thin ceramic-like shell from HSQ annealed in the presence of carbonyl molecules. These new Si-core–thin shell nanoparticles have unique, extremely narrow emission peaks at room temperature.

Results and discussion

Preparation of samples

In this work we used both commercial HSQ (c-HSQ) resist purchased from Dow Corning (type XR-1541), obtained as 10 wt% solution in methyl isobutyl ketone and our in-house prepared HSQ (in-HSQ) prepared by the procedure described in ref. 23 and 26. Immediately after synthesis, the in-HSQ was heated up to 50 °C for 2 hours in order to evaporate the majority of the HCl *via* the glass valve. Afterwards, the in-HSQ was immediately transferred to various solvents (*i.e.* acetone, methyl isobutyl ketone (MIBK), and ethyl acetate). In the main body of this article we will describe the modification of in-HSQ with MIBK (in-HSQ-MIBK), since this was the first studied modifying molecule. The other molecules have been selected to test the concept of thin shell Si-NC creation.

The in-HSQ in the modifying solvent was mixed on a hot plate (20 RPM, 30 °C) for 48 hours in a closed vessel. During this treatment, the mixture was exposed to light in order to promote the solvent to get into the polymer HSQ structure and possibly bind to it (when the samples were mixed in the dark or for a shorter period of time, 3–6 times fewer Si-NCs with narrow emission were produced). The modified c-HSQ and modified in-HSQ were spin coated on a Si-wafer substrate carrier and dried in air at room temperature for 12 hours prior to the annealing in order to evaporate the excess solvent.

The sample on the carrier was transferred to a furnace and annealed in a slightly reducing 5% H_2 : 95% Ar atmosphere. The samples were heated at a rate of 17 °C min^{-1} until the desired temperature of 1000 °C was reached, and then they were annealed at this temperature for 1 hour. After annealing, the modified HSQ formed a porous organosilicon film (for a TEM image of the film see Fig. S1 in the ESI†) in which we identified Si-NCs with ultranarrow linewidth using luminescence micro-spectroscopy. The resulting film appeared to be highly porous and could be scratched out from the carrier surface by applying some force, or by exposure to HF vapours. The origin and structure of the organic film and the Si-NCs with thin shells is discussed below. The samples were stable in air without losing their optical properties (up to 4 years in the case of in-HSQ-MIBK).

ATR-FTIR characterization of Si-NCs embedded in-HSQ-MIBK

For exploring the chemical structure of Si-NCs embedded in annealed modified HSQ, attenuated total reflectance Fourier transform infrared spectroscopy (ATR-FTIR) was used. A Nicolet FTIR spectrometer from Thermo Scientific equipped with a DTGS KBr detector and built-in germanium crystal ATR optics was employed for these measurements. Before measurements, the crystal was thoroughly rinsed with ethanol and dried. FTIR spectra were recorded as changes in absorption as a function of wavenumber in the range 600 cm^{-1} to 4000 cm^{-1} (4 cm^{-1} resolution). The spectrum consisted of 100 scans and a silicon surface was used as the background. The spectra of Si-NCs in-HSQ-MIBK before annealing and after annealing are displayed in Fig. 1.

Comparison of the FTIR spectra in Fig. 1 shows that some peaks shifted, changed their intensity, or disappeared after annealing. The FTIR spectra of the non modified HSQ are shown at Fig. S2D and E in the ESI.† From the comparison with our in-HSQ-MIBK, we can see that the peaks belonging to the HSQ structure are still present in both the annealed and non-annealed samples. For example, the peak around 800 cm^{-1} is assigned to the Si–O–Si bending mode in the literature,[26,32] and bands around 850 cm^{-1} probably belong to Si–O–Si frequencies that are identical for modified and unmodified HSQ. The peak around 2260 cm^{-1} corresponds to Si–H vibrations according to the literature[26] and is present in all non-annealed forms of HSQ used in this study. The most intense band at around 1123 cm^{-1} (Si–O–Si stretching) in in-HSQ-MIBK almost corresponds to the original HSQ frequencies, but we start to observe modifications caused most probably by the presence of MIBK in in-HSQ-MIBK on the lower energy shoulder. The major of the differences in FTIR spectra between the original unmodified HSQ and in-HSQ-MIBK are concentrated in two spectral regions: 1100–1750 cm^{-1} and 2800–3500 cm^{-1}.

We will now look more carefully at the FTIR spectra of the annealed in-HSQ-MIBK sample. The band at 1000–1200 cm^{-1} consists of 3 peaks. The vibration at 1023 cm^{-1} has been tentatively assigned to the Si–O–Si stretching mode. The precise position of the other two FTIR bands is hard to identify. We estimate their

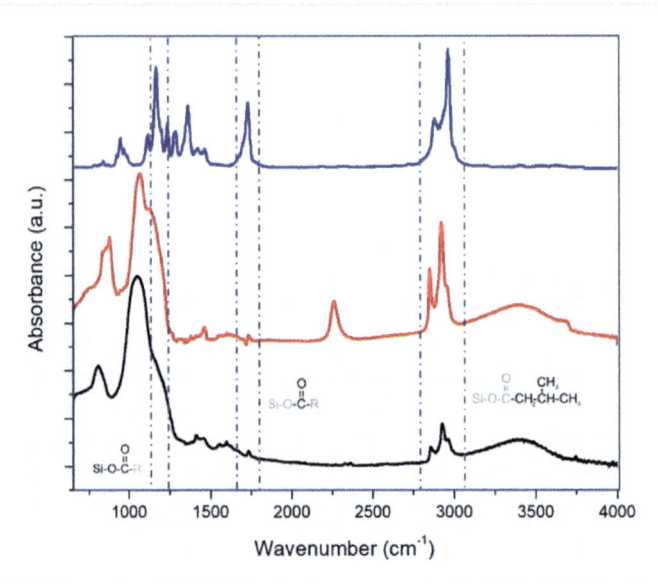

Fig. 1 ATR-FTIR spectra of MIBK (blue line) non-annealed in-HSQ-MIBK (red line) and in-HSQ-MIBK after thermal annealing (black line) on a silicon wafer.

positions to be approximately 1050 cm^{-1} and 1190 cm^{-1}. These two frequencies may belong to the Si–O–(C=O)–R group (where R stands for a carbon chain, as discussed below), since they are both well described in the ceramic materials literature[33–36] in the expected spectral range.

The frequency of the –(C=O)– group in MIBK (blue line in Fig. 1) has a value of 1716 cm^{-1} (for comparison see the IR spectrum in the NIST database[37]). For the annealed in-HSQ-MIBK, we observed a peak in the same spectral region with a frequency of 1740 cm^{-1} which, according to the literature[34,35] is within the limits of the interval of 1715–1770 cm^{-1} for frequencies belonging to –(C=O)– groups in silicon rich ceramics. The peak at 1466 cm^{-1} might belong to a C–H$_x$ group since it is in the expected frequency range for these groups.

In Fig. 1 we observed three significant peaks: 2855 cm^{-1}, 2930 cm^{-1} and 2960 cm^{-1} for both, annealed and non-annealed in-HSQ-MIBK. All frequencies except 2930 cm^{-1} can be assigned to vibrations in the MIBK molecule and they probably belong to the vibrations of the carbon chain, since the CH$_x$ bonds have vibration peaks in this zone. We observed small changes in peak intensities for annealed and non-annealed HSQ-MIBK, suggesting structural changes in the carbon chain during the annealing as well as in the drying process.

The FTIR spectrum of in-HSQ modified with acetone is shown in the ESI in Fig. S3† and that of in-HSQ modified with ethyl acetate is shown in Fig. S4.† The spectra indicate that again, the main peaks of HSQ and modified HSQ before annealing are quite similar, suggesting that the HSQ is still preserved. Interesting shifts of peak for the –(C=O)– group were observed: the wavenumber shifted from 1750 cm^{-1} in pure acetone to 1730 cm^{-1} before annealing, and to 1735 cm^{-1} after annealing of in-HSQ-acetone (see Fig. S3†). For ethyl acetate modified in-HSQ (Fig. S4†) we observed splitting of this peak into 1712 cm^{-1} and 1727 cm^{-1} bands for non-annealed in-HSQ-ethyl acetate which shifted to wavenumbers of 1710 cm^{-1} and 1735 cm^{-1} for the annealed sample. Whereas in the original ethyl acetate molecule this vibration had a wavenumber of 1742 cm^{-1}. That led us to the suspicion that this molecule can bind to HSQ *via* more than one binding site, which is also supported by the presence of a peak at around 3400 cm^{-1} which is typical for vibrations of –OH groups. The suspicion is supported by single dot spectroscopy, where we observed broader full width at half maxima (FWHM) PL peaks from single Si-NCs prepared from in-HSQ-ethyl acetate.

The peak at around 3400 cm^{-1} was observed for all HSQ modified with MIBK, acetone and ethyl acetate, but this peak was not present in pure HSQ or the pure modifying molecules, and therefore we are presuming that the creation of an –OH bond is a result of bonding of the modification molecules to the HSQ polymer. The main conclusion from these FTIR results is that the HSQ-modifying organic molecules still remain in some form in the sample after annealing. Based on the FTIR results, we hypothesise that Si–O–(C=O)–R,[36] where R stands for a carbon chain, was created on the surface of the Si-NCs and formed a preceramic-like shell as discussed below.

TEM study of Si-NCs prepared from in-HSQ-MIBK

For TEM observations, Si-NCs were released from the in-HSQ-MIBK sample using HF vapours. The sample on a Si-wafer was held for 3 seconds above a droplet of

HF, and the disruption of the film was clearly seen by the naked eye. Subsequently, the sample was collected from the wafer with a droplet of ethanol. The Si-NCs with chemical residua of the film were transferred onto a ~20 nm silicon nitride membrane and were studied using a bright-field TEM JEM-2100F microscope. Fig. 2a, which is a forward Fourier transform of the area with a nanoparticle image from Fig. S1b,† clearly shows peaks corresponding to Si {2$\bar{2}$0} (5.2 nm^{-1}) and {111} (3.2 nm^{-1}) crystallographic planes, confirming the presence of silicon nanocrystallites (weak peaks corresponding to {311} (6.1 nm^{-1}) planes can also be discerned). Another high resolution TEM image with Si {111} (0.31 nm) lattice fringes more clearly visible is presented in Fig. 2b.

For most of the individually standing small Si-NCs, we could also observe a thin shell (1–2 nm) around the nanocrystal core (Fig. 2c and d). The bright-field contrast is believed to be the atomic number contrast of the Si core with the shell consisting of the light O, C and H elements, as discussed below. As we can see from TEM images, the nanocrystal core is of the size of around 2–4 nm in diameter and the shell thickness is around 1–2 nm.

Pre-ceramic shell formation on Si-NCs during annealing of modified HSQ

The HSQ polymer is stable up to 900 °C,[25,28] when annealed in a slightly reducing 5% H$_2$: 95% Ar atmosphere. But as we observed from FTIR spectra of the modified in-HSQ before and after annealing (see Fig. 1, S3 and S4†) the organic character of the sample seems to be preserved even after annealing. Since the annealing of our samples was done at a temperature of 1000 °C, one would expect that all of the

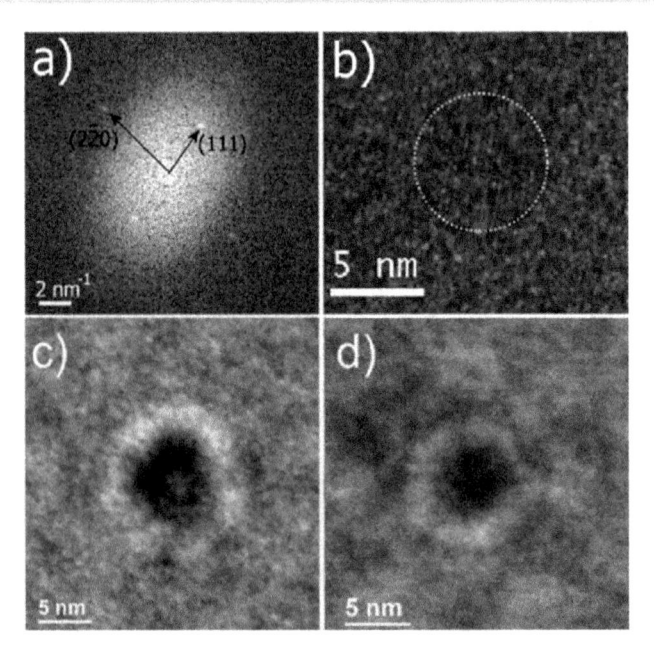

Fig. 2 Bright field TEM images of Si-NCs released from the matrix: (a) Fourier transform of a HRTEM image showing {220} and {111} Si planes from a typical nanoparticle in S1b;† (b) a high resolution TEM Si-NC image with Si {111} plane lattice fringes visible, (c and d) examples of Si-NCs showing a thin shell present on the silicon crystalline core.

organic molecules would be fully carbonized, but our results contradict this. When we investigated the literature for annealing of organosilanols and silicon based ceramics and glass of less complex polymers than HSQ, we found supporting evidence for the persistence of organic molecules even at high annealing temperatures. The persistence of organic molecules at higher annealing temperatures actually scales up with the complexity of the organosilanol molecules used as carriers. For linear Si–O structures, like single standing TEOS molecules, the organic molecules are preserved up to temperatures of 400 °C.[38,39] For more complex linear structures, which contain Si–O ring structures, the organic character of the sample is observed for annealing temperatures up to 600 °C.[39–43] Finally for Si–O structures arranged in an intertangled 3D structure, which are slightly less ordered than HSQ, the organic character is preserved at temperatures up to 700–800 °C (ref. 42–45) or even in some cases close to 1000 °C.[46,47] For each of those Si–O structures, the loss of organic character corresponds to the temperature of disintegration of the carrier Si–O structure. As was mentioned above, HSQ is fairly stable up to a temperature of 1000 °C in an inert atmosphere, therefore the organic molecules can be protected in the HSQ polymer at lower temperatures and they get released only when the HSQ starts to disintegrate.

In an ideal case, the individual HSQ units are closely interconnected, forming a well-arranged structure consisting of a minimum of 50 or more HSQ units. For the explanation of the formation of Si-NCs, we will address this well arranged HSQ as a HSQ parcel. When a Si-NC is being created from the HSQ parcel, the oxygen atoms present in the HSQ molecule are pushed to the border of the HSQ parcel, while silicon atoms cluster in the centre of the HSQ parcel and form the crystalline silicon core of the Si-NC. For a newly created Si-NC, there are approximately 4–8 extra oxygen atoms for each Si surface atom of the newly created Si-NCs, which have been pushed out of the centre of the HSQ parcel. Under normal annealing conditions of nonmodified HSQ, they will create silicon oxide surface bonds[23] on a Si-NC and form a thick SiO_2 matrix, but it seems the presence of MIBK changes this situation dramatically.

It is energetically unfavourable to have MIBK in the centre of a Si crystal, and therefore, these molecules were most probably pushed together with the excess oxygen to the surface of newly created Si-NCs. It should be noted that sources of oxygen were limited, and it originated solely from HSQ. Therefore, not all of the MIBK molecules were carbonized in the annealing process in a slightly reducing 5% H_2 : 95% Ar atmosphere. This also means that the reaction was self-controlled. Since there was a surplus of MIBK molecules for each Si surface bond, the non-carbonized MIBK molecules could bond to the Si surface atoms of a new Si-NC. The MIBK could bond to Si-NCs *via* one oxygen atom and a Si–O-ketone chain surface layer was probably formed. The presence of such bonds is supported by FTIR (all typical frequencies are present). Also, Si–O–C groups are typically present in silicon oxide ceramics,[35,43,48] or silicon oxycarbide glasses.[41,42,46,47] According to the literature[35] this bond can be created at temperatures of 500 °C and ceramic films containing these groups have high thermal and mechanical stability.[35] The annealing time of our samples was quite long, so the system had sufficient time for creation of spherical nanoparticles with homogenous surface shells (see Fig. 2).

Based on this hypothesis, we anticipated that the shell on the Si-NCs would be even thinner when acetone was used for modification of HSQ and, similarly, that the shell would be thicker when ethyl acetate was used. The FTIR study of the in-HSQ-ethyl acetate indicates that the bonding was inhomogeneous which is reflected by the inhomogeneous PL FWHM of individual Si-NCs.

The amount of Si-NCs created from the modified in-HSQ is not large, since the in-HSQ polymer was not perfect, and a reasonably big zone of good quality in-HSQ is needed for the creation of core–shell Si-NCs. The drying procedure of the thin layers seems to play an important role in core–shell Si-NC formation, since even fewer nanocrystals were produced when we tried to anneal bulk modified HSQ. Unfortunately, the current low production yields of these samples, prevents their mass characterization. Despite this fact, their narrow PL FWHMs at room temperature are unprecedented, as shown below.

Since we annealed organic molecules which can serve as a source of carbon, we should address the question of the presence of carbon dots in our sample. We cannot exclude the presence of carbon dots based on the FTIR spectra, but the optical measurements described below point out that carbon dots are not present.

Optical measurements of individual nanocrystals

The used micro-photoluminescence setup can be used to measure photoluminescence spectra of individual quantum dots or fluorophores at room and low temperatures. It consists of an optical inverted microscope system (Zeiss Axio observer Z1m) connected to a spectrograph (Andor Shamrock 500i) and an EMCCD camera (Andor, iXon3). The microscope is equipped with a micromanipulator ITGCA2 for precise positioning of the sample. The sample is excited by a continuous wave laser with an emission wavelength of 405 nm. The laser beam is directed at the sample in a dark field geometry and the emission signal is collected using an objective lens and led through a set of filters and through a spectrograph to the camera. For room temperature measurements, an objective Nikon EC Epiplan Neofluar 100x/0.70 NA was used. The measurements were controlled through Andor Solis 4.22 software and the data were then processed in Origin8. All spectra have been corrected for the system response and background.

The luminescence signal from the in-HSQ-MIBK Si-NCs exhibited similarities with that of individual Si-NCs in a silicon oxide shell. Firstly, the effect of blinking was observed as shown in Fig. 3, where we can clearly distinguish ON and OFF states from the PL intensity time trace. The blinking trace was extracted from a series of photoluminescence images with a time bin of 2 s. Secondly, the photoluminescence peak positions of the in-HSQ-MIBK Si-NCs were found to vary from 1.7 eV to 2.4 eV with the majority of dots having a peak position around 2 eV, as can be seen in Fig. 4. The spectral range of the in-HSQ-MIBK Si-NCs corresponds to the spectral range of Si-NCs prepared by other methods, such as thermal oxidation.[15]

At the same time, we observed green emitting nanocrystals at the tail of the distribution (see Fig. 4), which suggests a different surface passivation of the Si-NCs than a silicon oxide layer. Usually green emission is observed for Si-NCs with hydrogen bonds on the surface but from FTIR it is clear that the corresponding frequency of Si–H (2250 cm^{-1}) was missing after annealing (see Fig. 1). According to the literature, Si-NCs with hydrogen bonds on the surface are green/blue emitting but immediately after exposure to oxygen they shift to red emission due to rapid oxidation.[49] A wide range of PL emission of individual Si-NCs also

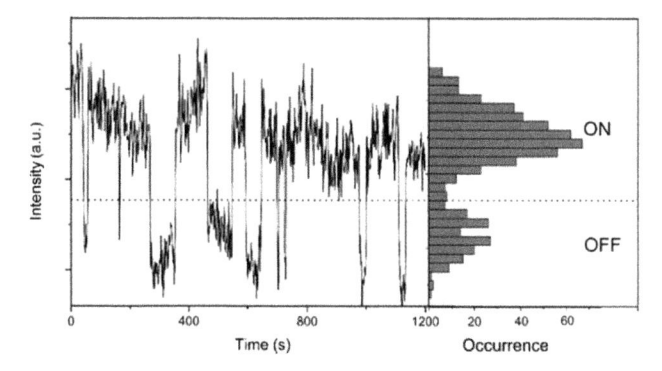

Fig. 3 Blinking trace of a in-HSQ-MIBK Si-NC exhibiting ON–OFF levels and statistics of intensity occurrence.

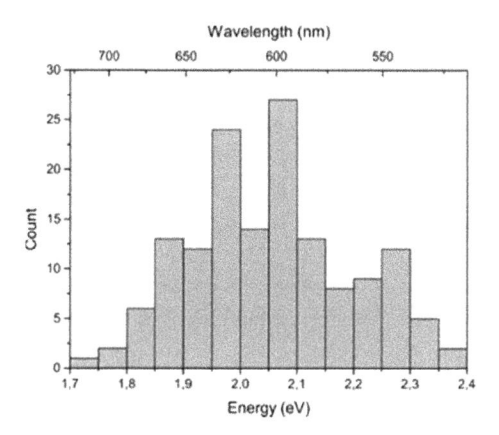

Fig. 4 Statistics of photoluminescence peak positions of individual MIBK-Si-NCs at room temperature.

suggests variation in size of the Si-NCs.[50] Prolongation of the annealing time with an increase of HSQ parcels will lead to bigger Si-NCs and a change in emission wavelengths to higher values.[23]

Examples of individual spectra from in-HSQ-MIBK Si-NCs are shown in Fig. 5. The spectra of individual Si-NCs have the same shape as those for Si-NCs embedded in oxide shells[15] but the linewidth clearly appears to be different *i.e.* much narrower.

The most significant change in the Si-NCs prepared from the in-HSQ-MIBK is their FWHM. If we look at the FWHM distribution in Fig. 6 and fit it with a Gaussian function, we obtain a FWHM of around 21 meV. This is significantly narrower than the linewidth of Si-NCs in a silicon oxide matrix (~150 meV)[15] and even narrower than the linewidth that was observed in the work of Kusova *et al.*,[14] in which they capped the surface of oxide passivated Si-NCs with methyl groups in xylene-based suspensions *via* photochemical reaction, with the resulting linewidth of single Si-NCs around 100 meV.

Another significant observation arose from the single dot PL measurements of Si-NCs prepared from in-HSQ-MIBK at 70 K.[51] In all of the studied Si-NCs prepared

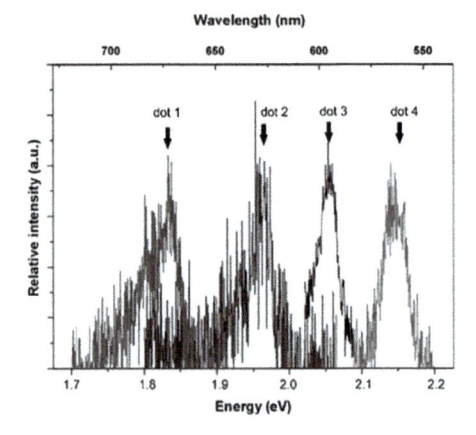

Fig. 5 Photoluminescence of single silicon nanocrystals prepared from in-HSQ modified by MIBK, measured at room temperature, with excitation of 405 nm.

from modified HSQ, we did not observe any TO or LO phonon lines, which are partially responsible for the broadening of PL FWHM of individual Si-NCs in a SiO$_2$ matrix. In our previous article,[51] we attributed the narrowing of the FWHM to the fact that the mechanical coupling of a nanocrystal to its matrix largely influences the PL linewidth due to exciton–phonon interaction. Our results unequivocally showed that previous theoretical models predicting necessary broadening due to the indirect band-gap structure of silicon were incorrect, having general implications for modelling and applications of silicon nanostructures.[51]

Based on the exciton–phonon coupling model, the thinner the shell is of a single Si-NC, the narrower the PL linewidth will be. This was confirmed by a single dot FWHM study of Si-NCs prepared from in-HSQ modified with acetone since we observed an even narrower linewidth for individual Si-NCs of around 17

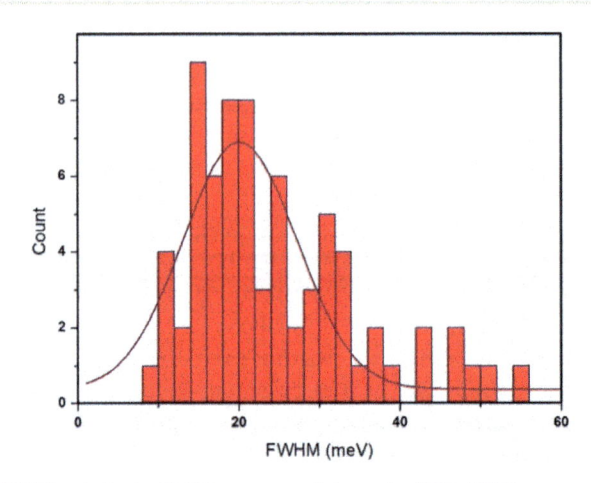

Fig. 6 PL FWHM of single Si-NCs prepared from in-HSQ-MIBK, measured at room temperature.

Table 1 Optical parameters of Si-NCs prepared from in-HSQ modified with acetone, methyl isobutyl ketone or ethyl acetate (for detailed data see ESI Fig. S5 and Fig. S6)

HSQ modifying molecule	FWHM (meV)	Central emission wavelength (eV)
Acetone	17	1.93
Methyl isobutyl ketone	21	1.97
Ethyl acetate	32	2.00

meV (for FWHM distribution and PL peak positions see the ESI Fig. S5,† where the narrowest dots had a FWHM of 10 meV, *i.e.* 3 nm at room temperature). In the case of the Si-NCs prepared from in-HSQ modified with ethyl acetate, the PL FWHM was around 32 meV (see ESI Fig. S6†), which is more than the values for the other two molecules used in this study. The FTIR results for the modification with ethyl acetate show the chemical non-uniformity of the surface shell which is also reflected by a broadening of the linewidth. The overview of the PL properties for all modifications used in this article are summarized in Table 1.

All of the single dot PL spectra of the Si-NCs were taken from the annealed modified in-HSQ layer, since they exhibited the same spectral features as single standing Si-NCs released from the film. It was just convenient in terms of time to measure Si-NCs in the non-disrupted sample layer, since we captured more single Si-NCs (up to 20 spectrally well resolved) within one PL measurement experiment. Spectral hopping was observed for around 3% of all PL spectra of individual Si-NCs. The FWHM of both PL peaks was the same within statistical error and the distance between the PL peaks was in the range of 0.03–0.09 eV. The spectral hopping was also observed in the case of the Si-NCs with a silicon oxide shell and it is considered as supporting evidence for the presence of a single PL emitting nanoparticle.

Since we annealed carbon rich samples, we compared the PL features of our core–shell Si-NCs to those of carbon dots. The FWHM of the single Si-NCs prepared from modified HSQ (see Table 1) is much narrower than that of carbon dots, whose reported narrowest FWHM is around 75 meV (ref. 17) with the majority having a linewidth > 100 meV.[17–19] It is also known that carbon dots, upon high excitation power, tend to have lower PL intensities[52] whereas our Si-NCs prepared from in-HSQ-MIBK showed an increase in PL even when illuminated with a 6200 W cm^{-2} 405 nm continuous laser[53] and showed the same dependence of PL on the excitation intensity as Si-NCs embedded in SiO_2. Our study,[51] in which we compared the PL behaviour of single Si-NCs in SiO_2, Si-NCs passivated with dodecane, and Si-NCs prepared from in-HSQ-MIBK at various temperatures, showed that the Si-NCs prepared from in-HSQ-MIBK follow the same trends as the other Si-NCs for lower temperatures and the main differences in the various FWHMs were solely attributed to their surface passivated layers. These optical measurements strongly support that we indeed observed Si-NCs with a core–shell layer structure.

Conclusions

We report a one step preparation technique for Si-NCs with a thin pre-ceramic shell. The individual Si-NCs showed ultranarrow luminescence linewidths at room temperature (with average FWHMs as follows for single Si-NCs prepared from in-HSQ modified with acetone ∼17 meV, MIBK ∼21 meV, and ethyl acetate ∼32 meV), which were significantly less than for other silicon nanocrystals and even CdSe quantum dots and carbon dots and, to our knowledge, are the narrowest linewidths reported to date for any quantum dot under ambient conditions. PL blinking and spectral hopping in individual nanocrystals were also detected, confirming that they were individual separated emitters. The FTIR indicated that the organic character of the samples was preserved during annealing. The Si-NCs prepared from in-HSQ-MIBK did not show any deterioration of luminescence for at least 4 years. One step preparation of silicon nanocrystals with a thin shell is the first step towards an inexpensive and accessible material for applications in solar cells and LEDs. Si-NCs with a thin shell featuring a narrow spectral line might also be found useful in biology as fluorescent markers as an alternative to toxic CdSe quantum dots.

Conflicts of interest

There are no conflicts to declare.

Acknowledgements

A. Fucikova acknowledges funding from the GACR 18-07977Y.

Notes and references

1 L. T. Canham, *Appl. Phys. Lett.*, 1990, **57**, 1046–1048.
2 S. Tiwari, F. Rana, H. Hanafi, A. Hartstein, E. F. Crabbé and K. Chan, *Appl. Phys. Lett.*, 1996, **68**, 1377.
3 M. Sailor and E. Lee, *Adv. Mater.*, 1997, **9**(10), 783–793.
4 A. Fucikova, J. Valenta, I. Pelant and V. Brezina, *Chem. Pap.*, 2009, **63**(6), 704–708.
5 A. Fucikova, J. Valenta, I. Pelant, M. H. Kalbacova, A. Broz, B. Rezek, A. Kromka and Z. Bakaeva, *RSC Adv.*, 2014, **4**(20), 10334–10342.
6 Z. Li and E. Ruckenstein, *Nano Lett.*, 2004, **4**(8), 1463–1467.
7 R. Walters, H. Atwater and G. Bourianoff, *Nat. Mater.*, 2005, **4**, 143–146.
8 Ch.-Y. Liu, Z. C. Holman and U. R. Kortshagen, *Nano Lett.*, 2009, **9**(1), 449–452.
9 D. Jurbergs, E. Rogojina, L. Mangolini and U. Kortshagen, *Appl. Phys. Lett.*, 2006, **88**, 233116.
10 F. Sangghaleh, I. Sychugov, Z. Yang, J. G. C. Veinot and J. Linnros, *ACS Nano*, 2015, **9**(7), 7097–7104.
11 M. A. Islam, M. H. Mobarok, R. Sinelnikov, T. K. Purkait and J. G. C. Veinot, *Langmuir*, 2017, **33**(35), 8766–8773.
12 J. Valenta, A. Fucikova, F. Vácha, F. Adamec, J. Humpolíčková, M. Hof, I. Pelant, K. Kůsová, K. Dohnalová and J. Linnros, *Adv. Funct. Mater.*, 2008, **18**, 2666–2672.
13 J. Valenta and I. Pelant, *Luminescence Spectroscopy of Semiconductors*, OUP Oxford, England, 2012.

14 K. Kůsová, O. Cibulka, K. Dohnalová, I. Pelant, J. Valenta, A. Fučíková, K. Žídek, J. Lang, J. Englich, P. Matějka, P. Štěpánek and S. Bakardjieva, *ACS Nano*, 2010, **4**(8), 4495–4504.

15 I. Sychugov, R. Juhasz, J. Valenta and J. Linnros, *Phys. Rev. Lett.*, 2005, **94**, 087405.

16 G. Schlegel, J. Bohenberger, I. Potapova and A. Mews, *Phys. Rev. Lett.*, 2002, **88**, 137401.

17 B. van Dam, H. Nie, B. Ju, E. Marino, J. M. J. Paulusse, P. Schall, M. Li and K. Dohnalová, *Small*, 2017, **13**, 1702098.

18 S. Ghosh, A. M. Chizhik, N. Karedla, M. O. Dekaliuk, I. Gregor, H. Schuhmann, M. Seibt, K. Bodensiek, I. A. T. Schaap and O. Schulz, *Nano Lett.*, 2014, **14**(10), 5656–5661.

19 X. Li, Z. Zhao and Ch. Pan, *Chem. Commun.*, 2016, **52**, 9406.

20 M. Falconieri, E. Borsella, R. D'Amato, F. Fabbri, E. Trave, V. Bello, G. Mattei, Y. Nie and D. Wang, *AIP Conf. Proc.*, 2010, **1275**, 50–57.

21 U. R. Kortshagen, R. M. Sankaran, R. N. Pereira, S. L. Girshick, J. J. Wu and E. S. Aydil, *Chem. Rev.*, 2016, **116**(18), 11061–11127.

22 G. B. Teh, N. Saravanan, R. D. Tilley, S. Ramesh and Y. S. Lim, *Z. Phys. Chem.*, 2009, **223**, 1417–1426.

23 C. M. Hessel, E. J. Henderson and J. G. C. Veinot, *Chem. Mater.*, 2006, **18**, 6139–6146.

24 Y. Kawakami, *React. Funct. Polym.*, 2007, **67**, 1137–1147.

25 R. H. Baney, M. I. A. Sakakibara and T. Suzuki, *Chem. Rev.*, 1995, **95**, 1409–1430.

26 E. J. Henderson, J. A. Kelly and J. G. C. Veinot, *Chem. Mater.*, 2009, **21**(22), 5426–5434.

27 T. K. Purkait, M. Iqbal, M. A. Islam, M. H. Mobarok, Ch. M. Gonzalez, L. Hadidi and J. G. C. Veinot, *J. Am. Chem. Soc.*, 2016, **138**, 7114–7120.

28 J. A. Kelly, E. J. Henderson and J. G. C. Veinot, *Chem. Commun.*, 2010, **46**(46), 8704–8718.

29 A. Angı, M. Loch, R. Sinelnikov, J. G. C. Veinot, M. Becherer, P. Lugli and B. Rieger, *Nanoscale*, 2018, **10**(22), 10337–10342.

30 B. O. Dabbousi, J. V. Rodriguez-Viejo, F. V. Mikulec, J. R. Heine, H. Mattoussi, R. Ober, K. F. Jensen and M. G. Bawendi, *J. Phys. Chem. B*, 1997, **101**(46), 9463–9475.

31 D. A. Hanifi, N. D. Bronstein, B. A. Koscher, Z. Nett, J. K. Swabeck, K. Takano, A. M. Schwartzberg, L. Maserati, K. Vandewal, Y. van de Burgt, A. Salleo and A. P. Alivisatos, *Science*, 2019, **363**, 1199–1202.

32 A. Coyopol, G. García-Salgado, T. Díaz-Becerril, H. Juárez, E. Rosendo, R. López, M. Pacio, J. A. Luna-López and J. Carrillo-López, *J. Nanomater.*, 2012, 368268, 7.

33 D. R. Anderson in *Analysis of Silicones*, ed. A. Lee Smith, Wiley-Interscience, New York, 1974, ch. 10.

34 L. J. Bellamy, *The infrared spectra of complex molecules*, Chapman and Hall, London, 3rd edn, 1975, ch. 20.

35 S. Jing, H.-J. Lee and C. K. Choi, *J. Korean Phys. Soc.*, 2002, **41**(5), 769–773.

36 R. Anderson, B. Arkles and G. L. Larson, *Silicon Compounds: Register and Review*, Petrarch Systems, Inc., Bristol, PA, 4th edn, S-7, 1987.

37 P. Linstrom, *NIST Chemistry WebBook, CAS Registry Number: 108-10-1*, National Institute of Standards and Technology, 1997.
38 P. Colombo, T. E. Paulson and C. G. Pantano, *J. Am. Ceram. Soc.*, 1997, **80**(9), 2333–2340.
39 F. Babonneau, L. Bois and J. Livage, *J. Non-Cryst. Solids*, 1992, **147**, 280–284.
40 X. Wang, F. Schmidt, D. Hanaor, P. H. Kamm, S. Li and A. Gurlo, *Addit. Manuf.*, 2019, **27**, 80–90.
41 A. K. Singh and C. G. Pantano, *J. Sol-Gel Sci. Technol.*, 1997, **8**, 371–376.
42 B. F. Sousa, I. V. P. Yoshida, J. L. Ferrari and M. A. Schiavon, *J. Mater. Sci.*, 2013, **48**, 1911–1919.
43 Y. Tanaka, C. Mori, N. Suzuki, T. Kasai, K. Iimura and T. Kato, Characterization of Si–O–C ceramics prepared by the pyrolysis of phenyl silicones, *Studies in Surface Science and Catalysis*, ed. Y. Iwasawa and N. O. H. Kunieda, Elsevier Science B. V., 2001, vol. 132, pp. 777–780.
44 D. S. Wang and C. G. Pantano, *J. Non-Cryst. Solids*, 1992, **147**, 115–122.
45 H. Bréquel, J. Parmentier, S. Walter, R. Badheka, G. Trimmel, S. Masse, J. Latournerie, P. Dempsey, C. Turquat, A. Desmartin-Chomel, L. Le Neindre-Prum, U. A. Jayasooriya, D. Hourlier, H.-J. Kleebe, G. D. Sorarù, S. Enzo and F. Babonneau, *Chem. Mater.*, 2004, **16**(13), 2585–2598.
46 C. G. Pantano, A. K. Singh and H. Zhang, *J. Sol-Gel Sci. Technol.*, 1999, **14**(1), 7–25.
47 J. V. Ryan and C. G. Pantano, *J. Vac. Sci. Technol., A*, 2007, **25**, 153–159.
48 *Handbook of Advanced Ceramics: Materials, Applications, Processing, and Properties*, ed. S. Somiya, Academic Press, 2nd edn, 2013, pp. 1025–1101.
49 Y. Kanemitsu, H. Mimura, T. Matsumoto and T. Nakamurad, *J. Lumin.*, 1997, **73–74**, 344–346.
50 G. Ledoux, J. Gong, F. Huisken, O. Guillois and C. Reynaud, *Appl. Phys. Lett.*, 2002, **80**, 4834.
51 I. Sychugov, A. Fucikova, F. Pevere, Z. Yang, J. G. C. Veinot and J. Linnros, *ACS Photonics*, 2014, **1**(10), 998–1005.
52 S. J. Zhu, Q. N. Meng, L. Wang, J. H. Zhang, Y. B. Song, H. Jin, K. Zhang, H. Sun, H. C. Wang and B. Yang, *Angew. Chem., Int. Ed.*, 2013, **52**, 3953–3957.
53 F. Pevere, I. Sychugov, F. Sangghaleh, A. Fucikova and J. Linnros, *J. Phys. Chem. C*, 2015, **119**(13), 7499–7505.

Faraday Discussions

PAPER

Critical assessment of wet-chemical oxidation synthesis of silicon quantum dots†

Jonathan L. Wilbrink,[abc] Chia-Ching Huang,[b] Katerina Dohnalova [bc] and Jos M. J. Paulusse [*ac]

Received 1st October 2019, Accepted 1st November 2019
DOI: 10.1039/c9fd00099b

Wet-chemical synthetic procedures are powerful strategies to afford fluorescent silicon quantum dots (Si QDs) in a versatile and scalable manner. However, development of Si QDs is still hampered by a lack of control over photoluminescence emission, in addition to synthesis and characterization complexities. The wet-chemical Si QD synthesis by oxidation of magnesium silicide (Mg_2Si) with bromine (Br_2) was revisited and a control reaction was carried out where the silicon source was omitted. Both reaction conditions resulted in substantial quantities of fluorescent material. Moreover, a comparative analysis of their optical properties (UV-vis/fluorescence) revealed no apparent differences. Other characterization techniques also confirmed the resemblance of the two materials as ^{1}H NMR, FTIR and XPS spectra were nearly identical for both samples. Elemental analysis revealed the presence of only 2 wt% silicon in the Si QD sample. No evidence was found for the formation of significant amounts of Si QDs *via* this wet-chemical procedure.

Introduction

Quantum dots (QDs), also coined semiconductor nanocrystals, are nanoparticles that possess unique properties on the quantum level that differ from bulk properties.[1,2] Depending on the particle size and shape, QDs display photoluminescence, following the quantum confinement model.[1] As a consequence, by varying QD dimensions, emission can be tuned over a broad spectral range. Owing to their superior optical properties, such as size-dependent fluorescence,[3–5] color definition[6] and resistance against photobleaching,[7,8] QDs are promising materials for a wide variety of applications, including photovoltaics,[9–11] lighting

[a]*Department of Biomolecular Nanotechnology, MESA+ Institute for Nanotechnology, TechMed Institute for Health and Biomedical Technologies, Faculty of Science and Technology, University of Twente, P.O. Box 217, Enschede, 7500 AE, The Netherlands. E-mail: J.M.J.Paulusse@utwente.nl*
[b]*Institute of Physics, University of Amsterdam, Science Park 904, 1098 XH Amsterdam, The Netherlands*
[c]*SpectriS-dot b.v., Wilsonstraat 34, 2131 PT Hoofddorp, The Netherlands*
† Electronic supplementary information (ESI) available. See DOI: 10.1039/c9fd00099b

and displays,[6,12] and biomedical imaging.[13-19] This application potential is, however, severely limited by the fact that conventional QDs typically contain toxic and/or scarce elements (*e.g.* Cd, In, Pb, Se or Te). Silicon quantum dots (Si QDs) signify an appealing, non-toxic and highly sustainable alternative.[20-22]

Synthesis strategies for Si QDs are typically divided into top-down and bottom-up approaches (and combinations).[16] In top-down approaches, Si QDs are prepared from larger macroscopic materials, *e.g.* silicon wafers, to construct nanoparticles, which can be done mechanically,[23] as well as by etching with strong acids, such as HF.[24] In addition, the group of Veinot and coworkers developed an effective route based on the thermal decomposition of hydrogen-terminated silsesquioxane to produce Si QDs within a silica matrix, followed by liberation of the Si QDs by HF etching.[4,25,26]

In bottom-up approaches, Si QDs are prepared from molecular precursors. This includes syntheses by physical means, such as condensation of silane gas into Si QDs *via* laser pyrolysis[27,28] or plasma synthesis.[3,29,30] Other strategies that have been reported to yield Si QDs include decomposition of precursors in supercritical fluids[31] and by microwave irradiation.[32-34] Bottom-up approaches by chemical means are typically wet-chemical syntheses, where reactions are carried out in solution-phase.[16,21,35,36] Strategies include reduction of $SiCl_4$ (or $SiBr_4$, and sometimes co-reaction with alkyltrichlorosilanes) using reducing agents, such as alkali metals (*e.g.* Na, K),[37-39] alkaline earth metals (*e.g.* Mg),[40] sodium naphthalide[41,42] or $LiAlH_4$,[43-49] and in molten salt syntheses using active Al species.[50,51] Synthetic routes towards Si QDs have also been based on the oxidation of silicon species, typically Zintl salts (XSi_n, where X is an alkali metal or alkaline earth metal). Common methods are the oxidation of these Zintl salts by Br_2,[52-56] NH_4Br,[34,57-59] or NH_4Cl.[60,61] Furthermore, Zintl salts have been reacted with $SiCl_4$ to yield Si QDs in a metathesis type synthesis.[62-64] Regarding synthesis strategies, several reviews are recommended for a more complete and detailed overview.[16,21,36,65]

Wet-chemical synthesis is generally more versatile, less laborious and highly scalable, while elevated temperatures can be avoided, although limited control over fluorescence emission remains a major challenge.[16] Moreover, the number of reliable wet-chemical synthetic strategies towards Si QDs is still limited, with complications concerning synthesis and characterization hampering research into Si QDs, as well as their commercialization.

Fluorescence emission generally correlates well with Si QD size,[66] although discrepancies between Si QD size and emission have been reported.[25] In particular for wet-chemical syntheses, primarily blue or green fluorescence emission is observed, regardless of the reported size. There is still debate as to whether quantum confinement is in fact the origin of photoluminescence in Si QDs.[67-69] Bulk silicon is an indirect bandgap material,[70] and even on the nanoscale it is often implicated that fluorescence emission is either inefficient or mostly governed by surface defects rather than quantum confinement effects.[9,71] Moreover, such defects are frequently reported,[9,69,71,72] *e.g.* due to nitrogen-containing species on Si QD surfaces[73] or particle oxidation.[72,74] Interestingly, even two seemingly comparable Si QD samples, albeit synthesized *via* different procedures, displayed very different optical properties, *i.e.* blue and red emission.[75] These differences were explained by the occurrence of nitrogen-containing impurities in the blue-emitting Si QDs. Emission of Si QDs was demonstrated by Wiggers and coworkers to change from red to blue upon exposure to air, confirming that blue fluorescence can be induced by oxidation.[74] Surface-related fluorescence is typically undesirable, as the emission is no longer tunable by adjusting the particle size, and such particles are more susceptible to photobleaching.[7] Although photoluminescence is frequently assigned to surface defects when the quantum confinement model cannot be applied,[71,75] this is not necessarily true in certain cases, as discussed below.

Interestingly, side reactions occurring during Si QD synthesis may also induce the (majority of) fluorescence and obscure results. Recently, Oliinyk *et al.* criticized[76] the preparation of Si QDs *via* a one-pot microwave synthesis employing 3-aminopropyltrimethoxysilane (APTMS) as the silicon source, as reported earlier by He and coworkers.[32] Their criticism focused on discrepancies concerning the XRD data when comparing with other Si QD literature,[77] the absence of XPS data, TEM images showing crystalline material that is inconsistent with silicon, and most strikingly, very comparable optical properties for materials prepared in the presence, as well as in the absence of the silicon source.

Purification of fluorescent nanomaterials has also been pointed out to be critical.[78–80] For instance, fluorescent by-products (*e.g.* small molecules) were frequently insufficiently removed.[78–82] Crystalline nanoparticles of molybdenum sulfide, as observed by TEM, were reported to display strong fluorescence emission,[83] although more recently, smaller impurities were demonstrated to be the actual source of fluorescence.[84] Hence, connecting the optical properties to a nanomaterial remains a considerable challenge, as also pointed out earlier by Nandi and coworkers,[80] who particularly stressed the importance of careful interpretation of TEM data.[79,85]

Our investigations into the preparation of Si QDs *via* the oxidation of Mg_2Si with Br_2, a well-established, facile reaction generally resulting in high product yields,[52,53] revealed that attempts to alter the optical properties by changing the reaction parameters, such as temperature, reaction time, concentration and reagents, remained unsuccessful. A lack of size-control (and thus control over the emission) was noted earlier, but remained unexplained.[52] This prompted us to investigate this reaction in more detail. An in-depth comparative study on these Si QDs is presented here, and a comparison is made with material resulting from a silicon-free synthesis, where the Zintl salt was omitted. Both materials were characterized using a number of different techniques, including ^{1}H-NMR spectroscopy, FTIR, TEM, elemental analysis, XPS, fluorescence spectroscopy, UV-vis spectroscopy, and SEC. Both products possessed highly comparable properties, hinting towards a mechanism that presumably does not (exclusively) produce Si QDs, but primarily fluorescent by-products.

Experimental

Materials and methods

Syntheses were carried out under a nitrogen atmosphere, unless stated otherwise. Glassware was dried before use. Solvents were dried over 4 Å molecular sieves for at least 24 h before use.

n-Butyllithium (*n*-BuLi, 2.5 M, in hexanes), chloroform-d (99.8 atom% D) and *n*-octane (≥97%) were obtained from Acros, bromine (Br_2, ≥99.99% trace metals basis), 9,10-diphenylanthracene, hydrochloric acid (HCl, 37%), magnesium silicide (Mg$_2$Si, ≥99% trace metals basis, −20 mesh), molecular sieves (0.3 nm, rods), sodium sulfate (Na$_2$SO$_4$, anhydrous), and sodium thiosulfate (Na$_2$S$_2$O$_3$, ≥98%, anhydrous) were obtained from Sigma Aldrich, chloroform, ethanol, ethyl acetate and methanol were obtained from VWR, TEM grids were obtained from EM Resolutions (Sheffield, United Kingdom), and XPS substrates from SSENS (Hengelo, The Netherlands).

Characterization

Nuclear magnetic resonance (NMR) spectroscopy was performed on a Bruker 400 MHz or Bruker 600 MHz spectrometer. Samples were measured in chloroform-d. Attenuated total reflectance (ATR) Fourier-transform infrared (FTIR) spectra were obtained by drop-casting samples from chloroform solution on a Bruker Alpha-p. Elemental analysis for carbon was performed on a Vario Micro Cube CHNS-Analyser from Elementar, for bromine on an 883 Plus IC from Metrohm, and for silicon on a Specord 50 Plus UV/vis spectrophotometer from Analytik Jena. Transmission electron microscopy (TEM) was performed on a Philips CM300ST-FEG transmission electron microscope operated at 300 kV. Samples were drop-cast from a dilute chloroform solution on holey carbon films on a 200 mesh copper substrate. X-ray photoelectron spectroscopy (XPS) was measured on a PHI Quantera SXM with a monochromatic Al Kα source (1486.6 eV). The carbon C 1s signal at 284.8 eV was taken as the reference binding energy. Samples were drop-cast from a chloroform solution on gold-coated glass substrates (20 nm gold). UV-vis spectra were recorded on a Shimadzu UV-2401PC UV-vis spectrophotometer. Absorbance was kept below 0.05 for quantum yield determinations to avoid inner filter effects. Ethyl acetate was employed as the solvent for UV-vis and fluorescence spectroscopy. Fluorescence was measured on a Varian Cary Eclipse fluorescence spectrophotometer. Size exclusion chromatography (SEC) measurements were performed on a Waters Alliance e2696, equipped with a 2475 fluorescence detector, and a 2998 photodiode array detector to determine absorbance. Samples were run on an Agilent PLgel 5 μm MIXED-D column, using HPLC-grade chloroform as the eluent. Before measurement, samples were filtered on 0.2 μm Whatman Spartan syringe filters.

Synthesis of Si QDs

Si QDs were synthesized, with only minor changes, according to earlier literature procedures.[52,53] Degassed *n*-octane (100 mL, 0.615 mol) was added to Mg$_2$Si (100 mg, 1.30 mmol), followed by the addition of Br$_2$ (0.54 mL, 10.5 mmol). Subsequently, the dispersion was stirred for 2 h, whereupon the color changed from deep orange to pale orange. The dispersion was heated to reflux for 18 h (or 60 h, isolated yield 37.8 mg). Then, the solvent was removed under reduced pressure, under an inert atmosphere. Fresh, degassed *n*-octane (100 mL, 0.615 mol) was added and the mixture was cooled on ice. *n*-Butyllithium in hexane (2.5 M, 2.09 mL, 5.22 mmol) was slowly added and run overnight for complete reaction, while the dispersion was allowed to warm up to room temperature. Unreacted *n*-butyllithium was quenched with excess methanol (10 mL), and left to react for 1 h. Thereafter, the dispersion was filtered and extracted against aqueous HCl (1 M, 100 mL, 1×) and distilled water (100 mL, 3×). During extraction, the organic phase was deep orange, with some insoluble materials, especially at the solvent–solvent interface, which were discarded. The organic phase was dried over Na_2SO_4 and filtered, and the solvent was removed under reduced pressure. The obtained product was a dark brown material (24.4 mg).

Control synthesis

The control synthesis was performed similarly to the procedure described above, with the only change being that the silicon source, Mg_2Si, was omitted during the synthesis. Initially, more coloration was observed during the reaction, due to the absence of the dark purple Mg_2Si. The obtained product was a dark brown material (24.5 mg).

Reaction between *n*-octane and Br_2

Br_2 (0.54 mL, 10.5 mmol) was added to degassed *n*-octane (100 mL, 0.615 mol) and reacted for 2 h at room temperature. Upon reaction, the orange color of the Br_2 slowly fades. Subsequently, samples were drawn and analyzed with ^{1}H-NMR spectroscopy.

Results and discussion

Synthesis of Si QDs by oxidation of Mg_2Si by Br_2

To investigate the structural and optical properties of the nanomaterials prepared *via* the oxidation of Mg_2Si by Br_2,[52–56] as depicted in Scheme 1, this synthesis was carried out with only minor changes: most importantly, commercial Mg_2Si was used instead of that synthesized from Mg and Si, solvent was removed at reflux temperature, rather than at room temperature, and surface passivation with *n*-butyllithium was carried out for 1 day, instead of for 2 days. The products of these reactions here are denoted as Si QDs (in the presence of Mg_2Si) and control (in the absence of Mg_2Si).

The products were first characterized using ^{1}H NMR spectroscopy, as shown in Fig. 1. The spectrum for Si QDs is in accordance with spectra published before. Signals at 0.88 ppm and 1.25 ppm were observed and have been assigned previously to $–CH_3$ and $–CH_2–$ protons of the butyl surface passivation, respectively.[86] Similar peak broadening was observed as previously reported.[53] However, no proof for silicon-bound $–CH_2–$ protons around 0.6 ppm (ref. 87) was found, although whether such signals can be observed for these kinds of material is still the subject of debate.[88] In comparison with the control, no obvious differences between the spectra are apparent. Signal intensity was low for both samples, even at high concentrations (>20 mg mL^{-1}), while no precipitation was observed.

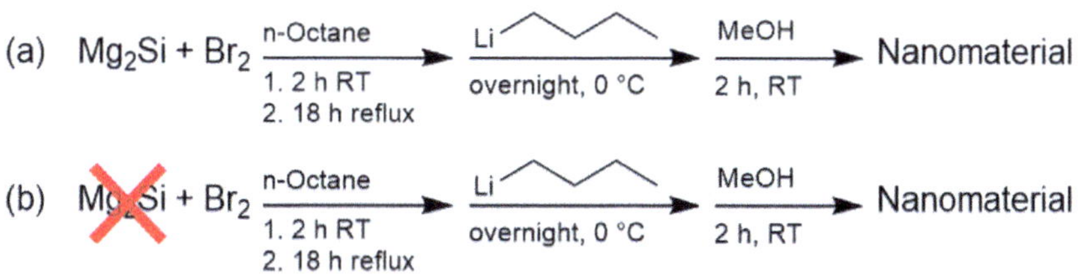

Scheme 1 Synthesis of Si QDs (a) and the control (b).

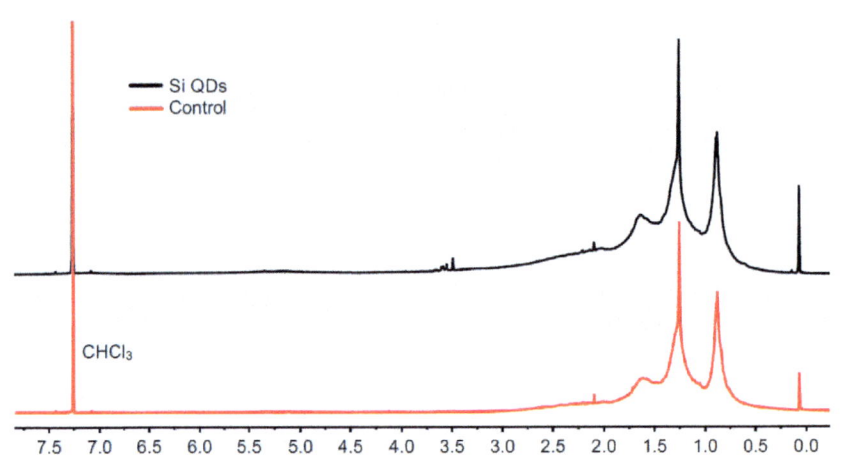

Fig. 1 ^{1}H NMR spectra from the Si QD synthesis (top) and the control synthesis (bottom) in chloroform-d.

The FTIR spectrum of the Si QDs (see Fig. 2) corresponds well with those from previous literature, although OH vibrations around 3425 cm^{-1} were not observed in previous reports,[52] or only to a minor extent.[53] Vibrations around 2927 cm^{-1} correspond to CH$_2$ and CH$_3$ groups, vibrations at 1456 cm^{-1} to CH$_2$ and CH$_3$ groups, and vibrations at 1377 cm^{-1} to CH$_3$ groups; all of them are characteristic for alkyl functionalities. Vibrations corresponding to Si–CH$_2$ bonds have been reported before to appear at approximately 1260 cm^{-1} and 1460 cm^{-1},[49,64,89,90] but are generally weak. Small vibrations were indeed observed at 1261 cm^{-1}, which might be related to Si–CH$_2$ vibrations, although it is not strong evidence for Si QDs, as it may also be assigned to other species. Vibrations at 1705 cm^{-1} are also present, which has been observed earlier.[52,53] However, the nature of these species remains unknown, as these vibrations were not assigned. The distinctive vibrations at 1073 cm^{-1} are attributed to Si–O bonds, as a result of (partial) oxidation of the Si QDs.[4,74] Notably, this signal was also observed in the control, which makes

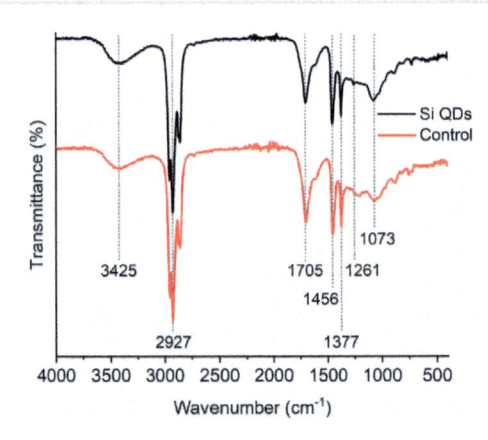

Fig. 2 FTIR spectra of Si QDs and the control.

the assignment of these vibrations as unique for Si QDs uncertain. Moreover, no discernible differences between the Si QDs and the control were observed.

The samples were further analyzed with TEM, as shown in Fig. 3. It can be observed that non-crystalline nanosized particles are present (size not quantified) in both cases. It should be mentioned that both TEM grids were evaluated until areas were found with high contrast, given that these images do not represent the full sample. However, this was performed in order to show the notable resemblance with Si QDs reported previously.[53] For the control, it can be excluded that these particles on TEM are Si QDs, indicating that mere particle observations using TEM do not infer the presence of Si QDs by default. Particles might be aggregates of organic matter, carbon dots, or a result of TEM sample preparation.[79,80,85] Fluorescence by quantum confinement effects needs to be confirmed, regardless of whether the Si QDs are amorphous[91] or crystalline.[92]

Elemental analysis was performed in order to examine the sample composition, and the presence of silicon in particular, as shown in Table S1.† A Si amount of 2.2 wt% was found for the Si QDs, while for the control a Si amount of 0.4 wt% was determined. In both samples, carbon (>60 wt%) and bromine (>3 wt%) were observed. In combination with the product yield, the reaction efficiency for silicon can be derived. At best, less than 1 mg of elemental Si from the starting material (Mg_2Si) can be traced back to the Si QDs, which is equal to 2.2% conversion of Mg_2Si. Considering the (slightly) higher Si content in Si QDs, conversion of Mg_2Si may have occurred, but is only marginal.

XPS analysis was performed to obtain more knowledge on the type of Si bonding in particular, and further confirm the presence of silicon (see Fig. 4 and S1–S4†). After deconvolution, a binding energy of 102.2 eV was observed for silicon in the Si QDs, and a binding energy of 102.4 eV was determined for the control. In the case of the control, Si arose from minor impurities, and therefore will be neglected. Binding energies around 102.2 eV were found for the Si QDs, indicating that no fully oxidized SiO_2 species were present, as binding energies would be observed around ∼103.5 eV.[25,56] However, XPS data for siloxanes shows similar binding energies to that for Si QDs,[93] whereas such materials consist primarily of Si–O bonds. In both the control and the Si QDs, it is not unlikely that similar oxygen-bound silicon material is present. However, most importantly, binding energies at ∼99.5 eV (depending on the oxidation state), corresponding to Si–Si bonds, are absent, though expected in the case of Si QDs.[25,77] Hence, the presence of appreciable amounts of Si QDs cannot be confirmed.

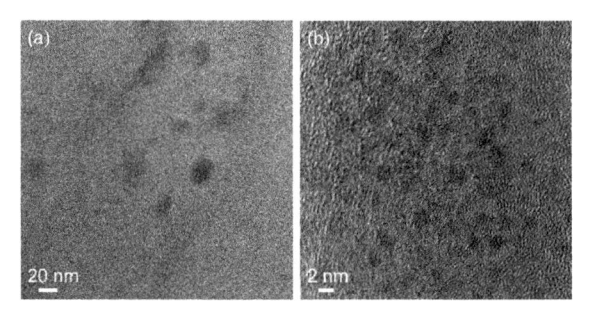

Fig. 3 TEM images of Si QDs (a) and the control (b).

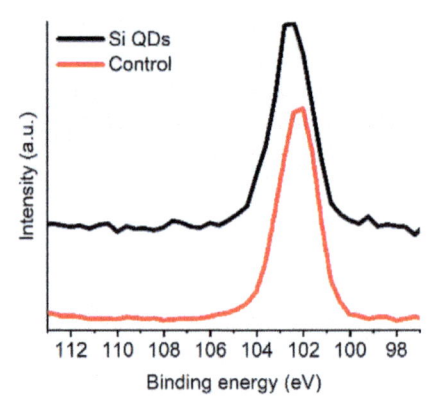

Fig. 4 XPS spectra Si QDs and the control (Si 2p scan).

The absorbance and fluorescence of Si QDs and the control are depicted in Fig. 5. The Si QDs revealed no characteristic features in the absorbance, as described earlier.[52,53] While it is known that QDs show features that correlate to their size, proposed for Si QDs as well,[94] the absence of features may also be an effect of a heterogeneous particle size distribution. However, the similarity with the control is of more concern, indicating no discernable differences.

The fluorescence properties of the samples were subsequently assessed, as shown in Fig. 5. Under visible light, both samples are colorless to pale yellow, while under UV illumination bright blue/green fluorescence can be observed, consistent with the emission observed in the excitation plots. For the Si QDs, the fluorescence spectra are comparable to those reported before,[52] although the emission observed for the Si QDs at lower excitation wavelengths is somewhat stronger. For excitation wavelengths between ∼260–320 nm, the emission wavelength is centered around ∼380 nm, while for higher excitation wavelengths, the emission was red-shifted and rapidly decreased in intensity. Thus, below excitation wavelengths of 320 nm, excitation-independent emission dominates, while for higher excitation wavelengths, excitation-dependent emission is stronger. The reported emission peak maximum (383 nm) in the literature is consistent with the maximum in our spectra. However, the most effective excitation wavelength was

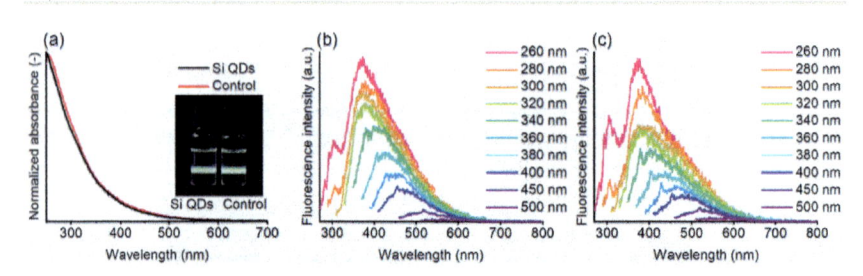

Fig. 5 Absorbance of Si QDs and the control, with an insert showing both samples under UV (excitation 405 nm) illumination (a), and excitation plots of Si QDs (b) and the control (c) in ethyl acetate, with the excitation wavelength ranging from 260 to 500 nm (see also Fig. S5†).

reported to be at ~320 nm, whilst in our case, the lowest excitation wavelength at 260 nm was the most efficient. The most plausible explanation would be that in the current work, more blue-emitting matter was produced due to slight experimental differences. For instance, shorter reaction times or slightly lower temperatures might possibly result in more blue-emitting material.[80] In Fig. S5,† the strongly overlapping emission of the Si QDs and the control can be observed, regardless of excitation wavelength. The fluorescence emission of the control cannot originate from Si QDs, yet still the spectra from Si QDs show the same identical emission spectra as the control, confirming that fluorescence emission in both the Si QDs and control samples is (mostly) determined by species other than Si QDs.

The quantum yield (QY) of the Si QDs, at an excitation wavelength of 350 nm, was determined *via* a relative method.[95] The QY was found to be 2.2%, which is comparable with previously published data.[53] For the control, the QY was somewhat lower at 1.1%.

By means of size exclusion chromatography (SEC) it was possible to separate the materials based on their size, as shown in Fig. 6, and evaluate the optical properties accordingly. For the Si QDs, it was observed that two populations are present, while only the smaller material showed significant fluorescence. Complete SEC traces of absorbance and fluorescence can be found in Fig. S6.† For fluorescence emission, it can be observed that for the Si QDs and the control, the fluorescence spectra were very comparable. In addition, both materials showed size-dependent fluorescence, which before was thought to be a result of quantum

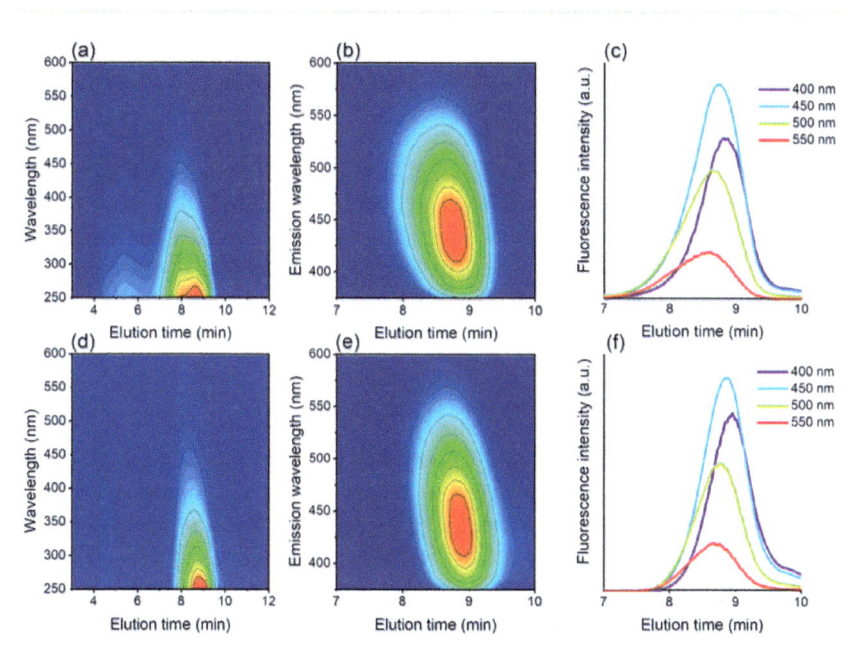

Fig. 6 SEC traces of absorbance (a) and fluorescence (b and c) of Si QDs, and absorbance (d) and fluorescence (e and f) of the control. The intensity of absorbance and fluorescence increases from blue to red in the contour plots. The emission intensities in (c) and (f) were measured at four different emission wavelengths.

confinement.[52] Emission at 400 nm was strongest at 8.87 min, while emission at 550 nm was strongest at an elution time of 8.60 min. Therefore, it can be concluded that solely size-dependent emission is not proof of Si QD emission.

There are, however, more notable deviations in the case of the absorbance spectra. It can be observed in Fig. 6 that for both samples, absorbance is observed over a much broader range (~3.5–10 min) than fluorescence (~8–9.5 min), indicating the presence of an emissive fraction at longer elution times, and a less- or non-emissive fraction at shorter elution times. The absorbance for Si QDs is much stronger than for the control at short elution times (~3.5–7 min), although also in the control absorbance was observed to a minor extent (see also Fig. S7†). Size fractionation of this material has been carried out, revealing enhanced silicon content (0.6 wt% vs. 0.2 wt%),[96] however, this species does not display strong fluorescence. Remarkably, a similar effect was also found in carbon dot synthesis, where size-separation was performed via dialysis.[79] Smaller materials (<1 kDa) showed strong fluorescence, while the larger materials (≥50 kDa) showed much weaker fluorescence. This again suggests that in the samples investigated here, the fluorescence emission is likely caused by small fluorescent materials other than Si QDs.

The findings that the reactions carried out in the presence and absence of Mg_2Si result in products with almost identical properties raises questions concerning the actual nature of the material formed. Therefore, determining the origin of these by-products was attempted. Direct exposure of Mg_2Si to Br_2 in the absence of a solvent with overnight stirring did not result in any noticeable reaction. Alternatively, elemental (insoluble) silicon may have been formed, as presumed for the Si QDs synthesis carried out in glyme.[52] In addition, it is known that reactions between Mg_2Si and aqueous HCl yield (higher) silanes,[97,98] assuming that aqueous HBr reacts in a similar fashion. Hereunder, it is shown that HBr is indeed formed during Si QD synthesis, although it remains unclear whether other species (besides octyl bromides and HBr) would form and how the reaction might proceed further.

After addition of Br_2 to the reaction mixture containing Mg_2Si and n-octane, the color of the Br_2 fades. This is also the case in the absence of Mg_2Si. Bromination of alkanes has been demonstrated before,[99–102] and results in the formation of alkyl bromides and HBr gas.[102] ^{1}H NMR spectroscopic analysis indeed confirms the formation of octyl bromides, as evidenced by signals at 3.99, 4.04 and 4.13 ppm, which were assigned (primarily) to three isomers of secondary octyl bromides (ratios not quantified, see Fig. S8†). Formation of considerable amounts of HBr gas was visually confirmed, as well as through the use of a pH indicator. Despite considerable evolution of HBr gas, it should be noted that the solubility of HBr in n-octane is also significant.[103] Formation of octyl bromides was observed before, but was then not considered to be the origin of strongly fluorescent material.[53] Experimentally, products of the reaction between n-octane and Br_2 were collected, possible residual Br_2 and HBr were removed by extraction against water, and the organic phase, containing n-octane and octyl bromides, was dried over Na_2SO_4. Subsequently, this mixture of n-octane and octyl bromides was heated to reflux, and fluorescent material was again obtained. This indicates that HBr and Br_2 are likely not essential for formation of fluorescent by-products upon heating, but rather that other, earlier formed species are, e.g. octyl bromides.

In addition to our critical assessment of Si QD synthesis by means of wet-chemical oxidation, we stress that related work on Si QDs should be critically assessed as well, and bottom-up syntheses in particular.

In our view, most essential in Si QD synthesis are two aspects. Firstly, it should be shown convincingly that (crystalline) Si QDs have been synthesized successfully, preferably using an appropriate combination of methods discussed below.[26,76] Secondly, it should be confirmed that these Si QDs are indeed the true origin of displayed fluorescence. Careful purification and characterization are therefore essential. Frequently, successful fabrication of (size-tunable) fluorescent Si QDs is claimed, almost entirely based on particles observed by TEM in combination with fluorescence spectra. However, as indicated, the presence of (crystalline) particles alone is not direct proof of the formation of fluorescing Si QDs.[79,85] Furthermore, it is not improbable that side products formed during synthesis might attach to the surface of the true QDs, further complicating observations.[80]

As long as the quantum confinement model is applicable, evidence of a direct relation between determined size and fluorescence emission would be of great value in this regard, for instance achieved by fluorescence spectroscopy on size-separated samples. Alternatively, differently sized particles may be prepared and their respective sizes and fluorescence emissions can be assessed.[4,104] Unfortunately, surface effects influencing fluorescence, and thus reducing or eliminating the effect of size on emission, especially towards smaller sizes, may complicate such characterization methods.[105]

Veinot and coworkers provided guidelines for the evaluation of Si QDs,[26] which are in line with a more general approach published on inorganic colloidal materials.[106] In fact, characterization can be performed almost inexhaustibly, and a combination of techniques will always be required due to the complex nature of Si QDs.[26] Most conveniently, information regarding crystallinity can be obtained using TEM[107,108] and XRD,[77] size is readily assessed by TEM and SAXS,[25,66] while the nature of Si-bonding can be determined using Raman spectroscopy,[109] FTIR[108,110] and XPS.[25] The optical properties can be ascertained through fluorescence and UV-vis spectroscopic analysis, while chemical composition can be determined by elemental analysis and, trivially but importantly, by reporting product yields. Further insights can be gained from techniques such as dynamic light scattering,[111] atomic force microscopy,[112] thermogravimetric analysis,[113] NMR spectroscopy,[77] fluorescence lifetime experiments,[114] ultracentrifugation,[115] size-selective precipitation,[105] and size exclusion chromatography.[28,116]

For purification, chromatographic separation based on size (e.g. SEC) or polarity (e.g. HPLC) are powerful tools for removing (fluorescent) by-products,[80,117] while dialysis may be convenient as well.[79] It has been shown that purification in the field of nanotechnology has been frequently performed only to a limited extent, if at all, leading to misinterpretations.[78] For instance, centrifugation or filtration alone is typically insufficient.[79] Dialysis, especially when using a molecular weight cut-off below 20 kDa, is not recommended, due to its limited ability to remove organic residues.[79]

Our studies indicate that the reaction solvents may not be as inert as generally considered, and can induce fluorescence (possibly in combination with other reactants), especially when heat or microwave radiation is applied. For instance, it has been shown that refluxing ethylene glycol readily results in fluorescent material,[84] as reported for poly(ethylene glycol) as well.[118] The combination of sodium and refluxing diglyme has also been demonstrated to lead to fluorescent by-products (suggested to be polymerized diglyme).[119] These by-products could also form during a Si QD synthesis where sodium naphthalide is reacted in refluxing glyme.[41] Gu et al. obtained fluorescent materials when heating a series of solvents (DMF, DMAc, xylene, n-hexane and cyclohexane) to 260 °C in an autoclave.[120] Generally, reaction temperatures in wet-chemical Si QD syntheses (e.g. glyme at 85 °C,[52,64] n-octane at 126 °C,[52,53] DMF at 153 °C (ref. 57 and 58)) and microwave-assisted Si QD syntheses (e.g. DMSO, DMF, acetonitrile at 160 °C,[33] glycerol at 180 °C[121] and DMF at 275 °C (ref. 34)) are slightly lower. However, it is important to rule out the formation and/or insufficient removal of such fluorescent materials when comparable conditions are employed. For example, microwave-assisted synthesis in DMF at 275 °C was confirmed to result in the formation of fluorescent by-products.[34] Hence the requirement for effective purification. Straightforward control experiments (without a source of silicon, in the case of Si QDs) would help to confirm that the observed fluorescence does not originate from by-products.[76,84]

Conclusions

In the assessment of the wet-chemical oxidation of Mg_2Si with Br_2 we observed the formation of fluorescent nanomaterials, even in the absence of a silicon precursor. We demonstrated that the materials formed in the presence and absence of Mg_2Si display very comparable optical properties. Not only were the optical properties similar, but other physicochemical properties, as determined using other characterization techniques, were found to be similar. In fact, only SEC characterization and elemental analysis revealed minute differences. However, no evidence was found supporting the formation of Si QDs in substantial amounts. No crystallinity corresponding to silicon was observed, no size-dependent emission different from the control without Mg_2Si was shown, and no proof of Si–Si bonds was observed. We suggest that wet-chemical syntheses, especially those involving heating, can lead to the formation of fluorescent by-products with similar optical properties to those of Si QDs. The presence of fluorescent impurities might not easily be distinguished or detected due to a generally high content of organic material in the samples and there being no need for the carbon material to be crystalline to show size-dependent properties.[78]

The optical properties observed in Si QD synthesis should be convincingly linked to the actual nanomaterial. This may be achieved by appropriate purification by chromatography and a combination of characterization techniques to determine Si QD crystallinity, size, bonds, optical properties, composition and product yields. Considerations regarding Si QDs may also be extended to other types of fluorescent nanoparticles.

Conflicts of interest

There are no conflicts to declare.

Acknowledgements

We thank Dr Rico Keim for performing TEM measurements and Mr Gerard Kip for performing XPS measurements.

References

1 A. D. Yoffe, *Adv. Phys.*, 2001, **50**, 1–208.
2 C. B. Murray, D. J. Norris and M. G. Bawendi, *J. Am. Chem. Soc.*, 1993, **115**, 8706–8715.
3 A. Gupta, M. T. Swihart and H. Wiggers, *Adv. Funct. Mater.*, 2009, **19**, 696–703.
4 C. M. Hessel, E. J. Henderson and J. G. C. Veinot, *Chem. Mater.*, 2006, **18**, 6139–6146.
5 D. Norris and M. Bawendi, *Phys. Rev. B: Condens. Matter Mater. Phys.*, 1996, **53**, 16338–16346.
6 Y. Shirasaki, G. J. Supran, M. G. Bawendi and V. Bulović, *Nat. Photonics*, 2013, **7**, 13–23.
7 M. Dasog, G. B. De Los Reyes, L. V. Titova, F. A. Hegmann and J. G. C. Veinot, *ACS Nano*, 2014, **8**, 9636–9648.
8 X. Gao, L. Yang, J. A. Petros, F. F. Marshall, J. W. Simons and S. Nie, *Curr. Opin. Biotechnol.*, 2005, **16**, 63–72.
9 Z. Ni, S. Zhou, S. Zhao, W. Peng, D. Yang and X. Pi, *Mater. Sci. Eng., R*, 2019, **138**, 85–117.
10 F. Priolo, T. Gregorkiewicz, M. Galli and T. F. Krauss, *Nat. Nanotechnol.*, 2014, **9**, 19–32.
11 P. V. Kamat, *J. Phys. Chem. C*, 2008, **112**, 18737–18753.
12 F. Maier-Flaig, J. Rinck, M. Stephan, T. Bocksrocker, M. Bruns, C. Kübel, A. K. Powell, G. A. Ozin and U. Lemmer, *Nano Lett.*, 2013, **13**, 475–480.
13 E. J. Henderson, A. J. Shuhendler, P. Prasad, V. Baumann, F. Maier-Flaig, D. O. Faulkner, U. Lemmer, X. Y. Wu and G. A. Ozin, *Small*, 2011, **7**, 2507–2516.
14 F. Erogbogbo, K. T. Yong, I. Roy, G. X. Xu, P. N. Prasad and M. T. Swihart, *ACS Nano*, 2008, **2**, 873–878.
15 U. Resch-Genger, M. Grabolle, S. Cavaliere-Jaricot, R. Nitschke and T. Nann, *Nat. Methods*, 2008, **5**, 763–775.
16 X. Cheng, S. B. Lowe, P. J. Reece and J. J. Gooding, *Chem. Soc. Rev.*, 2014, **43**, 2680–2700.
17 X. Michalet, F. F. Pinaud, L. A. Bentolila, J. M. Tsay, S. Doose, J. J. Li, G. Sundaresan, A. M. Wu, S. S. Gambhir and S. Weiss, *Science*, 2005, **307**, 538–544.
18 I. L. Medintz, H. T. Uyeda, E. R. Goldman and H. Mattoussi, *Nat. Mater.*, 2005, **4**, 435–446.
19 K. D. Wegner and N. Hildebrandt, *Chem. Soc. Rev.*, 2015, **44**, 4792–4834.
20 L. M. Kustov, P. V. Mashkin, V. N. Zakharov, N. B. Abramenko, E. Y. Krysanov, L. A. Aslanov and W. Peijnenburg, *Environ. Sci.: Nano*, 2018, **5**, 2945–2951.
21 M. Dasog, J. Kehrle, B. Rieger and J. G. C. Veinot, *Angew. Chem., Int. Ed.*, 2016, **55**, 2322–2339.
22 P. Reiss, M. Carrière, C. Lincheneau, L. Vaure and S. Tamang, *Chem. Rev.*, 2016, **116**, 10731–10819.
23 A. S. Heintz, M. J. Fink and B. S. Mitchell, *Adv. Mater.*, 2007, **19**, 3984–3988.
24 K. Sato, H. Tsuji, K. Hirakuri, N. Fukata and Y. Yamauchi, *Chem. Commun.*, 2009, 3759–3761.

25 C. M. Hessel, D. Reid, M. G. Panthani, M. R. Rasch, B. W. Goodfellow, J. Wei, H. Fujii, V. Akhavan and B. A. Korgel, *Chem. Mater.*, 2012, **24**, 393–401.
26 R. J. Clark, M. Aghajamali, C. M. Gonzalez, L. Hadidi, M. A. Islam, M. Javadi, M. H. Mobarok, T. K. Purkait, C. J. T. Robidillo, R. Sinelnikov, A. N. Thiessen, J. Washington, H. Yu and J. G. C. Veinot, *Chem. Mater.*, 2017, **29**, 80–89.
27 X. Li, Y. He, S. S. Talukdar and M. T. Swihart, *Langmuir*, 2003, **19**, 8490–8496.
28 G. Ledoux, J. Gong, F. Huisken, O. Guillois and C. Reynaud, *Appl. Phys. Lett.*, 2002, **80**, 4834–4836.
29 L. Mangolini, E. Thimsen and U. Kortshagen, *Nano Lett.*, 2005, **5**, 655–659.
30 F. Kunze, S. Kuns, M. Spree, T. Hülser, C. Schulz, H. Wiggers and S. M. Schnurre, *Powder Technol.*, 2019, **342**, 880–886.
31 J. D. Holmes, K. J. Ziegler, R. C. Doty, L. E. Pell, K. P. Johnston and B. A. Korgel, *J. Am. Chem. Soc.*, 2001, **123**, 3743–3748.
32 Y. Zhong, F. Peng, F. Bao, S. Wang, X. Ji, L. Yang, Y. Su, S. T. Lee and Y. He, *J. Am. Chem. Soc.*, 2013, **135**, 8350–8356.
33 S. P. Pujari, H. Driss, F. Bannani, B. Van Lagen and H. Zuilhof, *Chem. Mater.*, 2018, **30**, 6503–6512.
34 T. M. Atkins, A. Thibert, D. S. Larsen, S. Dey, N. D. Browning and S. M. Kauzlarich, *J. Am. Chem. Soc.*, 2011, **133**, 20664–20667.
35 B. Song and Y. He, *Nano Today*, 2019, **26**, 149–163.
36 B. F. P. McVey, S. Prabakar, J. J. Gooding and R. D. Tilley, *ChemPlusChem*, 2017, **82**, 60–73.
37 J. R. Heath, *Science*, 1992, **258**, 1131–1133.
38 A. Kamyshny, V. Zakharov, M. Zakharov, A. Yatsenko, S. Savilov, L. Aslanov and S. Magdassi, *J. Nanopart. Res.*, 2011, **13**, 1971–1978.
39 L. A. Aslanov, V. N. Zakharov, A. V. Pavlikov, S. V. Savilov, V. Y. Timoshenko and A. V. Yatsenko, *Russ. J. Coord. Chem.*, 2013, **39**, 427–431.
40 S. Semlali, B. Cormary, M. L. De Marco, J. Majimel, A. Saquet, Y. Coppel, M. Gonidec, P. Rosa and G. L. Drisko, *Nanoscale*, 2019, **11**, 4696–4700.
41 R. K. Baldwin, K. A. Pettigrew, E. Ratai, M. P. Augustine and S. M. Kauzlarich, *Chem. Commun.*, 2002, **17**, 1822–1823.
42 Q. Li, T. Y. Luo, M. Zhou, H. Abroshan, J. Huang, H. J. Kim, N. L. Rosi, Z. Shao and R. Jin, *ACS Nano*, 2016, **10**, 8385–8393.
43 R. D. Tilley and K. Yamamoto, *Adv. Mater.*, 2006, **18**, 2053–2056.
44 R. D. Tilley, J. H. Warner, K. Yamamoto, I. Matsui and H. Fujimori, *Chem. Commun.*, 2005, 1833–1835.
45 J. Wang, S. Sun, F. Peng, L. Cao and L. Sun, *Chem. Commun.*, 2011, **47**, 4941–4943.
46 X. Cheng, R. Gondosiswanto, S. Ciampi, P. J. Reece and J. J. Gooding, *Chem. Commun.*, 2012, **48**, 11874–11876.
47 J. P. Wilcoxon, G. A. Samara and P. N. Provencio, *Phys. Rev. B: Condens. Matter Mater. Phys.*, 1999, **60**, 2704–2714.
48 K. Linehan and H. Doyle, *Small*, 2014, **10**, 584–590.
49 M. Rosso-Vasic, E. Spruijt, B. Van Lagen, L. De Cola and H. Zuilhof, *Small*, 2008, **4**, 1835–1841.
50 N. Lin, Y. Han, L. Wang, J. Zhou, J. Zhou, Y. Zhu and Y. Qian, *Angew. Chem., Int. Ed.*, 2015, **54**, 3822–3825.
51 A. Shavel, L. Guerrini and R. A. Alvarez-Puebla, *Nanoscale*, 2017, **9**, 8157–8163.

52 K. A. Pettigrew, Q. Liu, P. P. Power and S. M. Kauzlarich, *Chem. Mater.*, 2003, **15**, 4005–4011.
53 L. Ruizendaal, S. P. Pujari, V. Gevaerts, J. M. J. Paulusse and H. Zuilhof, *Chem.–Asian J.*, 2011, **6**, 2776–2786.
54 Q. Liu and S. M. Kauzlarich, *Mater. Sci. Eng., B*, 2002, **96**, 72–75.
55 M. Imamura, J. Nakamura, S. Fujimasa, H. Yasuda, H. Kobayashi and Y. Negishi, *Eur. Phys. J. D*, 2011, **63**, 289–292.
56 C. W. Hsu, D. Septiadi, C. H. Lai, P. Chen, P. H. Seeberger and L. De Cola, *ChemPlusChem*, 2017, **82**, 660–667.
57 X. Zhang, D. Neiner, S. Wang, A. Y. Louie and S. M. Kauzlarich, *Nanotechnology*, 2007, **18**, 095601.
58 B. M. Nolan, T. Henneberger, M. Waibel, T. F. Fässler and S. M. Kauzlarich, *Inorg. Chem.*, 2015, **54**, 396–401.
59 D. Neiner, H. W. Chiu and S. M. Kauzlarich, *J. Am. Chem. Soc.*, 2006, **128**, 11016–11017.
60 B. Cho, S. Baek, H. G. Woo and H. Sohn, *J. Nanosci. Nanotechnol.*, 2014, **14**, 5868–5872.
61 D. Rodríguez Sartori, C. R. Lillo, J. J. Romero, M. Laura Dellarciprete, A. Miñán, M. Fernández Lorenzo De Mele and M. C. Gonzalez, *Nanotechnology*, 2016, **27**, 475704.
62 D. Mayeri, B. L. Phillips, M. P. Augustine and S. M. Kauzlarich, *Chem. Mater.*, 2001, **13**, 765–770.
63 R. A. Bley and S. M. Kauzlarich, *J. Am. Chem. Soc.*, 1996, **118**, 12461–12462.
64 C. S. Yang, R. A. Bley, S. M. Kauzlarich, H. W. H. Lee and G. R. Delgado, *J. Am. Chem. Soc.*, 1999, **121**, 5191–5195.
65 B. Ghosh and N. Shirahata, *Sci. Technol. Adv. Mater.*, 2014, **15**, 014207.
66 Y. Yu, G. Fan, A. Fermi, R. Mazzaro, V. Morandi, P. Ceroni, D. M. Smilgies and B. A. Korgel, *J. Phys. Chem. C*, 2017, **121**, 23240–23248.
67 K. Dohnalová, A. Fučíková, C. P. Umesh, J. Humpolíčková, J. M. J. Paulusse, J. Valenta, H. Zuilhof, M. Hof and T. Gregorkiewicz, *Small*, 2012, **8**, 3185–3191.
68 X. Wen, P. Zhang, T. A. Smith, R. J. Anthony, U. R. Kortshagen, P. Yu, Y. Feng, S. Shrestha, G. Coniber and S. Huang, *Sci. Rep.*, 2015, **5**, 12469.
69 R. Sinelnikov, M. Dasog, J. Beamish, A. Meldrum and J. G. C. Veinot, *ACS Photonics*, 2017, **4**, 1920–1929.
70 T. Takagahara and K. Takeda, *Phys. Rev. B*, 1992, **46**, 15578–15581.
71 D. C. Hannah, J. Yang, P. Podsiadlo, M. K. Y. Chan, A. Demortière, D. J. Gosztola, V. B. Prakapenka, G. C. Schatz, U. Kortshagen and R. D. Schaller, *Nano Lett.*, 2012, **12**, 4200–4205.
72 K. Dohnalová, A. N. Poddubny, A. A. Prokofiev, W. D. De Boer, C. P. Umesh, J. M. Paulusse, H. Zuilhof and T. Gregorkiewicz, *Light: Sci. Appl.*, 2013, **2**, e47.
73 M. Dasog, Z. Yang, S. Regli, T. M. Atkins, A. Faramus, M. P. Singh, E. Muthuswamy, S. M. Kauzlarich, R. D. Tilley and J. G. C. Veinot, *ACS Nano*, 2013, **7**, 2676–2685.
74 A. Gupta and H. Wiggers, *Nanotechnology*, 2011, **22**, 055707.
75 J. Fuzell, A. Thibert, T. M. Atkins, M. Dasog, E. Busby, J. G. C. Veinot, S. M. Kauzlarich and D. S. Larsen, *J. Phys. Chem. Lett.*, 2013, **4**, 3806–3812.
76 B. V. Oliinyk, D. Korytko, V. Lysenko and S. Alekseev, *Chem. Mater.*, 2019, **31**, 7167.

77 A. N. Thiessen, M. Ha, R. W. Hooper, H. Yu, A. O. Oliynyk, J. G. C. Veinot and V. K. Michaelis, *Chem. Mater.*, 2019, **31**, 678–688.
78 J. B. Essner and G. A. Baker, *Environ. Sci.: Nano*, 2017, **4**, 1216–1263.
79 J. B. Essner, J. A. Kist, L. Polo-Parada and G. A. Baker, *Chem. Mater.*, 2018, **30**, 1878–1887.
80 N. C. Verma, A. Yadav and C. K. Nandi, *Nat. Commun.*, 2019, **10**, 2391.
81 Y. Song, S. Zhu, S. Zhang, Y. Fu, L. Wang, X. Zhao and B. Yang, *J. Mater. Chem. C*, 2015, **3**, 5976–5984.
82 C. J. Reckmeier, J. Schneider, Y. Xiong, J. Häusler, P. Kasák, W. Schnick and A. L. Rogach, *Chem. Mater.*, 2017, **29**, 10352–10361.
83 V. Štengl and J. Henych, *Nanoscale*, 2013, **5**, 3387–3394.
84 L. Jiang and H. Zeng, *Nanoscale*, 2015, **7**, 4580–4583.
85 S. Khan, A. Sharma, S. Ghoshal, S. Jain, M. K. Hazra and C. K. Nandi, *Chem. Sci.*, 2017, **9**, 175–180.
86 L. Ruizendaal, *Functional Silicon Nanoparticles*, Wageningen University, 2011.
87 W. Biesta, B. Van Lagen, V. S. Gevaert, A. T. M. Marcelis, J. M. J. Paulusse, M. W. F. Nielen and H. Zuilhof, *Chem. Mater.*, 2012, **24**, 4311–4318.
88 I. M. D. Höhlein, J. Kehrle, T. K. Purkait, J. G. C. Veinot and B. Rieger, *Nanoscale*, 2015, **7**, 914–918.
89 J. H. Warner, A. Hoshino, K. Yamamoto and R. D. Tilley, *Angew. Chem., Int. Ed.*, 2005, **44**, 4550–4554.
90 A. Shiohara, S. Hanada, S. Prabakar, K. Fujioka, T. H. Lim, K. Yamamoto, P. T. Northcote and R. D. Tilley, *J. Am. Chem. Soc.*, 2010, **132**, 248–253.
91 N. M. Park, C. J. Choi, T. Y. Seong and S. J. Park, *Phys. Rev. Lett.*, 2001, **86**, 1355–1357.
92 W. L. Wilson, P. F. Szajowski and L. E. Brus, *Science*, 1993, **262**, 1242–1244.
93 M. Ouyang, C. Yuan, R. J. Muisener, A. Boulares and J. T. Koberstein, *Chem. Mater.*, 2000, **12**, 1591–1596.
94 R. Intartaglia, K. Bagga, F. Brandi, G. Das, A. Genovese, E. Di Fabrizio and A. Diaspro, *J. Phys. Chem. C*, 2011, **115**, 5102–5107.
95 J. R. Lakowicz, *Principles of fluorescence spectroscopy*, Springer Science & Business Media, 2013.
96 Reaction conditions differed from the syntheses presented in this article and concern higher Mg$_2$Si concentrations.
97 A. Stock and C. Somieski, *Ber. Dtsch. Chem. Ges.*, 1916, **49**, 111–157.
98 H. J. Emeléus and A. G. Maddock, *J. Chem. Soc.*, 1946, 1131–1134.
99 Y. Manabe, Y. Kitawaki, M. Nagasaki, K. Fukase, H. Matsubara, Y. Hino, T. Fukuyama and I. Ryu, *Chem.–Eur. J.*, 2014, **20**, 12750–12753.
100 M. S. Kharasch, W. Hered and F. R. Mayo, *J. Org. Chem.*, 1941, **6**, 818–829.
101 G. Egloff, R. E. Schaad and C. D. Lowry, *Chem. Rev.*, 1931, **8**, 1–80.
102 X. Wang, R. F. Liu, J. Cork, Y. Gu, D. C. Upham, B. Laycock and E. W. McFarland, *Ind. Eng. Chem. Res.*, 2017, **56**, 9411–9418.
103 E. R. Boedeker and C. C. Lynch, *J. Am. Chem. Soc.*, 1950, **72**, 3234–3236.
104 C. M. Hessel, E. J. Henderson and J. G. C. Veinot, *J. Phys. Chem. C*, 2007, **111**, 6956–6961.
105 M. L. Mastronardi, F. Maier-Flaig, D. Faulkner, E. J. Henderson, C. Kübel, U. Lemmer and G. A. Ozin, *Nano Lett.*, 2012, **12**, 337–342.
106 C. J. Murphy and J. M. Buriak, *Chem. Mater.*, 2015, **27**, 4911–4913.

107 M. G. Panthani, C. M. Hessel, D. Reid, G. Casillas, M. José-Yacamán and B. A. Korgel, *J. Phys. Chem. C*, 2012, **116**, 22463–22468.
108 Y. Yu, C. M. Hessel, T. D. Bogart, M. G. Panthani, M. R. Rasch and B. A. Korgel, *Langmuir*, 2013, **29**, 1533–1540.
109 C. M. Hessel, J. Wei, D. Reid, H. Fujii, M. C. Downer and B. A. Korgel, *J. Phys. Chem. Lett.*, 2012, **3**, 1089–1093.
110 L. Mangolini and U. Kortshagen, *Adv. Mater.*, 2007, **19**, 2513–2519.
111 K. Kůsová, O. Cibulka, K. Dohnalová, I. Pelant, J. Valenta, A. Fučíková, K. Žídek, J. Lang, J. Englich, P. Matějka, P. Štěpánek and S. Bakardjieva, *ACS Nano*, 2010, **4**, 4495–4504.
112 M. Rosso-Vasic, E. Spruijt, Z. Popović, K. Overgaag, B. Van Lagen, B. Grandidier, D. Vanmaekelbergh, D. Domínguez-Gutiérrez, L. De Cola and H. Zuilhof, *J. Mater. Chem.*, 2009, **19**, 5926–5933.
113 F. Hua, M. T. Swihart and E. Ruckenstein, *Langmuir*, 2005, **21**, 6054–6062.
114 M. Jakob, A. Aissiou, W. Morrish, F. Marsiglio, M. Islam, A. Kartouzian and A. Meldrum, *Nanoscale Res. Lett.*, 2018, **13**, 338.
115 J. B. Miller, A. R. Van Sickle, R. J. Anthony, D. M. Kroll, U. R. Kortshagen and E. K. Hobbie, *ACS Nano*, 2012, **6**, 7389–7396.
116 L. Pitkänen and A. M. Striegel, *TrAC, Trends Anal. Chem.*, 2016, **80**, 311–320.
117 Y. Shen, M. Y. Gee, R. Tan, P. J. Pellechia and A. B. Greytak, *Chem. Mater.*, 2013, **25**, 2838–2848.
118 M. Chen, W. Wang and X. Wu, *J. Mater. Chem. B*, 2014, **2**, 3937–3945.
119 D. E. Harwell, J. C. Croney, W. Qin, J. T. Thornton, J. H. Day, E. K. Hajime and D. M. Jameson, *Chem. Lett.*, 2003, **32**, 1194–1195.
120 J. Gu, X. Li, D. Hu, Y. Liu, G. Zhang, X. Jia, W. Huang and K. Xi, *RSC Adv.*, 2018, **8**, 12556–12561.
121 H. L. Ye, S. J. Cai, S. Li, X. W. He, W. Y. Li, Y. H. Li and Y. K. Zhang, *Anal. Chem.*, 2016, **88**, 11631–11638.

Faraday Discussions

DISCUSSIONS

Synthesis and functionalisation of silicon nanostructures: general discussion

Paola Ceroni, Carina Crucho, Anna Fucikova, Ankit Goyal, Yao He, Ali Reza Kamali, Katerina Kusova, Giacomo Morselli, Katerina Dohnalova, Jos Paulusse, Holger Stephan, Wei Sun, Ming Lee Tang and Han Zuilof

DOI: 10.1039/D0FD90004D

Anna Fucikova opened a general discussion of the paper by Han Zuilhof: Really nice article. Could the authors please tell us if they tried to dry their samples and redissolve them, and did they maintain their optical properties?

Han Zuilof replied: Thank you for the compliment. No, we did not try to dry up our samples, as evaporating the 2.5 ml of DMSO in which the particles were dispersed would take at least 2 days of heating at 80 °C and 1 mbar. During this time, oxidation of the particles (especially the nanoparticles capped with smaller silanols) will take place, which would also affect the optical properties. For the nanoparticles capped with decyldimethylsilanol, this experiment could perhaps be tried, but it was not attempted for the current work.

Carina Crucho said: A few years ago, I developed an approach to track silica surface modification and quantify functional group coverage using only solution NMR on silica nanoparticles.[1] Looking at your NMR spectrum, the signals are quite visible. I think it would be interesting to do a quantification and calculate the number of molecules per nanoparticle. Knowing the number of Si–H bonds, it would be possible to see how many Si–H groups had reacted and whether the ligands were monolayered or adsorbed on the surface. The broadening of the NMR peaks is due to restricted mobility of the species.

1 C. I. C. Crucho, C. Baleizão and J. P. S. Farinha, *Anal. Chem.*, 2017, **89**(1), 681–687, DOI: 10.1021/acs.analchem.6b03117.

Han Zuilof answered: First of all, that method is really nice – great work! It would indeed be very interesting to be able to quantify the number of molecules per nanoparticle as obtained *via* our method. While quantification might be somewhat limited in accuracy, this would certainly be of interest for future studies!

Ming Lee Tang remarked: Hello. Nice to meet you virtually – I've read many of the papers from your group. Too bad we couldn't meet in person. (1) Have you done XRD to verify that this microwave synthesis does indeed produce Si? I see there is TEM data but I am just curious. (2) Do silanols without flanking dimethyl groups work for this functionalization, *i.e.* R–CH$_2$SiOH, not R–CMe$_2$SiOH presented here?

Han Zuilof responded: Thanks for your comments! XRD was performed on the nanoparticles produced using this microwave method; those data were published in a previous paper from our group.[1] We also considered using silanols without flanking dimethyl groups due to the reduced steric hindrance, but did not include them in the present paper. Our main concern with the use of such "unflanked" silanols was the stability, as the Si–H bond present in such silanols might also react with another silanol or environmental oxygen, leading to multilayers and oxidation.

1 S. P. Pujari, H. Driss, F. Bannani, B. van Lagen and H. Zuilhof, *Chem. Mater.*, 2018, **30**, 6503–6512.

Katerina Dohnalova remarked: First, I would like to voice great collegial appreciation for the fact that the authors are contributing to the difficult and complex research field of the bottom-up synthesis of Si nanoparticles. They are offering an interesting synthesis and capping routes, which seem very up-scalable and "simple". Nevertheless, I am of the opinion that the authors do not offer any evidence that the emissive material is truly the Si nanoparticles. It is beyond doubt that the authors produced a material that emits brightly, and they show Si content in the solution, but there is no evidence that it is the silicon content that emits. Independent of that, there are also further doubts because (i) none of the existing theoretical models suggest that crystalline (or amorphous) Si nanoparticles of ∼6 nm mean diameter could have visible emission with fast radiative rates, and (ii) recently many critical papers have appeared proving that bottom-up Si nanoparticle synthesis routes are "plagued" with highly emissive carbon nanomaterial by-products (including the cited paper by Oliinyk *et al.*[1]), especially when the synthesis route involves organic material and some form of heating (including microwaves). Given this landscape, I am of the opinion (open to debate) that the presented evidence does not show that brightly emissive Si nanoparticles can be produced using the proposed method and would like to ask for (1) supporting evidence for this claim (such as correlative optical-Raman/AFM/STM/SEM microscopy, or similar), plus (2) a repeated microwave synthesis without the Si precursor (as the simplest step).

1 B. V. Oliinyk, D. Korytko, V. Lysenko and S. Alekseev, *Chem. Mater.*, 2019, **31**, 7167.

Han Zuilof replied: First of all, thanks for the comments – they point at a recent discussion on the nature of the NPs produced by these wet-chemical methods induced by Oliinyk *et al.*[1], and we think both points strike at the heart of this scientific discussion.

The complete characterization of the particles used in this study was published previously by our group.[2] Since the synthesis of the particles is not the main focus of the current paper, such extensive characterization was not repeated here. We did perform a control reaction without the silicon compound, and the optical properties of this mixture were significantly different (about 50 nm blue-shifted). In addition, the XPS of our NPs does not display N, thereby excluding that we are looking at $C_wN_xSi_yO_z$ species with a high value of x. Since we form significant amounts of NPs in our method (typically, 100s of mgs of isolated NPs per experiment), significant sources of elements are needed to get to a substantial contribution to the composition of the NPs. Of course, all this primarily only shows that the species formed with the Si source present is different to those formed without its presence, but this at least indicates that the fluorescent species contains silicon and no nitrogen. In addition, upon exposure to ambient air, the fluorescence of the silicon mixture displays a large blue-shift, and a white precipitate (confirmed to be SiO_2 by XPS) forms in the solution. If the solution is exposed to air or water for longer times, the fluorescence drops in intensity as the precipitate sinks to the bottom of the cuvette. Therefore, we would argue that the fluorescent material in our study is indeed a silicon species. If the fluorescent material was carbon nanoparticles, the fluorescent properties would likely not change readily or upon removal of SiO_2. Finally, it is evident that more work is clearly needed in this field, also in view of the still wide-ranging variation in optical properties and deviations between experiment and theory. Collaborative efforts between synthetic chemists, spectroscopists and theoreticians are likely to be the most fruitful to clarify the issues at hand here.

1 B. V. Oliinyk, D. Korytko, V. Lysenko and S. Alekseev, *Chem. Mater.*, 2019, **31**, 7167–7172.
2 S. P. Pujari, H. Driss, F. Bannani, B. van Lagen and H. Zuilhof, *Chem. Mater.*, 2018, **30**, 6503–6512.

Ali Reza Kamali opened a general discussion of the paper by Ankit Goyal: In this work, silicon nanoparticles with particle sizes of less than 20 mm were ball-milled with attrition ball-milling equipment using WC balls for 20 h. It is mentioned that the temperature is raised to 500 °C during the ball-milling process. I have three questions: (1) How could the temperature inside the ball-milling jar be estimated? (2) In my opinion, 500 °C is very difficult to reach during normal ball-milling processes. One simple consequence of such a high temperature is that the high pressures caused by the expansion of gas inside the milling jar lead to either an explosion or a leakage problem. Can you comment on this please? (3) What was the level of W and C contamination from the WC milling media, and what was its effect on the optical performance of the Si nanoparticles?

Ankit Goyal responded: Thank you for the questions. We used ~20 micrometer particles as a starting powder. (1) Local temperatures at the surface of the sub-microscopic particles could reach up to 500 °C. In existing commercial or modified home-made ball-mills (such as ours), there is no provision for measuring the actual temperature inside the mill. Even if one could measure the temperature inside the chamber, maybe using IR detectors, it is very difficult to measure sub-microscopic temperatures during a continuous random process. 500 °C is a theoretical estimate. (2) In commercial ball-mills where sealed jars are used, it is very difficult to mill silicon or any hard materials. As you say, that could lead to a very high temperature and cause an explosion, but in our case, we have a modified ball-mill where we have a provision for continuous cooling to avoid too much heat and continuous gas flow so the milling jar is not pressurized at any moment during the process. (3) I haven't seen any traces of tungsten in the sample. I saw SiC formation in the FTIR so I would assume some carbon would have come from the milling media, but exact quantification wasn't done. I assume it would be less than one part per thousand (ppt). Carbon caused the formation of a Si–C surface on the silicon core, which could be one possible reason for the multi-phononic PL peak we saw in the PL spectrum, but it could also be due to the presence of metallic impurities so this part is still to be figured out.

Holger Stephan opened a general discussion of the paper by Giacomo Morselli: Have you carried out dynamic light scattering (DLS) measurements with ammonium-terminated silicon nanocrystals suspended in methanol/water in order to get information about the size distribution and zeta potential?

Giacomo Morselli replied: We performed both the analyses suggested by the kind reader. However, in an aqueous solution of KCl (1 mM) we observed aggregation (DLS size about 100 nm and zeta potential +14 mV), a slight quenching and blue-shift over time, probably due to the oxidation of the surface. This is the reason why we aimed to further functionalize the surface with amide bonds (to carboxymethyl PEG or other biological molecules), capping all the amino groups and protecting the surface. Moreover, amino-functionalised SiNCs are reported to be cytotoxic, and this is another reason to carry out the amide coupling. We didn't perform the DLS measurements in methanol, in which the nanocrystals seem to be more stable (well-suspended and maintaining their photoluminescence after more than one year). This could be interesting and we thank the reader very much for the suggestion.

Holger Stephan remarked: Are coupling reactions of ammonium-terminated silicon nanocrystals with biological molecules such as peptides also planned? This would be very interesting.

Giacomo Morselli answered: Yes, we are planning to functionalise SiNCs with peptides and, more generally, each molecule that we aim to link to the SiNC surface could be interesting for ameliorating their properties and widening their applications.

Carina Crucho asked: I wonder if the authors tried to do a quantification of the amine groups with the ninhydrin test? This is an important result. It is also important for the complete characterization of the material. In addition, this is widely used in the characterization of amine functionalized silica nanoparticles, as for example in the work developed by Sun *et al.*[1] The authors could then compare the results with the ones in the paper by Stephan and coworkers from the present Faraday Discussion (DOI: 10.1039/c9fd00091g).

1 Y. Sun, F. Kunc, V. Balhara, B. Coleman, O. Kodra, M. Raza, M. Chen, A. Brinkmann, G. P. Lopinski and L. J. Johnston, *Nanoscale Adv.*, 2019, **1**, 1598–1607, DOI: 10.1039/c9na00016j.

Giacomo Morselli responded: We performed a quantification of the amino groups from the NMR spectra, comparing with a standard at a known concentration. About 2 chlorosilanes out of 3 reacted with the trityl amine. This is due to the steric hindrance of the triphenylmethyl group.

Carina Crucho commented: I work with silica nanoparticles and I use quantification by solution-state ^{1}H-NMR to quantify organic moieties (hydrophobic and hydrophilic) attached to the nanoparticle surface.[1] I know from experience that in the case of amine groups, as the protonated amines have solubility in water, it is possible to see the NMR peaks of the protons attached to the carbon next to the amine group or next to the silicon atom (even without dissolution of the silica matrix) if the NMR spectra are recorded in D_2O. In Fig. 1 of the paper (DOI: 10.1039/c9fd00089e), spectra (a) and (b) were recorded in different NMR solvents, so we can't compare. I wonder if the authors tried different NMR solvents or mixtures of NMR solvents, for example, DMSO and D_2O.

1 C. I. C. Crucho, C. Baleizão and J. P. S. Farinha, *Anal. Chem.*, 2017, **89**(1), 681–687, DOI: 10.1021/acs.analchem.6b03117.

Giacomo Morselli replied: Unfortunately, the tritylamine functionalised SiNCs and the deprotected ones could not be suspended in the same solvent. So, we could only assess the fading of the aromatic groups in the second spectrum. We thank the reader for the suggestion.

Carina Crucho remarked: I wonder if the authors considered using zeta potential measurements to monitor changes on the particle surface?

Giacomo Morselli answered: We performed zeta potential measurements in 1 mM KCl (aq), yielding a potential of +14 mV (DLS measurements confirmed slight aggregation of the nanoparticles in the same solution). Moreover, a slight quenching and blue-shift was observed in aqueous solution over time, probably due to the oxidation of the surface. This is the reason why we aimed to further functionalize the surface with amide bonds (to carboxymethyl PEG or other biological molecules), capping all the amino groups and protecting the surface.

Ankit Goyal remarked: I have some questions regarding the photophysical properties of the synthesized SiNCs: 1. The PL spectrum shown in Fig. 4(a) of the paper (DOI: 10.1039/c9fd00089e) looks very broad (~520 nm to 1000 nm) which suggests a broad size distribution. Have you measured the size distribution or done some size separation to see size dependant emission? What does the PLE look like for the obtained PL spectrum, and please mention the excitation wavelength used to obtain this PL spectrum.

Giacomo Morselli responded: The emission spectrum looks very broad because it is reconstructed and the band shape is affected by this procedure: due to the position of the emission of the sample, we had to record it with two different detectors, one for visible light (excitation at 452 nm) and one for the NIR spectral range (excitation at 610 nm), and then combine the two spectra. Unfortunately, the maximum lies exactly in the middle of the sensitivity zones for the two detectors, resulting in a hollow at about 750 nm. With our analysis software we could "flatten" the zone, but there should be a maximum instead. The PLE spectrum is the green line in Fig. 8 of the paper (DOI: 10.1039/c9fd00089e).

Ankit Goyal commented: 2. In Fig. 5 of the paper (DOI: 10.1039/c9fd00089e), the absorption spectrum also looks shifted along with the PL of the SiNCs. Could you please comment on how oxidation happened in water but not in ethanol? I assume "A" is absorbance here?

Giacomo Morselli replied: It is really difficult to assess if the absorbance is shifted: the sample was more dilute than the one in ethanol and so the curve is lower. Anyway, the oxidation also occurs in ethanol, but at a much slower rate.

Ankit Goyal remarked: 3. You mentioned that upon excitation at 340 nm, 70% of light is absorbed by pyrene and 30% of light is absorbed by the silicon core. Could you please comment on how you quantified light absorbance here?

Giacomo Morselli responded: The absorption spectrum of pyrene-functionalised silicon nanocrystals closely matches the absorption spectrum of the pyrene chromophore and the pristine silicon nanocrystals, demonstrating no significant interaction between the chromophores and the silicon core in the ground state. Therefore, superimposing the absorption spectra of pyrene-free silicon nanocrystals and pyrene-functionalised silicon nanocrystals with the same silicon concentration (Fig. 6 of the paper, DOI: 10.1039/c9fd00089e), the difference in the absorbance at 340 nm is due to the pyrene moieties, which is 70% of the total.

Ankit Goyal commented: 4. You have mentioned that: "The same value (3.7 ns) corresponds to the rise time in PL emission of the SiNCs at 700 nm upon excitation at 340 nm, which proves the occurrence of energy transfer from the pyrenes to the silicon core." This is quite unclear. In which figure do you see emission at 700 nm?

Giacomo Morselli responded: The emission of the silicon core at 700 nm is visible in Fig. 7a and b (dashed green line) of the paper (DOI: 10.1039/c9fd00089e).

Holger Stephan opened a general discussion of the paper by Yao He: Have you investigated how stably europium is bound within Eu@SiNPs in cell medium in the presence of challenging proteins?

Yao He replied: Thanks for your valuable comments. In our case, the Eu@SiNP nanothermometer maintains stable fluorescence intensity for 48 h of storage in cell media (DMEM and RPMI 1640).

Holger Stephan commented: Do you believe that the Eu@SiNP-based nanothermometer can also be used *in vivo*, for example to differentiate between normal and highly proliferative cells?

Yao He responded: The Eu@SiNP-based nanothermometer is useful for *in vitro* applications but it is not suitable *in vivo* due to its blue emission. To address this issue, we plan to design new kinds of NIR-emitting nanothermometers in the future, which would be available for *in vivo* applications.

Wei Sun remarked: Prof. He is a very well-known expert in the field of bio-applications of silicon nanostructures. This work is also innovative in its nano-thermometer use. I would just like to draw his attention to the fact that a recent report has different thoughts on the synthesis, optical properties and definition of SiNPs synthesized using APTMS and trisodium citrate dihydrate (Na$_3$Cit) under microwave heating.[1] The authors showed that even without a Si source, similar luminescent nanomaterials were also generated under the reaction conditions. The paper by de Cola and coworkers in this issue of *Faraday Discussions* (DOI: 10.1039/c9fd00127a) has also reported similar concerns. Personally, I think the luminescence obviously did not originate from the band gap of SiQDs, and the NPs synthesized seem to be a composite of Si, C, N and O, as seen from the XPS results where Si appeared to be mostly in its higher oxidation states, very different from the Si(0) nanoparticles synthesized from thermal disproportionation or non-thermal plasma. Nevertheless, since the particles contained Si and performed well in the demonstrated applications, I appreciate the work, so this is just something interesting with respect to the description and definition of "SiNPs" that I would like to mention.

1 B. V. Oliinyk, D. Korytko, V. Lysenko and S. Alekseev, *Chem. Mater.*, 2019, **31**, 7167–7172.

Yao He answered: We agree with these opinions that the luminescence did not originate from the band gap of SiQDs, and the NPs synthesized seem to be a composite of Si, C, N, and O. Based on the optical characterization (*e.g.*, Fig. 4 and 5, DOI: 10.1039/c9fd00088g) and previous reports,[1–4] we deduce that the luminescence of Eu@SiNPs is contributed by both the quantum-size effect and surface properties (*e.g.*, surface defects and surface ligands). On the other hand, we thank you for your kind reminder about the recent report regarding the difference between SiNPs and CNPs (carbon NPs). We actually noted the paper several months ago, and then made a systematic comparison between SiNPs (synthesized using APTMS and trisodium citrate dihydrate) and CNPs (synthesized without a Si source). Based on our experimental data, both SiNPs and CNPs produce similar luminescence; however, they show distinctive differences in terms of optical stability (*e.g.*, pH and storage stability) and biological effects (*e.g.*, intracellular distribution). Moreover, XPS and element analysis provide convincing evidence that the SiNPs contain Si element, consistent with the above comment (*i.e.*, "the NPs synthesized seem to be a composite of Si, C, N, and O").

We really appreciate these precious comments and suggestions, thanks again!

1 X. Shen, B. Song, B. Fang, X. Yuan, Y. Li, S. Wang, S. Ji and Y. He, *Nano Res.*, 2019, **12**, 315–322.
2 S. Ciampi, J. B. Harper and J. Gooding, *Chem. Soc. Rev.*, 2010, **39**, 2158–2183.
3 M. Dasog, J. Kehrle, B. Rieger and J. G. C. Veinot, *Angew. Chem., Int. Ed.*, 2016, **55**, 2322–2339.
4 J. M. Lauerhaas and M. J. Sailor, *Science*, 1993, **261**, 1567–1568.

Ming Lee Tang opened a general discussion of the paper by Anna Fucikova: Hi there. I read your 2014 *ACS Photonics* paper[1] and was very surprised to see the narrow linewidth of the silicon nanoparticles in your report since my group routinely measures broad photoluminescence linewidths. Here you show similar experimental observations – very intriguing. You include TEM, but is there XRD data showing that nanocrystalline Si is present?

1 I. Sychugov, A. Fucikova, F. Pevere, Z. Yang, J. G. C. Veinot and J. Linnros, *ACS Photonics*, 2014, **1**, 998–1005.

Anna Fucikova responded: Thanks for the question. We were surprised by the narrow linewidths as well, therefore we tried to do XRD but there were too few nanoparticles present in the sample so we could not get reliable data. Although the synthesis produces perfect nanocrystals, it does not produce a sufficient amount of them for mass characterization methods. Therefore, we decided to test our idea differently: we prepared a set of samples which only differed in the length of the molecule and had a ketone group in the molecule (we used acetone, methyl isobutyl ketone, and ethyl acetate) – our results are summarized in this *Faraday Discussions* paper (DOI: 10.1039/c9fd00093c). We confirmed that the shorter the molecule, the thinner the shell and the narrower the linewidth, which is in agreement with our theory from the 2014 *ACS Photonics* paper.[1] The crystallinity of the nanoparticles was confirmed with TEM but also indirectly *via* optical characterization of the single silicon nanocrystals. They have the same properties as silicon nanocrystals with a silicon oxide shell except for the linewidth. They blink, spectral hopping is observed, they respond similarly to higher excitation power, and their luminescence properties respond in the same way to lowering the temperature except that we do not observe any phonon replicas in the spectra of single nanocrystals. This might contribute to the even narrower linewidth and also indicates that the shell is uniform and rigid.

1 I. Sychugov, A. Fucikova, F. Pevere, Z. Yang, J. G. C. Veinot and J. Linnros, *ACS Photonics*, 2014, **1**, 998–1005.

Paola Ceroni remarked: Q1: Would it be possible to liberate your SiNCs from the surrounding matrix and make measurements by dispersing them in solution?

Anna Fucikova responded: Thank you for this question. We successfully liberated the silicon nanocrystals from the matrix and the photoluminescence study of the individual nanocrystals showed the same photoluminescence linewidth distribution as the nanocrystals in the matrix. Unfortunately the sample concentration was not sufficient to perform the measurements in solution.

Paola Ceroni asked: Q2: Do you know the emission quantum yield of your particles?

Anna Fucikova answered: Since the amount of sample produced is so low, we could not measure the absolute quantum yield. We have just a rough estimate based on a comparison of the spectral blinking of individual nanocrystals with thin shells and silicon nanocrystals with silicon oxide shells: our samples blinked

2–3 times less than the ones with silicon oxide shells. Based on this, the QY is higher than 30% as a minimum.

Paola Ceroni remarked: Q3: What is the experimental evidence for spectral hopping? How can you estimate a 3% spectral hopping?

Anna Fucikova responded: The spectral hopping was observed by photoluminescence spectral studies of single nanocrystals. More than 200 nanocrystals were spectrally analysed for each sample, and out of these spectra of individual nanocrystals, around 3% showed spectral hopping.

Ali Reza Kamali opened a general discussion of the paper by Jos M. J. Paulusse: In this research, the synthesis of Si QDs by oxidation of magnesium silicide (Mg_2Si) with bromine (Br_2) is reported. Also, a control sample is prepared without the involvement of the Si source. The reaction involving Mg_2Si produced 24.4 mg of a dark brown material. On the other hand, the reaction without the presence of Mg_2Si also produced 24.5 mg of a dark brown material. The Si amount in the Si QDs and the control sample was found to be 2.2 and 0.4 wt%, respectively. 1. Considering that there was no Si source (Mg_2Si) involved in the preparation of the control sample, what is the source of the Si observed in the elemental analysis of the control sample?

Jos Paulusse responded: The detected Si content is low. We think the source of Si in the control reaction is possibly grease or lubricants on the glassware/equipment used (though care was taken to avoid the use of silicone-based grease).

Ali Reza Kamali asked: 2. Which kind of elemental analysis was used to evaluate the silicon content of the samples?

Jos Paulusse replied: The Wurzschmitt digestion was used for digestion of silicon, followed by UV-Vis analysis.

Ali Reza Kamali asked: 3. How was the formation of crystalline silicon confirmed? Which characterization technique was used to confirm this?

Jos Paulusse responded: The formation of crystalline silicon was not confirmed. TEM was used to detect any crystalline silicon, but no crystalline materials were observed.

Katerina Kusova remarked: This publication questions the wet-chemical synthesis of SiQDs by the oxidation of Mg_2Si with Br_2 and calls for a critical assessment of other synthesis techniques. I have more of a comment than a question here. When I measured a sample prepared by the solution-phase reduction of $SiCl_4$ with $LiAlH_4$, I compared the PL of the sample and that of a pure solvent. The PL was about 100x higher in the sample, suggesting that the observed PL came from SiQDs. However, a more thorough analysis (spectra, decays, several excitation wavelengths) showed that the PL of the SiQD sample and that of the pure solvent were exactly the same, this time proving that the product of the synthesis only amplified the (very weak) PL of the solvent. Do the authors have experience with a similar problem?

Jos Paulusse replied: We indeed have indications that the reduction synthesis also produces small molecule materials with blue fluorescence emission. These molecules are possibly based on only several silicon atoms instead of larger clusters.

Conflicts of interest

There are no conflicts to declare.

Faraday Discussions

PAPER

The influence of hydrofluoric acid etching processes on the photocatalytic hydrogen evolution reaction using mesoporous silicon nanoparticles†

Sarah A. Martell, Ulrike Werner-Zwanziger and Mita Dasog *

Received 30th September 2019, Accepted 16th October 2019
DOI: 10.1039/c9fd00098d

H_2 has been identified as one of the potential energy vectors that can provide a sustainable energy supply when produced through solar-driven water-splitting reaction. Si is the second most abundant element in the Earth's crust and can absorb a significant fraction of the solar spectrum while presenting little toxicity risk, making it an attractive material for photocatalytic H_2 production. Hydrogen-terminated mesoporous Si (mp-Si) nanoparticles can be utilized to effectively drive the hydrogen evolution reaction using UV-to-visible light. In this work, the response of the photocatalytic activity of mp-Si nanoparticles to a series of HF acid treatments was investigated. A two-step magnesiothermic reduction method was used to prepare crystalline mp-Si nanoparticles with a specific surface area of 573 m² g⁻¹. The HF etching process was optimized as a function of the amount of acid added and the reaction time. The reaction time did not influence the H_2 evolution rate substantially. However, the amount of HF used did have a significant effect on the photocatalytic activity. In the presence of ≥1.0 mL HF acid per 0.010 g of Si, morphological damage was observed using electron microscopy. N_2 adsorption measurements indicated that the pore size and surface area were also altered. Solution-phase $^{19}F\{^1H\}$ NMR studies indicated the formation of SiF_5^- and SiF_6^{2-} when larger volumes of HF were used. Both factors, morphological damage and the presence of byproducts in the pores, likely result in a lowering of the photocatalytic H_2 evolution rate. Under the optimized HF treatment conditions (0.5 mL of HF per 0.010 g of Si), a H_2 evolution rate of 1398 ± 30 μmol g⁻¹ h⁻¹ was observed.

Introduction

H_2 is an energy carrier that can be combusted without the release of CO_2 and is also an important chemical feedstock.[1,2] Industrially, it is predominantly produced *via*

Department of Chemistry, Dalhousie University, 6274 Coburg Road, Halifax, NS, Canada. E-mail: mita. dasog@dal.ca

† Electronic supplementary information (ESI) available: N_2 adsorption isotherms, pore size distribution plots, Kubelka–Munk function of hydrogen-terminated mp-Si nanoparticles, and ssNMR spectra. See DOI: 10.1039/c9fd00098d

steam reforming of natural gas which is estimated to generate 8.6 tons of CO_2 per ton of H_2.[3,4] As an alternative to this high carbon-footprint process, solar-driven water reduction systems have been explored extensively to sustainably produce H_2.[5-7] Coupled photovoltaic (PV)-electrolyzer systems,[8-10] integrated photoelectrochemical (PEC) cells,[11-13] and photobiological systems[14-16] have all been demonstrated to enable solar-to-hydrogen conversion. Recent techno-economic analysis has shown that the direct conversion of water to H_2 using dispersed particulate photocatalysts may be a simple and cost-effective alternative.[17,18]

The common semiconductors investigated for particulate photocatalytic water-splitting are metal oxides that absorb in the UV-region or nitrides, oxynitrides, or doped metal oxides that absorb visible light.[19-25] While bulk Si has been extensively explored for water-splitting reactions in PV-electrolyzer and PEC systems,[26-28] its application in particulate systems has been limited. This is due to the low energy difference between the conduction band-edge of Si and the water reduction potential, thus limiting efficient electron transfer from Si particles to the solution.[29] However, quantum confinement increases the bandgap of Si (1.3–2.0 eV) from its bulk value (1.1 eV), which shifts the conduction band-edge position and allows for efficient electron transfer to drive H_2 formation.[30] Recent studies on photocatalytic H_2 generation using particulate Si have focused on mesoporous nanostructures, because the porosity increases the number of catalytic sites and enhances light absorption *via* light trapping mechanisms.[29,31,32] Due to these properties, mesoporous Si (mp-Si) generally outperforms conventional Si nanoparticles in H_2 generation efficiency.

Along with porosity, the surface chemistry of Si also plays an important role in determining the efficiency of the photocatalytic reactions. Surface oxides can introduce trap states where photo-generated charge-carriers can recombine instead of performing fuel forming reactions.[31] One way to mitigate this issue is to treat Si with HF to remove the surface oxides.[33-35] However, the relationship between the etching routine and the observed photocatalytic activity, if any, has not been explored. Optimizing the etching conditions may enable the realization of superior photocatalytic activity using mp-Si nanoparticles. In this study, the influence of the amount of HF and the etching time on the photocatalytic H_2 generation using mp-Si nanoparticles is reported. The porous nanoparticles were prepared *via* magnesiothermic reduction of SiO_2 and treated with different volumes of 48% HF solution ranging between 0.2 and 5.0 mL to yield hydrogen-terminated surfaces. These nanoparticles were explored for photocatalytic H_2 generation under 100 mW cm^{-2} of broadband illumination. The structural properties of the etched mp-Si nanoparticles were studied using electron microscopy, N_2 adsorption, and solution- and solid-state NMR (ssNMR) techniques.

Results

The precursor silica was synthesized through a base-catalyzed Stöber sol–gel method which yielded an amorphous product as indicated by the broad peak at 22° in the powder X-ray diffraction (XRD) pattern (Fig. 1A).[36] Scanning electron microscopy (SEM) analysis showed spherical nanoparticles with an average diameter of 296 ± 11 nm (Fig. 1B). SiO_2 nanoparticles were reacted with Mg powder initially at 650 °C for 0.5 h followed by 300 °C for 6 h under Ar atmosphere. We recently demonstrated that this two-step annealing process results in

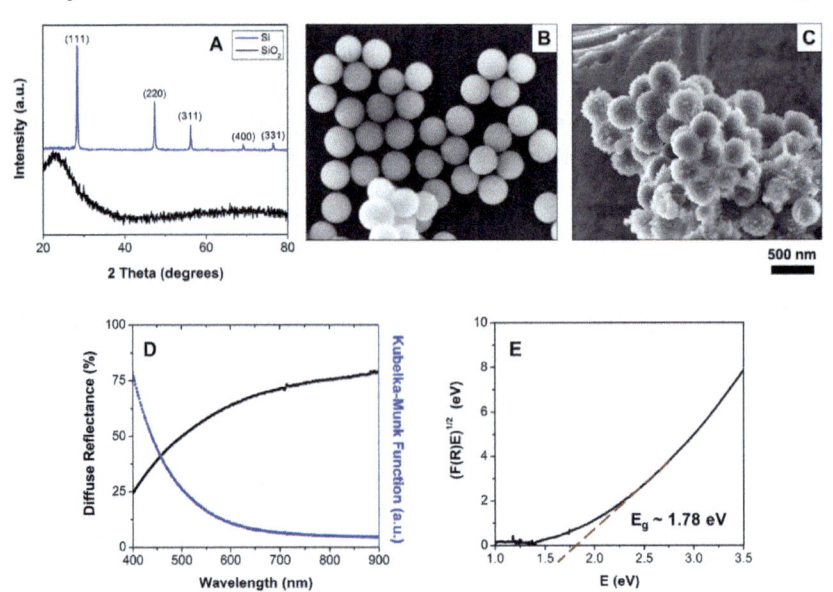

Fig. 1 (A) Powder XRD pattern of SiO$_2$ and the mp-Si nanoparticles, SEM images of (B) SiO$_2$ and (C) the mp-Si nanoparticles, (D) diffuse reflectance and Kubelka–Munk function and (E) Tauc plot of the mp-Si nanoparticles.

crystalline and high surface area mp-Si nanoparticles without a large amount of sintering.[37] This is crucial as sintering can increase the distance that the carriers have to travel and introduce grain boundaries with trap states, both of which can lead to carrier recombination. After the reaction, the resulting product was cooled to room temperature and then treated with 1.0 M HCl solution to remove MgO and any unreacted Mg. After 4 h of HCl reaction, a brown-solid was collected via filtration and dried overnight. The powder XRD pattern of the isolated product showed characteristic reflections at 28.5°, 47.4°, 56.2°, 69.2°, and 76.6° corresponding to the (111), (220), (311), (400), and (331) planes of diamond structured Si (Fig. 1A).[37] The SEM image showed mp-Si particles which are mostly spherical with a rough surface (Fig. 1C). The specific surface area of the mp-Si nanoparticles was determined to be 573 ± 32 m^2 g^{-1} using the Brunauer–Emmett–Teller (BET) method (Fig. S1A, ESI†).[38] The average pore diameters were found to be 10 and 28 nm using the Barrett–Joyner–Halenda (BJH) model (Fig. S1B†).[39] The diffuse reflectance spectrum and corresponding Kubelka–Munk function of the mp-Si nanoparticles is shown in Fig. 1D. The bandgap (E_g) was estimated to be 1.78 eV from the Tauc plot (Fig. 1E).

The mp-Si nanoparticles were treated with a mixture of EtOH, H$_2$O, and varying volumes of 48% HF to yield hydrogen-terminated surfaces. The attenuated total reflectance infrared (ATR-IR) spectra of the mp-Si nanoparticles showed characteristic Si–H$_x$ stretches at ∼2080 cm^{-1} for all the etched samples (Fig. S2†).[30] All the particles showed some degree of oxidation as evidenced by the presence of Si–O–Si stretches at ∼1100 cm^{-1},[30] however, the intensity of these stretches decreased with increasing amount of HF. After treatment with HF, the hydrogen-terminated mp-Si nanoparticles were isolated, rinsed with DI-water, and quickly transferred to a Pyrex reaction vessel for photocatalysis experiments. The amount of H_2 evolved was quantified using gas chromatography (GC) and is summarized in Fig. 2. A representative H_2 evolution trend over the course of 5 h is shown in Fig. 2A. The unetched mp-Si nanoparticles had an average H_2 evolution rate of 838 ± 22 μmol h^{-1} g^{-1} of Si in 5 h. The hydrogen-terminated mp-Si nanoparticles had an average H_2 evolution rate of 815 ± 26, 1194 ± 18, 1060 ± 24, 775 ± 36, and 516 ± 45 μmol h^{-1} g^{-1} of Si when 0.2, 0.5, 1.0, 2.0, and 5.0 mL of 48% HF was used for the etching process, respectively. The apparent quantum efficiencies (AQE) of the mp-Si nanoparticles were calculated at 420 nm. The values were found to be 2.3 ± 0.3, 2.1 ± 0.4, 3.1 ± 0.4, 2.5 ± 0.4, 1.9 ± 0.3, and 1.2 ± 0.4% when no HF, 0.2, 0.5, 1.0, 2.0, and 5.0 mL of 48% HF was used for the etching process, respectively.

The morphology of the mp-Si nanoparticles before and after treatment with HF acid was investigated using transmission electron microscopy (TEM). Similar to the SEM image (Fig. 1C), the TEM image of the unetched sample (Fig. 3) mostly contained porous spherical nanoparticles. In the presence of 0.2 mL of HF, no morphological damage was observed, and the sample mostly consisted of intact nanoparticles. More broken nanoparticles were observed when 0.5 mL of HF was used. However, in the presence of 1.0, 2.0, and 5.0 mL of HF, significant damage was observed and none of the original spherical particles were visible. Instead, smaller non-spherical pieces (<250 nm) and particles ranging between 6 and 10 nm were observed. The absorption properties (based on the Kubelka–Munk function) of the mp-Si nanoparticles treated with 1.0–5.0 mL were similar (Fig. S3†). Particles treated with 0.2 mL were like the unetched particles and showed higher absorption between 300 and 550 nm compared to the 1.0–5.0 mL set whereas those etched with 0.50 mL had slightly higher absorption between 300 and 375 nm. The specific surface areas of the HF treated mp-Si nanoparticles were 560 ± 21, 428 ± 25, 243 ± 18, 149 ± 30, and 132 ± 24 m^2 g^{-1} when 0.2, 0.5, 1.0, 2.0, and 5.0 mL of HF was used, respectively (Fig. S4†). The mp-Si pore diameter decreased with increasing volume of HF. The particles treated with 0.2 mL of HF had a broad pore size distribution with an average diameter of around 18 nm (Fig. S5†). With increasing volume of HF, pores with ~4 nm diameters began to form and the relative number of pores between 15 and 30 nm in diameter decreased (Fig. S5†).

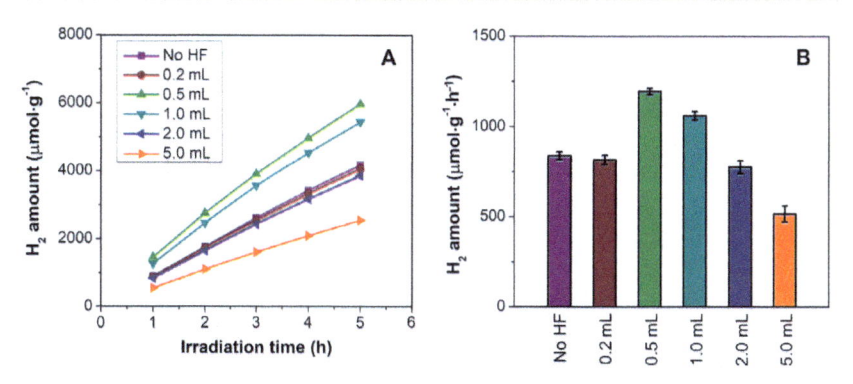

Fig. 2 (A) Representative reaction time course of photocatalytic H_2 evolution and (B) average H_2 evolution rates observed with mp-Si nanoparticles treated with varying volumes of HF under broadband illumination with 100 mW cm^{-2} power density.

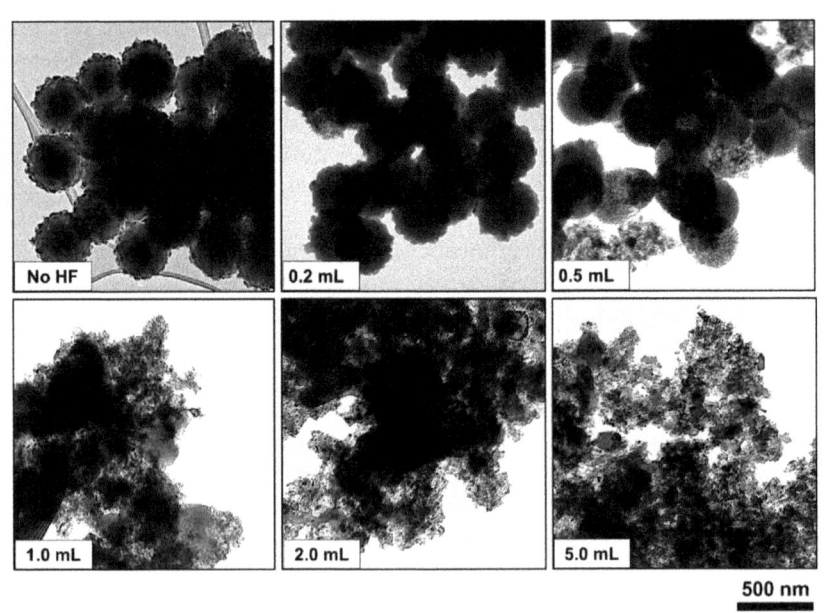

Fig. 3 TEM images of mp-Si nanoparticles treated with varying volumes of 48% HF solution.

To complement the TEM and N_2 adsorption studies which provided information on the morphological differences in the mp-Si nanoparticles and investigate if any of the HF etching products or byproducts were contained within the pores of the mp-Si, NMR studies were performed on nanoparticles dispersed in acetone-d_6. The $^{19}F\{^1H\}$ NMR spectra showed a broad peak between −178 and −185 ppm for all samples. The peak width of the signal increased with increasing volume of HF (Fig. 4B). HF signals have previously been reported to be broad peaks with chemical shifts observed here and line widths changing as a function of their concentration.[40–42] When 0.2 and 0.5 mL of HF was used, a second broad peak was observed around −151 ppm (Fig. 4A) which is assumed to be HF in a different chemical environment. For volumes ≥0.5 mL of HF, two sharp signals between −151 and −152 ppm were observed which is consistent with the presence of SiF_5^- (smaller peak) and SiF_6^{2-} (larger peak).[43,44] The intensity of both SiF_5^- and SiF_6^{2-} signals increases with increasing volume of HF. ^1H ssNMR experiments on hydrogen-terminated silicon nanoparticles have previously reported chemical shifts ranging between 6 and 3 ppm for the H atoms.[45] The ^1H ssNMR shown in Fig. S6† exhibits peaks between 7 and 0 ppm which likely result from the Si–H and Si–OH groups.[45,46] The intensities of the signals between 3 and 0 ppm reflect the increased Si–H_x species with increasing HF amount, while those between 7 and 3 ppm are already present in the unetched sample and likely belong to Si–OH. In the solution-phase ^1H NMR experiments, the signal for hydrogen-terminated mp-Si nanoparticles was found to shift with differing HF volume (Fig. 4C). The ^1H NMR peak shifted from approximately 4.7 to 3.3 ppm when the HF volume changed from 5.0 to 0.2 mL, respectively. A quartet with a chemical shift ~3.6 ppm was observed in all spectra, corresponding to ethanol –CH_2– protons.[47]

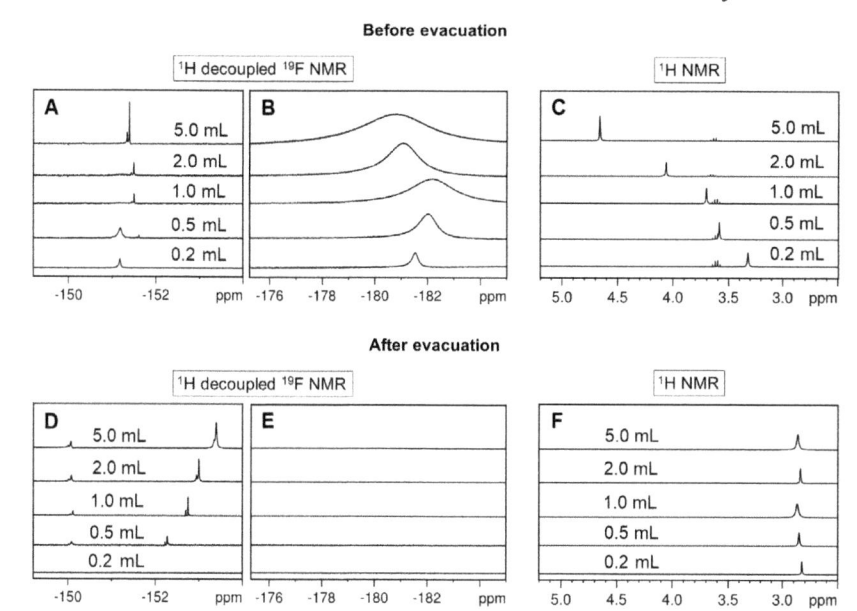

Fig. 4 (A, B, D and E) ^{19}F{^{1}H} NMR and (C and F) ^{1}H NMR spectra of mp-Si nanoparticles treated with HF before and after evacuation.

To remove the potentially trapped or adsorbed molecular species from the nanoparticles, the mp-Si was evacuated for 1 h and further investigated using NMR. ^{19}F{^{1}H} NMR spectra of the evacuated samples showed that the HF had been successfully removed through the disappearance of the peaks between −178 and −182 and at −151 ppm (Fig. 4D and E). However, multiple sharp signals were observed with chemical shifts ranging between −150 and −154 ppm which could be due to SiF_5^- and SiF_6^{2-} in multiple chemical environments (Fig. 4D). In the ^{1}H NMR spectra of the evacuated samples, the signals resulting from the hydrogen-terminated Si surface did not show much variation in chemical shift and all appeared between 3.0 and 2.8 ppm (Fig. 4F). This indicates that HF was the major species responsible for the downfield shifting of the ^{1}H NMR signal when higher amounts of HF were used.

The unetched mp-Si nanoparticles and those treated with 0.20–1.0 mL of HF, were analyzed using ^{29}Si{^{1}H} CP/MAS ssNMR. This technique allows detection of Si atoms that are directly bound to protons (e.g. Si–H$_x$ and Si–OH species), as well as those within close proximity of protons. The unetched mp-Si nanoparticles showed multiple overlapping signals ranging from −50 to −109 ppm (Fig. S7†). Signals from Si Q-sites ((Si–O)$_{4-x}$Si(OH)$_x$) are observed between −60 and −120 ppm.[48] The signal for elemental Si is typically found between −60 and −85 ppm.[45,49] Although the absence of nearby protons suppresses the signals for the crystalline Si-core in the ^{29}Si{^{1}H} CP/MAS ssNMR spectra, they are observed in the directly excited ^{29}Si ssNMR spectra at −80.9 ± 1 ppm. It is likely that the ^{29}Si NMR spectrum of the unetched nanoparticles has signals resulting from Q-sites and elemental Si atoms that are in proximity to -OH protons. The peaks at −5.1 and −20.4 ppm have previously been reported to result from (OH)$_n$Si(-Si)$_m$(-OSi)$_{4-n-m}$ species.[50] The chemical identity of the species at −50 ppm is currently unclear. The 0.2, 0.5, and

1.0 mL etched samples had peaks around −95 ppm, likely corresponding to Si–H$_x$ moieties (Fig. S7†). However, the Q-site Si atoms also have chemical shifts in that range complicating the determination of definitive assignments.[49] The overlay plot of the unetched and etched samples shows that the width of this peak in the etched samples is narrower (Fig. S8†). The signals between −61 and −65 ppm are typical of Q^0-site Si atoms.[48] There were two peaks between −5 and −25 ppm in the etched samples. Signals in this region have previously been assigned to incompletely etched surfaces such as HSi(–SiH$_3$)$_2$–OSiH$_3$ or HO(H)Si(–SiH$_3$)$_2$ but can also correspond to the suboxide species observed in unetched samples.[50]

The evacuated mp-Si nanoparticles were investigated for photocatalytic H$_2$ evolution using a similar set-up to that used before and the comparative rates before and after evacuation are shown in Fig. 5. The unetched samples had similar H$_2$ evolution rates before and after evacuation, which were 838 ± 22 and 843 ± 21 μmol h^{-1} g^{-1} of Si, respectively. The highest increase in evolution rate after evacuation was observed for the 0.5 mL HF etched sample, which had a value of 1398 ± 30 μmol h^{-1} g^{-1} of Si. There was also a slight increase with the 0.2 and 1.0 mL HF etched samples which had H$_2$ evolution rates of 920 ± 23 and 1156 ± 32 μmol h^{-1} g^{-1} of Si, respectively. The 2.0 mL (812 ± 40 μmol h^{-1} g^{-1}) and 5.0 mL (592 ± 41 μmol h^{-1} g^{-1}) samples had H$_2$ production rates similar to those of the samples before evacuation was performed.

The reaction of mp-Si nanoparticles with 48% HF solution was optimized as a function of etching time. A 0.5 mL volume was used for these studies and the particles were reacted for 1, 5, 10, and 15 min. The photocatalytic H$_2$ evolution rate was found to be 1340 ± 50, 1398 ± 30, 1395 ± 32, and 1403 ± 27 μmol h^{-1} g^{-1} of Si for 1, 5, 10, and 15 min, respectively.

Discussion

The ∼300 nm mp-Si nanoparticles were synthesized using a previously reported magnesiothermic reduction method which resulted in a porous 3-D network of

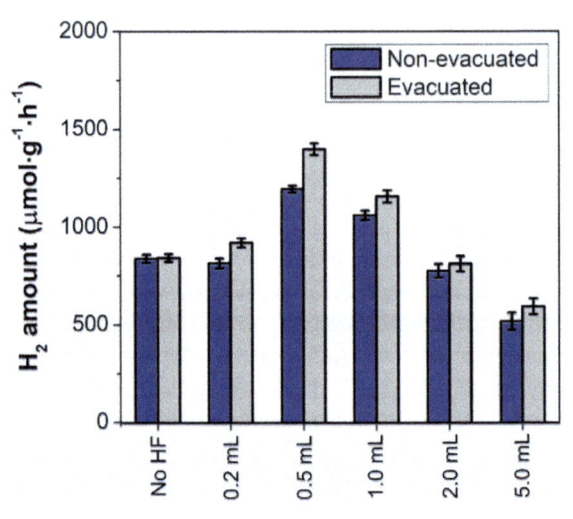

Fig. 5 Average H$_2$ evolution rates observed with mp-Si nanoparticles treated with varying volumes of HF under broadband illumination before and after the evacuation step.

nano-sized (3–10 nm) primary Si particles.[51] The small size of the primary particles reduces the carrier diffusion lengths and increases the bandgap, both of which should allow for better carrier transfer to generate H_2 from water.[29,31,52] The as-prepared (unetched) particles had a photocatalytic H_2 evolution rate of 838 ± 22 μmol h^{-1} g^{-1} of Si which is higher than that of previously reported mp-Si generated *via* magnesiothermic reduction.[32] This shows that the two-step reduction method results in a higher performing material. Both IR (Fig. S2†) and ^{29}Si ssNMR (Fig. S7†) characterization showed the presence of oxide species in the unetched mp-Si nanoparticles. To further improve the H_2 evolution rates, attempts were made to remove the oxide using HF acid according to the following reaction:[35]

$$SiO_2 + 4HF \rightarrow SiF_4 + 2H_2O$$

Volumes of 48% HF varying between 0.2 mL and 5.0 mL were investigated for the oxide etching step. This should result in mp-Si nanoparticles with hydrogen-terminated surfaces, and this was confirmed by the presence of Si–H$_x$ stretches in the IR (Fig. S2†) and ^{1}H NMR (Fig. 4C and S6†) spectra. However, oxide species were detected in all the etched samples at some concentration in both the IR (Fig. S2†) and the ^{29}Si ssNMR spectra (Fig. S7†). This shows that either the etching process is incomplete, or some degree of oxidation occurs during sample preparation. ^{19}F{^{1}H} NMR characterization of the etched samples showed that residual HF was present in all the samples (Fig. 4B). This had a de-shielding effect on the chemical shift of the hydrogen atoms attached to mp-Si nanoparticles. As the amount of HF present decreased the chemical shift of the hydrogen-terminated mp-Si nanoparticles also decreased from 4.7 to 3.3 ppm (Fig. 4C). In the presence of ≥0.5 mL of HF, formation of the secondary reaction byproducts (as shown in the reactions below) SiF_5^- and SiF_6^{2-} was observed using ^{19}F{^{1}H} NMR (Fig. 4A):

$$SiF_4 + HF \rightarrow HSiF_5$$

$$SiF_4 + 2HF \rightarrow H_2SiF_6$$

The intensity of the penta- and hexa-fluorosilicate signals in the ^{19}F{^{1}H} NMR spectra increased with increasing HF volume. The volume of HF acid used did not significantly influence the absorption properties of the mp-Si nanoparticles (Fig. S3†).

The photocatalytic H_2 evolution rate using etched mp-Si nanoparticles was investigated. Despite having a negligible amount of residual byproducts, Si etched with 0.2 mL of HF did not have the highest activity. This is likely due to a higher concentration of oxide species, which result in carrier recombination. This is supported by the higher intensity of Si–O–Si stretches in the IR spectrum and the lower AQE. The mp-Si nanoparticles etched with 0.5 mL of HF resulted in the best performing photocatalyst in this study. This is likely due to the optimum combination of oxide removal and byproduct amounts. With increasing HF volume (*i.e.* ≥1.0 mL), the H_2 evolution rate decreased. This decrease could be due to (i) morphological damage, as observed in the TEM images (Fig. 3) or (ii) blocking of the catalytic sites due to the presence of a higher concentration of byproducts such as HF, SiF_5^-, or SiF_6^{2-}. The change in morphology when

≥1.0 mL of HF was used resulted in reduced specific surface area and formation of smaller pores, which can lower the number of catalytic sites and limit the diffusion of reactants and products. Both of these can result in reduced charge-carrier transfer (as reflected in the AQE measurements) and H_2 formation.

To study the effect of reaction byproducts, the mp-Si nanoparticles were evacuated after the etching process. The trend for the photocatalytic H_2 evolution rate after evacuation was the same as that before. The 0.2, 0.5, and 1.0 mL etched samples showed an increase in the rate, but the others remained similar to those before. The $^{19}F\{^1H\}$ NMR spectra showed the disappearance of all the peaks corresponding to HF, however, SiF_5^- and SiF_6^{2-} remained which were found to be in two different chemical environments. It is possible that the applied vacuum redistributes the fluorosilicate species within the pores of the mp-Si nanoparticles instead of effectively removing it. When in contact with water at neutral pH (such as through photocatalysis reaction), these fluorosilicate species will undergo hydrolysis to form SiO_2 as shown below:[43]

$$SiF_5^- + 2H_2O \rightarrow SiO_2 + 5F^- + 4H^+$$

$$SiF_6^{2-} + 2H_2O \rightarrow SiO_2 + 6F^- + 4H^+$$

Formation of the oxide can introduce trap states, and block pores and catalytic sites, which can all reduce H_2 evolution rates.[31,53,54] It is likely that both the change in particle morphology and formation of deleterious byproducts lowers the H_2 evolution rates, however, the current techniques do not allow us to confidently differentiate or quantify these contributions. The H_2 evolution rates did not change significantly with different etching times.

Conclusions

In this study, the influence of the HF etching routine on the H_2 evolution using mp-Si photocatalyst was investigated. HF treatment was performed to remove the oxide from Si particles and improve the catalytic performance. The highest H_2 evolution rate (1398 ± 30 µmol g^{-1} h^{-1}) was observed when 0.5 mL per 0.010 g of Si was used. A lower volume (0.2 mL) of HF did not remove enough of the oxide in mp-Si, whereas the higher volumes (≥1.0 mL) introduced reaction byproducts and morphological changes, all of which led to reduced H_2 evolution rates. This shows that care must be taken when an HF etching step is involved, especially with porous nanoparticles that can trap reaction byproducts. Reaction byproducts and changes in morphology can lower the catalytic rates more than the initial oxide itself.

Experimental

Materials

Ammonium hydroxide (NH_4OH, 28%) and magnesium −325 mesh powder (Mg, 99.8%) were purchased from Alfa Aesar. Tetraethyl orthosilicate (TEOS, 99.9%), hydrochloric acid (HCl, 37%), calcium chloride ($CaCl_2$) acetone-d_6, methanol, ethanol (95%), and benzene were purchased from Sigma-Aldrich. Hydrofluoric acid (HF, 48%) was purchased from ACP. All reagents were used as received without further purification.

Synthesis of Stöber SiO$_2$ nanoparticles

The precursor SiO$_2$ nanoparticles were synthesized using a sol–gel process. TEOS (30.0 mL) was added to 95% EtOH (700 mL) followed by the addition of aqueous 28% NH$_4$OH (60 mL) solution. The mixture was left to stir at room temperature for 18 h (ambient humidity was recorded at 60%). The white product was collected through vacuum filtration. The product was washed twice with 95% EtOH and re-collected through centrifugation at 3300 rpm for 15 min. It was then dried overnight in an oven at 100 °C.

Synthesis of mp-Si nanoparticles

Mp-Si was synthesized through a previously optimized two-step magnesiothermic reduction reaction. SiO$_2$ nanoparticles (0.220 g) and −325 mesh Mg powder (0.196 g) were mixed together using a mortar and pestle until a homogeneous gray powder was obtained. The powder was transferred to a ceramic reaction boat which was placed in a quartz tube under Ar atmosphere. The mixture was initially heated to 650 °C at a ramp rate of 10 °C min^{-1} and held at that temperature for 0.5 h, followed by 300 °C for 4 h. The reaction product was treated with HCl (1.0 M, 25.0 mL) for 6 h (stirred at 300 rpm). The mp-Si was then collected through suction filtration and washed with distilled H$_2$O (150 mL). The mp-Si was then dried overnight in an oven at 100 °C.

HF acid etching procedure

Caution: HF can be extremely dangerous and must be handled with great care. Distilled H$_2$O (5.00 mL) and EtOH (5.00 mL, 95%) were added to the mp-Si nanoparticles (10 mg). The mixture was sonicated for 10 s to improve dispersion. The desired amount of HF (0.2 mL, 0.5 mL, 1.0 mL, 2.0 mL or 5.0 mL; 48%) was added with stirring. The mixture was stirred for 5 min at 300 rpm. It was then separated into 2 aliquots and centrifuged for 6 min at 3300 rpm. The supernatant was decanted, and the etched mp-Si was transferred to the appropriate vessel for immediate analysis or experimental usage. A supersaturated solution of CaCl$_2$ was available at all times for neutralization purposes.

Photocatalytic hydrogen evolution experiments

The photocatalytic experiments were performed in a gas-tight Pyrex cell. The etched mp-Si nanoparticles were rinsed with 5.0 mL of DI-water and transferred to the reaction vessel which was flushed with N$_2$ for 0.5 h. 10.0 mL of H$_2$O and 5.0 mL of MeOH (sacrificial agent) were bubbled with N$_2$ for 0.5 h and transferred to the reaction flask containing the mp-Si nanoparticles under N$_2$ flow. A broadband LED (Thorlabs, SOLIS-2C) was used as the light source and the photocatalytic reactions were carried out at 100 mW cm^{-2} power density. The illumination power was calibrated using a silicon photodiode (Thorlabs). Gas from the reaction vessel headspace was analyzed periodically for H$_2$ quantification using gas chromatography with a thermal conductivity detector (Agilent 7890A) and N$_2$ was used as a carrier gas. The experiments were performed in triplicate to ensure reproducibility. Apparent quantum efficiency (AQE) measurements were performed by irradiating the catalyst solution for 1 h with a 420 nm LED source (Thorlabs). The efficiency was calculated using the equation

$$\text{AQE} = \frac{2 \times \text{number of evolved } H_2 \text{ molecules}}{\text{number of incident photons}} \times 100$$

Materials characterization techniques

Scanning electron microscopy (SEM) images were collected using a Hitachi S-4700 electron microscope. The samples were drop-cast from aqueous dispersions onto a silicon wafer. Transmission electron microscopy (TEM) images were obtained using a FEI TechaniTM 12 electron microscope with an accelerating voltage of 120 kV. TEM samples were prepared by drop-casting nanoparticle suspensions onto a 200 mesh carbon coated copper grid. The images were analyzed using ImageJ software (version 1.52a).

The XRD patterns were collected on a Rigaku Ultima IV X-ray diffractometer employing $Cu_{K\alpha}$ radiation (λ = 1.54 Å). The samples were placed on a zero-background Si wafer and spectra were collected at 3 counts per s.

N_2 adsorption measurements were performed using an ASAP 2020 analyser (Micromeritics, Norcross, GA, USA).

Diffuse reflectance measurements were performed on a Cary-5000 spectrometer with an integrating sphere and the Kubelka–Munk function was obtained using the relationship:

$$\frac{k}{s} = \frac{(1-R)^2}{2R}$$

where k is the absorption coefficient, s is the scattering coefficient, and R is reflectance.

Solution phase NMR spectra were collected on a Bruker AV-300 and AV-500C spectrometer. Acetone-d_6 was used to disperse the hydrogen-terminated mp-Si NMR samples. Benzene was added as an internal standard. Samples were prepared in 0.5 mm borosilicate glass NMR tubes. 64 scans were acquired for both ^1H and $^{19}F\{^1H\}$ (no baseline subtraction) NMR experiments. The spectra were recorded at 298 K. ^1H and ^{29}Si magic angle spinning (MAS) ssNMR experiments were carried out on a Bruker Avance NMR spectrometer with a 16.4T magnet (139.2 MHz ^{29}Si Larmor frequencies). The samples were spun at 6.5 kHz in rotors of 4 mm diameter. The ^{29}Si ssNMR spectra were accumulated using $^{29}Si\{^1H\}$ cross-polarization (CP) at 4 ms CP contact times acquiring between 14 400 and 3600 scans using a 1 s recycle delay. The chemical shift scale was referenced externally against Kaolin at −91.0 ± 0.5 ppm as the secondary chemical shift standard relative to TMS. ^1H ssNMR spectra were acquired with background suppression[55] adding 16 scans and referenced externally based on the ^{29}Si resonance using the IUPAC conversion factor.[56]

Conflicts of interest

There are no conflicts to declare.

Acknowledgements

The authors acknowledge funding from the Natural Sciences and Engineering Research Council of Canada (RGPIN 2017-05143), Canada Foundation for Innovation, and Dalhousie University. SM thanks Nova Scotia Graduate Scholarship and NSERC for funding. P. Scallion and Clean Technologies Research Institute are thanked for access and assistance with the SEM analysis and Dr P. Li for TEM. Dr S. MacQuarrie and C. Forbes are thanked for assistance with the N_2 isotherm measurements.

References

1 N. S. Lewis, *Science*, 2016, **351**, 353.
2 K. Mazloomi and C. Gomes, *Renew. Sustain. Energy Rev.*, 2012, **16**, 3024–3033.
3 B. Chen, Z. Liao, J. Wang, H. Yu and Y. Yang, *Int. J. Hydrogen Energy*, 2012, **37**, 3191–3200.
4 J. Dufour, D. P. Serrano, J. L. Gálvez, J. Moreno and C. García, *Int. J. Hydrogen Energy*, 2009, **34**, 1370–1376.
5 S. Chu, W. Li, Y. Yan, T. Hamann, I. Shih, D. Wang and Z. Mi, *Nano Futures*, 2017, **1**, 022001.
6 S. Y. Tee, K. Y. Win, W. S. Teo, L. D. Koh, S. Liu, C. P. Teng and M. Y. Han, *Adv. Sci.*, 2017, **4**, 1600337.
7 J. R. McKone, N. S. Lewis and H. B. Gray, *Chem. Mater.*, 2014, **26**, 407–414.
8 J. T. Davis, J. Qi, X. Fan, J. C. Bui and D. V. Esposito, *Int. J. Hydrogen Energy*, 2018, **43**, 1224–1238.
9 J. Jia, L. C. Seitz, J. D. Benck, Y. Huo, Y. Chen, J. W. D. Ng, T. Bilir, J. S. Harris and T. F. Jaramillo, *Nat. Commun.*, 2016, **7**, 1–6.
10 O. Khaselev, A. Bansal and J. A. Turner, *Int. J. Hydrogen Energy*, 2001, **26**, 127–132.
11 T. Yao, X. An, H. Han, J. Q. Chen and C. Li, *Adv. Energy Mater.*, 2018, **8**, 1–36.
12 C. Jiang, S. J. A. Moniz, A. Wang, T. Zhang and J. Tang, *Chem. Soc. Rev.*, 2017, **46**, 4645–4660.
13 P. Peerakiatkhajohn, J.-H. Yun, S. Wang and L. Wang, *J. Photonics Energy*, 2016, **7**, 012006.
14 M. L. Ghirardi, A. Dubini, J. Yu and P. C. Maness, *Chem. Soc. Rev.*, 2009, **38**, 52–61.
15 A. Melis and M. R. Melnicki, *Int. J. Hydrogen Energy*, 2006, **31**, 1563–1573.
16 E. Eroglu and A. Melis, *Bioresour. Technol.*, 2011, **102**, 8403–8413.
17 B. A. Pinaud, J. D. Benck, L. C. Seitz, A. J. Forman, Z. Chen, T. G. Deutsch, B. D. James, K. N. Baum, G. N. Baum, S. Ardo, H. Wang, E. Miller and T. F. Jaramillo, *Energy Environ. Sci.*, 2013, **6**, 1983–2002.
18 M. R. Shaner, H. A. Atwater, N. S. Lewis and E. W. McFarland, *Energy Environ. Sci.*, 2016, **9**, 2354–2371.
19 Z. Wang, C. Li and K. Domen, *Chem. Soc. Rev.*, 2019, **48**, 2109–2125.
20 T. Jafari, E. Moharreri, A. S. Amin, R. Miao, W. Song and S. L. Suib, *Molecules*, 2016, **21**, 900.
21 K. Maeda and K. Domen, *J. Phys. Chem. Lett.*, 2010, **1**, 2655–2661.
22 M. Ge, J. Cai, J. Iocozzia, C. Cao, J. Huang, X. Zhang, J. Shen, S. Wang, S. Zhang, K. Q. Zhang, Y. Lai and Z. Lin, *Int. J. Hydrogen Energy*, 2017, **42**, 8418–8449.

23 R. D. Tentu and S. Basu, *Curr. Opin. Electrochem.*, 2017, **5**, 56–62.
24 M. Matsuoka, M. Kitano, M. Takeuchi, K. Tsujimaru, M. Anpo and J. M. Thomas, *Catal. Today*, 2007, **122**, 51–61.
25 M. Ni, M. K. H. Leung, D. Y. C. Leung and K. Sumathy, *Renew. Sustain. Energy Rev.*, 2007, **11**, 401–425.
26 L. Ji, M. D. Mcdaniel, S. Wang, A. B. Posadas, X. Li, H. Huang, J. C. Lee, A. A. Demkov, A. J. Bard, J. G. Ekerdt and E. T. Yu, *Nat. Nanotechnol.*, 2015, **10**, 84–90.
27 Q. Huang, Z. Ye and X. Xiao, *J. Mater. Chem. A*, 2015, **3**, 15824–15837.
28 D. M. Powell, M. T. Winkler, H. J. Choi, C. B. Simmons, D. B. Needleman and T. Buonassisi, *Energy Environ. Sci.*, 2012, **5**, 5874–5883.
29 H. Song, D. Liu, J. Yang, L. Wang, H. Xu and Y. Xiong, *ChemNanoMat*, 2017, **3**, 22–26.
30 M. Dasog, J. R. Thompson and N. S. Lewis, *Chem. Mater.*, 2017, **29**, 7002–7013.
31 F. Dai, J. Zai, R. Yi, M. L. Gordin, H. Sohn, S. Chen and D. Wang, *Nat. Commun.*, 2014, **5**, 1–11.
32 H. Sun, J. Chen, S. Liu, D. K. Agrawal, Y. Zhao, D. Wang and Z. Mao, *Int. J. Hydrogen Energy*, 2019, **44**, 7216–7221.
33 R. T. Howe and D. S. Soane, *J. Electrochem. Soc.*, 1994, **141**, 270–274.
34 V. Passi, U. Sodervall, B. Nilsson, G. Petersson, M. Hagberg, C. Krzeminski, E. Dubois, B. Du Bois and J. P. Raskin, *Microelectron. Eng.*, 2012, **95**, 83–89.
35 D. M. Knotter, *J. Am. Chem. Soc.*, 2000, **122**, 4345–4351.
36 K. Nozawa, H. Gailhanou, L. Raison, P. Panizza, H. Ushiki, E. Sellier, J. P. Delville and M. H. Delville, *Langmuir*, 2005, **21**, 1516–1523.
37 S. A. Martell, Y. Lai, E. Traver, J. Macinnis, D. D. Richards, S. Macquarrie and M. Dasog, *ACS Appl. Nano Mater.*, 2019, **2**, 5713–5719.
38 S. Brunauer, P. H. Emmett and E. Teller, *J. Am. Chem. Soc.*, 1938, **60**, 309–319.
39 S. Storck, H. Bretinger and W. F. Maier, *Appl. Catal., A*, 1998, **174**, 137–146.
40 D. K. Hindermann and C. D. Cornwell, *J. Chem. Phys.*, 1968, **48**, 2017–2025.
41 M. Karg, G. Scholz, R. König and E. Kemnitz, *Dalton Trans.*, 2012, **41**, 2360–2366.
42 S. Wiemers-Meyer, M. Winter and S. Nowak, *Phys. Chem. Chem. Phys.*, 2016, **18**, 26595–26601.
43 W. F. Finney, E. Wilson, A. Callender, M. D. Morris and L. W. Beck, *Environ. Sci. Technol.*, 2006, **40**, 2572–2577.
44 Q. Ji, L. Zhou and H. A. Nasr-El-Din, *SPE Lat. Am. Caribb. Pet. Eng. Conf. Proc.*, 2014, vol. 2, pp. 1465–1484.
45 M. P. Hanrahan, E. L. Fought, T. L. Windus, L. M. Wheeler, N. C. Anderson, N. R. Neale and A. J. Rossini, *Chem. Mater.*, 2017, **29**, 10339–10351.
46 K. Chen, M. Abdolrhamani, E. Sheets, J. Freeman, G. Ward and J. L. White, *J. Am. Chem. Soc.*, 2017, **139**, 18698–18704.
47 H. E. Gottlieb, V. Kotlyar and A. Nudelman, *J. Org. Chem.*, 1997, **62**, 7512–7515.
48 M. Mägi, E. Lippmaa, A. Samoson, G. Engelhardt and A. R. Grimmer, *J. Phys. Chem.*, 1984, **88**, 1518–1522.
49 A. N. Thiessen, M. Ha, R. W. Hooper, H. Yu, A. O. Oliynyk, J. G. C. Veinot and V. K. Michaelis, *Chem. Mater.*, 2019, **31**, 678–688.
50 R. A. Faulkner, J. A. DiVerdi, Y. Yang, T. Kobayashi and G. E. Maciel, *Materials*, 2013, **6**, 18–46.
51 Y. Lai, J. R. Thompson and M. Dasog, *Chem.–Eur. J.*, 2018, **24**, 7913–7920.

52 M. Shao, L. Cheng, X. Zhang, D. D. D. Ma and S. T. Lee, *J. Am. Chem. Soc.*, 2009, **131**, 17738–17739.
53 N. L. Fry, G. R. Boss and M. J. Sailor, *Chem. Mater.*, 2014, **26**, 2758–2764.
54 A. Walcarius and L. Mercier, *J. Mater. Chem.*, 2010, **20**, 4478–4511.
55 D. G. Cory and W. M. Ritchey, *J. Magn. Reson.*, 1988, **80**, 128–132.
56 R. K. Harris, E. D. Becker, S. M. Cabral De Menezes, R. Goodfellow and P. Granger, *Pure Appl. Chem.*, 2001, **73**, 1795–1818.

Faraday Discussions

PAPER

Low temperature radical initiated hydrosilylation of silicon quantum dots

Timothy T. Koh, [†a] Tingting Huang,[†a] Joseph Schwan,[b] Pan Xia,[c] Sean T. Roberts, [*d] Lorenzo Mangolini [*bc] and Ming L. Tang [*ac]

Received 20th December 2019, Accepted 13th January 2020
DOI: 10.1039/c9fd00144a

The photophysics of silicon quantum dots (QDs) is not well understood despite their potential for many optoelectronic applications. One of the barriers to the study and widespread adoption of Si QDs is the difficulty in functionalizing their surface, to make, for example, a solution-processable electronically-active colloid. While thermal hydrosilylation of Si QDs is widely used, the high temperature typically needed may trigger undesirable side-effects, like uncontrolled polymerization of the terminal alkene. In this contribution, we show that this high-temperature method for installing aromatic and aliphatic ligands on non-thermal plasma-synthesized Si QDs can be replaced with a low-temperature, radical-initiated hydrosilylation method. Materials prepared *via* this low-temperature route perform similarly to those created *via* high-temperature thermal hydrosilylation when used in triplet fusion photon upconversion systems, suggesting the utility of low-temperature, radical-initiated methods for creating Si QDs with a range of functional behavior.

Introduction

Silicon is both non-toxic and earth-abundant, making it a highly attractive material for quantum-confined semiconductor structures. However, compared to those of direct gap chalcogenide quantum dots (QDs), the electronic states of silicon QDs remain a subject of sizable debate.[1–5] This is not surprising as the electronic properties of QDs are greatly affected by their surface,[6–9] and yet surface functionalization of Si QDs is a key step in transforming these materials into solution-processable optoelectronics. The strong dependence of the electronic

---

[a]*Department of Chemistry, University of California Riverside, Riverside, CA 92521, USA. E-mail: mltang@ucr.edu*

[b]*Department of Mechanical Engineering, University of California Riverside, Riverside, CA 92521, USA. E-mail: lmangolini@engr.ucr.edu*

[c]*Materials Science & Engineering Program, University of California Riverside, Riverside, CA 92521, USA*

[d]*Department of Chemistry, The University of Texas at Austin, Austin, TX 78712, USA. E-mail: roberts@cm.utexas.edu*

† Equal contribution.

properties of Si QDs on their surface ligands stems from Si's ability to form strong, covalent bonds to surface-bound molecules. This is distinct from lead and cadmium-based QDs, which tend to ionically coordinate molecules to their surfaces. In principle, the covalently bonded nature of Si QD surfaces creates the opportunity to rationally employ well-understood, widely-used organo-silicon chemistry to functionalize Si QDs to achieve a range of desired optical and electronic properties.

Thermal hydrosilylation is the most widely used method to functionalize Si QDs. Unfortunately, this reaction has many limitations, including use of high temperatures (150–170 °C), substrate limitations due to the need to incorporate a terminal alkene, and the need for a large excess of reagents to complete functionalization in a reasonable timeframe. In particular, this need for a great excess of alkene or alkyne ligands is undesirable when designer ligands are to be incorporated on Si QD surfaces. As such, it is of practical interest to find conditions to hydrosilylate Si QDs at low temperatures in high yield to broaden the range of ligands that can be attached to Si QD surfaces.

In this work, we examine a radical initiator commonly used for polymerization. While radical initiators have been explored to functionalize both porous and single crystal Si,[10–15] no systematic investigation has been reported for Si QDs. Carroll et al. showed that hydrosilylation of Si QDs can be achieved at 140 °C when initiated by 1,1′-azobis-(cyclohexanecarbonitrile) (ABCN) in neat alkene.[16] Similarly, 2,2′-azobis(2-methylpropionitrile) (AIBN)[17–19] and 1,1′-azobis(cyclohexanecarbonitrile) (ACHN)[20] were used at lower temperatures. In all these cases, linear aliphatic molecules were bound to the Si QD surface, and the terminal alkene was used in great excess, usually as a neat solvent. The ability to extend this radical based method to forming covalent bonds between aromatic ligands and Si QDs, similar to forming Si–C bonds homogeneously in solution, will allow greater synergy between functional organic materials and Si QDs.

Here, we show that AIBN catalyzed hydrosilylation of Si QDs with aromatic ligands can be achieved at temperatures as low as 60 °C with aliphatic and aromatic alkenes. This is important for multi-excitonic processes like photon upconversion, where Dexter energy transfer occurs between the QD and the surface-bound ligands.[21–27] Since the efficiency of this short-range energy transfer decays exponentially with increasing distance between the QD donor and molecular acceptor, it is imperative to have both species within 1 nm of each other.[28] We show here that the photon upconversion quantum yield (QY) with anthracene-functionalized Si QD light absorbers (Si:9EA) is comparable for larger Si QDs created by both thermal and radical-initiated hydrosilylation. For smaller Si QDs, the photon upconversion QY is ~3.3%, lower than that achieved with thermal methods.[29]

Results and discussion

Fig. 1 shows our light harvesting scheme in which non-thermal plasma synthesized silicon QDs are functionalized with octadecene (ODE) and 9-vinylanthracene (9VA) ligands under one of two conditions: (1) thermally at high temperature (170 °C), or (2) at a relatively low temperature (60 °C) with an AIBN radical initiator. In doing so, Si QDs functionalized with 9-ethylanthracene (9EA) and octadecane (ODA) are created, abbreviated to Si:9EA. These Si:9EA QDs are

Fig. 1 (Left) Non-thermal plasma-synthesized silicon quantum dots (Si QDs) were hydrosilylated with 9-vinyl anthracene (9VA) and octadecene, to give Si QDs functionalized with 9-ethylanthracene (9EA) and octadecane (Si:9EA). (Right) Si:9EA can photosensitize triplet excited states in diphenylanthracene (DPA) via triplet energy transfer (TET). Photon upconversion occurs via triplet–triplet annihilation (TTA) between pairs of DPA molecules in their triplet excited states, producing violet light.

colloidally stable in non-polar solvents like toluene or hexane due to the presence of the solubilizing ODA groups. Excitation of Si:9EA QDs creates an excitonic state with triplet character that photosensitizes the surface bound 9EA molecules.[29] This allows for triplet energy transfer (TET) between Si:9EA and diphenylanthracene (DPA) emitter molecules in solution. Photon upconversion is observed when DPA emits from its singlet state following triplet–triplet annihilation between two DPA molecules in their lowest excited triplet state, T_1. Note that direct photoexcitation of a mixture of Si QDs and DPA with light comprised of photons with energy smaller than DPA's 3.2 eV optical gap, such as the 488 nm, 532 nm and 640 nm light used here, does not result in photon upconversion in the absence of surface-bound 9EA on the Si QDs.

Preparation of anthracene functionalized Si QDs

Table 1 outlines the synthetic details. Non-thermal plasma synthesized Si QDs[30] were directly transferred into an air-free glove box to store under nitrogen for future use. Thermal hydrosilylation of Si QDs with 9VA and ODE was performed as previously reported.[29] For the AIBN-initiated radical hydrosilylation of the Si QDs, 0.21 mg of Si QDs were mixed with 0.5 mg of AIBN. Subsequently, either 0.2 or 1.0 mol equivalents of ODE were added, corresponding to 0.03 or 0.144 mL,

Table 1 Synthetic conditions employed for radical-initiated hydrosilylation of Si QDs compared to the conditions previously reported for thermal hydrosilylation. Photoluminescence (PL) spectra were measured in toluene at room temperature with the excitation source set to 532 nm. Photon upconversion measurements were performed with 5.2 mM DPA in toluene at room temperature. Upconversion QYs are reported using a 488 nm CW laser in the linear regime.

Synthesis	AIBN radical initiated: 60 °C, 15 h						Thermal: 170 °C, 3 h			
Sample	RI	RII	RIII	RIV	RV	RVI	T1.5	T2	T3	T4
ODE (mol eq.)	0.2		1.0				4.6			
9VA/ODE (%)	1.25	1.5	0.5	1.5	7.8	15.7	1.50	2.0	3.0	4.0
PL λ_{MAX} (nm)	776	777	774	792	800	804	771	776	784	785
PLQY (%)	12.0	18.3	4.5	2.1	8.4	3.1	11.8	10.4	6.9	5.0
Upconversion QY (%)	0.89	1.13	0.61	0.58	0.25	0.11	0.82	1.57	0.42	0.19

respectively. As shown in Table 1, 9VA was added in different amounts by mole percentage, and the solution was topped up with toluene to the final volume of 0.56 mL. This solution was sealed, stirred and heated in a nitrogen glovebox at 60 °C for 15 hours. In Table 1 and the remainder of this manuscript, Si QDs functionalized via radical-initiated or thermal hydrosilylation are labelled as R and T, respectively. There are about 2–3 anthracene ligands per QD.

In the AIBN initiated hydrosilylation, the cloudy solution turned clear, indicating Si QD functionalization. Methanol was used to separate functionalized QDs from free 9VA precursor molecules. 1.7 mL of methanol were added to the reaction solution to precipitate out Si:9EA by centrifugation at 14 000 rpm for 20 min. The precipitated Si:9EA was redispersed in 0.5 mL of toluene, followed by addition of 1.5 mL of methanol and centrifugation at 14 200 rpm for an additional 20 min. This cleaning procedure was repeated three times in total. These cleaning steps resulted in full removal of free 9VA and ODE molecules in solution, as shown by UV-vis absorption spectra of the supernatant which did not show any traces of anthracene. The second pellet was redispersed in toluene, split into two portions, and then precipitated out a third time with 0.75 mL of methanol. One portion was used for upconversion measurements by redispersing it in 0.4 mL of 5.2 mM DPA solution in toluene, while the other portion was saved for characterization.

Fig. 2 shows the electronic absorption and attenuated total reflection infrared (ATR-IR) spectra of Si:9EA colloids synthesized by the AIBN radical-initiated method after cleaning. There is no visible difference compared to the thermally hydrosilylated Si:9EA that we previously reported.[29] For example, vibronic features corresponding to surface-bound anthracene molecules were observed following both synthetic methods. In Fig. 2b, strong C–H absorption from ODA at $\sim$3000 cm^{-1} and C=C aromatic stretching bands from 1600–1700 cm^{-1} are observed, confirming covalent binding of 9EA and ODA to the Si QD surface.

An increase in the amount of 9VA during hydrosilylation bathochromically shifts the photoluminescence (PL) of the Si:9EA QDs. This can be seen in Fig. 3, which plots the PL collected from Si:9EA in toluene following excitation with a 532 nm CW laser. In Fig. 3, we have also highlighted the 9EA triplet energy of

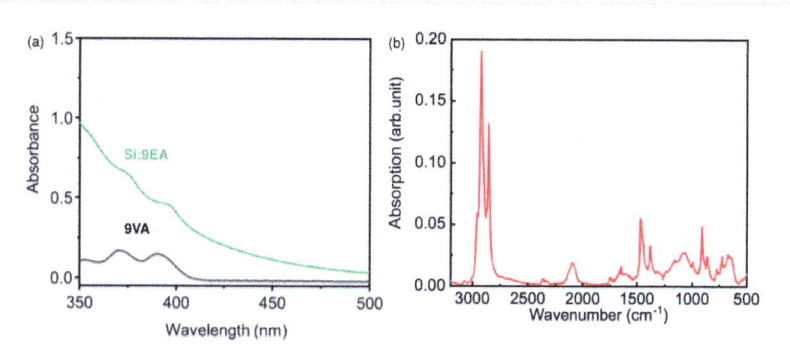

Fig. 2 (a) Absorption spectra of 9VA and Si:9EA QDs hydrosilylated at 60 °C with AIBN radical initiator (green). Vibronic features corresponding to the molecular absorption of the anthracene precursor are observable (black). (b) Attenuated total reflection infra-red (ATR-IR) spectra of Si:9EA showing that octadecane and 9-ethylanthracene (ODA and 9EA) are covalently bound to the surface of the Si QDs.

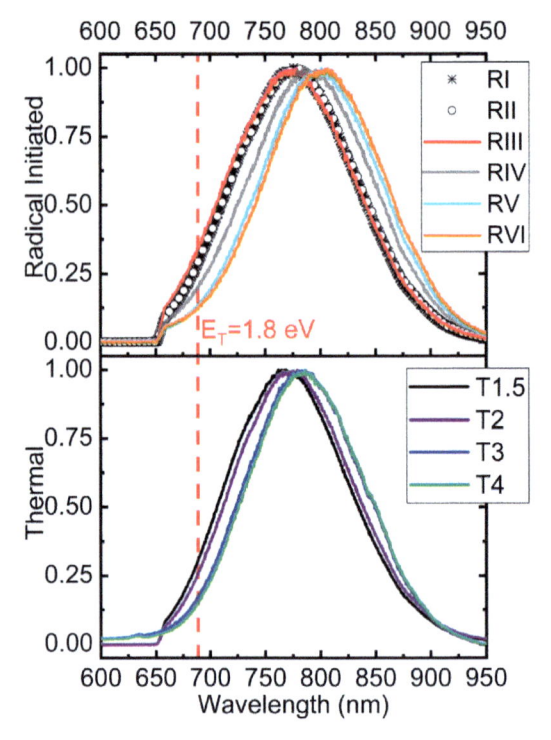

Fig. 3 PL from Si QDs red shifts as the 9VA precursor is increased during hydrosilylation, independent of whether it was initiated by AIBN (top) or thermally performed (bottom). See Table 1 for sample information. Si:9EA was dissolved in toluene and excited using a 532 nm CW laser with a power density of 830 mW cm^{-2} at the sample.

1.8 eV, which falls on the high energy side of the Si QD PL spectrum. Thus, the bathochromic shift could stem from QD size heterogeneity as smaller QDs with a wider bandgaps will energetically favor triplet energy transfer to surface-bound anthracene, leading to quenching of their emission. Consistent with this hypothesis, we observed a narrowing of the Si emission lineshape with increasing 9EA loading. Comparing T1.5 to T4, or the thermally hydrosilylated silicon QDs with 9VA at 1.5% and 4%, respectively, the FWHM of the Si QD PL decreases from 0.27 to 0.24 eV. Similarly, comparing the AIBN initiated radical hydrosilylation samples III to IV with 0.5 and 15.7 mol% of 9VA with respect to ODE, the FWHM decreases from 0.28 to 0.25 eV with 532 nm excitation, too. When the PL traces in Fig. 3 are normalized to their maxima, we also see a decrease in Si QD emission QY with increasing surface concentration of 9VA. Table 1 shows that Si:9EA QD PL decreases by over a factor of two as the amount of surface-bound 9EA increases for the thermal hydrosilylation samples T1.5, T2, T3 and T4. The QY trend is less clear for the AIBN radical-initiated functionalization, possibly due to complication by other surface states, but a net QY decrease with increasing 9EA concentration was seen for the samples we investigated. We note that both radical-initiated and thermal hydrosilylation lead to similar PL shifts with similar 9VA/ODE ratios.

Fig. 4 and 5 (corresponding to samples T2 and RII, respectively) show that the Si PL emitted by Si:9EA blue shifts as the power density of the CW 488 nm, 532 nm

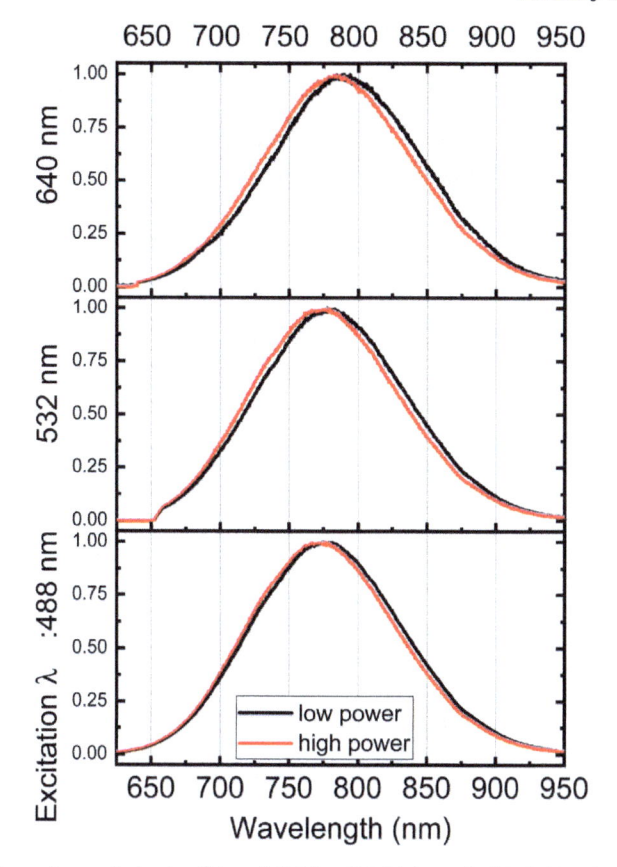

Fig. 4 PL from thermally hydrosilylated Si QDs, Si:9EA (sample T2, Table 1) shows a blue shift as the power density of the 488 nm, 532 nm and 640 nm CW lasers is increased.

and 640 nm lasers increases. This trend is summarized in Table 2. This blue shift could again be attributed to the inhomogeneous size distribution. Gregorkiewicz *et al.* suggested that at high excitation density, smaller Si QDs preferentially emit light as larger Si QDs possess higher extinction cross-sections, making them more susceptible to absorption saturation.[31] Alternatively, there might be a possibility of bi-excitons forming on Si QDs.[32] The fact that blue-shifted emission with increasing excitation density is not observed in direct gap QDs may arise from the long >500 μs lifetimes of Si QDs, compared to the short ~10 ns lifetime of cadmium chalcogenide QDs, or the <10 μs for lead chalcogenide QDs. Time-resolved experiments are needed to differentiate these and other possible scenarios.

Photon upconversion

As shown in Table 1 and Fig. 6, both thermal and radical-initiated hydrosilylation result in photon upconversion QYs between 1.3% and 1.5% for Si QDs whose emission peaks at 780 nm. This PL maximum was obtained for thermally hydrosilylated Si QDs with ODE only when excited with a 488 nm CW laser. Compared to the direct gap chalcogenide QDs, the indirect gap nature of Si QDs

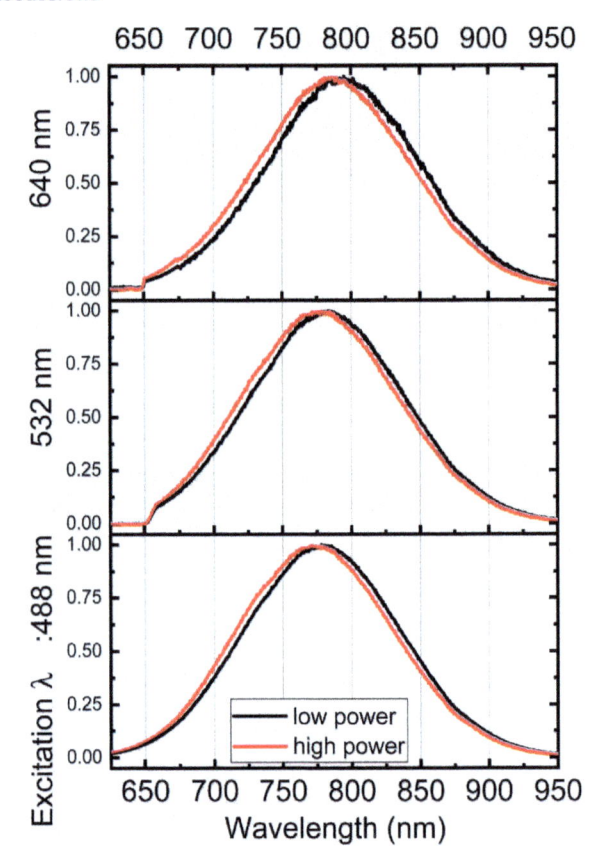

Fig. 5 PL from Si:9EA prepared by AIBN-initiated hydrosilylation of Si QDs (sample RII, Table 1) shows a blue shift as the power density of the 488 nm, 532 nm and 640 nm CW lasers is increased.

makes it impossible to index particle size using electronic absorption spectroscopy. On the other hand, the PL of Si QDs is affected by the synthetic methodology, thus in this work we compare the Si QD size by comparing the PL when thermally hydrosilylated with ODE only is carried out. The photon upconversion QY was determined with a R6G standard as described previously.[29] The photon upconversion QY for the samples in Table 1 is not high as the large size of the Si QDs causes their bandgaps to be too small relative to the T_1 level of 9EA of $\sim$1.8 eV

Table 2 As the excitation density of the CW lasers is increased by 10×, a slight blue shift of the PL of Si:9EA is observed for all excitation wavelengths. Samples T2 (Fig. 4) and RII (Fig. 5) were dissolved in toluene. All measurements were performed at RT

Sample	λ_{EXC} (nm)	488		532		640	
	Power (mW cm^{-2})	100	1800	85	830	18	130
RII	PLQY (%)	32.0	17.2	33.6	18.3	9.6	7.6
	PL λ_{MAX} (nm)	779	774	783	778	795	787
T2	PLQY (%)	16.0	9.9	21.8	11.8	21.2	13.0
	PL λ_{MAX} (nm)	778	774	781	776	791	785

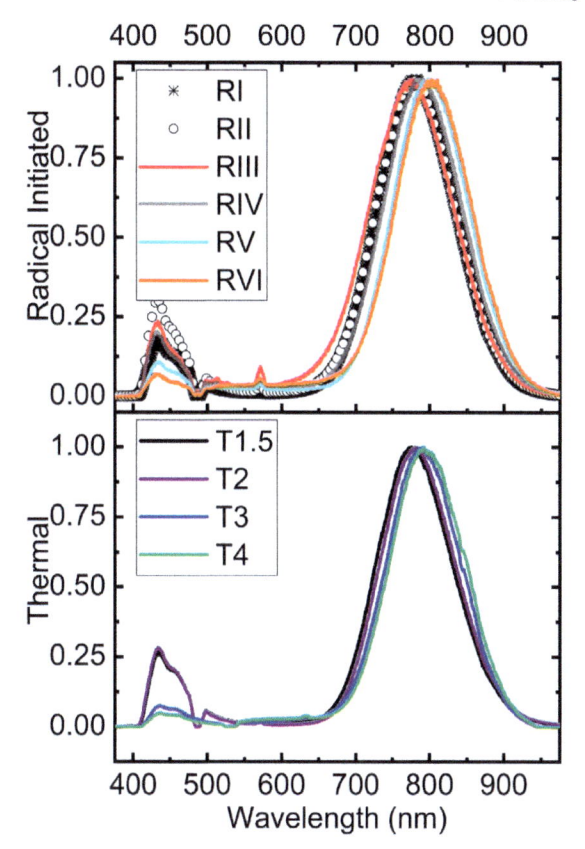

Fig. 6 The photon upconversion observed when Si:9EA colloids from Table 1 were dissolved in 5.2 mM DPA and excited with a 488 nm CW laser.

to allow efficient energy transfer. This makes triplet energy transfer (TET in Fig. 1) a thermodynamically uphill process. Using Si QDs of a smaller size that emit maximally at 700 nm, a photon upconversion QY of 3.34% was obtained. This is a factor of 2 lower than the value we previously reported.[29] This suggests that the heating step at 170 °C may remove some surface states deleterious for triplet energy transfer.

Conclusions

We have shown that functionalization of non-thermal plasma-synthesized silicon QDs can be performed *via* radical-initiated hydrosilylation at low temperature with a substantially lower concentration of ligands. The photon upconversion quantum yields here are a barometer of the efficiency of triplet energy transfer between the Si QD and the surface-bound anthracene, a measure of the quality of the organic–inorganic surface. For larger sized Si QDs, both hydrosilylation methods perform equally well. However, for smaller QDs, the high-temperature thermal hydrosilylation method gives higher photon upconversion QYs compared to the low-temperature radical-initiated route, suggesting size dependent surface reactivity. The functionalization of Si QDs needs be further developed to enhance electronic communication between the Si semiconductor core and conjugated ligands on its surface.

Conflicts of interest

There are no conflicts to declare.

Acknowledgements

M. L. T. acknowledges National Science Foundation (NSF) grant OISE-1827087 and the Alfred P. Sloan Foundation. L. M. acknowledges support from the NSF under award number 1351386 (CAREER). STR acknowledges support from the NSF (CHE-1610412), the Robert A. Welch Foundation (Grant F-1885), and Research Corporation for Science Advancement (Grant #24489).

References

1 W. de Boer, D. Timmerman, K. Dohnalova, I. N. Yassievich, H. Zhang, W. J. Buma and T. Gregorkiewicz, Red spectral shift and enhanced quantum efficiency in phonon-free photoluminescence from silicon nanocrystals, *Nat. Nanotechnol.*, 2010, **5**, 878–884.
2 K. Dohnalova, A. N. Poddubny, A. A. Prokofiev, W. de Boer, C. P. Umesh, J. M. J. Paulusse, H. Zuilhof and T. Gregorkiewicz, Surface brightens up Si quantum dots: direct bandgap-like size-tunable emission, *Light: Sci. Appl.*, 2013, **2**, e47.
3 W. de Boer, D. Timmerman, I. Yassievich, A. Capretti and T. Gregorkiewicz, Reply to 'Absence of redshift in the direct bandgap of silicon nanocrystals with reduced size', *Nat. Nanotechnol.*, 2017, **12**, 932–933.
4 F. Trojanek, K. Neudert, M. Bittner and P. Maly, Picosecond photoluminescence and transient absorption in silicon nanocrystals, *Phys. Rev. B: Condens. Matter Mater. Phys.*, 2005, **72**, 075365.
5 J. W. Luo, S. S. Li, I. Sychugov, F. Pevere, J. Linnros and A. Zunger, Absence of redshift in the direct bandgap of silicon nanocrystals with reduced size, *Nat. Nanotechnol.*, 2017, **12**, 930–932.
6 N. P. Brawand, M. Voros and G. Galli, Surface dangling bonds are a cause of B-type blinking in Si nanoparticles, *Nanoscale*, 2015, **7**, 3737–3744.
7 A. Puzder, A. J. Williamson, J. C. Grossman and G. Galli, Surface control of optical properties in silicon nanoclusters, *J. Chem. Phys.*, 2002, **117**, 6721–6729.
8 A. Puzder, A. J. Williamson, J. C. Grossman and G. Galli, Surface chemistry of silicon nanoclusters, *Phys. Rev. Lett.*, 2002, **88**, 097401.
9 G. Allan, C. Delerue and M. Lannoo, Nature of luminescent surface states of semiconductor nanocrystallites, *Phys. Rev. Lett.*, 1996, **76**, 2961–2964.
10 J. M. Buriak, Organometallic chemistry on silicon and germanium surfaces, *Chem. Rev.*, 2002, **102**, 1271–1308.
11 C. Chatgilialoglu, Organosilanes as radical-based reducing agents in synthesis, *Acc. Chem. Res.*, 1992, **25**, 188–194.
12 M. R. Linford and C. E. D. Chidsey, Alkyl monolayers covalently bonded to silicon surfaces, *J. Am. Chem. Soc.*, 1993, **115**, 12631–12632.

13 M. R. Linford, P. Fenter, P. M. Eisenberger and C. E. D. Chidsey, Alkyl Monolayers on Silicon Prepared from 1-Alkenes and Hydrogen-Terminated Silicon, *J. Am. Chem. Soc.*, 1995, **117**, 3145–3155.
14 M. Miyano, Y. Kitagawa, S. Wada, A. Kawashima, A. Nakajima, T. Nakanishi, J. Ishioka, T. Shibayama, S. Watanabe and Y. Hasegawa, Photophysical properties of luminescent silicon nanoparticles surface-modified with organic molecules *via* hydrosilylation, *Photochem. Photobiol. Sci.*, 2016, **15**, 99–104.
15 R. Mazzaro, A. Gradone, S. Angeloni, G. Morselli, P. G. Cozzi, F. Romano, A. Vomiero and P. Ceroni, Hybrid Silicon Nanocrystals for Color-Neutral and Transparent Luminescent Solar Concentrators, *ACS Photonics*, 2019, **6**, 2303–2311.
16 G. M. Carroll, R. Limpens and N. R. Neale, Tuning Confinement in Colloidal Silicon Nanocrystals with Saturated Surface Ligands, *Nano Lett.*, 2018, **18**, 3118–3124.
17 A. Angi, M. Loch, R. Sinelnikov, J. G. C. Veinot, M. Becherer, P. Lugli and B. Rieger, The influence of surface functionalization methods on the performance of silicon nanocrystal LEDs, *Nanoscale*, 2018, **10**, 10337–10342.
18 F. Sangghaleh, I. Sychugov, Z. Y. Yang, J. G. C. Veinot and J. Linnros, Near-Unity Internal Quantum Efficiency of Luminescent Silicon Nanocrystals with Ligand Passivation, *ACS Nano*, 2015, **9**, 7097–7104.
19 Z. Y. Yang, C. M. Gonzalez, T. K. Purkait, M. Iqbal, A. Meldrum and J. G. C. Veinot, Radical Initiated Hydrosilylation on Silicon Nanocrystal Surfaces: An Evaluation of Functional Group Tolerance and Mechanistic Study, *Langmuir*, 2015, **31**, 10540–10548.
20 R. T. Anderson, X. Zang, R. Fernando, M. J. Dzara, C. Ngo, M. Sharps, R. Pinals, S. Pylypenko, M. T. Lusk and A. Sellinger, Direct Conversion of Hydride- to Siloxane-Terminated Silicon Quantum Dots, *J. Phys. Chem. C*, 2016, **120**, 25822–25831.
21 Z. Y. Huang, X. Li, M. Mahboub, K. M. Hanson, V. M. Nichols, H. Le, M. L. Tang and C. J. Bardeen, Hybrid Molecule-Nanocrystal Photon Upconversion Across the Visible and Near-Infrared, *Nano Lett.*, 2015, **15**, 5552–5557.
22 Z. Y. Huang and M. L. Tang, Designing Transmitter Ligands That Mediate Energy Transfer between Semiconductor Nanocrystals and Molecules, *J. Am. Chem. Soc.*, 2017, **139**, 9412–9418.
23 M. Wu, D. N. Congreve, M. W. B. Wilson, J. Jean, N. Geva, M. Welborn, T. Van Voorhis, V. Bulovic, M. G. Bawendi and M. A. Baldo, Solid-state infrared-to-visible upconversion sensitized by colloidal nanocrystals, *Nat. Photonics*, 2016, **10**, 31–34.
24 L. Nienhaus, M. F. Wu, N. Geva, J. J. Shepherd, M. W. B. Wilson, V. Bulovic, T. Van Voorhis, M. A. Baldo and M. G. Bawendi, Speed Limit for Triplet-Exciton Transfer in Solid-State PbS Nanocrystal-Sensitized Photon Upconversion, *ACS Nano*, 2017, **11**, 7848–7857.
25 B. Joarder, N. Yanai and N. Kimizuka, Solid-State Photon Upconversion Materials: Structural Integrity and Triplet-Singlet Dual Energy Migration, *J. Phys. Chem. Lett.*, 2018, **9**, 4613–4624.
26 K. Okumura, K. Mase, N. Yanai and N. Kimizuka, Employing Core-Shell Quantum Dots as Triplet Sensitizers for Photon Upconversion, *Chem.–Eur. J.*, 2016, **22**, 7721–7726.

27 C. Mongin, S. Garakyaraghi, N. Razgoniaeva, M. Zamkov and F. N. Castellano, Direct observation of triplet energy transfer from semiconductor nanocrystals, *Science*, 2016, **351**, 369–372.
28 X. Li, Z. Y. Huang, R. Zavala and M. L. Tang, Distance-Dependent Triplet Energy Transfer between CdSe Nanocrystals and Surface Bound Anthracene, *J. Phys. Chem. Lett.*, 2016, **7**, 1955–1959.
29 P. Xia, E. K. Raulerson, D. Coleman, C. S. Gerke, L. Mangolini, M. L. Tang and S. T. Roberts, Achieving spin-triplet exciton transfer between silicon and molecular acceptors for photon upconversion, *Nat. Chem.*, 2020, **12**, 137–144.
30 L. Mangolini and U. Kortshagen, Plasma-assisted synthesis of silicon nanocrystal inks, *Adv. Mater.*, 2007, **19**, 2513–2519.
31 E. M. L. D. de Jong, H. Rutjes, J. Valenta, M. T. Trinh, A. N. Poddubny, I. N. Yassievich, A. Capretti and T. Gregorkiewicz, Thermally stimulated exciton emission in Si nanocrystals, *Light: Sci. Appl.*, 2018, **7**, 17133.
32 M. C. Beard, K. P. Knutsen, P. R. Yu, J. M. Luther, Q. Song, W. K. Metzger, R. J. Ellingson and A. J. Nozik, Multiple exciton generation in colloidal silicon nanocrystals, *Nano Lett.*, 2007, **7**, 2506–2512.

Faraday Discussions

PAPER

Modulating donor–acceptor transition energies in phosphorus–boron co-doped silicon nanocrystals *via* X- and L-type ligands†

Gregory F. Pach, Gerard M. Carroll, Hanyu Zhang and Nathan R. Neale *

Received 8th October 2019, Accepted 6th January 2020
DOI: 10.1039/c9fd00106a

In this work, we explore the effect of ligand binding groups on the visible and NIR photoluminescent properties within phosphorus–boron co-doped silicon nanocrystals (PB:Si NCs) by exploiting both the X-type (covalent) and L-type (Lewis donor molecule) bonding interactions. We find that the cooperative nature of both X- and L-type bonding from alkoxide/alcohol, alkylamide/alkylamine, and alkylthiolate/alkylthiol on PB:Si NCs results in photoluminescence (PL) energy blue shifts from the as-synthesized, hydride-terminated NCs (PB:Si–H) in excess of 0.4 eV, depending on the surface termination. These PL blue shifts appear greatest in the most strongly confined samples with diameters <4 nm where the surface-to-volume ratio is high and, therefore, the ligand effects are most pronounced. A correlation between the donor group strength (either X-type or L-type) and the degree of D–A state modulation is found, and the proportion of the PL blue shift from the X- and L-type interactions is quantified. Raman spectroscopy is used to provide additional evidence of the strength of the L-type donor groups. Additionally, we probe how the nature of the ligand chemistry affects the radiative lifetime and PL efficiency and find that the ligands do not significantly change the D–A emission dynamics, and all samples retain the long 50–130 μs lifetimes characteristic of these transitions. Finally, we describe three mechanisms that operate to affect the D–A recombination energies: (1) X-type ligands that modulate the PB:Si–X NC wavefunction; (2) L-type ligands that perturb the donor and acceptor states *via* a molecular orbital theory picture; and (3) X- and L-type ligands that cause a dielectric increase around the PB:Si NC core, which provides Coulomb screening and modulates the donor and acceptor states even further.

Chemistry and Nanoscience Center, National Renewable Energy Laboratory, 15013 Denver West Parkway, Golden, Colorado 80401, USA. E-mail: nathan.neale@nrel.gov

† Electronic supplementary information (ESI) available. See DOI: 10.1039/c9fd00106a

Introduction

Quantum-confined semiconducting nanocrystals (NCs) have drawn considerable interest for light emitting applications due to their size-tunable optical properties and high quantum yields.[1,2] The ubiquity of Si in the electronics industry, as well as its low toxicity, makes Si NCs particularly interesting candidates for near-infrared (NIR) emission applications, in which the majority of materials contain heavy metals and/or use toxic synthetic precursors.[3,4] However, two limitations of Si are its indirect, rather than direct, band gap that is 1.1 eV in bulk. The indirect optical gap results in poor visible light absorption, though it is possible to modulate the absorption coefficient through quantum confinement.[5] For some applications the large Stokes shift afforded by this property is actually desirable, for example in luminescent solar concentrators.[6] In addition, though quantum confinement pushes the band gap emission to the visible region,[7] several studies have shown that the incorporation of impurity dopants into semiconducting NCs can be used to further tune the optical properties of NC ensembles.[8–10] For example, for two decades it has been known that doping Si NCs with substitutional phosphorus or boron dopant atoms results in sub-band gap emission due to the interaction of the exciton with the dopant state.[11,12]

To this end, nonthermal plasma synthesis of semiconductor NCs provides a particular synthetic advantage for dopant incorporation by circumventing the "self-purification effect" in which impurity atoms are expelled from the NC core to the surface.[13,14] This is because nonthermal plasma synthesis is a kinetic growth process occurring out of equilibrium that can irreversibly trap dopants within the core. This process has been widely used to dope Si NCs with phosphorus and boron impurities.[15–20] Expanding upon these single dopant systems, a recent publication from our group has also demonstrated the ability to use nonthermal plasma synthesis to simultaneously incorporate both boron and phosphorus dopants into co-doped Si NCs (PB:Si).[21] Delerue described the emission properties of these PB:Si NCs as purely governed by donor–acceptor (D–A) transitions attributed to charge-compensating phosphorus–boron (P–B) pairs.[22] Consequently, our results showed that the peak emission energy of these plasma-synthesized PB:Si NCs is consistently red-shifted several hundred meV from that of intrinsic Si NCs of the same diameter, below the bulk 1.1 eV band gap at lightly quantum-confined NCs (diameters > 4 nm). The results from plasma-synthesized PB:Si NCs are in line with many prior studies on co-doped PB:Si NCs fabricated using thermal precipitation from phosphoborosilicate glass.[23–28]

More generally, the large surface-to-volume ratio of the nanostructures results in the surface playing a critical role in dictating the NC optoelectronic properties, and the ability to control these properties *via* controlled surface chemistry is now widely established.[29–31] In Si NCs, there are three known methods of controlling the optoelectronic properties: (1) quantum confinement to tune the band edge emission as originally demonstrated by Brus *et al.* in 1993;[32] (2) supra-band gap oxide and/or nitrogen-related defect state emission described by the seminal work of Tilley, Kauzlarich, Veinot and coworkers in 2013;[33] and (3) intentional surface chemistry manipulation. The latter method can be broken up into several sub-categories including (i) Lewis acid–base interactions *via* donor molecule coordination to either polarized Si surface atoms or surface heteroatoms,[20,34] (ii) coupling of π-conjugated ligand molecular orbitals to the Si NC wavefunction,[35,36] and (iii) covalent surface group modification including halogens *Si–X (where *Si denotes a surface Si atom; X = Cl, Br, I)[37] or other main group ligands *Si–ER (where E = CH$_2$, NH, O, S; R = alkyl).[38,39] Theoretical frameworks exist for surface interactions on the Si NCs based on changing the dielectric matrix described by Luo and Zunger,[40] or altering the bonding structure at the Si NC surface introduced by Galli and Kimerling.[41]

For covalent ligand attachment to Si NCs, surface functionalization is conventionally performed *via* a radical reaction involving the removal of a surface hydrogen atom (H˙) or silyl group (˙SiH$_3$) and substitution with a covalent ligand bond.[42,43] In our recent studies, we demonstrated the use of a variety of surface terminations on Si NCs, including alkyls (from reaction with 1-dodecene), alkoxides (1-dodecanol), secondary amides (1-dodecylamine), and thiolates (1-dodecanethiol).[38,39] We proposed that the ligand head groups (the only atoms that change *via* this chemistry) interact with the Si NC electronic wavefunction and, depending on the ligand, modulate the peak photoluminescence (PL) emission energy by up to ~150 meV for small NC sizes (~3 nm). In addition to covalent functionalization, we have pioneered Lewis acid–base interactions that either create or perturb the localized surface plasmonic resonance (LSPR) in Si NCs.[20,34] To date there are no studies describing whether these Lewis acid–base interactions can also be employed to control the emission properties in Si NCs, since previously all such interactions have been on doped NCs that do not exhibit band edge luminescence.

In this report, we explore the effect of ligand binding groups on the visible and NIR photoluminescent properties within PB:Si NCs by exploiting both the X-type (covalent) and L-type (Lewis donor molecule) bonding interactions classically described by Green.[44] We explore both the individual and cooperative nature of X- and L-type bonding, which results in PL energy shifts from the as-synthesized, hydride-terminated NCs (PB:Si–H) in excess of 0.4 eV, depending on the surface termination. We probe the emission energies as a function of size and find a correlation between the donor group strength (either X-type or L-type) and the degree of D–A state modulation, and additionally quantify the proportion of the PL blue shift from the X- and L-type interactions. We also probe how the nature of the ligand chemistry affects the radiative lifetime and PL efficiency as well as vibrational states within the PB:NCs *via* Raman spectroscopy. Finally, we describe three mechanisms that operate to affect the D–A recombination energies based on these experiments.

Results

All PB:Si NCs in this study were synthesized using a capacitively-coupled 13.56 MHz nonthermal plasma with process gases SiH$_4$, B$_2$H$_6$, and PH$_3$ and additional carrier gases Ar and H$_2$ (see Table S1† for process conditions of all PB:Si NCs used in this study). As-synthesized co-doped PB:Si–H NCs are terminated with surface hydride groups that are readily distinguishable *via* FTIR (Fig. S1†). Surface silicon hydride groups *SiH (ν = 2087 cm^{-1}), *SiH$_2$ (ν = 2108 cm^{-1}), and *SiH$_3$ (ν = 2138 cm^{-1}) are present, where the asterisk (*) denotes a surface atom, as well as surface dopants such as *PH$_x$ (ν = 2276 cm^{-1}); larger size NCs additionally exhibit *BH$_x$ (ν ~ 2500 cm^{-1}).[21] Notably, LSPR is absent in the FTIR spectra of these hydride-terminated PB:Si (PB:Si–H) NCs, which can be attributed to near-perfect charge compensation between charges introduced from donor and acceptor impurities.[21] The NC size is determined *via* XRD and the Scherrer broadening of the (111) diffraction peak (Fig. S2†) correlates well with our experimentally determined PL sizing curve.[43] Additional evidence of dopant atom incorporation can also be seen through a continual shift of the (111) diffraction peak to higher angles with increasing NC size, representing a smaller lattice constant consistent with a greater number of substitutional boron sites.[20]

The nanoparticle surface can be modified post-synthesis to render PB:Si NCs colloidally dispersible. Covalent functionalization of PB:Si NCs follows our previously reported procedure for intrinsic Si NCs[38,43] by heating hydride-terminated PB:Si–H NCs in neat reactant (1-dodecene, 1-dodecanethiol, 1-dodecylamine, or 1-dodecanol) at 140–150 °C for ~4 h with the addition of 1,1′-azobis(cyclohexanecarbonitrile) radical initiator (see Experimental for full details). Hydride (as-synthesized) and alkyl (from reaction with 1-dodecene) groups bind solely *via* X-type interactions at the surface of the PB:Si–X NCs (Fig. 1), where the X-type formalism refers to a covalent bond between a ligand and a metal (in this case silicon).[44] In addition to conventional X-type ligation, highly Lewis-acidic boron sites reside at the NC surface in a three-coordinate arrangement, allowing acid–base interactions.[20,45–47] It has been shown that boron atoms are primarily incorporated on the NC surface as opposed to the core, due to the reduced solubility of boron within the NC core caused by the mismatch in atomic radii between B and Si.[45] As is evident from the spontaneous solubility of as-synthesized boron-doped silicon (B:Si) NCs in strongly polar solvents such as methanol, *N*-methylpyrrolidone (NMP), and dimethylsulfoxide

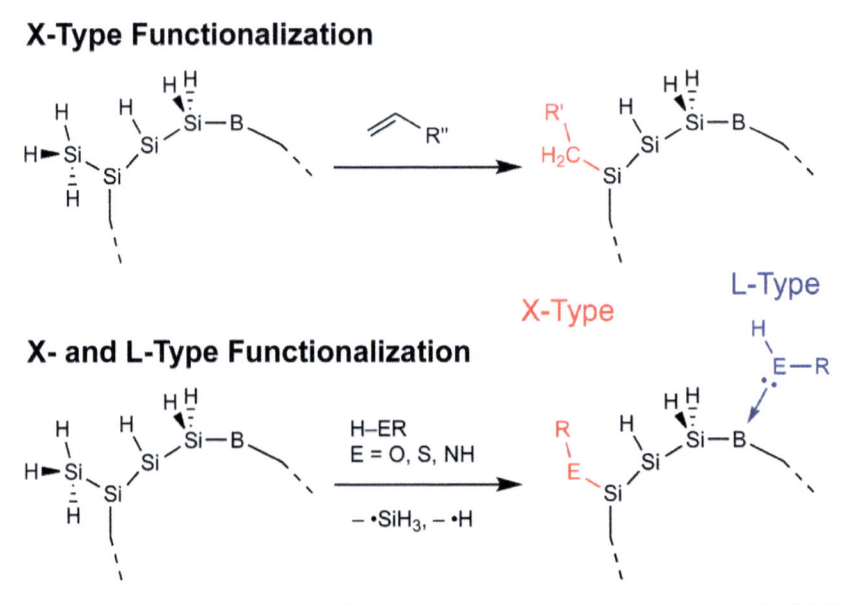

Fig. 1 Schematic illustration of the X- and L-type functionalization mechanisms for PB:Si NCs. X-type functionalization involves the formation of a covalent bond with a surface *Si atom, whereas L-type functionalization is Lewis donor molecule coordination to a Lewis acidic surface *B atom.

(DMSO), the electron pair from a filled O 2p orbital in the donor molecule bonds *via* dative interaction with the empty 2p orbital of the surface boron atoms.[21] This type of bonding interaction is classified by Green as L-type, where both electrons are provided by ligand interactions.[44] Three of the surface ligands studied in this report (1-dodecanethiol, 1-dodecylamine, and 1-dodecanol) have Lewis basic head groups that bond *via* such an L-type interaction with surface *B in addition to X-type bonding. We annotate these species as L→PB:Si–X throughout this work.

Fig. 2 summarizes the experimental PL results and details how the emission energy can be modified by changing the surface termination of PB:Si NCs (see Fig. S3† for all emission profiles). The expected peak emission energy for alkyl-terminated intrinsic Si NCs as a function of the NC diameter (D_{NC}) can be approximated using an empirically developed $D_{NC}^{-1.69}$ dependence (Fig. 2b, black dashed trace).[43] The peak emission energy of the PB:Si–H NCs (Fig. 2b, gray points) is consistently ~400 meV red-shifted from that of the intrinsic Si NC trend across all sizes within the confinement regime, owing to lower energy D–A transitions, and follows a slightly different power dependence of $D_{NC}^{-2.09}$.[21] We find that the PB:Si–H NCs synthesized in this study still follow this dependence, showing the reproducibility of the plasma growth process (Fig. S4†). Subsequent functionalization and solubilization of PB:Si–X and L→PB:Si–X NCs blue-shifts the emission relative to the as-synthesized PB:Si–H NCs for all functional groups studied. The greatest shift is induced by cooperative alcohol/alkoxide terminations that display changes in the emission energy of over 400 meV. Unlike intrinsic Si NCs with alkoxide surface chemistry that exhibit size-dependent band edge emission (where stronger *Si–OR effects are observed for smaller, more quantum-confined NCs[38]), the change in the D–A emission energy for each L→PB:Si–X functionalized NC across the entire size range studied here (2.7–5.1 nm) is only ~50 meV (Fig. S5†). The fact that the smallest shift results from X-type-only interactions through alkyl (*Si–CH$_2$R) bonding suggests that Lewis acid–base interactions at the NC surface also influence the PB:Si NC electronic structure of the donor and acceptor states.

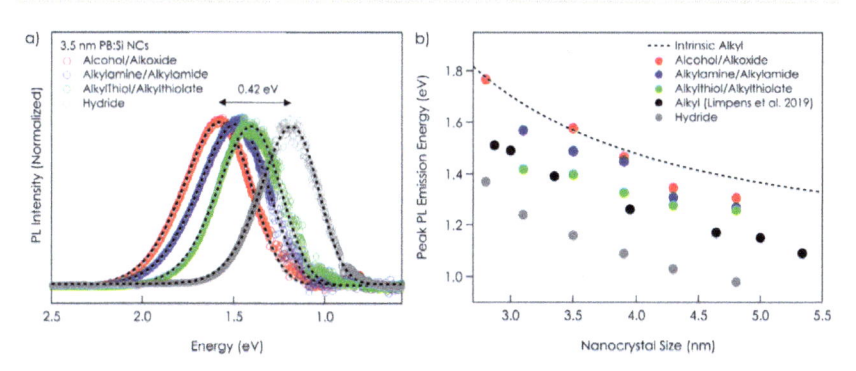

Fig. 2 (a) Representative shifts in the peak PL emission energy from 3.5 nm RHO→PB:Si–OR (red), RH$_2$N→PB:Si–NHR (blue), and RHS→PB:Si–SR (green) NCs. All traces are fit with a lognormal (black dashed traces). (b) Summary of the peak PL energies for a variety of PB:Si NC sizes and surface chemistries. Peak PL energies from alkyl-functionalized PB:Si–CH$_2$R NCs are included from our previous study.[21]

We next attempt to quantify the individual effects of the covalently bound X-type and coordinated L-type ligands. Similar to Fujii's original report of the dispersion of PB:Si NCs in methanol,[26] we find that PB:Si–H NCs directly disperse in a solution of toluene and 1-dodecanol or 1-dodecylamine at room temperature (Fig. S6†). In contrast, similar experiments with 1-dodecanethiol in toluene do not induce spontaneous suspension. We suspect that these solubility differences result from two effects. First, the weaker nature of the Lewis acid–base complex with thiols (*i.e.*, RHS→*B) relative to that with alcohols and alkylamines does not provide sufficient energy for solvation.[47] Second, the PL spectra from the functionalized L→PB:Si–X NCs produced *via* either the spontaneous solubilization process or the radical-initiated thermal functionalization route align nearly exactly (Fig. S7a†). We therefore hypothesize that the dispersion process in protic reagents such as alcohols and amines does *not* interact solely *via* Lewis acid–base chemistry, and that the reaction to generate X-type ligands also occurs under ambient conditions. Our previous study of the mechanism of radical-initiated surface reactions on nonthermal plasma-synthesized intrinsic Si NCs showed that abstraction of surface silyl groups, *SiH$_3$, can occur spontaneously at room temperature,[43] which can explain our observation here. In addition, we find that FTIR spectra from B:Si NCs solvated in DMSO (termed "DMSO–B:Si" in our prior work[20]) exhibit a strong absorption at ∼2240 cm^{-1} characteristic of the so-called 'back-bonded' Si–H stretch, where a high frequency ∼2240 cm^{-1} Si–H vibration results from the *Si atom also being bound to a highly electron-withdrawing oxygen atom, (O)*SiH$_x$. Since coordination of DMSO *via* an L-type interaction as (CH$_3$)$_2$(O)S=O→PB:Si cannot be the source of this IR absorption feature—there is no strongly affected *SiH$_x$ vibration *via* this L-type interaction since it bonds to a surface boron atom—we posit that this high energy Si–H stretch results from hydro- and/or silylsilylation across the sulfoxide S=O bond in DMSO to give an X-type ligand (CH$_3$)$_2$(O)(Y)S–O–*SiH$_x$ (Y = H for hydrosilylation and SiH$_3$ for silylsilylation), in addition to an L-type interaction. See Fig. S8† for full description of the reaction with DMSO.

Additional confirmation that the shifts in PL energy in Fig. 2 must be attributed to both X-type and L-type interactions is derived from an experiment in which 3.5 nm PB:Si NCs are successively functionalized with covalent and then dative ligands. PB:Si NCs are first functionalized with covalent alkyl groups using the radical-initiated thermal method. After collecting the PL spectrum of the resulting PB:Si–CH$_2$R NCs, a small amount of 1-dodecanol is added to the toluene solution in the cuvette. As shown in Fig. S7b,† the PL spectrum of the resulting alcohol/alkyl RHO→PB:Si–CH$_2$R NCs blue-shifts by ∼50 meV from the alkyl-only PB:Si–CH$_2$R NCs. These data provide clear evidence that both types of surface chemistry interactions affect the energy of the D–A recombination in PB:Si NCs. In addition, this experiment allows us to quantify the shift derived from alcohol L-type coordination to surface *B atoms at ∼50 meV. Since the PL energy of 3.5 nm diameter alcohol/alkoxide RHO→PB:Si–CH$_2$R is blue-shifted by ∼200 meV relative to alkyl-only PB:Si–CH$_2$R (Fig. 2), this additionally means that X-type ligation with alkoxide results in ∼150 meV PL blue-shift relative to alkyl. Thus, the effect of X-type ligation on D–A transition energies in PB:Si NCs appears to be similar to that in our prior report on the effect of this ligand head group on the band edge emission energies in intrinsic Si NCs for small NC sizes (<4 nm).[38]

To probe possible changes to the radiative mechanism, we characterize the radiative lifetimes of L→PB:Si–X NCs using time-resolved photoluminescence (TRPL) spectroscopy. Fig. 3a shows representative TRPL traces for 3.5 nm L→PB:Si–X NCs. The effective lifetime is retrieved by fitting a single exponential to the dominant, long-lived component of the decay. Fig. 3b summarizes the lifetimes of several different sized functionalized samples with alkyl, alkylamine/alkylamide, alcohol/alkoxide, and alkylthiol/alkylthiolate surface terminations (see Fig. S9† for all TRPL data and fits). The lifetimes of all the L→PB:Si–X NCs, irrespective of the ligands, are found to be slightly faster than those found for alkyl-terminated intrinsic Si NCs of the same size (Fig. 3b, black dashed trace), consistent with Fujii and our prior reports.[21,26,48] While D–A recombination is generally expected to be a slow, tunneling process, the slightly faster effective lifetime is most likely due to a convolution of different radiative decay pathways.[49] From the shape of the TRPL decay, it is evident that multiple decay pathways indeed exist. These traces can be fit using a multi-exponential fit (Fig. S10†) that can quantify the various distinct decay pathways. Another of our studies has revealed that functionalizing Si NCs with alkylthiolate ligands can induce mid-gap trap states that contribute to Shockley–Read–Hall (SRH)-type recombination.[39] This is reflected in the large amplitude decay within the first 20 μs in the RHS→PB:Si–SR TRPL trace as seen in Fig. 3a and all RHS→PB:Si–SR traces in Fig. S9.† In agreement with an additional alkylthiol/alkylthiolate-derived recombination pathway, the photoluminescent quantum yield (PLQY) spectra of L→PB:Si–X samples exhibit a vast disparity in the radiative recombination efficiency of alcohol/alkoxide- and alkylamine/alkylamide-terminated PB:Si NC ensembles (both 12–13%) compared with alkylthiol/alkylthiolate terminations (2%) (Fig. S11†).

To further help elucidate the different surface environments in L→PB:Si–X and PB:Si–X NCs, we use Raman spectroscopy to examine the nature of the local boron mode with different surface functionalizations. Collecting the Raman spectrum for RH_2N→PB:Si–NHR NCs proved difficult due to excess ligand that formed a solid at room temperature and induced overwhelming scattering of the

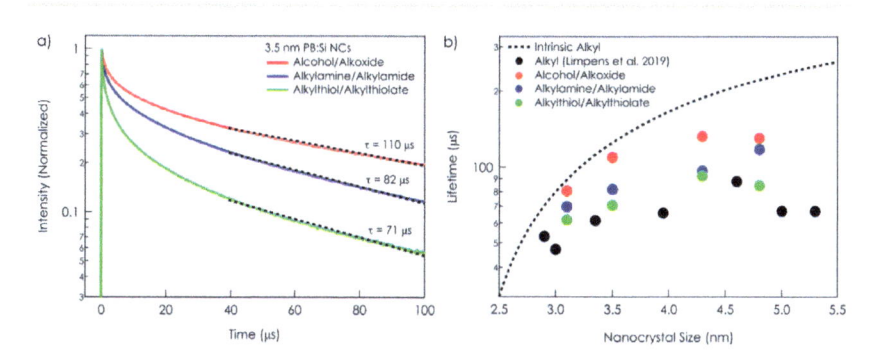

Fig. 3 (a) Representative TRPL traces from 3.5 nm RHO→PB:Si–OR (red), RH_2N→PB:Si–NHR (blue), and RHS→PB:Si–SR (green) NCs. Effective lifetimes are calculated by fitting the tail of the decay to a single exponential (black dashed traces). (b) Summary of all TRPL lifetimes acquired for multiple NC sizes and surface terminations (see Fig. S9† for all TRPL traces). Lifetimes from alkyl-terminated PB:Si–CH_2R NCs are included from our previous study.[21]

Raman signal. RH$_2$N→PB:Si–NHR NCs are therefore not included in this analysis. As shown in Fig. 4, the two Raman modes of the as-synthesized PB:Si–H NCs at 498 cm^{-1} and ~600 cm^{-1} are attributed to the silicon LO phonon mode (broadened and made asymmetric by Fano interference[50]) and boron-related local vibrational modes, respectively. The LO phonon mode of silicon shifts by 18 cm^{-1} to 516 cm^{-1} upon functionalization. This may be related to the reconstruction of surface hydrides upon thermal treatment during ligation, which has been hypothesized to also enhance the PLQY.[51] Assignment of the B-related local vibrational mode is derived from previous Raman studies on PB:Si NC systems.[46,52] Somogyi *et al.* modeled the vibrational density of states (VDOS) within PB:Si NCs and showed a large density of boron states extending from 600 to 800 cm^{-1} that could be attributed to disperse boron environments within PB:Si NCs (*i.e.*, surface or core states, proximity to phosphorus atoms, coordination environment, *etc.*).[52] Generally, the high-energy tail of this broad peak is attributed to boron at the surface, and thus can provide a picture of the boron environment at the PB:Si surface. Upon surface functionalization, the boron mode is uniquely affected by the identity of the ligand binding group. For the RHO→PB:Si–OR and RHS→PB:Si–SR NCs, the boron mode shifts and narrows; a signature of different local environments. In contrast, for the alkyl-terminated PB:Si–CH$_2$R NCs it remains as a broad peak, indicating that the Lewis acid–base interaction with surface boron atoms *B is responsible for the narrowing. The direction of the energy shift is also ligand dependent. Whereas RHO→PB:Si–OR results in a shift to lower energies relative to H–PB:Si, RHS→PB:Si–SR shows just the opposite. This shift is likely related to the strength of the acid–base interactions at the surface of the L→PB:Si–X NCs. In summary, the shift and sharpening of the B-related local Raman mode observed within RHO→PB:Si–OR and RHS→PB:Si–SR NCs is clearly attributed to L-type interactions at the NC surface.

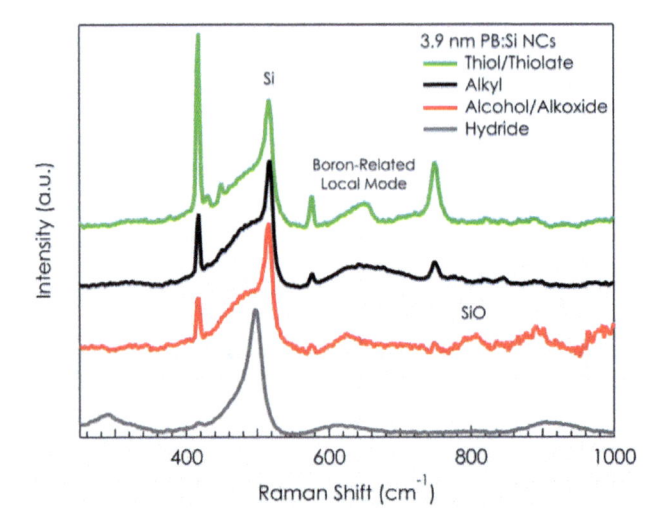

Fig. 4 Raman spectra of 3.9 nm RHS→PB:Si–SR (green), PB:Si–CH$_2$R (black), RHO→PB:Si–OR (red), as well as PB:Si–H (gray) NCs. Strong, sharp peaks at 415–450, 577, and 750 cm^{-1} can all be attributed to the sapphire substrate (see Fig. S12† for blank sapphire Raman spectrum).

Discussion

The mechanism accounting for the blue-shifted emission energy upon functionalization of PB:Si NCs is multi-faceted. To fully explain this picture, the different radiative mechanisms between PB:Si NCs and intrinsic Si NCs must first be understood. Delerue recently employed atomistic tight-binding calculations to model the specific energy levels created by phosphorus and boron impurities within PB:Si NCs.[22] The results show that close, but not nearest-neighbor, P–B pairs account for the D–A transitions observed experimentally. Furthermore, the extent of the PL red-shift is largely dependent on the number of P–B pairs within the NC, where even a single P–B pair shows a large impact on the radiative energy. It's important to note that phosphorus and boron impurities are only optically active when in a silicon substitutional site (*i.e.*, in a four-coordinate arrangement). Therefore, trivalent boron atoms on the NC surface do not contribute to the effective number of P–B pairs. Based on Delerue's calculations, we estimate that between 5 and 10 optically active P–B pairs within our PB:Si–H NCs account for the 400 meV shift in emission energy from that of intrinsic Si NCs. The fact that such large shifts are seen for a relatively small number of optically active dopants is a result of the dielectric confinement effect, in which partially unscreened Coulomb interactions create much deeper donor and acceptor levels than what is typically observed for impurity states in bulk Si.[22] It follows that different dielectric media surrounding the PB:Si NC will impact the level of confinement (greater Coulomb interactions for a greater mismatch in the dielectric constant) and subsequently the D–A energy levels. The changing emission energies from PB:Si–H to L→PB:Si–X and PB:Si–CH$_2$R NCs can therefore be partially explained by the changing dielectric medium between NC ensembles measured as free-standing powders (PB:Si–H) and in colloidal solutions (L→PB:Si–X and PB:Si–CH$_2$R). The more closely matched dielectric constants of the Si core with its ligands and solvent shell (compared with argon glove box atmosphere and hydride surface atoms for PB:Si–H) should increase the Coulomb interaction (decrease the Coulomb screening) and create shallower D–A energy levels, thereby blue-shifting the emission from PB:Si–H NCs.

In addition to dielectric effects, our prior study has shown that there is interaction between covalent X-Type ligands and the Si NC wavefunction[38] that is undoubtedly responsible for some of the PL blue shift upon functionalization. This X-type ligand effect on the degree of exciton localization occurs not just for Si NCs but also metal chalcogenide quantum dots, where exchanging hard carboxylate or amide ligands with softer species such as sulfide-, selenide-, or telluride-based ligands results in a red shift. For further reading on these effects, see ref. 53–61.

Two other possibilities for the PL blue shift have been proposed in prior studies on PB:Si NCs synthesized using thermal precipitation from phosphoborosilicate glass. Such PB:Si NCs exhibit a similar blue shift in the emission energy after particles are HF-etched from the SiO$_2$ matrix and suspended in solution.[27,62] This blue shift has been attributed to both increased quantum confinement (due to HF etching) and fewer P–B pairs (again, resulting from etching), not to any ligand effects.[27,62] However, we do not believe these mechanisms to be operative for nonthermal plasma-synthesized PB:Si NCs, since dopants would have to be removed from the NC core or, alternatively, the local bonding environment would have to be changed in such a way to render substitutional dopants "inactive" (*i.e.*, trivalent arrangements). Instead, we find that the number of substitutional boron sites in the NC core is unchanged based on our observation that surface functionalization does not modify the (111) diffraction peak position or width (see XRD data in Fig. S13†). This is to be expected, as there is insufficient driving force at ambient temperature to expel NC dopant atoms from the core for covalent semiconductors such as silicon. This high metastability of doped covalent semiconductor NCs such as silicon is a key difference from more ionic semiconductor NCs, such as metal chalcogenides, where impurities are more easily expelled.[13,14]

Next, we propose that a 3rd mechanism operates to induce a blue shift in L→PB:Si–X NCs relative to solely X-type PB:Si–H and PB:Si–CH$_2$R NCs. In the modeling work by Delerue,[22] D–A states within PB:Si NCs are still relatively delocalized (characterized by the participation ratio, or degree of coupling between D–A states and Si NC band states), particularly for acceptor states. This is consistent with the experiments here, since otherwise we would not observe a size-dependent shift in the D–A transition energy or effects from ligands on core-residing D–A states without some degree of state coupling. This coupling allows for orbital overlap and the formation of P–B pairs between surface boron and core-residing phosphorus atoms. Invoking molecular orbital (MO) theory,[30] the functional group that coordinates with surface boron sites in P–B pairs will greatly affect the specific energy of the acceptor and donor states (Fig. 5). If this

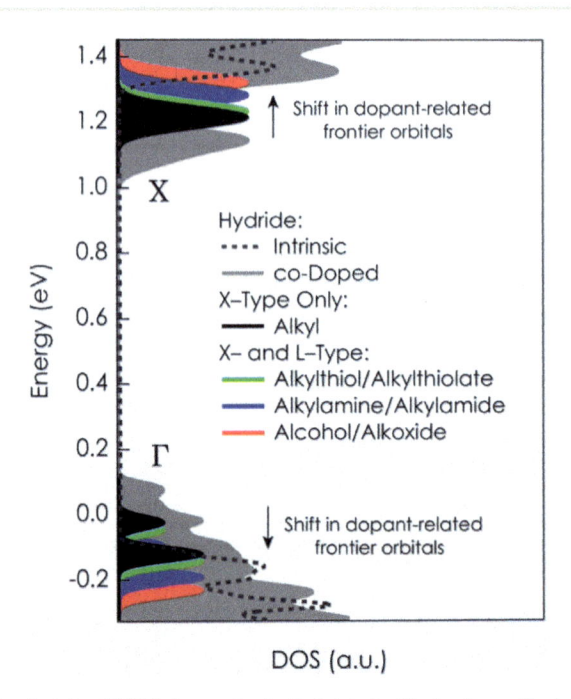

Fig. 5 Density of states (DOS) shown for both intrinsic (dashed gray line) and co-doped (solid gray) hydrogen-terminated Si NCs, as calculated by Delerue.[22] The frontier orbitals associated with both the HOMO and LUMO levels shift upon functionalization with L- and/or X-type ligands.

MO picture is valid, the extent of the PL energy shift should be proportional to the electron donating ability of the bonding atom. Experimentally, we find the alcohol (HOR) to induce the greatest shift in the acceptor level, followed by the alkylamine (H$_2$NR) and then the alkylthiol (HSR), perfectly in line with the electron donating ability of each functional group. As the energy of the acceptor level moves closer to the valence band, the donor level moves higher to compensate. These combined effects cause the D–A transition to be at a higher energy than without the L-type interaction, exactly what we observe experimentally for small <4 nm diameter L→PB:Si–X NCs (see Fig. S7b†). Similar effects have been observed for II–VI NCs, where the exchange of electron-donating L-type oleylamine ligands with electron-withdrawing, Lewis acidic Z-type M(oleate)$_2$ (M = Zn, Cd) relaxes quantum confinement, leading to an emission red shift.[63]

We note that at smaller D_{NC} < 4 nm, we find that the D–A energy levels within alkoxide terminated NCs become degenerate with the bands, as evidenced by both the emission energy and the NC lifetimes. As D_{NC} increases, the effect of both X-type and L-type surface coordination becomes weaker due to the shrinking surface-to-volume ratio of the NC; thus, the core-residing P–B pairs are less affected by the surface chemistry/dielectric environment. This is evidenced by the convergence of the emission energies of X-type and L-type L→PB:Si–X NCs at D_{NC} = 4.8 nm (see Fig. 2b).

Conclusions

This work shows that the D–A transitions in PB:Si NCs respond strongly to the surface chemistry. Effects from covalent X-type surface bonding interactions in alkyl PB:Si–CH$_2$R NCs are found to increase the D–A recombination energy by ∼200 meV relative to the as-prepared hydride-terminated PB:Si–H NCs, resulting from changes in the potential of dopant-related states. A secondary and equally strong perturbation of the D–A energy levels occurs through L-type Lewis base coordination of surface boron atoms by HOR, H$_2$NR, and HSR for L→PB:Si–X NCs (X = OR, NHR, SR). This L-type effect can be described by a molecular orbital model, with stronger donor groups leading to a higher energy donor level (closer to the conduction band) and a lower energy acceptor level (closer to the valence band). For both X-type and L-type ligands, an increase in the dielectric environment around the PB:Si NCs results in less Coulomb screening and, therefore, donor and acceptor states that are closer to the conduction and valence bands, respectively, compared with as-grown PB:Si–H NCs.

These results map out the strong effects of ligands on the D–A levels in PB:Si NCs, all of which result in higher emission energies. For NIR applications, modulating the energetics in the opposite direction (lower energy transitions) is desirable. The work presented here details the challenges that must be tackled to achieve lower energy emission. Donors and acceptors with states further from the bands than P and B should be developed. Surface structures with a highly mismatched dielectric environment to the underlying PB:Si NC core are crucial to ensuring poor Coulomb screening and localization of the states. Similarly, ligands that interact to stabilize the surface but provide weaker Lewis acid–base interactions, thereby weakly perturbing the donor and acceptor levels, should be targeted in future research. Finally, all of these challenges must be overcome while simultaneously achieving high PLQY values, which are far from optimized in Si NCs. Still, the future is bright for co-doped PB:Si NCs, which continue to demonstrate fascinating fundamental photophysics while also holding tantalizing promise for future applications.

Experimental

General methods

All solvents were purchased from Sigma-Aldrich or Fisher, ACS grade or better unless otherwise noted. Toluene and 1-dodecene were distilled from sodium under nitrogen. 1-Dodecylamine, 1-dodecanol, and 1-dodecanethiol were dried under vacuum at 100 °C for 24 hours. All experiments with PB:Si NCs were conducted in an argon-atmosphere glove box, with the exception of powder X-ray diffraction (performed in air; total Si NC air exposure time ≤1 h).

Nanoparticle growth

All PB:Si–H NCs in this study were grown *via* a non-thermal plasma synthesis method using a capacitively-coupled 13.56 MHz RF plasma. Using an optimized ratio of process gases $SiH_4/PH_3/B_2H_6$ and carrier gases (Ar, H_2), the NC size was controlled by varying the pressure within the reactor tube. Table S1† summarizes the synthetic methods for all PB:Si NCs reported. Hydrogen-terminated PB:Si NCs (PB:Si–H) were collected downstream from the plasma on a 400-mesh stainless steel filter and transferred *via* load-lock to an inert-atmosphere argon-filled glovebox for collection and subsequent functionalization/characterization.

CAUTION! Singly doped and co-doped Si NCs evolve toxic gas (PH_3 and/or B_2H_6) upon air exposure. Rigorous air-free handling is required. Allow samples to air oxidize for at least 1 h in exhaust ventilation before removing into lab ambient conditions.

Surface functionalization

As-prepared, hydride-terminated PB:Si–H NCs were functionalized *via* hydrosilylation. Approximately 1 mL of each ligand (in this study either 1-dodecene, 1-dodecanethiol, 1-dodecanol, or 1-dodecylamine) was added to 5 mg of PB:Si–H NC powder with the addition of ~2 mg 1,1′-azobis(cyclohexanecarbonitrile) (ABCN) radical initiator. The solution was heated at 140–150 °C for ~4 h to ensure saturated ligation. After cooling to ambient temperature, a few mL of dry toluene was added to dilute the sample, and the solution was filtered through a 0.45 μm PTFE filter.

Fourier-transform infrared (FTIR) spectroscopy

FTIR absorbance measurements were conducted on a Bruker Alpha FTIR spectrometer using a diffuse reflectance infrared Fourier transform spectrometer (DRIFTS) attachment with a resolution of 4 cm^{-1}. Reflective gold-coated polished Si wafers were used for background measurements, and PB:Si–H NCs were deposited by mechanically pressing powder directly onto the same substrates. Spectra were baseline-corrected using the concave rubberband correction method (2 iterations).

Photoluminescence (PL), and time-resolved photoluminescence (TRPL) spectroscopy

PL measurements were conducted using a home-built system. Samples in toluene solutions were excited using a Thorlabs fiber-coupled 405 nm LED pulsed at 10 Hz using a Thorlabs DC2200 LED driver. Visible detection was made using an Ocean Optics Ocean FX spectrometer whereas NIR detection was made using an Ocean Optics NIRQuest spectrometer. Spectra were stitched at the crossover point between the two detectors (940–1000 nm) using LabVIEW program developed in-house. Detector calibration was done using an Ocean Optics HL-2000-HP black-body lamp.

TRPL spectra were collected using a home-built setup in which a Nd:YAG laser was used for photoexcitation and subsequently tuned to the desired wavelength (450 nm) using an optical parametric oscillator (OPO) (Spectra Physics Quanta Ray and GWU Premi Scan). PL spectra were collected in an off-axis, front-face geometry using a Si avalanche photodiode (APD, Thorlabs APD430A). A combination of 700 to 700 nm longpass filters was used to block the excitation light and the signal from the APD was digitized using a fast oscilloscope (Tektronix DPO7254).

Absolute PL quantum yield (PLQY) measurements were conducted on PB:Si NCs in toluene solution using a fiber-coupled integrating sphere (Labsphere 3P-GPS). A 532 nm collimated laser diode (Thorlabs CPS532-C2) was used as an excitation source. The sphere was fibre-coupled, using a Thorlabs M35L02 fibre (Ø 1000 µm), to a Princeton HRS-300-SS spectrograph including a PYL100F Si CCD.

X-ray diffraction

XRD patterns were taken on a Bruker D8 Discover diffractometer using Cu Kα radiation ($\lambda = 1.54$ Å). PB:Si NCs were dispersed in toluene slurries on Si-based zero diffraction plates. NC sizes were determined from the XRD spectra by use of the Scherrer-broadening of the (111) diffraction peaks. The NC size was calculated using the equation $D_{NC} = k\lambda/w \cos(c)$, where D_{NC} is the diameter of the NC, k is a shaping factor of 1.1, λ is the X-ray wavelength (1.54 Å), and w and c are the width and center of the Si (111) diffraction peak in radians, respectively.

Raman spectroscopy

Raman spectra of all PB:Si NCs are obtained *via* an InVia Renishaw confocal Raman microscope with a 532 nm laser and 50× objective lens. The samples are sealed in an air-free holder with a 300 µm-thick transparent sapphire optical window. The data are collected at an output laser power of ∼900 µW with 2 second integration time and 5 accumulations.

Conflicts of interest

There are no conflicts to declare.

Acknowledgements

We would like to thank Fernando Urias-Cordero for his help in nanocrystal growth, as well as Lance Wheeler for helpful discussion. This work was authored by the National Renewable Energy Laboratory, operated by the Alliance for Sustainable Energy, LLC, for the U.S. Department of Energy (DOE) under Contract No. DE-AC36-08GO28308. Funding provided by U.S. Department of Energy Office of Science Chemical Sciences, Geosciences, and Biosciences Solar Photochemistry Program. The views expressed in the article do not necessarily represent the views of the DOE or the U.S. Government. The U.S. Government retains and the publisher, by accepting the article for publication, acknowledges that the U.S. Government retains a nonexclusive, paid-up, irrevocable, worldwide license to publish or reproduce the published form of this work, or allow others to do so, for U.S. Government purposes.

Notes and references

1 J. M. Caruge, J. E. Halpert, V. Wood, V. Bulović and M. G. Bawendi, *Nat. Photonics*, 2008, **2**, 247.
2 D. V. Talapin, J.-S. Lee, M. V. Kovalenko and E. V. Shevchenko, *Chem. Rev.*, 2010, **110**, 389.
3 H. Lu, G. M. Carroll, N. R. Neale and M. C. Beard, *ACS Nano*, 2019, **13**, 939.
4 R. Mazzaro, F. Romano and P. Ceroni, *Phys. Chem. Chem. Phys.*, 2017, **19**, 26507.
5 B. G. Lee, J.-W. Luo, N. R. Neale, M. C. Beard, D. Hiller, M. Zacharias, P. Stradins and A. Zunger, *Nano Lett.*, 2016, **16**, 1583.
6 F. Meinardi, S. Ehrenberg, L. Dhamo, F. Carulli, M. Mauri, F. Bruni, R. Simonutti, U. Kortshagen and S. Brovelli, *Nat. Photonics*, 2017, **11**, 177.
7 M. L. Mastronardi, F. Maier-Flaig, D. Faulkner, E. J. Henderson, C. Kübel, U. Lemmer and G. A. Ozin, *Nano Lett.*, 2012, **12**, 337.
8 D. M. Kroupa, B. K. Hughes, E. M. Miller, D. T. Moore, N. C. Anderson, B. D. Chernomordik, A. J. Nozik and M. C. Beard, *J. Am. Chem. Soc.*, 2017, **139**, 10382.
9 H. Lu, G. M. Carroll, X. Chen, D. K. Amarasinghe, N. R. Neale, E. M. Miller, P. C. Sercel, F. A. Rabuffetti, A. L. Efros and M. C. Beard, *J. Am. Chem. Soc.*, 2018, **140**, 13753.
10 H. Zhang, R. Zhang, K. S. Schramke, N. M. Bedford, K. Hunter, U. R. Kortshagen and P. Nordlander, *ACS Photonics*, 2017, **4**, 963.
11 M. Fujii, S. Hayashi and K. Yamamoto, *J. Appl. Phys.*, 1998, **83**, 7953.
12 A. Mimura, M. Fujii, S. Hayashi, D. Kovalev and F. Koch, *Phys. Rev. B*, 2000, **62**, 12625.
13 G. Galli, *Nature*, 2005, **436**, 32–33.
14 S. C. Erwin, L. Zu, M. I. Haftel, A. L. Efros, T. A. Kennedy and D. J. Norris, *Nature*, 2005, **436**, 91.
15 X. D. Pi, R. Gresback, R. W. Liptak, S. A. Campbell and U. Kortshagen, *Appl. Phys. Lett.*, 2008, **92**, 123102.
16 D. J. Rowe, J. S. Jeong, K. A. Mkhoyan and U. R. Kortshagen, *Nano Lett.*, 2013, **13**, 1317.
17 N. J. Kramer, K. S. Schramke and U. R. Kortshagen, *Nano Lett.*, 2015, **15**, 5597.
18 B. T. Diroll, K. S. Schramke, P. Guo, U. R. Kortshagen and R. D. Schaller, *Nano Lett.*, 2017, **17**, 6409.
19 R. Limpens and N. R. Neale, *Nanoscale*, 2018, **10**, 12068.

20 R. Limpens, G. F. Pach, D. W. Mulder and N. R. Neale, *J. Phys. Chem. C*, 2019, **123**, 5782.
21 R. Limpens, G. F. Pach and N. R. Neale, *Chem. Mater.*, 2019, **31**, 4426.
22 C. Delerue, *Phys. Rev. B*, 2018, **98**, 045434.
23 N. X. Chung, R. Limpens, A. Lesage, M. Fujii and T. Gregorkiewicz, *Phys. Status Solidi A*, 2016, **213**, 2863.
24 M. Fujii, K. Toshikiyo, Y. Takase, Y. Yamaguchi and S. Hayashi, *J. Appl. Phys.*, 2003, **94**, 1990.
25 M. Fujii, Y. Yamaguchi, Y. Takase, K. Ninomiya and S. Hayashi, *Appl. Phys. Lett.*, 2004, **85**, 1158.
26 Y. Hori, S. Kano, H. Sugimoto, K. Imakita and M. Fujii, *Nano Lett.*, 2016, **16**, 2615.
27 H. Sugimoto, M. Fujii, K. Imakita, S. Hayashi and K. Akamatsu, *J. Phys. Chem. C*, 2013, **117**, 11850.
28 R. Limpens, M. Fujii, N. R. Neale and T. Gregorkiewicz, *J. Phys. Chem. C*, 2018, **122**, 6397.
29 D. M. Kroupa, M. Vörös, N. P. Brawand, B. W. McNichols, E. M. Miller, J. Gu, A. J. Nozik, A. Sellinger, G. Galli and M. C. Beard, *Nat. Commun.*, 2017, **8**, 15257.
30 N. C. Anderson, M. P. Hendricks, J. J. Choi and J. S. Owen, *J. Am. Chem. Soc.*, 2013, **135**, 18536.
31 J. Owen, *Science*, 2015, **347**, 615.
32 K. A. Littau, P. J. Szajowski, A. J. Muller, A. R. Kortan and L. E. Brus, *J. Phys. Chem.*, 1993, **97**, 1224.
33 M. Dasog, Z. Yang, S. Regli, T. M. Atkins, A. Faramus, M. P. Singh, E. Muthuswamy, S. M. Kauzlarich, R. D. Tilley and J. G. C. Veinot, *ACS Nano*, 2013, **7**, 2676.
34 L. M. Wheeler, N. R. Neale, T. Chen and U. R. Kortshagen, *Nat. Commun.*, 2013, **4**, 2197.
35 H. Li, Z. Wu, T. Zhou, A. Sellinger and M. T. Lusk, *Phys. Chem. Chem. Phys.*, 2014, **16**, 19275.
36 T. Zhou, R. T. Anderson, H. Li, J. Bell, Y. Yang, B. P. Gorman, S. Pylypenko, M. T. Lusk and A. Sellinger, *Nano Lett.*, 2015, **15**, 3657.
37 M. Dasog, K. Bader and J. G. C. Veinot, *Chem. Mater.*, 2015, **27**, 1153.
38 G. M. Carroll, R. Limpens and N. R. Neale, *Nano Lett.*, 2018, **18**, 3118.
39 G. M. Carroll, R. Limpens, G. F. Pach, Y. Zhai, M. C. Beard and N. R. Neale, 2019, in preparation.
40 J.-W. Luo, P. Stradins and A. Zunger, *Energy Environ. Sci.*, 2011, **4**, 2546.
41 L. D. Negro, S. Hamel, N. Zaitseva, J. H. Yi, A. Williamson, M. Stolfi, J. Michel, G. Galli and L. C. Kimerling, *IEEE J. Sel. Top. Quantum Electron.*, 2006, **12**, 1151.
42 J. M. Buriak, *Chem. Mater.*, 2014, **26**, 763.
43 L. M. Wheeler, N. C. Anderson, P. K. B. Palomaki, J. L. Blackburn, J. C. Johnson and N. R. Neale, *Chem. Mater.*, 2015, **27**, 6869.
44 M. L. H. Green, *J. Organomet. Chem.*, 1995, **500**, 127.
45 J. Ma, S.-H. Wei, N. R. Neale and A. J. Nozik, *Appl. Phys. Lett.*, 2011, **98**, 173103.
46 K. Nomoto, H. Sugimoto, A. Breen, A. V. Ceguerra, T. Kanno, S. P. Ringer, I. P. Wurfl, G. Conibeer and M. Fujii, *J. Phys. Chem. C*, 2016, **120**, 17845.
47 L. M. Wheeler, N. J. Kramer and U. R. Kortshagen, *Nano Lett.*, 2018, **18**, 1888.
48 H. Sugimoto, M. Yamamura, R. Fujii and M. Fujii, *Nano Lett.*, 2018, **18**, 7282.

49 A. Marchioro, P. J. Whitham, K. E. Knowles, T. B. Kilburn, P. J. Reid and D. R. Gamelin, *J. Phys. Chem. C*, 2016, **120**, 27040.
50 D. M. Sagar, J. M. Atkin, P. K. B. Palomaki, N. R. Neale, J. L. Blackburn, J. C. Johnson, A. J. Nozik, M. B. Raschke and M. C. Beard, *Nano Lett.*, 2015, **15**, 1511.
51 L. Mangolini and U. Kortshagen, *Adv. Mater.*, 2007, **19**, 2513.
52 B. Somogyi, E. Bruyer and A. Gali, *J. Chem. Phys.*, 2018, **149**, 154702.
53 R. D. Harris, S. Bettis Homan, M. Kodaimati, C. He, A. B. Nepomnyashchii, N. K. Swenson, S. Lian, R. Calzada and E. A. Weiss, *Chem. Rev.*, 2016, **116**, 12865.
54 M. T. Frederick and E. A. Weiss, *ACS Nano*, 2010, **4**, 3195.
55 M. T. Frederick, V. A. Amin, L. C. Cass and E. A. Weiss, *Nano Lett.*, 2011, **11**, 5455.
56 M. B. Teunis, S. Dolai and R. Sardar, *Langmuir*, 2014, **30**, 7851.
57 C. Giansante, I. Infante, E. Fabiano, R. Grisorio, G. P. Suranna and G. Gigli, *J. Am. Chem. Soc.*, 2015, **137**, 1875.
58 H. Zhang, B. Hu, L. Sun, R. Hovden, F. W. Wise, D. A. Muller and R. D. Robinson, *Nano Lett.*, 2011, **11**, 5356.
59 J. J. Buckley, E. Couderc, M. J. Greaney, J. Munteanu, C. T. Riche, S. E. Bradforth and R. L. Brutchey, *ACS Nano*, 2014, **8**, 2512.
60 B. K. Hughes, D. A. Ruddy, J. L. Blackburn, D. K. Smith, M. R. Bergren, A. J. Nozik, J. C. Johnson and M. C. Beard, *ACS Nano*, 2012, **6**, 5498.
61 K. J. Schnitzenbaumer, T. Labrador and G. Dukovic, *J. Phys. Chem. C*, 2015, **119**, 13314–13324.
62 H. Sugimoto, M. Fujii, K. Imakita, S. Hayashi and K. Akamatsu, *J. Phys. Chem. C*, 2013, **117**, 6807.
63 Y. Zhou, F. Wang and W. E. Buhro, *J. Am. Chem. Soc.*, 2015, **137**, 15198.

Faraday Discussions

PAPER

Ab initio studies of the optoelectronic structure of undoped and doped silicon nanocrystals and nanowires: the role of size, passivation, symmetry and phase

Stefano Ossicini, *[ab] Ivan Marri,[b] Michele Amato,[c] Maurizia Palummo,[d] Enric Canadell[e] and Riccardo Rurali[e]

Received 16th September 2019, Accepted 22nd October 2019
DOI: 10.1039/c9fd00085b

Silicon nanocrystals and nanowires have been extensively studied because of their novel properties and their applications in electronic, optoelectronic, photovoltaic, thermoelectric and biological devices. Here we discuss results from ab initio calculations for undoped and doped Si nanocrystals and nanowires, showing how theory can aid and improve comprehension of the structural, electronic and optical properties of these systems.

1 Introduction

The scaling down of Si structures to nanometer size has opened up new chances to overcome the inability of bulk Si to be an efficient light emitter. In low-dimensional Si-based nanosystems, such as porous-Si (a quantum sponge made up of Si-nanostructures), Si-nanowires (Si-NWs) and Si-nanocrystals (Si-NCs), the possibility of achieving efficient visible photoluminescence (PL) has been clearly demonstrated.[1-7] In these nanostructures the low-dimensionality causes zone folding of the conduction band minimum of bulk Si, thus introducing a quasi-direct band gap. Furthermore, the quantum confinement (QC) effect induced by size and dimensionality reduction of the systems enlarges the energy band gap enabling light emission in the visible range.[8] QC can also enhance the spatial

---

[a] Dipartimento di Scienze e Metodi Dell'Ingegneria, Centro Interdipartimentale En&Tech, Universitá di Modena e Reggio Emilia, Via Amendola 2 Pad. Morselli, I-42125 Reggio Emilia, Italy. E-mail: stefano.ossicini@unimore.it

[b] Centro S³, CNR-Istituto di Nanoscienze, Via Campi 213/A, I-41125 Modena, Italy

[c] Laboratoire de Physique des Solides (LPS), CNRS, Université Paris Sud, Université Paris-Saclay, Centre Scientifique D'Orsay, F91405, Orsay Cedex, France

[d] Dipartimento di Fisica and INFN, Universitá di Roma Tor Vergata, Via Della Ricerca Scientifica 1, 00133 Roma, Italy

[e] Institut de Ciencia de Materials de Barcelona (ICMAB-CSIC), Campus de Bellaterra, 08193 Bellaterra, Barcelona, Spain

localization of electron and hole wave functions and their overlap, and thus the probability of their recombination. Moreover, size reduction offers the possibility of modulating the electronic and optical properties and therefore of increasing solar light harvesting, opening up new routes in the development of efficient nanostructured Si-based solar cell devices.[9]

Si-NCs have attracted, from the beginning, much interest because they exhibit very bright visible PL, which is tunable with respect to their dimensions, and sample dependent high quantum yields that can be enhanced by careful surface passivation.[10] Optical gain in Si-NCs has been observed and studied[11–14] and multiple exciton generation effects in excited carrier dynamics after the absorption of a single high-energy photon have been proven in a range of Si-NCs, revealing efficient carrier photogeneration that could increase solar cell performance.[15–22] Indeed full exploitation of the size, shape and surface termination[23–32] of Si-NCs together with their low toxicity and good bio-compatibility has boosted their application to microelectronics,[33–36] photonics,[37–39] non-linear optics,[40] photovoltaics,[9,41,42] thermoelectrics[43] and nano- and biomedicine.[44–49]

Si-NWs are particularly appealing thanks to bottom-up growth, which allows the limits of conventional lithography-based top-down design to be overcome. Si-NWs with very small diameters (of the order of a few nm) have been successfully grown and several applications have been demonstrated in the fields of electronic devices, nonvolatile memories, photonics, photovoltaics, biological sensors and thermoelectrics.[5,50–52] In particular, in these one-dimensional structures the electronic and optical properties strongly depend not only on the diameter, but also on the growth orientation, surface termination and crystal phase.[53,54]

An additional degree of freedom in semiconductor materials design is given by the introduction of impurities. Controlled doping is at the heart of the modern Si-based semiconductor industry. The presence of dopants remarkably changes the optical and electronic properties of the host material. In particular, B and P impurities introduce very shallow levels in the band gap of bulk Si (above the valence band and below the conduction band, respectively) and can then be efficiently ionized at room temperature, dramatically increasing the conductivity. Doping in Si nanostructures can be used to alter in a controllable way the electronic, optical and transport properties of nanomaterials. As a consequence, intentional doping with n- and p-type impurities can be exploited to design and realize novel devices. The possibility of enhancing the electrical conductivity of nanosized systems has been attempted, for instance, by fabricating porous-Si from n- or p-doped bulk Si by means of electrochemical etching.[55] However, although the etching process does not remove the impurities from the system,[56] a very low conductivity was measured, even for the larger mesoporous samples. This suggests that the ionization of the impurities at room temperature may be strongly quenched with respect to the bulk. Therefore the possibility of generating free charge carriers from impurity states can be limited by size effects. As for the optical properties, the introduction of an isoelectronic impurity in Si nanostructures can modify the indirect nature of the band gap in bulk Si. However, Si does not possess proper isoelectronic impurities that can strongly localize excitons at room temperature and enhance the PL intensity. An alternative approach is given by codoping. *i.e.* the formation of Si nanosystems with the same number of n- and p-type impurities.[57] The first attempts to dope Si nanostructures started about two decades ago. The results obtained revealed that the doped Si nanostructures showed different properties with respect to the doped Si-bulk.[58–60]

From an experimental point of view, several factors contribute to making the interpretation of the measurements on these systems a difficult task. First of all, independent of the fabrication technique, the degree of reproducibility of a single individual NC is very low. For instance, samples show a large variation in Si-NC size, a magnitude that is by itself difficult to determine. In this case it is possible that the measured quantity does not correspond exactly to the mean size but instead to the most responsive Si-NCs.[61] Moreover, Si-NCs synthesized by different techniques often show different properties in terms of their size and shape and in the structure of the interface. Finally, in solid nanocrystal arrays, some collective effects caused by electron, photon, and phonon transfer between Si-NCs can render the interpretation of the experimental results more complicated.

In this context the role of theoretical modeling and simulations is extraordinarily important. In particular, *ab initio* approaches represent a unique and very powerful instrument to predict and design the properties of novel molecules, materials, and devices with an accuracy that complements experimental characterization. Thus the importance of theoretical investigation lies not only in the interpretation of experimental results but also in the possibility of predicting the structural, electronic, optical and transport properties of unexplored materials.

In this work we will describe recent and new results obtained by *ab initio* simulations for doped and codoped Si-NCs and Si-NWs. We will mainly consider B and P as dopant atoms due to their high solid solubility in silicon and because they are the most commonly used impurities at the experimental level. The whole discussion will be conducted taking the corresponding results for undoped nanostructures as a reference to underline the differences and novelties introduced by the presence of impurities. In particular, together with previous outcomes, we will present new results related to the doping of Si nanocrystals with different group-III and group-V species, to the band structures of cubic Si-NWs and to the comparison of the stability and segregation of several dopants in cubic and hexagonal Si-NWs. It is worthwhile to note that new calculations have been explicitly performed in order to achieve better comparison with respect to previous results obtained through the use of several computational packages. In particular we note that different codes applied to the same nanosystem calculate energy eigenvalues that differ by less than 0.05 eV.

The paper is organized in the following way: in Section 2 we briefly describe the *ab initio* models used for the calculations of the structural, electronic and optical properties of the Si nanostructures. In Section 3 we present results regarding freestanding and matrix-embedded Si-NCs, with a particular focus on the theoretical calculations for undoped (see Subsections 3.1) and doped NCs. We will consider mainly B and P impurities (see Subsections 3.2), but a section has been devoted to other p- and n-doping atomic species (see Subsections 3.3). Section 4 is instead devoted to Si-NWs, with a subsection dedicated to theoretical results for NWs with different phases (Subsection 4.3). Conclusions are presented in Section 5.

2 Computational methods

Computational modelling of the structural, electronic and optical properties of complex systems is accessible nowadays, thanks to the impressive development of both theoretical approaches and computing facilities. Surfaces, interfaces, nanostructures, and even biological systems can now be studied within *ab initio* methods. Even within the Born–Oppenheimer approximation, which allows decoupling of the ionic and electronic dynamics, solving the Schrödinger equation which governs the physics of all those systems is a formidable task due to the many-body long-range electronic coulomb interactions. It is hence necessary to resort to further approximations. Density functional theory (DFT) avoids dealing directly with the many-body equation by mapping the interacting system into a fictitious non-interacting system, which is then described by self-consistent single particle equations.[62,63] Although DFT is a ground state theory based on relations that represent a fictitious auxiliary system with no physical meaning, it is often used to calculate band structures and their eigenvalues are often interpreted as one electron excitation energies corresponding to the excitation spectra of the system upon removal or addition of an electron. The level of accuracy is often related to the approximation used to describe the exchange correlation energy (E_{xc}). Here we adopt the local density approximation (LDA) where E_{xc} of the system is replaced with that of a uniform electron gas with the same density. DFT-LDA is computationally cheap and permits investigation of systems of a realistic size. The agreement with experiments is often remarkable, but sometimes only qualitative; the electronic gaps of semiconductors are, as a matter of fact, always systematically underestimated within DFT-LDA. The Green's function approach should be the formally correct approach to obtain electronic addition and removal energies, through the solution of the so-called quasi-particle (QP) equation. This is formally similar to the KS equation but instead of the exchange–correlation potential a non-local, energy dependent and non-Hermitian self-energy (usually estimated within the GW approximation) operator appears. This physically means that excitations are described in terms of a particle of finite lifetime which represents the extra electron (and/or the extra hole added to the system) plus its screened interaction with the electrons of the system.[64] Nevertheless, due to the complexity and high-computational cost of solving this QP equation, what is usually done is to estimate in a perturbative way the exchange correlation potential, V_{xc}, and correct the KS energies. To resolve this problem one can use, on top of DFT-LDA, the Green's function approach (in the so-called GW approximation), which maps the many-body electronic problem to a system of quasi-particles. The main consequence of adopting this approach is the opening of the band gap by amounts that, for nanostructures, are strongly size-dependent.

As for the optical properties, one can perform first-principles calculations of the dielectric response taking into account many-body interactions (namely excitonic and local-field effects) through the solution of the Bethe–Salpeter equation (BSE)[65] always within the many-body Green's function approach. However, it is worth noting that results obtained through LDA are still interesting for several reasons: (i) the inclusion of many-body effects is computationally very demanding and thus limited to small nanocrystals and nanowires,[53,66,67] (ii) an almost complete compensation of self-energy and excitonic effects[68–70] has been observed in silicon based nanostructures, thus rendering the KS based description significant, (iii) very often, one is interested in the trends of a specific property whose behaviour remains similar going from an independent-particle approximation to many-particle corrections. Thus, when not stated otherwise, the results of this paper have mainly been obtained using DFT-LDA approaches.

In all the performed supercell calculations the optimized structures have been achieved by relaxing all the atomic positions and cell parameters. Several different packages have been used: QUANTUM ESPRESSO,[71,72] SIESTA,[73] ABINIT,[74] VASP[75] and YAMBO code.[76]

3 Silicon nanocrystals

3.1 Undoped isolated and matrix embedded Si-NCs with differing passivation

Si-NCs have largely been investigated in the last two decades by the scientific community for different technological applications. Research activities have been mainly focused on the study of the microscopic properties of these systems, and in particular of the mechanisms that rule their electronic and optical properties. Despite the relevant efforts dedicated to the study of these systems, important issues remain unsolved. For instance, experiments have not totally clarified the microscopic origin of the PL peak in Si-NCs; sometimes it has been assigned to transitions between states localized inside the Si-NC, and other times to transitions between defect and/or interface states.[33] In this context, theory can play a crucial role and can support experimental investigation by solving fundamental controversies. In order to investigate PL in Si-NCs, we have considered two different types of surface termination for our model. For the first case the dangling bonds at the surface of the Si-NCs were passivated with H atoms. Although Si-NCs usually experience oxidation, there have been experiments where NC surfaces were passivated with H.[77] For the second case we performed studies on the same Si-NCs, but this time with dangling bonds capped by OH terminations. NCs have always been modelled using supercell methods, in which a large region of vacuum is adopted to separate points at the surface of the NC and its images, thus avoiding effects induced by spurious interactions between replicas. We consider both spherical-like and faceted Si-NCs. The first are obtained by cutting Si atoms outside a sphere from the Si bulk, whereas the faceted NCs result from a shell-by-shell construction procedure, in which one starts from a central atom and adds shells of atoms successively. Fig. 1 shows the obtained results for the highest occupied molecular orbital (HOMO)–lowest unoccupied molecular orbital (LUMO) gaps for both the hydrogenated and hydroxide Si-NCs. Investigation of the spatial localization of the HOMO and LUMO for the Si-NCs passivated with H shows that these states are not related to the H atoms. Therefore, the dependence of the HOMO–LUMO gap on the Si-NC diameter is a direct consequence of the QC effect (the NC shape does not have any influence) and thus the observed PL is directly related to the states localized inside the Si-NCs.[60,78–80] For the OH-terminated Si-NCs, instead, the changes in the HOMO–LUMO gap are strongly dependent on the oxidation degree (see Table 1) at the NC interface. In fact, in these NCs, the Si atoms at the interface have a different number of nearby O atoms. By defining the oxidation degree Ω as the ratio between the number of O atoms and the number of Si atoms bonded to them, we can compare the behavior of the HOMO–LUMO gap with the oxidation degree (see second and third columns of Table 1). What emerges is a strong correlation between the calculated gap values and Ω (a larger oxidation degree results in a larger gap), consequently the QC effect dependence from the size is still there (NCs with similar Ω show gaps with a clear dependence on the diameter), but it is strongly modulated by the presence of oxidation. It is worthwhile to note that

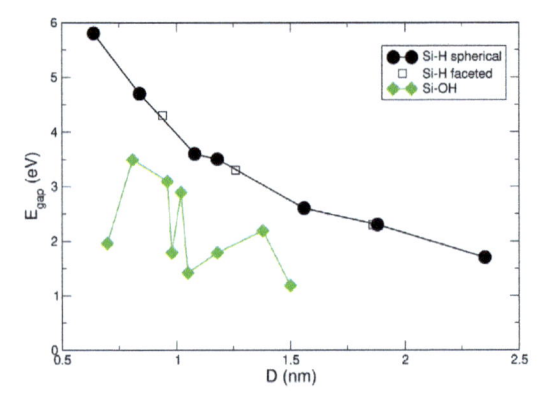

Fig. 1 HOMO–LUMO gaps for hydrogenated (black dots and black squares) and hydroxide (green diamonds) Si-NCs as function of NC diameter.

Table 1 Energy gaps (HOMO–LUMO) for the OH-terminated Si-NCs as a function of the diameter and oxidation degree Ω

d (nm)	Ω	E_{gap}
0.70	1.6	1.97
0.81	3.0	3.50
0.96	3.0	3.05
0.98	1.5	1.89
1.02	2.8	2.93
1.05	1.2	1.50
1.18	2.0	1.83
1.38	3.0	2.26
1.50	1.5	1.19

a size-dependent experimental study of the Si core-level shifts for Si-NCs embedded in a SiO$_2$ matrix showed that the shell around the Si-NCs consists of three different Si suboxide states, where for NCs with a diameter of less than 3 nm the presence of suboxides with higher oxidation is favored.[81]

Thus these results suggest that, in the case of Si-NCs embedded in SiO$_2$ matrices, for diameters above a certain threshold, about 2–3 nm, the emission peak of the Si-NCs simply follows the QC model, while interface states assume a crucial role only for very small Si-NCs.[82–89]

3.2 Single B and P doped and codoped isolated and embedded Si-NCs

In bulk Si the level of doping lies in the interval 10^{13} to 10^{18} cm^{-3}. In a Si-NC of about 2 nm in diameter, which possess more than 200 Si atoms, the introduction of a single impurity atom corresponds to a doping level of about 10^{20} cm^{-3}, which is a quite different situation. Despite the large number of works dedicated to the study of doped and codoped Si-NCs, important unsolved issues still exist. Effective dopant locations and the occurrence of self-purification effects (the tendency of the Si-NC to expel dopant atoms towards its surface), as well as the role played by the chemical environment of the Si-NC are topics still under discussion.[56,58,59,90–97]

In H-terminated Si-NCs the dopant atom could be either inside the NC or in a surface position. In order to define the most stable impurity location, we have probed all the possible substitutional sites, moving from the center of the Si-NC to the surface. For each configuration, calculations of formation energies (FE) have been performed. As is known, this is a crucial quantity that allows understanding of the energy cost of creating a defect in a host crystal.[98–100] The results for the FE show that the neutral B impurities tend to migrate towards a subsurface position: the position directly below the interface Si atoms linked to H. This is explained by considering that such positions are the only ones that allow significant atomic relaxation around the impurity. For P impurities this behavior depends on the Si-NC diameter, in fact P atoms prefer to stay at the subsurface positions in Si-NCs with diameters less than 2 nm, whereas for larger nanocrystals the Si-NC center is the most energetically favorable position. These results are not only confirmed by several calculations,[91,97,101] but are also supported by experiments,[102–104] thus showing the possibility of incorporating B and P as impurities even in small Si-NCs.[60]

Calculations on matrix-isolated singly doped Si-NCs clearly indicate that for n-type doping the impurity tends to settle in the Si-NC core. Meanwhile, in the case of p-type doping interfacial sites are favored. Moreover, doping of the SiO_2 region is unlikely to occur. In particular, a very high FE is required to move a P dopant from the Si-NC core to the silica.[105] This is in agreement with the observation that for Si-NCs embedded in a SiO_2 matrix, the matrix provides a strong barrier to P diffusion, inducing P segregation in the Si rich region.[106,107] By contrast, still consistent with experiments,[108] the diffusion of B toward the matrix is significantly easier. Once more the structural relaxation around the impurity plays a significant role. P atoms maintain a sort of tetrahedral coordination in all cases, with all the bonds showing approximately constant characteristic lengths with Si and/or O. Instead, when placed at the interface with two O atoms, B atoms tend to form, a strong antibonding interaction with one of the neighbors, causing repulsion to a large distance. Experimental results based on atom probe tomography and proximity histograms support these conclusions.[109–111]

With regards to the electronic and optical properties for B and P singly doped Si-NCs, one has to consider that in Si bulk the extra hole associated with the B impurity creates an acceptor level slightly above the valence band maximum in energy, whereas for a P impurity the extra electron creates a donor state slightly below the conduction band minimum. Fig. 2 and 3 show our results for the position in energy of the calculated impurity state for the H-Si-NCs singly doped with B (red squares) or P (green squares). The position of these impurity levels is compared with the energy of the LUMO for the undoped NCs. In both figures the zero point for the energy is set to the HOMO level of the corresponding undoped Si-NCs. The binding energy (BE) of the acceptor is defined as the energy difference between the impurity level and the HOMO, whereas the BE of the donor is given by the energy difference between the LUMO and the impurity level. Looking at the plots, it is clear that the BE strongly depends on the NC size. Going to smaller NCs, the impurity levels, unlike in the bulk case, are localized well inside the Si band gap and thus necessitate a larger activation energy. As a consequence, the optical properties for singly-doped Si-NCs show intense absorption peaks, due to transitions that involve the impurity states in the absorption. Only for the larger

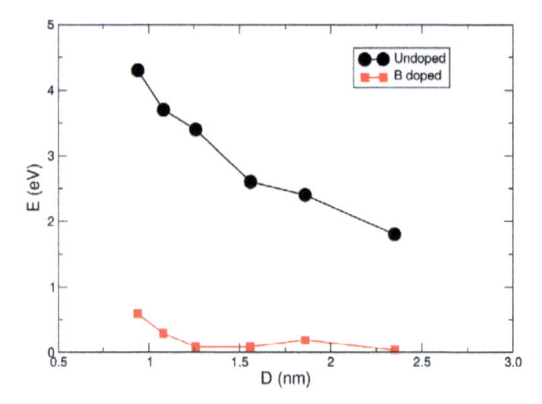

Fig. 2 Position in energy, with respect to the HOMO (set equal to zero), of the B related impurity state (red squares) and the LUMO state for the undoped Si-NCs (black dots), as a function of the NC diameter.

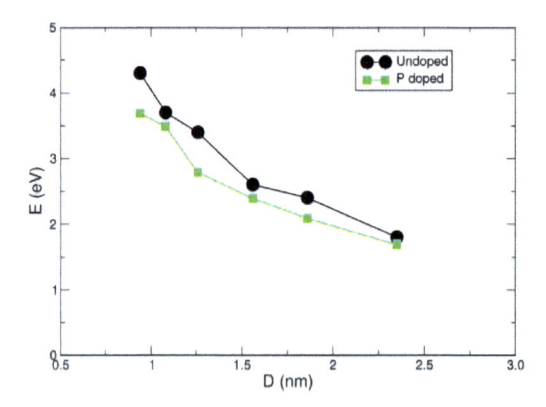

Fig. 3 Position in energy, with respect to the HOMO (set equal to zero), of the P related impurity state (green squares) and the LUMO state for the undoped Si-NCs (black dots), as a function of the NC diameter.

NCs (diameters of the order of 3 nm) are the absorption spectra of the singly doped case very similar to those of the undoped case.[60,80]

In the case of B and P codoped NCs the simultaneous doping strongly reduces the FE with respect to both B or P singly doped cases, independent of their size.[66,92] This reduction is a consequence of the carrier compensation in codoped Si-NCs, that recover, around the impurities, an almost T_d configuration. Furthermore, the FE of the codoped Si-NCs depends on the distance between the two impurities. A reduction in the distance between the two impurities results in a reduction in the FE. In particular, the minimum of the FE is obtained when the impurities are located at the nearest neighbor locations at the surface of the Si-NCs. Notably, it has been observed that codoping is an effective means for promoting segregation and stability of the B and P impurities in the Si-NC interface region.[109] The fact that Si-NCs can be more easily codoped than singly-doped explains the feasibility of easily growing[57,112] B and P codoped colloidal Si-NCs.

As for the optoelectronic properties, Fig. 4 shows our calculated energy gaps for B and P codoped Si-NCs in comparison with the calculated gaps for the corresponding undoped cases. We see that the codoping induces a true lower energy gap. As shown in the inset of Fig. 4, this is due to the presence of both an acceptor (now fully occupied and localized on the B atom) and a donor state (now empty and localized on the P atom) within the undoped Si-NC band gap. In the codoped case the PL emission has thus originated from a donor–acceptor pair recombination mechanism. This mechanism is at the root of the possibility to tune the PL in codoped Si-NCs. This tuning has been experimentally observed and is clearly related to the presence of donor–acceptor pair recombination.[113–117] Notably, as a consequence of donor–acceptor pair recombination, evidence of carrier multiplication in codoped Si-NCs has been reported.[114]

3.3 Other group-III and group-V dopants

As previously underlined there are several difficulties concerning impurity doping in Si-NCs, *i.e.* self-purification effects, large impurity formation energies and difficulty tracking the locations of dopant elements. Thus, while doping of Si-NCs with boron and phosphorus has been studied in detail, doping with other group-III and group-V species has been investigated in very few cases. To our knowledge there are no experimental works regarding other group-III elements, whereas for group-V dopants, with the exception of a study concerning the role of nitrogen in the formation of matrix-embedded Si-NCs,[118] only arsenic has attracted interest. In particular the structural and optical properties of As-doped Si-NCs embedded in silicon dioxide matrices have been investigated using Rutherford backscattering spectrometry, transmission electron microscopy and photoluminescence[119,120] and very recently using atom probe tomography.[121] It turns out that As shows similar behavior to phosphorus both regarding PL and the efficient incorporation of impurities in the core of Si-NCs. In this section we briefly present results of *ab initio* simulations concerning formation energies, changes in structural features, atomic location and optical properties, for Al (group-III) and N and As (group-V) doped Si-NCs, comparing the obtained trends with those for B

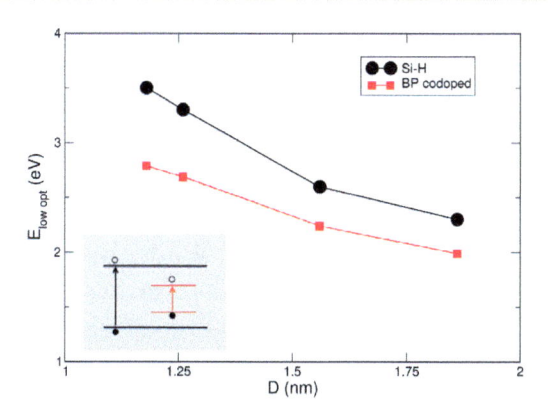

Fig. 4 Comparison between the HOMO–LUMO gaps of the undoped (black dots) and codoped (red squares) hydrogenated Si-NCs as a function of the NC diameter. The inset shows a schematic illustration of the energy level structures for both cases.

and P doped nanocrystals of the same size. All these new calculations have been performed for free standing Si-NCs capped by hydrogen atoms with the impurity located at the center of the NCs, whereas for the case of Si-NCs embedded in a SiO$_2$ matrix we have varied the atomic position of the dopant species.

The calculated formation energies are similar for the group-III dopants (Al *vs.* B), whereas for the group-V dopants they are larger for N with respect to P and As. This is a consequence of a larger rearrangement of Si atoms near the impurity for the N case, owing to the small dimensions of N. It is interesting to note that for the Al impurity, as in the case with B, we found a C_{3v} symmetry around the dopant, due to the presence of one longer and three equally shorter Si-impurity distances, whereas in the case of As and P we found four similar impurity-Si bond lengths. The case of N represents again an exception, since the similarity between the four impurity-Si bond lengths is not so clear. Thus it seems that the structural features induced by the presence of the impurity are linked directly to the their valence (trivalent *vs.* pentavalent atoms). In the case of Si-NCs embedded in a silicon dioxide matrix, again, for p-type doping the behavior of Al is similar to that of B (the sites at the interface with the matrix are favoured), and for n-type doping the behavior of N is similar to that of P (there is a strong tendency for the impurity to be located in the Si-NC core).

Concerning the computed optoelectronic properties, the Al impurity (like B) induces shallow acceptor levels in the band gap of the Si-NCs, whereas N, differently from P, induces deep donor levels. Moreover, as a consequence of the quantum confinement effect the energy position of these levels strongly depends on the Si-NC diameter. For smaller diameters the acceptor and donor levels are located deeper in the nanocrystal band gap. Regarding the calculated optical spectra, in small Si-NCs the Al-related absorption peaks are more intense than those related to the B impurity, but they become a similar intensity for larger nanocrystals. For group-V impurities the deeper donor character of the N impurity is clearly reflected in the optical spectra, which show impurity-related absorption peaks higher in energy with respect to those of the other dopants.

4 Silicon nanowires

4.1 Cubic-diamond Si nanowires

Because of their natural compatibility with silicon based technologies, cubic-diamond (3C) Si-NWs have being extensively theoretically studied and several experiments have already characterized their main structural and electronic properties.[5,6,50–52,54,122] It has been possible to fabricate, for example, single-crystal Si-NWs with diameters as small as 1 nm and lengths of a few tens of micrometers. Ultrathin NWs show a substantial blue-shift when decreasing the size as revealed by scanning tunneling spectroscopy measurements.[123] Si-NWs grow along well-defined crystalline directions; depending on the size, one can fabricate ⟨110⟩, ⟨111⟩ and ⟨112⟩ NWs.[124] In particular, surface passivation is required to obtain semiconducting ultrathin NWs.

Fig. 5 reports our DFT-LDA calculations for the band structures of Si-NWs of differing orientation and size. All the dangling bonds at the NW surface are capped by H atoms. We note that (i) in all cases, except for thin ⟨111⟩ NWs, we find a direct band gap at Γ. The direct nature of the fundamental gap is due to the folding of bulk bands, (ii) moreover, as a consequence of QC effects, the electronic

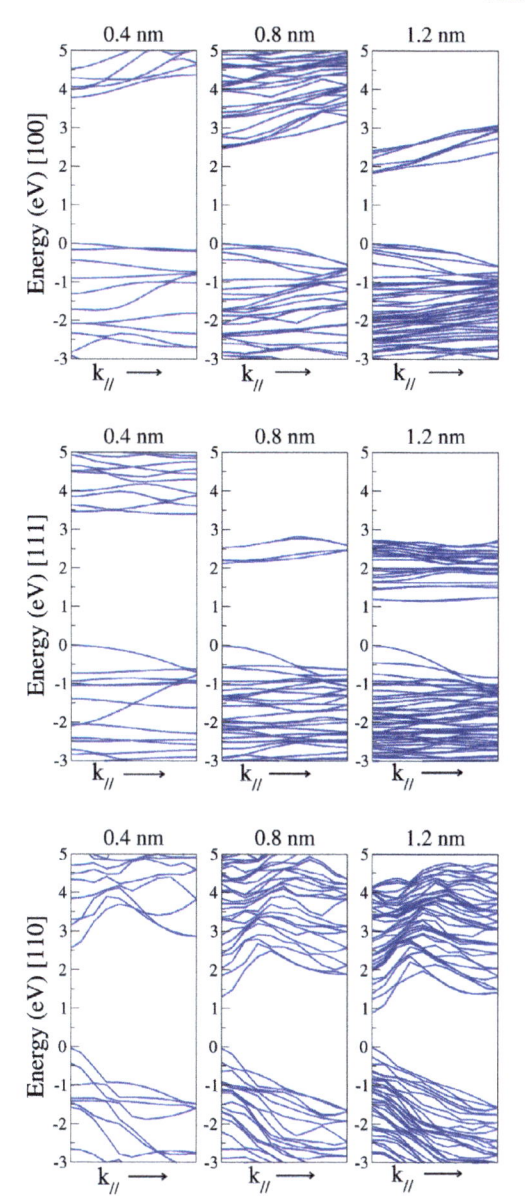

Fig. 5 Band structure at the DFT-LDA level for Si-NWs grown along the [100] (top panel), [111] (central panel) and [110] (lower panel) directions. In each panel the plots for different wire cross sections, $l = 0.4$ nm (left), $l = 0.8$ nm (center), $l = 1.2$ nm (right) are shown. The zero point for the energy is set to the top of the valence band.

gaps are much larger with respect to bulk Si and (iii) the electronic properties vary changing the growth orientation of wires. This is a result of the anisotropic behaviour of these low dimensional systems. Indeed the gap for the ⟨100⟩ direction is larger than that of the ⟨111⟩ and ⟨110⟩ directions.

With regards to the optoelectronic properties, we found that the GW corrections to the DFT-LDA band gaps are much larger than those in the bulk case and depend on the size and orientation of the wire. The excitonic effects are very strong, originating from a very large binding energy for the electron–hole bound states. Indeed the wires are almost transparent when the light is polarized in the direction perpendicular to the axis of the wire.[53] This is due to the microscopic fields, induced by the external perturbation, which screen the electric field in the NWs.

Table 2 reports our calculated DFT-LDA, quasi-particle (self-energy, GW calculations) and excitonic (BSE calculations) gaps for all the Si-NWs in Fig. 5. An almost complete compensation of the self-energy and excitonic contributions, concerning the absorption onset, is observed. This fact confirms again the validity of the DFT-LDA description, especially regarding trends of a specific quantity.

4.2 Doped cubic-diamond Si nanowires

Doping in 3C-Si-NWs has been examined in depth from theoretical and experimental points of view,[125–137] focusing mainly on B and P single doping. Experimentally, it has been clearly demonstrated that, with the current growth methods, it is possible to incorporate group III and V impurities into the wire and to obtain both n-type and p-type wires. On the other hand, theoretical investigations have shown that B and P impurities give rise to energetic levels deep in the band gap, and that they prefer to segregate towards the surface of the wire in order to minimize mechanical stress and distortions. The simultaneous addition of B and P impurities (codoping) has also been demonstrated to be a fundamental tool for modifying the electronic and optical features of NWs.[138] Unlike the case of B and P singly doped systems, very few theoretical studies[125,131] have addressed the analysis of B and P codoped Si-NWs, and moreover, they have been mainly focused on very thin NWs (with diameters of up to 1.6 nm). We investigated the FE and the band structure for ⟨110⟩ oriented B–P codoped H terminated Si-NWs with a diameter of 2.4 nm using SIESTA code.[139] The results show that the impurities tend to get closer together and to occupy edge positions, as a consequence of greater freedom of mechanical relaxations. P atoms will have a preference to sit in inner subsurface sites (as in the case of codoped Si-NCs). The simultaneous addition of B and P only slightly modifies the energy band gap value with respect to the pure wire (see Table 3) and the introduction of impurities does not change the dispersion of the electronic states too much. Analysis of the wave function

Table 2 DFT, quasi-particle and excitonic gaps for Si-NWs with differing orientation and size. All values are in eV

Wire size	Orientation	E_{gap} (DFT)	E_{gap} (QP)	E_{gap} (EXC)
0.4 nm	[100]	3.8	6.2	4.4
	[111]	3.5	5.7	4.1
	[110]	2.5	4.2	3.0
0.8 nm	[100]	2.5	4.2	3.0
	[111]	2.2	3.5	2.7
	[110]	1.4	3.0	1.8
1.2 nm	[100]	1.8	3.1	2.3
	[111]	1.2	2.3	1.9
	[110]	1.0	2.1	1.7

Table 3 DFT-LDA band gaps (in eV) for the pure H terminated Si-NW ($\langle 110 \rangle$ direction and a diameter of 2.4 nm) and for different codoped P–B configurations

Pure	B core–P core	B core–P edge	P core–B edge	B edge–P edge
0.97	0.91	0.81	0.80	0.96

localization of the band edges shows that the top of the valence band is located at the B impurity, while the bottom of the conduction band is located at P. The calculated reduction in the band gaps is thus due to donor–acceptor transitions as in the case of Si-NCs, even if in the case of NWs, due to the fact that QC effects act only in the directions perpendicular to the growth axis, this reduction is much smaller. Only for edge–edge configuration are the valence band maximum and conduction band minimum both localized on Si, whereas the impurity states are localized in the bands.

4.3 Hexagonal Si nanowires

The theoretical and experimental investigation of novel phases in Si-NWs has attracted huge interest in the last years.[140–146] In particular, Si-NWs showing the hexagonal-diamond (2H) phase have been the object of intense study because of their potential for adding new functionalities to 3C Si-NWs. The electronic, optical and transport properties of 2H-Si-NWs[147–153] have been under scrutiny, highlighting how this phase can offer a further degree of property modulation with respect to the versatility of 3C-Si-NWs. For example, it has been shown that 2H-Si-NWs are characterized by a more pronounced QC effect than 3C-Si-NWs and by a more enhanced optical absorption in the visible region.[149] For very small diameters, 2H-Si-NWs have a larger band gap with respect to 3C-Si ones, but the difference vanishes for diameters in the order of 4 nm and then it changes sign for larger diameters. For very large NWs, when the QC effects become negligible, the band gaps converge to the bulk value, which is 0.1–0.2 eV larger for 3C than for 2H. Indeed, different from the 3C case, the band gap of 2H-Si-NWs is indirect for the smallest considered diameter, but it becomes direct for the other cases.[149]

4.4 Doped hexagonal Si-NWs

Much less attention has been dedicated to the behaviour of dopants in 2H-Si-NWs. In a recent study[153] we performed *ab initio* DFT calculations to describe the properties of p-type and n-type dopants located at the innermost sites in 2H-Si-NWs, comparing our results with those obtained for 3C-Si-NWs. All the surface dangling bonds were saturated by H atoms. Our findings for the middle and large diameter NWs (beyond the QC regime) can be summarized as follows: (i) p-dopants have a lower formation energy when they are in 2H wires with respect to 3C ones, (ii) n-type dopants, in first approximation, do not show any preference of phase. The origin of this differing behaviour was ascribed to the different symmetry that is induced when a dopant is inserted in one specific phase. While p-type dopants preserve the C_{3v} symmetry of the host 2H lattice, n-type dopants show a clear tendency to occupy T_d lattice sites.

Here, we expand on our previous results by investigating the structural and electronic properties of group III and V dopants in ultrathin Si-NWs (with a diameter of 2 nm). We performed *ab initio* DFT calculations, the details of which are the same as in ref. 153.

As a first step, in order to investigate the structural stability of an impurity in one phase with respect to the other, we performed calculations of the FE, within the scheme reported in ref. 153. Results from these calculations are reported in Table 4. The table clearly shows that, for both group III and V impurities, the

Table 4 Calculated formation energies (E_{form}) (in eV) for group III (B, Al and Ga) and V (N, P and As) impurities in 3C-Si- and 2H-Si-NWs

Group	Impurity	3C-NWs	2H-NWs
III	B	−6.338	−6.230
	Al	−2.609	−2.637
	Ga	−1.727	−1.754
V	N	−4.201	−4.158
	P	−5.059	−4.908
	As	−2.541	−2.423

Table 5 Nearest neighbour distances for group III impurities in 3C-Si- and 2H-Si-NWs. For the 2H systems, the first three bond lengths correspond to bonds between atoms in the basal plane of the hexagonal cell. The fourth bond length is that along the c direction instead

Impurity	NW structure	Bond length (Å)
B	Cubic-diamond (3C)	2.048
		2.048
		2.048
		2.055
	Hexagonal-diamond (2H)	2.052
		2.052
		2.052
		2.063
Al	Cubic-diamond (3C)	2.394
		2.394
		2.394
		2.424
	Hexagonal-diamond (2H)	2.413
		2.414
		2.414
		2.397
Ga	Cubic-diamond (3C)	2.396
		2.396
		2.396
		2.444
	Hexagonal-diamond (2H)	2.420
		2.421
		2.421
		2.395

difference in the FE between the two phases does not exceed 0.1 eV. Although the stability of the impurities is slightly larger in the cubic phase, this means that there is not a clear preference of one phase with respect to the other. This finding is in contrast to what was found in the case of middle and large diameter NWs.[153] This differing behaviour can be related to the different structural reorganization around the impurity that one can observe in ultrathin NWs. In the bulk system and large diameter NWs, the host crystal, due to its symmetry, may favor or not favor the stability of a given impurity; in the case of a small NW the major degree of geometrical relaxation is an ideal chemical environment for all types of dopant. In other words, each impurity can achieve structural stability regardless of the phase. This is more evident if one looks at the first neighbour distance around the impurity after relaxation reported in Tables 5 and 6.

Interestingly, the local symmetry around the impurity in an ultrathin NW does not change too much from one phase to the other. This is in contrast to what we found in large diameter NWs in which the rigidity of the host lattice has a much pronounced influence on determining dopant symmetry. When the size of the NW shrinks down to a few nm, the potential for the Si atoms to accommodate stress is much larger in both the phases and hence the role of the symmetry of the host lattice is less pronounced. This is the reason why, as is the case with Si-NCs, group III atoms have C_{3v} symmetry (three short bonds and one longer bond with

Table 6 Nearest neighbour distances for group V impurities in 3C-Si- and 2H-Si-NWs. For the 2H systems, the first three bond lengths correspond to bonds between atoms in the basal plane of the hexagonal cell. The fourth bond length is that along the c direction instead

Impurity	NW structure	Bond length (Å)
N	Cubic-diamond (3C)	2.000
		2.004
		2.004
		2.005
	Hexagonal-diamond (2H)	1.984
		1.984
		1.985
		2.057
P	Cubic-diamond (3C)	2.332
		2.332
		2.333
		2.328
	Hexagonal-diamond (2H)	2.345
		2.345
		2.346
		2.347
As	Cubic-diamond (3C)	2.435
		2.439
		2.435
		2.427
	Hexagonal-diamond (2H)	2.449
		2.449
		2.450
		2.444

first neighbours) in both 3C- and 2H-Si-NWs (see Table 5), while when group V atoms are present, independent of the phase, a T_d configuration is adopted (in which the four bond lengths are practically the same, as shown in Table 6, with the exception of N, the smallest impurity).

Calculations of the electronic structure have been carried out for all the dopants. Here we show, in particular, the outcomes for the B and P impurities, which are the most common dopants for both 3C- and 2H-NWs. These results are reported in Fig. 6 and 7. With respect to the band structures of the corresponding undoped NWs[149] we note the presence of the impurity states near the band edges (the B atom state is located near the valence band, while that related to the P atom is near the conduction band). Comparing the two plots for the doped structures, there are a few interesting differences that can be pointed out: (i) the impurity state in the 2H phase is always flatter than that in 3C, as a consequence of the more pronounced QC effect of the former and (ii) the activation energy of the B dopant is larger in the case of 2H-Si-NWs with respect to the 3C phase (70 meV

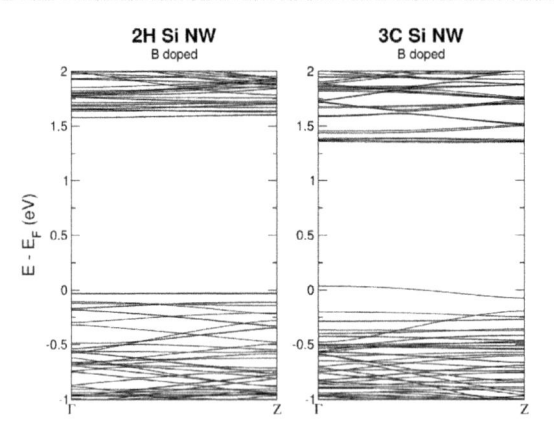

Fig. 6 Band structures of B doped 2H-Si-NWs (left) and B doped 3C-Si-NWs (right). The zero point for the energy is set to the Fermi energy of the system.

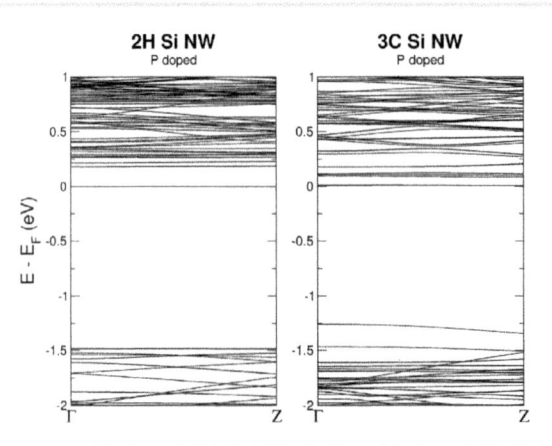

Fig. 7 Band structures of P doped 2H-Si-NWs (left) and P doped 3C-Si-NWs (right). The zero point for the energy is set to the Fermi energy of the system.

compared to 163 meV, respectively). In the case of P we observed opposite behaviour as the activation energy amounts to 180 meV in 2H-Si-NW, compared to the 95 meV calculated for the 3C systems.

5 Conclusions

The properties of undoped and n- and p-doped Si-NCs and Si-NWs obtained through *ab initio* calculations have been discussed. We showed that first-principles calculations can be a powerful tool for understanding the effects induced by substitutional doping on the structural, electronic and optical properties of free-standing and matrix-embedded Si nanostructures. The preferential positioning of the dopants and their effects on the structural properties with respect to the undoped cases, as a function of the nanocrystal diameter and termination, have been characterized through total-energy considerations. The localization of the acceptor and donor related levels in the band gaps of the Si-NCs and Si-NWs, together with the impurity activation energies, have been discussed as a function of the nanostructure size. The dopant induced differences in the optical properties with respect to the undoped cases have been presented. The role played by interface effects, orientation growth and phases has been outlined. We have studied B and P codoped nanocrystals and nanowires showing that if carriers are perfectly compensated, the Si nanostructures undergo minor structural distortion around the impurities, inducing a significant decrease in the impurity formation energies with respect to the singly doped case. Due to codoping, additional peaks were introduced in the absorption spectra, giving rise to a size- and dopant localization-dependent red-shift of the optical spectra. These results give a strong indication that it is possible to efficiently tune the optical properties of silicon nanostructures with doping.

Conflicts of interest

There are no conflicts to declare.

Acknowledgements

S. O. acknowledges support/funding from University of Modena and Reggio Emilia under project "FAR2017INTERDISC". S. O. and I. M. thank the Super-Computing Interuniversity Consortium CINECA for support and high-performance computing resources under the Italian Super-Computing Resource Allocation (ISCRA) initiative, PRACE for awarding us access to the resource MARCONI HPC cluster based in Italy at CINECA. I. M. acknowledges support/funding from European Union H2020-EINFRA-2015-1 and H2020-INFRAEDI-2018-1 programs under grant agreement No. 676598 and No. 824143, project MaX-MAterials at the eXascale. M. A. greatly acknowledges the Transnational Access Programme of the HPC-EUROPA3 (project HPC17PB9IZ). Some of the high-performance computing (HPC) resources for this project were granted by the Institut du development et des ressources en informatique scientifique (IDRIS) under the allocation A0040910089 *via* GENCI (Grand Equipment National de Calcul Intensif). This work was supported by the ANR HEXSIGE project (ANR-17-CE030-0014-01) of the French Agence Nationale de la Recherche. M. P.

acknowledges INFN for financial support through the National project Nemesys. We also acknowledge financial support by the Ministerio de Economía, Industria y Competitividad (MINECO) and MICIU (Ministerio de Ciencia y Universidades) under Grants FEDER-MAT2017-90024-P, FIS2015-64886-C5-4-P and PGC2018-096955-B-C44-P, the Severo Ochoa Centres of Excellence Program under Grant SEV-2015-0496 and the Generalitat de Catalunya under Grant 2017 SGR 1506.

Notes and references

1 L. T. Canham, *Appl. Phys. Lett.*, 1990, **57**, 1046–1048.
2 H. Takagi, H. Ogawa, Y. Yamazaki, A. Ishikazi and T. Nakagiri, *Appl. Phys. Lett.*, 1990, **56**, 2379–2381.
3 S. Ossicini, F. Bernardini and A. Fasolino, *Phys. Rev. Lett.*, 1994, **72**, 1044–1047.
4 O. Bisi, S. Ossicini and L. Pavesi, *Surf. Sci. Rep.*, 2000, **38**, 1–126.
5 R. Rurali, *Rev. Mod. Phys.*, 2010, **82**, 427–449.
6 M. Amato, M. Palummo, R. Rurali and S. Ossicini, *Chem. Rev.*, 2014, **114**, 1371–1412.
7 A. R. Goñi, L. R. Muniz, J. S. Reparaz, M. I. Alonso, M. Garriga, A. F. Lopeandia, J. Rodríguez-Viejo, J. Arbiol and R. Rurali, *Phys. Rev. B: Condens. Matter Mater. Phys.*, 2014, **89**, 045428.
8 I. Pelant, *Phys. Status Solidi A*, 2011, **208**, 625–630.
9 F. Priolo, F. Gregorkiewicz, M. Galli and T. F. Krauss, *Nat. Nanotechnol.*, 2014, **1**, 19–32.
10 D. Jurbergs, E. Rogojina, L. Mangolini and U. Kortshagen, *Appl. Phys. Lett.*, 2006, **88**, 233116.
11 L. Pavesi, L. Dal Negro, C. Mazzoleni, G. Franzó and F. Priolo, *Nature*, 2000, **408**, 440–444.
12 L. Dal Negro, M. Cazzanelli, L. Pavesi, S. Ossicini, D. Pacifici, G. Franzó, F. Priolo and F. Iacona, *Appl. Phys. Lett.*, 2003, **82**, 4636–4638.
13 J. Ruan, P. M. Fauchet, L. Dal Negro, M. Cazzanelli and L. Pavesi, *Appl. Phys. Lett.*, 2003, **83**, 5479–5481.
14 E. Degoli, R. Guerra, F. Iori, R. Magri, I. Marri, O. Pulci, O. Bisi and S. Ossicini, *C. R. Phys.*, 2009, **10**, 575–586.
15 M. C. Beard, K. P. Knutsen, P. Yu, J. M. Luther, Q. Song, W. K. Metzger, R. J. Ellingson and A. J. Nozik, *Nano Lett.*, 2007, **7**, 2506–2512.
16 D. Timmermann, L. Izeddin, P. Stallinga, I. N. Yassevich and T. Gregorkiewicz, *Nat. Photonics*, 2008, **2**, 105–109.
17 M. T. Trinh, R. Limpens, W. D. A. M. de Boer, J. M. Schins, L. D. A. Siebbeles and T. Gregorkiewicz, *Nat. Photonics*, 2012, **6**, 316–321.
18 M. Govoni, I. Marri and S. Ossicini, *Phys. Rev. B: Condens. Matter Mater. Phys.*, 2011, **84**, 075215.
19 M. Govoni, I. Marri and S. Ossicini, *Nat. Photonics*, 2012, **6**, 672–679.
20 I. Marri, M. Govoni and S. Ossicini, *J. Am. Chem. Soc.*, 2014, **136**, 13257–13266.
21 I. Marri, M. Govoni and S. Ossicini, *Beilstein J. Nanotechnol.*, 2015, **6**, 343–352.
22 I. Marri, M. Govoni and S. Ossicini, *Sol. Energy Mater. Sol. Cells*, 2016, **145**, 162–169.
23 J. D. Holmes, K. J. Ziegler, R. C. Doty, L. E. Pell, K. P. Johnston and B. A. Korgel, *J. Am. Chem. Soc.*, 2001, **123**, 3743–3748.

24 J. G. C. Veinot, *Chem. Commun.*, 2006, 4160–4168.
25 M. L. Mastronardi, F. Maier-Flaig, D. Faulkner, E. J. Henderson, C. Küble, U. Lemmer and G. A. Ozin, *Nano Lett.*, 2012, **12**, 337–342.
26 C. M. Hessel, D. Reid, M. G. Panthani, M. R. Rasch, B. W. Goodfellow, J. W. Wei, H. Fujii, V. Akhavan and B. A. Korgel, *Chem. Mater.*, 2012, **24**, 393–401.
27 L. Mangolini, *J. Vac. Sci. Technol., B: Nanotechnol. Microelectron.: Mater., Process., Meas., Phenom.*, 2013, **31**, 020801.
28 O. Wolf, M. Dasog, Z. Yang, I. Balberg, J. C. G. Veinot and O. Millo, *Nano Lett.*, 2013, **13**, 2516–2521.
29 M. Dasog, G. B. De Los Reyes, L. V. Titova, F. Hegmann and J. G. C. Veinot, *ACS Nano*, 2014, **8**, 9636–9648.
30 K. Dohnalová, T. Gregorkiewicz and K. Kůsová, *J. Phys.: Condens. Matter*, 2014, **26**, 173201.
31 Y. Yu, C. E. Rowland, R. Schaller and B. A. Korgel, *Langmuir*, 2015, **31**, 6886–6893.
32 D. Kim, J. M. Zuidema, J. Kang, Y. Pan, L. Wu, D. Warther, B. Arkles and M. J. Sailor, *J. Am. Chem. Soc.*, 2016, **138**, 15106–15109.
33 S. Ossicini, L. Pavesi and F. Priolo, *Springer Tracts Mod. Phys.*, 2003, **194**, 1–282.
34 D. V. Talapin, J. S. Lee, M. V. Kovalenko and E. V. Shevchenko, *Chem. Rev.*, 2010, **110**, 389–458.
35 D. P. Puzzo, E. J. Henderson, M. G. Helander, Z. Wang, G. A. Ozin and Z. Lu, *Nano Lett.*, 2011, **11**, 1585–1590.
36 M. Dasog, J. Kehrle, B. Rieger and J. C. G. Veinot, *Angew. Chem., Int. Ed.*, 2016, **55**, 2322–2339.
37 N. Daldosso and L. Pavesi, *Laser Photonics Rev.*, 2009, **3**, 508–534.
38 L. Kriachtchev, S. Ossicini, F. Iacona and F. Gourbilleau, *Int. J. Photoenergy*, 2012, 872576.
39 R. Sinelnikov, M. Dasog, J. Beamish, A. Meldrum and J. C. G. Veinot, *ACS Photonics*, 2017, **4**, 1920–1929.
40 S. V. Makarov, M. I. Petrov, U. Zywietz, V. A. Milichko, D. A. Zuev, N. Lopanitsyna, A. Kuksin, I. Mukhin, G. Zograf, E. Ubyivovk, D. Smirnova, S. Starikov, B. N. Chichkov and Y. S. Kivshar, *Nano Lett.*, 2017, **17**, 3914–3918.
41 M. A. Green, *Prog. Photovolt. Res. Appl.*, 2001, **9**, 123–135.
42 A. J. Nozik, *Nano Lett.*, 2010, **10**, 2735–2741.
43 T. Claudio, N. Stein, D. G. Stroppa, B. Klobes, M. M. Koz, P. Kudejova, N. Petermann, H. Wiggers, G. Schierning and R. P. Hermann, *Phys. Chem. Chem. Phys.*, 2014, **16**, 25701–25709.
44 J.-H. Park, L. Gu, G. von Maltzahn, E. Ruoslahti, S. N. Bhatia and M. J. Sailor, *Nat. Mater.*, 2009, **8**, 331–336.
45 L. Gu, D. J. Hall, Z. Qin, E. Anglin, J. Joo, D. J. Mooney, S. B. Howell and M. J. Sailor, *Nat. Commun.*, 2013, **4**, 2326.
46 L. T. Canham, *Nanomedicine*, 2013, **8**, 1573–1576.
47 J. Joo, X. Liu, V. R. Kotamrju, E. Ruoslahti, Y. Nam and M. J. Sailor, *ACS Nano*, 2015, **9**, 6233–6241.
48 L. Ostrovska, A. Broz, A. Fucikova, T. Belinova, H. Sugimoto, T. Kanno, M. Fujii, J. Valenta and M. Hubalek Kabalcova, *RSC Adv.*, 2016, **6**, 63403–63413.

49 B. F. P. McVey, S. Prabakar, J. J. Goodig and R. D. Tilley, *ChemPlusChem*, 2017, **82**, 60–73.
50 Y. Li, F. Qian, J. Xiang and C. M. Lieber, *Mater. Today*, 2006, **9**, 18–27.
51 V. Schmidt, J. V. Wittermann and U. Gösele, *Chem. Rev.*, 2010, **110**, 361–388.
52 Y. Wang, T. Wang, P. Da, M. Xu, H. Wu and G. Zheng, *Adv. Mater.*, 2013, **25**, 5177–5195.
53 M. Bruno, M. Palummo, A. Marini, R. Del Sole and S. Ossicini, *Phys. Rev. Lett.*, 2007, **98**, 036807.
54 M. Amato and R. Rurali, *Prog. Surf. Sci.*, 2016, **91**, 1–28.
55 A. G. Cullis, L. T. Canham and P. D. J. Calcott, *J. Appl. Phys.*, 1997, **82**, 909–965.
56 G. Polisski, D. Kovalev, G. Dollinger, T. Sulima and F. Koch, *Phys. B*, 1999, **273–274**, 951–954.
57 M. Fujii, Y. Yamaguchi, Y. Takase, K. Ninomiya and S. Hayashy, *Appl. Phys. Lett.*, 2005, **87**, 211919.
58 B. L. Oliva-Chatelain, T. M. Ticich and A. R. Barron, *Nanoscale*, 2016, **8**, 1733–1745.
59 E. Arduca and M. Perego, *Mater. Sci. Semicond. Process.*, 2017, **62**, 156–170.
60 I. Marri, S. Ossicini and E. Degoli, *Prog. Surf. Sci.*, 2017, **92**, 3705–3748.
61 G. M. Credo, M. D. Mason and S. K. Buratto, *Appl. Phys. Lett.*, 1999, **74**, 1978–1981.
62 P. Hohenberg and W. Kohn, *Phys. Rev. B: Solid State*, 1964, **136**, 864–871.
63 W. Kohn and L. J. Sham, *Phys. Rev. A*, 1965, **140**, 1133–1138.
64 L. Hedin, *Phys. Rev. A*, 1965, **139**, 796.
65 G. Onida, L. Reining and A. Rubio, *Rev. Mod. Phys.*, 2002, **74**, 601–659.
66 F. Iori, E. Degoli, R. Magri, I. Marri, G. Cantele, D. Ninno, F. Trani, O. Pulci and S. Ossicini, *Phys. Rev. B: Condens. Matter Mater. Phys.*, 2007, **76**, 085302.
67 M. Bruno, M. Palummo, A. Marini, R. Del Sole, V. Olevano, A. N. Kholod and S. Ossicini, *Phys. Rev. B: Condens. Matter Mater. Phys.*, 2005, **72**, 153310.
68 M. Luppi and S. Ossicini, *Phys. Rev. B: Condens. Matter Mater. Phys.*, 2005, **71**, 035340.
69 R. Guerra, I. Marri, R. Magri, L. Martin-Samos, O. Pulci, E. Degoli and S. Ossicini, *Phys. Rev. B: Condens. Matter Mater. Phys.*, 2009, **79**, 155320.
70 R. Guerra, I. Marri, R. Magri, L. Martin-Samos, O. Pulci, E. Degoli and S. Ossicini, *Superlattices Microstruct.*, 2009, **46**, 246–252.
71 P. Giannozzi, S. Baroni, N. Bonini, M. Calandra, R. Car, C. Cavazzoni, D. Ceresoli, G. L. Chiarotti, M. Cococcioni, J. Dabo, A. Dal Corso, S. Fabris, G. Fratesi, S. de Gironcoli, R. Gebauer, U. Gerstmann, C. Gougoussis, A. Kokalj, M. Lazzeri, L. Martin-Samos, N. Marzari, F. Mauri, R. Mazzarello, S. Paolini, A. Pasquarello, L. Paulatto, C. Sbraccia, S. Scandolo, G. Sclauzero, P. Seitsonen, A. Smogunov, P. Umari and R. M. Wentzcovitch, *J. Phys.: Condens. Matter*, 2009, **21**, 395502.
72 P. Giannozzi, O. Andreussi, T. Brumme, O. Bunau, M. B. Nardelli, M. Calandra, R. Car, C. Cavazzoni, D. Ceresoli, M. Cococcioni, *et al.*, *J. Phys.: Condens. Matter*, 2017, **29**, 465901.
73 J. M. Soler, E. Artacho, J. D. Gale, A. García, J. Junquera, P. Ordejón and D. Sánchez-Portal, *J. Phys.: Condens. Matter*, 2002, **14**, 2745–2779.
74 X. Gonze, J.-M. Beuken, R. Caracas, F. Detraux, M. Fuchs, G.-M. Rignanese, L. Sindic, M. Verstaete, G. Zerah, F. Jollet, M. Torrent, A. Roy, M. Mikami, P. Ghosez, J.-P. Raty and D. C. Allan, *Comput. Mater. Sci.*, 2002, **25**, 478–492.

75 G. Kresse and J. Furthmüller, *Comput. Mater. Sci.*, 1996, **6**, 15–50.
76 D. Sangalli, A. Ferretti, H. Miranda, C. Attaccalite, I. Marri, E. Cannuccia, P. Melo, M. Marsili, F. Paleari, A. Marrazzo, *et al.*, *J. Phys.: Condens. Matter*, 2019, **31**, 325902.
77 Y. Q. Wang, S. Smirani and G. G. Ross, *Nano Lett.*, 2004, **4**, 2041–2045.
78 R. Guerra, E. Degoli and S. Ossicini, *Phys. Rev. B: Condens. Matter Mater. Phys.*, 2009, **80**, 155332.
79 L. E. Ramos, E. Degoli, G. Cantele, S. Ossicini, D. Ninno, J. Furthmüller and F. Bechstedt, *J. Phys.: Condens. Matter*, 2007, **19**, 466211.
80 L. E. Ramos, E. Degoli, G. Cantele, S. Ossicini, D. Ninno, J. Furthmüller and F. Bechstedt, *Phys. Rev. B: Condens. Matter Mater. Phys.*, 2008, **78**, 235310.
81 S. Kim, M. C. Kim, S. H. Choi, K. J. Kim, H. N. Hwang and C. C. Hwang, *Appl. Phys. Lett.*, 2007, **92**, 103113.
82 M. V. Wolkin, J. Jorne, J. M. Fauchet, G. Allan and C. Delerue, *Phys. Rev. Lett.*, 1999, **82**, 000197.
83 L. Dorigoni, O. Bisi, F. Bernardini and S. Ossicini, *Phys. Rev. B: Condens. Matter Mater. Phys.*, 1996, **53**, 4557–4564.
84 M. Sykora, L. Mangolini, R. D. Schaller, U. Kortshagen, D. Jurbergs and V. I. Klimov, *Phys. Rev. Lett.*, 2008, **100**, 067401.
85 M. Luppi and S. Ossicini, *J. Appl. Phys.*, 2003, **94**, 2130–2132.
86 N. Daldosso, M. Luppi, S. Ossicini, E. Degoli, R. Magri, G. Dalba, D. Fornasini, R. Grisenti, F. Rocca, L. Pavesi, S. Boninelli, F. Priolo, C. Spinella and F. Iacona, *Phys. Rev. B: Condens. Matter Mater. Phys.*, 2003, **68**, 085327.
87 Z. Zhou, L. Brus and R. Friesner, *Nano Lett.*, 2003, **3**, 163–167.
88 R. Guerra and S. Ossicini, *Phys. Rev. B: Condens. Matter Mater. Phys.*, 2010, **81**, 245307.
89 G. Seguini, C. Castro, S. Schamm-Chardon, G. BenAssayag, P. Pellegrino and M. Perego, *Appl. Phys. Lett.*, 2013, **103**, 023103.
90 D. V. Melnikov and J. R. Chelikowsky, *Phys. Rev. Lett.*, 2004, **92**, 046802.
91 G. Cantele, E. Degoli, E. Luppi, R. Magri, D. Ninno, G. Jadonisi and S. Ossicini, *Phys. Rev. B: Condens. Matter Mater. Phys.*, 2005, **72**, 113303.
92 S. Ossicini, E. Degoli, F. Iori, E. Luppi, R. Magri, G. Cantele, F. Trani and D. Ninno, *Appl. Phys. Lett.*, 2005, **87**, 173120.
93 G. M. Dalpian and J. R. Chelikowsky, *Phys. Rev. Lett.*, 2006, **96**, 226802.
94 T. L. Chan, M. L. Tiago, E. Kaxiras and J. R. Chelikowsky, *Nano Lett.*, 2008, **8**, 596–600.
95 D. J. Norris, A. L. Efros and S. C. Erwin, *Science*, 2008, **319**, 1766–1779.
96 J. Li, S. H. Wei, S. S. Li and J. B. Xia, *Phys. Rev. B: Condens. Matter Mater. Phys.*, 2008, **77**, 113304.
97 J. R. Chelikowsky, M. M. G. Alemany, T. L. Chan and G. M. Dalpian, *Rep. Prog. Phys.*, 2011, **74**, 046501.
98 S. B. Zhang and J. E. Northrup, *Phys. Rev. Lett.*, 1991, **67**, 2339–2342.
99 J. E. Northrup and S. B. Zhang, *Phys. Rev. B: Condens. Matter Mater. Phys.*, 1993, **47**, 6791–6794.
100 C. Freysoldt, B. Grabowski, T. Hickel, J. Neugebauer, G. Kresse, A. Janotti and C. G. Van de Walle, *Rev. Mod. Phys.*, 2014, **86**, 253–305.
101 J. H. Eom, T. L. Chan and J. R. Chelikowsky, *Solid State Commun.*, 201, **150**, 130–132.

102 X. D. Pi, R. Gresback, R. W. Liptak, S. A. Campbell and U. Kortshagen, *Appl. Phys. Lett.*, 2008, **92**, 123102.
103 H. Sugimoto, M. Fujii, M. Fukuda, K. Imakita and S. Hayashi, *J. Appl. Phys.*, 2011, **110**, 063528.
104 N. J. Kramer, K. S. Schramke and U. R. Kortshagen, *Nano Lett.*, 2015, **15**, 5597–5603.
105 R. Guerra and S. Ossicini, *J. Am. Chem. Soc.*, 2014, **136**, 4404–4409.
106 M. Perego, G. Seguini and M. Fanciulli, *Surf. Interface Anal.*, 2013, **45**, 386–389.
107 M. Perego, G. Seguini, E. Arduca, J. Frascaroli, D. De Salvador, M. Mastromatteo, A. Carnera, G. Nicotra, M. Scuderi, C. Spinella, G. Impellizzeri, C. Lenardi and E. Napolitani, *Nanoscale*, 2015, **7**, 14469–14475.
108 M. Xie, D. Li, L. Chen, F. Wang, X. Zhu and D. Yang, *Appl. Phys. Lett.*, 2013, **102**, 123108.
109 K. Nomoto, H. Sugimoto, A. Breen, A. V. Ceguerra, T. Kanno, S. P. Ringer, I. Perez-Würfl, G. Conibeer and M. Fujii, *J. Phys. Chem. C*, 2016, **120**, 17845–17852.
110 R. Khelifi, D. Mathiot, R. Gupta, D. Muller, M. Roussel and S. Duguay, *Appl. Phys. Lett.*, 2013, **102**, 013116.
111 M. Frégnaux, R. Khelifi, D. Muller and D. Mathiot, *J. Appl. Phys.*, 2014, **116**, 143505.
112 M. Fujii, Y. Yamaguchi, Y. Takase, K. Ninomiya and S. Hayashi, *Appl. Phys. Lett.*, 2004, **85**, 1158–1160.
113 T. Nakamura, S. Adachi, M. Fujii, H. Sugimoto, K. Miura and S. Yamamoto, *Phys. Rev. B: Condens. Matter Mater. Phys.*, 2015, **91**, 165424.
114 N. X. Chung, R. Limpens, A. Lesage, M. Fujii and T. Gregorkiewicz, *Phys. Status Solidi A*, 2016, **11**, 2863–2866.
115 D. Li, Y. Jiang, P. Zhang, D. Shan, J. Xu, W. Li and K. Chen, *Appl. Phys. Lett.*, 2017, **110**, 233105.
116 O. Ashkenazi, D. Azulay, I. Balberg, S. Kano, H. Sugimoto, M. Fujii and O. Millo, *Nanoscale*, 2017, **9**, 17884–17892.
117 H. Sugimoto, M. Yamamur, R. Rujii and M. Fujii, *Nano Lett.*, 2018, **18**, 7282–7288.
118 V. Mulloni, P. Bellutti and L. Vanzetti, *Surf. Sci.*, 2005, **585**, 137–143.
119 E. Sun, F.-H. Su, C.-H. Chen, H.-L. Tsai, J.-R. Yang and M.-J. Chen, *J. Lumin.*, 2010, **130**, 1485–1488.
120 D. Puglia, G. Sombrio, R. dos Reis and H. Boudinov, *Mater. Res. Express*, 2018, **5**, 036201.
121 R. Demoulin, M. Roussel, S. Duguay, D. Muller, D. Mathiot, P. Pareige and E. Talbot, *J. Phys. Chem. C*, 2019, **123**, 7381–7389.
122 M. Palummo, C. Hogan and S. Ossicini, *Phys. Chem. Chem. Phys.*, 2015, **17**, 29085–29089.
123 D. Ma, C. Lee, F. Au, S. Tong and S. Lee, *Science*, 2003, **299**, 1874–1877.
124 Y. Wu, Y. Cui, L. Huynh, C. J. Barrelet, D. Bell and C. M. Lieber, *Nano Lett.*, 2004, **4**, 433–436.
125 H. Peelaers, B. Partoens and F. M. Peeters, *Nano Lett.*, 2006, **6**, 2781–2784.
126 B. Schoeters, O. Leenaerts, G. Pourtois and B. Partoens, *J. Appl. Phys.*, 2015, **118**, 104306.
127 M. V. Fernández-Serra, C. Adessi and X. Blase, *Phys. Rev. Lett.*, 2006, **96**, 166805.

128 M. Diarra, Y.-M. Niquet, C. Delerue and G. Allan, *Phys. Rev. B: Condens. Matter Mater. Phys.*, 2007, **75**, 045301.
129 C. R. Leao, A. Fazzio and A. J. R. da Silva, *Nano Lett.*, 2008, **8**, 1866–1871.
130 R. Rurali and X. Cartoixà, *Nano Lett.*, 2009, **9**, 975–979.
131 M. Palummo, F. Iori, R. Del Sole and S. Ossicini, *Superlattices Microstruct.*, 2009, **46**, 234–239.
132 R. Kagimura, R. W. Nunes and H. Chacham, *Phys. Rev. Lett.*, 2005, **95**, 115502.
133 N. Fukata, *Adv. Mater.*, 2009, **21**, 2829–2832.
134 E. Koren, N. Berkovitch and Y. Rosenwaks, *Nano Lett.*, 2010, **10**, 1163–1167.
135 S. G. Pavlov, R. Eichholz, N. V. Abrisimov, B. Redlich and H. W. Hubers, *Appl. Phys. Lett.*, 2011, **98**, 061102.
136 N. Fukata, J. Kaminaga, R. Takiguchi, R. Rurali, M. Dutta and K. Murakami, *J. Phys. Chem. C*, 2013, **117**, 20300–20307.
137 L. Yang, L. Zhihong, L. Xiaoxiang, S. Zhihua, W. Yanan, L. Rui, W. Dunwei, J. Jie, L. Joon Hwan, W. Haiyan, Y. Qingkai and B. Jiming, *Nanoscale*, 2015, **7**, 1601–1605.
138 N. Fukata, J. Chen, T. Sekiguchi, S. Matsuhita, T. Oshima, N. Uchida, K. Murakami, T. Tsurui and S. Ito, *Appl. Phys. Lett.*, 2008, **93**, 203106.
139 M. Amato, M. Palummo, R. Rurali and S. Ossicini, *J. Phys. D: Appl. Phys.*, 2014, **47**, 394013.
140 L. Vincent, G. Patriarche, G. Hallais, C. Renard, C. GardÃĺs, D. Troadec and D. Bouchier, *Nano Lett.*, 2014, **14**, 4828–4836.
141 L. Vincent, D. Djomani, M. Fakfakh, C. Renard, B. Belier, D. Bouchier and G. Patriarche, *Nanotechnology*, 2018, **29**, 125601.
142 H. I. T. Hauge, M. A. Verheijen, S. Conesa-Boj, T. Etzelstorfer, M. Watzinger, D. Kriegner, I. Zardo, C. Fasolato, F. Capitani, P. Postorino, S. Kölling, A. Li, S. Assali, J. Stangl and E. P. A. M. Bakkers, *Nano Lett.*, 2015, **15**, 5855–5860.
143 H. I. T. Hauge, S. Conesa-Boj, M. A. Verheijen, S. Koelling and E. P. A. M. Bakkers, *Nano Lett.*, 2017, **17**, 85–90.
144 J. Tang, J.-L. Maurice, F. Fossard, I. Florea, W. Chen, E. Johnson, M. Foldyna, L. Yu and P. R. i Cabarrocas, *Nanoscale*, 2017, **9**, 8113–8118.
145 Z. He, J.-L. Maurice, Q. Li and D. Pribat, *Nanoscale*, 2019, **11**, 4846–4853.
146 J. Carrete, M. López-Suárez, M. Raya-Moreno, A. S. Bochkarev, M. Royo, G. K. H. Madsen, X. Cartoixá, N. Mingo and R. Rurali, *Nanoscale*, 2019, **11**, 16007–16016.
147 C. Rödl, T. Sander, F. Bechstedt, J. Vidal, P. Olsson, S. Laribi and J.-F. Guillemoles, *Phys. Rev. B: Condens. Matter Mater. Phys.*, 2015, **92**, 045207.
148 T. Kaewmaraya, L. Vincent and M. Amato, *J. Phys. Chem. C*, 2017, **121**, 5820–5828.
149 M. Amato, T. Kaewmaraya, A. Zobelli, M. Palummo and R. Rurali, *Nano Lett.*, 2016, **16**, 5694–5700.
150 F. Fabbri, E. Rotunno, L. Lazzarini, D. Cavalcoli, A. Castaldini, N. Fukata, K. Sato, G. Salviati and A. Cavallini, *Nano Lett.*, 2013, **13**, 5900–5906.
151 S. Dixit and A. K. Shukla, *J. Appl. Phys.*, 2018, **123**, 224301.
152 F. Fabbri, E. Rotunno, L. Lazzarini, N. Fukata and G. Salviati, *Sci. Rep.*, 2014, **4**, 3603.
153 M. Amato, S. Ossicini, E. Canadell and R. Rurali, *Nano Lett.*, 2019, **19**, 866–876.

Faraday Discussions

PAPER

The red and blue luminescence in silicon nanocrystals with an oxidized, nitrogen-containing shell†

Pavel Galář,[a] Tomáš Popelář,[a] Josef Khun,[b] Irena Matulková,[c] Ivan Němec,[c] Kateřina Dohnalova Newell,[d] Alena Michalcová,[b] Vladimír Scholtz[b] and Kateřina Kůsová[*a]

Received 27th September 2019, Accepted 22nd October 2019
DOI: 10.1039/c9fd00092e

Traditionally, two classes of silicon nanocrystals (SiNCs) are recognized with respect to their light-emission properties. These are usually referred to as the "red" and the "blue" emitting SiNCs, based on the spectral region in which the larger part of their luminescence is concentrated. The origin of the "blue" luminescence is still disputed and is very probably different in different systems. One of the important contributions to the discussion about the origin of the "blue" luminescence was the finding that the exposure of SiNCs to even trace amounts of nitrogen in the presence of oxygen induces the "blue" emission, even in originally "red"-emitting SiNCs. Here, we obtained a different result. We show that the treatment of "red" emitting, already oxidized SiNCs in a water-based environment containing air-related radicals including nitrogen-containing species as well as oxygen, diminishes, rather than induces the "blue" luminescence.

1 Introduction

Silicon nanocrystals (SiNCs) are an efficient light-emitting material with prospective applications ranging from biological imaging through light-emitting devices to photovoltaics.[2–5] As is expected in a quantum-dot-based system, their optical properties are tunable with size. However, it is not only the maximum of the photoluminescence (PL) which changes with size as a result of the bandgap opening.[6–13] Unlike in other quantum-dot-based systems, the decay lifetimes also

---

[a]*Institute of Physics, Czech Academy of Sciences, Cukrovarnická 10, Prague 6, 162 00, Czech Republic. E-mail: kusova@fzu.cz*
[b]*University of Chemistry and Technology, Technická 3, Prague, 166 28, Czech Republic*
[c]*Department of Inorganic Chemistry, Faculty of Science, Charles University, Hlavova 7, Prague 2, 128 43, Czech Republic*
[d]*Institute of Physics, University of Amsterdam, Science Park 904, 1098 XH Amsterdam, The Netherlands*
† Electronic supplementary information (ESI) available. See DOI: 10.1039/c9fd00092e

depend on the size of the SiNCs[12,14–17] as the efficiency of the emission is tuned by a changing level of confinement. Naturally, the quantum yield also changes.[12,18]

"Bare" H-terminated SiNCs are strongly prone to oxidation under an ambient atmosphere,[19] implying that their surface always needs to be terminated with a non-silicon-based chemical species. In addition to the size, the type of surface termination also strongly affects the optical properties.[20–23] Probably the most typical surface chemistries include natural or intentional oxidation[6,10,24] and long-alkyl-chain attachment *via* hydrosilylation,[9,11,25,26] exhibiting emission in the red and near-infrared spectral region with radiative lifetimes on the order of 10–100 μs. However, a wide range of other types of surface terminations are possible.

The optical properties introduced so far apply only to one class of SiNCs. It is generally recognized that in addition to the "red" emitting SiNCs described above, a second class of SiNCs exists: the "blue" emitting ones with much faster emission rates. The blue PL band can either co-exist with the red band,[27,28] or it can stand completely on its own.[1,29] Since fast emission in the blue spectral region under UV illumination is quite a common occurrence, there are as many models explaining the physical origin of the blue band as there are proposed mechanisms for the switch between the red and the blue types of emission. Given the prevalence of the blue emission across various materials, it is actually quite possible that different explanations are valid for different samples and more than one of the proposed models are necessary to explain the wealth of the observed types of behavior.

Starting with oxidized SiNCs, the blue band has been proposed to be connected with the crystalline core of the SiNCs,[30] possibly with the non-thermalized direct Γ_{15}–Γ'_{25} transition[31–33] (see Fig. 1a), whereas other studies see it completely as a result of oxidation and SiO_2 defects in the surface oxide.[28,34] The blue-green emission of organically capped SiNCs was attributed to the crystalline core,[35] but a different study saw its origin in a surface defect or impurity.[1] As for the switch from the red to the blue PL, a discontinuity in the spectral tunability with decreasing size starting from the "red" emission is usually observed,[1,8,13,17,28] even though here an exception to the rule can be found.[7]

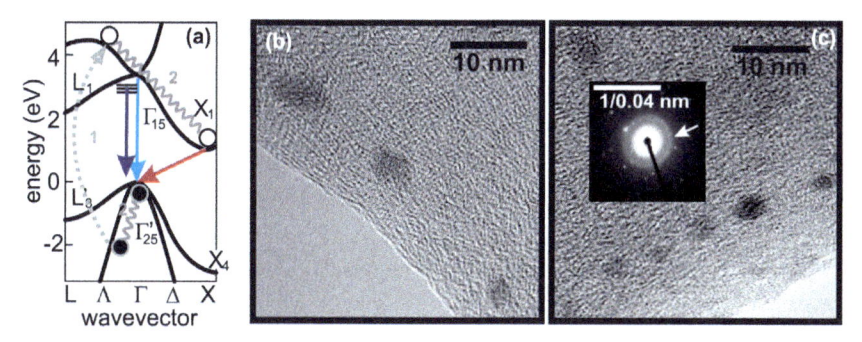

Fig. 1 (a) The band structure of bulk silicon with the transitions discussed in the text highlighted. Number (1) designates the excitation and number (2) the thermalization event. (b and c) Bright-field TEM images of SiNCs before (b) and after (c) surface modification. Individual NCs can be easily discerned. The corresponding selected-area electron diffraction pattern of the surface-modified SiNCs in (c) confirms the presence of randomly oriented NCs in the investigated spot.

Apart from the effect of oxidation,[28] an induced switch between red and blue emission has been ascribed to the presence of nitrogen "impurities".[1] In that work, SiNCs were prepared by a top-down approach (annealing of an Si-rich material, HSQ) and using bottom-up syntheses (Na_4Si_4/NH_4Br reaction and $SiCl_4$ reduction). Even though all of the samples had basically the same size and the same surface termination, the emitted PL was the "red" type and the "blue" type of emission for the top-down and bottom-up approaches, respectively. The incorporation of nitrogen and oxygen "impurities" was identified as the process responsible for this difference; the mixing of the "red-emitting" SiNCs with nitrogen-containing compounds even induced them to have "blue" PL. As a result of that work, nitrogen is seen as an undesirable element, which can always induce "blue" PL in SiNCs.

There is some evidence in the literature, though, showing that this is not necessarily always true. One body of work constitutes SiNCs prepared *via* the reduction of $SiCl_4$, surface-terminated using N-linked aromatic compounds (*e.g.* diphenylamine, aniline), where high quantum yields were achieved.[29] The maximum PL of the surface-terminated SiNCs could be tuned using different types of surface termination, with aniline-terminated SiNCs emitting in the orange spectral region.[29] Fast PL lifetimes were reported, with no information on a possible long component. Similarly, the attachment of N-containing porphyrin molecules active as antennae on "red-emitting" dodecyl-capped SiNCs did not induce a switch of the red PL to blue PL either.[36] Looking at older publications, films of SiNCs treated in N_2 plasma, resulting in nitride-terminated SiNCs, were also investigated. The presence of the nitrogen in the nitride-rich SiNC films, just at the surface of the SiNCs, did not induce blue PL.[37,38] Of particular interest is the comparison of such nitride-terminated SiNCs with oxidized ones,[38] which hints at the co-existence of a blue and a red PL band in the oxidized SiNCs, but also at its complete absence of the blue PL band in the nitride-terminated SiNCs. However, the optical experiments performed in that study were a bit rudimentary.

In this contribution, we study the effect of the treatment of oxidized SiNCs in a nitrogen-rich environment, focusing on the blue emission. The original, untreated oxidized SiNCs exhibit a weak blue PL band in addition to the standard red band. Despite the fact that nitrogen builds into the surface passivation layer of the SiNCs in the form of nitrites and/or nitrates, it is observed to cause the disappearance rather than the emergence of the blue band. This finding confirms that the presence of nitrogen close to SiNCs does not necessarily induce blue PL in SiNCs.

2 Methods

The studied SiNCs were prepared by a standardized electrochemical etching treatment followed by oxidation of the surface under ambient-conditions.[17] In brief, crystalline silicon wafers were exposed to a solution of HF/EtOH in an electrochemical cell while current was passed through the etching solution (please note that HF can cause serious burns to the skin with significant complications due to fluoride toxicity). After two hours of etching, the resulting thin layer was washed with ethanol and let to dry in an ambient atmosphere at slightly elevated humidity ($\approx 50\%$). Mechanical pulverization of the electroetched layer yielded a powder of oxidized SiNCs, which had already been characterized in great detail.[17,32,39,40]

In order to carry out the surface-modification treatment, SiNC powder was suspended in deionized water at a concentration of about 1 mg SiNCs/2 ml water and was treated in an ultrasonic pulse homogenizer for 1 hour. Then, the sample was placed in a quartz cuvette. The surface-modification treatment was performed in water using non-thermal plasma generated by electrical discharge in the point-to-plane arrangement and in the regime of transient spark. Transient spark discharge is described in ref. 41, and specific details will be published elsewhere. In brief, an electrode was inserted into the cuvette so that a discharge was generated between the electrode and the water surface under passing current. Processes induced by non-thermal plasma are related to the presence of high energy electrons. Their collisions with air molecules surrounding the water level produce reactive oxygen and nitrogen species, charged molecules and UV and visible photons under the conditions of the experiment. The radical species are dissolved in the water and the high energy molecules/atoms cause the radicalization of the surface part of the water. Moreover, the radicals from the air and the water induce secondary chemical processes in the water medium. This procedure yields the so-called plasma-activated water (PAW, for the details of PAW preparation and analysis see our previous studies[42,43]), in our case, rich in nitrates NO_3 and nitrites NO_2. Although in many cases PAW is also rich in peroxide, H_2O_2, under the conditions of our experiment, peroxides are not present. The surface of the SiNCs was modified in PAW under the plasma discharge while the suspension was magnetically stirred for about 30 min. The resulting suspension of modified SiNCs was characterized by the methods detailed below. The samples used for EDS and FTIR measurements were washed additionally with deionized water to remove the remaining radicals. PAW for the luminescence measurements was separated from the modified sample using filtration with centrifuge filtration units. The PAW was checked for the absence of red PL to confirm that no SiNCs were present.

The microstructure of the samples and the EDS spectra were observed using a TEM Jeol 2200FS (200 kV) equipped with an Oxford Instruments EDS detector. As a support for the nanoparticles, holey carbon 300 mesh copper grids were used.

For FTIR measurements, the samples were deposited onto an Al plate and dried for three and half days in air, which guaranteed that all volatile substances had already evaporated. The FTIR spectra were recorded using the ATR (Ge crystal) technique using a ThermoScientific Nicolet iN10 FTIR Microscope with a resolution of 4 cm^{-1} and the Norton-Beer strong apodization function in the spectral region of 675–4000 cm^{-1}, where the average spot size was 100 × 100 µm^2. Several different spots of each of the evaporated deposits were measured to confirm the homogeneity of samples.

For temporally resolved PL measurement, an optical setup composed of a femtosecond pulsed laser, Pharos SP – 1.5 mJ – 200 (150 fs), a harmonics generator HiRO (Light Conversion) and a streak camera (Hamamatsu) coupled with an imaging spectrospectrometer was employed. An excitation wavelength of 343 nm and a repetition rate of 1 kHz were used, and the impinging excitation intensity was 65 µJ cm^{-2} (1.2 × 10^{14} photons per cm^2). The suspensions of SiNCs were placed in (luminescence-free) quartz cuvettes and were magnetically stirred during the measurements, and we verified that the PL properties of both the samples were not influenced by the laser beam. The angle between the excitation beam and the detection path was 90°. The PL of the second, modified sample was measured just after that of the unmodified one with minimal delay between the two and under the same excitation/detection conditions. We have also thoroughly verified that the intensity readings of our apparatus are well-repeatable, which implies that the PL intensities reported here are fully comparable across the two samples. The acquired PL maps were analyzed following a dedicated procedure outlined in ref. 44, ensuring high-quality fits. In brief, both the PL onset edge and the decay itself were taken into account using a convolution of a decay function (single-exponential, stretched-exponential) and the (known) shape of the incoming laser pulse. Four spectrally neighboring PL decays (that is a spectral region of about 1.5 nm) were averaged to increase the signal-to-noise ratio and the fitting in the temporal domain was performed for all of these averaged PL decays ($\approx$130 decays per one spectrum). The background level (I_0 from eqn (1) introduced later) was determined as the mean of the values before the onset of PL and its homogeneity was checked in order to accurately discriminate possible weak signals from the background. At first, the data were fitted assuming only a simple single-exponential decay to find out the position of the onset edge, and in a second run, a more precise decay function was selected while the onset times were fixed to the values obtained from the first run. Spectrally uncorrected data were used for the fitting procedure, and a spectral correction was applied to the spectral part (decay amplitudes) after the fitting.

The transmittance spectra were collected using a Libra Biochrom S60 spectrophotometer.

For the calculations of the phononless radiative rates, the Si crystalline cores were built using the Vesta editor by making a cut-off from a bulk crystal along the (100) (110) and (111) crystal planes. In this way, approximately spherical SiNCs were obtained. All SiNCs had crystalline cores. The surface of the SiNCs had different types of facets with –SiX$_3$, –SiX$_2$ and –SiX binding sites. The –SiX$_3$ species were not chemically stable and were removed. The (100) facets were reconstructed. The resulting SiNCs were slightly tetrahedral and were 100% covered by hydrogen. OH passivation was 100%, and the mixed –OH and –O–NO$_2$-capped nanoparticle had –O–NO$_2$ attached to 49 surface sites and OH to the remaining 55 sites. Calculations were performed using density functional theory (DFT) as implemented in the CP2K package Quickstep,[45] with the generalized gradient approximation functional of Perdew–Burke–Ernzerhof.[46] The core was approximated by the Goedecker–Teter–Hutter pseudopotential potential.[47] For the basis set, we used the short-range Gaussian double-zeta valence polarized basis set DVZP-MOLOPT.[45] Plane wave cutoff of the integration grid was set to 400 Ry and the self-consistency convergence to 10^{-6}. The bandgap energies appeared to be shifted by a constant value of 0.5 eV, which is a known issue of ground state DFT.[48]

3 Results

3.1 Surface modification and changes in chemistry

In our study, we used porous-silicon-based SiNCs oxidized under an ambient atmosphere at slightly elevated humidity, which had already been studied in great detail. In brief, the size of the crystalline core of the light-emitting SiNCs was around 2.5–3 nm,[39] and the prevailing surface species was oxide.[39] The PL characteristics were dominated by a broad red PL band positioned around 650 nm with microsecond decay times, even though a fast blue band was also present.[17,32] The porous-silicon origin of the SiNCs, on the one hand, causes the presence of a broad distribution of sizes of NCs and crystallites in the samples, including crystallites with sizes of ≈ 50 nm.[40] On the other hand, it ensures that the small light-emitting SiNCs have crystalline cores, since they originate from a single-crystalline Si wafer. It might not be so straightforward to achieve such high concentration of particles with crystalline cores in bottom-up preparation approaches, such as in various types of chemical syntheses.

The water-based suspension of the oxidized SiNCs underwent surface modification by the application of non-thermal plasma under ambient conditions (atmosphere and temperature). The non-thermal plasma discharge ionizes the air surrounding the SiNC suspension and causes the generation of water-related reactive species. The air-originating radicals dissolve in the water and might react with the water-related species. The complex series of chemical reactions leads to a cocktail of reactive species dissolved in water, resulting in the so-called plasma-activated water (PAW). The type and the concentration of the reactive species in the PAW depend on the conditions of the plasma discharge; in our case, the prevailing species are NO_2^- (≈ 30 mg l^{-1}) and NO_3^- (≈ 5 g l^{-1}), but H_2O_2 is not present. High concentrations of these radicals caused very low pH values (1–2) in the PAW in our experiment. The details of the modification procedure are beyond the scope of the current contribution.

As can easily be expected from the absence of peroxides the PAW treatment does not affect the crystalline cores of the original SiNCs because the treatment conditions are too mild to possibly destroy a crystalline core. The presence of small crystalline particles in both the original sample and in the sample after the surface modification are documented *via* the TEM measurements in Fig. 1b and c. The SAED diffraction pattern of the surface-modified sample in the inset of Fig. 1c shows a clearly defined circle, typical of very small randomly oriented crystalline nanoparticles.[24,49] However, the already mentioned presence of larger crystallites originating from the etching preparation procedure causes residual diffraction of the silicon lattice.

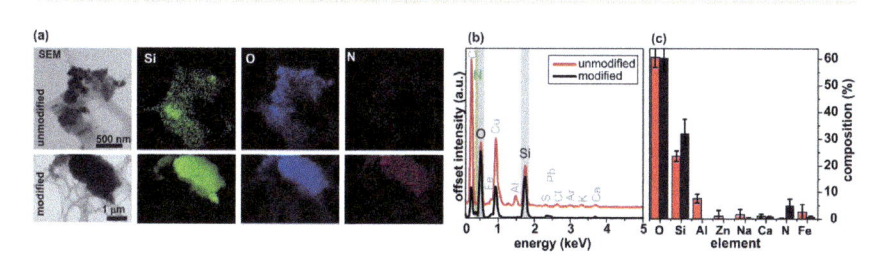

Fig. 2 (a) Elemental analysis of the unmodified (top row) and surface-modified (bottom row) SiNCs. The four pairs of images correspond to the electron-microscopy image (left), and the EDS maps of the Si, O and N content (right). Before the modification, the nitrogen content in the sample was very low, and after the modification, the N content significantly rose in the areas containing Si. (b) Examples of two EDS spectra. (c) Elemental composition averaged over several samples and spots and compensated for the signal of the substrate. The error bars represent the standard deviations of the average values. Color coding is the same as in (b).

The non-thermal plasma treatment of the oxidized SiNCs results in a change in the surface-passivation layer of the SiNCs. Fig. 2 compares electron micrographs and the energy dispersive X-ray spectroscopy (EDS) elemental maps of the original and surface-modified SiNCs, respectively. As expected, the most important elements present in the sample before the treatment were silicon and oxygen (see also the EDS spectra in Fig. 2b). When probed, trace amounts of nitrogen could also be detected. After treatment in the nitrite- and nitrate-rich PAW, silicon and oxygen still remained the most important elements in the modified sample, however, the amount of incorporated nitrogen dramatically increased (Fig. 2a).

Since the acquired EDS spectra from Fig. 2a tend to contain "flakes" of impurity material (such as Cu from the grids, see Fig. S1 in the ESI†), EDS spectra from two different samples were averaged over several spots and compensated for the signal of the carbon substrates. The resulting elemental composition is shown in Fig. 2c. Considering the sizes of the SiNCs under study, a rough estimate of 0.4 surface bonds to be passivated per one silicon atom in the crystalline core can be made.[56] The different observed silicon-to-oxygen ratio in the unmodified sample was a result of CO_2 being adsorbed by the sample during the measurement. The other elements present in trace amounts are residual impurities originating from the preparation procedure (Al – from the duralumin substrate of the electrochemical etching cell, Fe from the stainless steel scalpel used for mechanical pulverization, and Na from NaOH used for cleaning of the etched wafers). The concentration of all of these impurities decreased after the modification, implying that they are clearly washed out during the radical-removal procedure after the modification treatment and thus not involved in the modifications presented in this study.

The important changes in the elemental composition of the modified sample are a small increase in the silicon-to-oxygen content ratio accompanied by a large increase in nitrogen content, implying that part of the oxidized surface of the unmodified SiNCs is replaced with a nitrogen-containing compound. The increase in the nitrogen content is significant, and it rises from (0.2 ± 0.2)% to as much as (5.0 ± 2.5)%. Just to give an idea of what this means, we can imagine a "model" SiNC composed of about 700 Si core atoms (3 nm in diameter), which has about 250 surface bonds to be passivated.[56] Presuming a simple case of a 1 : 1 passivation of these surface bonds with bridging oxygen bonds Si–O–Si and with hydroxyl groups –OH, the hypothetical NC has 160 surface oxygen atoms. If 5% of this NC is supposed to be made up of nitrogen, about 40 of these surface oxygen atoms would have to be replaced with nitrogen atoms.

The chemical nature of the incorporated nitrogen was revealed by FTIR measurements in ATR mode, as shown in Fig. 3. The oxidized unmodified SiNC sample has already been studied in detail.[39] The prominent surface features are bridging Si–O–Si bonds at 1000–1250 cm^{-1} and 798 cm^{-1} accompanied by silanol Si–OH (846 and 943 cm^{-1}) and back-oxidized hydride O_x–SiH$_y$ (875 and 2255 cm^{-1}) groups. The PAW treatment led to the disappearance of the back-oxidized hydride groups whereas silanol groups were preserved. More importantly, some changes in the bridging-oxygen region are observed, suggesting reorganization in the surface oxides. However, the most marked feature of the SiNCs after the treatment was the emergence of a new FTIR band between 1250 and 1560 cm^{-1}. In accordance with the studies of the interactions of HNO_3 with SiO_2 and an organic self-assembled layer-coated Ge crystal under dry and humid

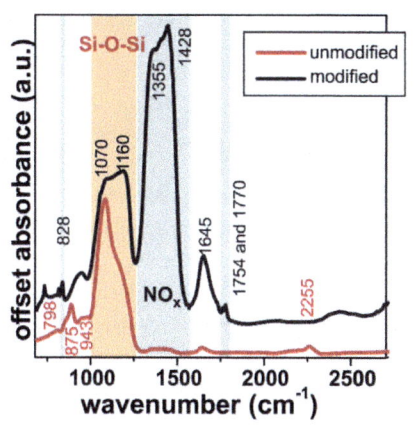

Fig. 3 FTIR-ATR measurements of the unmodified and modified samples and the assignments of the observed FTIR peaks (v stands for stretching, δ for deformation, s for symmetric, and a for antisymmetric) with the corresponding references used for the assignments noted.

Bond	Wavenumber	Ref
v_s Si–O–Si	≈795 cm^{-1}	50,51
v_a Si–O–Si	≈975–1130 cm^{-1}	50–52
v'_a Si–O–Si	≈1130–1260 cm^{-1}	50,52
δ O$_x$–SiH$_y$	875 cm^{-1}	51,52
v O$_x$–SiH$_y$	2262 cm^{-1}	51,52
v Si–OH	≈ 940 cm^{-1}	53
v_2 NO$_3^-$	828 cm^{-1}	54
v_3 NO$_3^-$	1300–1500 cm^{-1}	55
$(v_1 + v_4)$ NO$_3^-$	1750–1770 cm^{-1}	54
δ H$_2$O	1641 cm^{-1}	55
v H$_2$O	3400 cm^{-1}	55

N$_2$ flow,[55] these peaks result from the presence of nitrate/nitrite ions. The shape of this nitrate peak, which is asymmetric but exhibits minimal levels of splitting, indicates that the trigonal symmetry of the nitrate anions is almost retained on the SiNC surface. This fact could be explained by a set of weak bonding interactions ···O–NO$_2$ to the SiNC surface (probably also including a weak bidentate bonding mode of nitrates), which induce only a minimal distortion of symmetry. Additionally, the splitting of the combination band $(v_1 + v_4)$ of the nitrate anions into two bands at 1770 and 1754 cm^{-1} can also confirm the existence of a weak bonding interaction of NO$_3^-$ with the surface of the SiNCs. The separation of the combination band of 24 cm^{-1} represents the border of the monodentate and chelating bidentate interactions of nitrate ions according to the literature.[54] The band recorded at 828 cm^{-1}, overlapped by the vibrational motion of silanol, can be also assigned to the v_2 NO$_3^-$ vibration. An alternative explanation of the shape of the 1250–1560 cm^{-1} FTIR band could be also related to the formation of nitrate–water complexes (discussed in ref. 55 and references therein) bound to the SiNC surface. The presence of water molecules is indicated by the H$_2$O bending mode recorded at 1645 cm^{-1}. Thus, FTIR not only confirms the increase in nitrogen content, it also shows that the changes are due to the PAW modification and the incorporation of nitrate–water complexes into the surface layer of the SiNCs.

3.2 The blue photoluminescence band

The incorporation of nitrogen-containing complexes into the shells of the modified SiNCs influences the emitted PL. Even though the "red" PL band is affected by the treatment to some extent (see Fig. 5a), contrary to the results of ref. 1, it does not disappear. Thus, it is not of special interest here.

In order to study the "blue" PL band, we carried out time-resolved PL spectroscopy of both the unmodified and modified samples. The excitation wavelength of 343 nm was chosen for a direct comparison of the blue bands of the two samples, since it lies just outside the absorption range of PAW, see Fig. S2 in the ESI.† Even a single look at the measured temporally and wavelength-resolved dependencies of the PL intensity presented in Fig. 4a and d reveals that there is a marked difference between the blue PL bands of the two samples. Interestingly, in accordance with the results presented in ref. 38, the band is observed to diminish rather than increase in intensity following the nitrogen-enriching treatment, see also Fig. 5d.

More insights into the dynamics of the blue PL bands can be gained using a proper analysis, which is unavoidable considering that nearly any material has some blue PL when excited in the UV region. Here, we employ a procedure put forward in ref. 44, which takes into account the whole PL decay including the

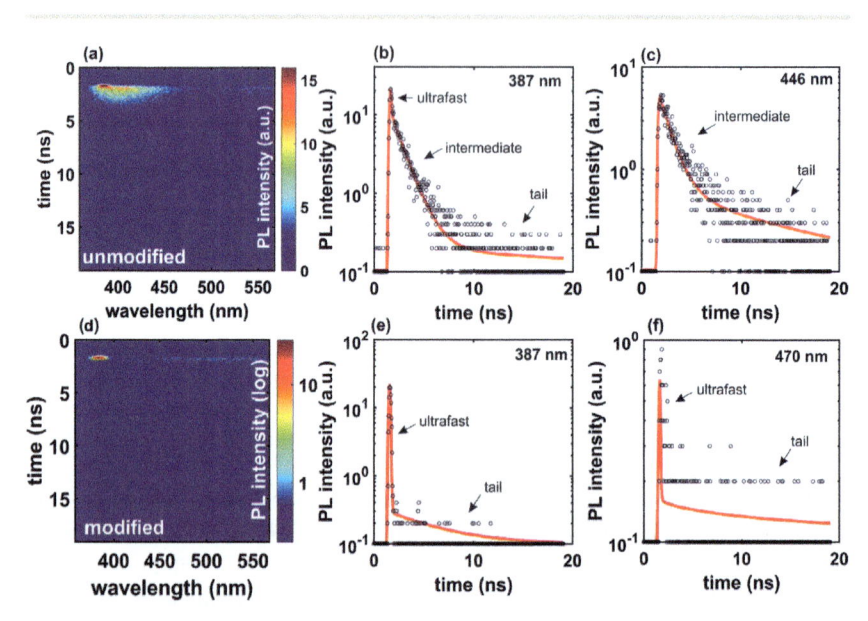

Fig. 4 Time-resolved PL measurements of the unmodified (top row) and surface-modified (bottom row) SiNCs (ultrafast pulsed excitation at 343 nm). The left columns show the measured PL maps (PL as a function of time and wavelength, please note that the PL intensity scale of the modified sample is logarithmic), the remaining two columns display examples of data fits: the measured data (black circles) and the corresponding fits (red curves) are plotted in a shifted logarithmic scale for two selected emission wavelengths.

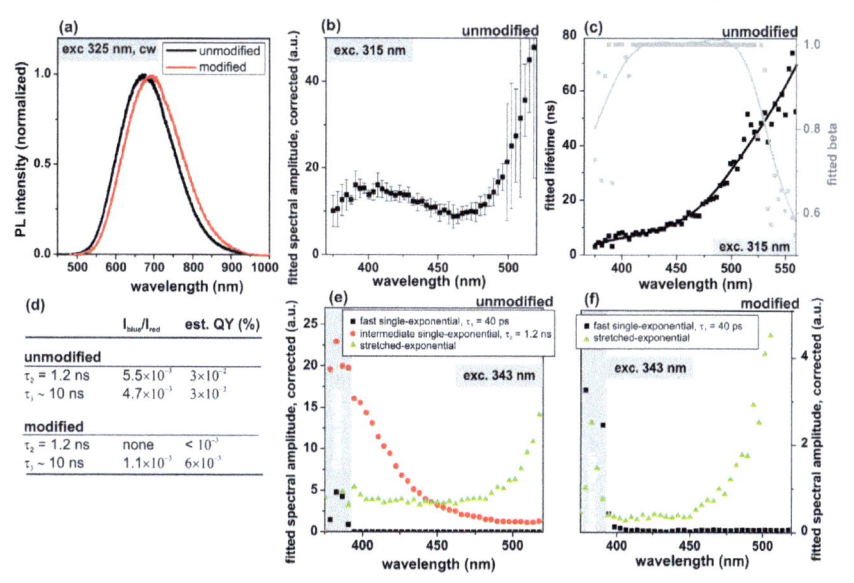

Fig. 5 (a) Comparison of normalized PL spectra of the red bands of the unmodified and modified SiNCs using 325 nm CW excitation. (b and c) Results of the fits of the PL of the unmodified SiNCs using pulsed 315 nm excitation (for the fits, see Fig. S3 in the ESI†). The curves represent spline-smoothing of the presented data to serve as guides-for-the-eye. (d) Integrated PL intensities for the second and third component and the corresponding rough estimates of quantum yields (see text for more details). (e and f) Spectral amplitudes of the fitted components of the PL of the unmodified and modified samples under pulsed 343 nm excitation, respectively. The corresponding PL maps are shown in Fig. 4.

onset edge. To do so, a convolution (★) of a (known) instrument response function $i_{irf}(t;\sigma)$ with the PL decay function (a single- or stretched-exponential characterized by lifetime τ and stretched parameter β including a proper normalization factor norm(τ,β)) is used as a fitting function:

$$i_{meas}(t,\lambda;\sigma,\tau_0,I_0,A(\lambda),\tau(\lambda),\beta(\lambda)) = \frac{A(\lambda)}{\text{norm}(\tau(\lambda),\beta(\lambda))}\exp\left\{-\left(\frac{t-\tau_0(\lambda)}{\tau(\lambda)}\right)^{\beta(\lambda)}\right\} \star i_{irf}(t;\sigma) + I_0, \quad (1)$$

where $A(\lambda)$ is the amplitude of the corresponding decay component, σ is the apparent width of the laser pulse in the chosen temporal window and I_0 is the background. The fitting was performed over the individual measured wavelength-resolved PL decays and the extracted characteristics $A(\lambda)$, $\tau(\lambda)$ and $\beta(\lambda)$ were checked for any underlying spectral dependencies. In contrast to simply focusing on the PL decay at a single wavelength or on the spectrally unresolved PL, this approach helps one determine if more than a single dynamic component is present and the additional information on its spectral dependence is beneficial for the identification of the underlying physical process.

Starting with the oxidized unmodified SiNCs, three different components were identified. The clear presence of these three components can be visualized when the measured PL decays are plotted on the offset logarithmic scale[44] (so that the smallest-PL-intensity data points, representing the measured zero, are also visible), as seen in the fit-and-data plots in Fig. 4b and c. The first component is ultrafast and it was fitted assuming a single-exponential decay with a lifetime set to the temporal difference between two consecutive datapoints ($\tau_1 = 40$ ps in the chosen temporal window), implying that the corresponding lifetime is ≤40 ps. The second component is also single-exponential, and was found to be emission-wavelength-independent with[57] $\tau_2 = 1.2$ ns. A weaker third component produced a slower-than-single-exponentially decaying tail and it was approximated with a stretched-exponential.

In order to suppress the ultrafast PL component, the two samples were also measured using the excitation wavelength of 315 nm. Examples of data-and-fit plots are given in Fig. S3 in the ESI† and the deduced amplitudes, lifetimes and β's are shown in Fig. 5b and c. This analysis confirms that the slower, stretched-exponential component is excitation wavelength independent and it also allows one to better characterize its spectral evolution, which peaks at ≈400 nm. Unfortunately, the properties of the intermediate component could not be accurately analyzed using the shorter excitation wavelength, because the blue-shift of the excitation wavelength causes a much more efficient excitation of the solvent, whose PL characteristics are far too similar to those of the SiNCs to allow for an accurate analysis.

As for the nitrogen-containing modified SiNCs, the PL intensity of their blue PL band is clearly diminished when compared to the unmodified sample. The intermediate $\tau_2 = 1.2$ ns component is completely missing, but a weak stretched-exponential tail persists, as seen in Fig. 4e and f. Despite what it might seem at first sight, the red curves in these plots fit the measured data very well. It is important to realize here that the plot is shown, contrary to standard representation, in a logarithmic vertical scale shifted by a small value to also include the zero values.[44] This representation provides a better visual cue as to how the signal approaches zero as opposed to a traditional logarithmic plot where the zero (or negative) values are simply omitted as a result of applying the logarithm. As the signal yielded by our apparatus is intrinsically digital, a tail of any PL decay, and that of a slower stretched-exponential one in particular, is made up by increasingly scarce photon-detection events one "level" above the zero. Such photon-detection events are exactly what can be seen in Fig. 4e and f. For a more in-depth discussion, readers should refer to ref. 44. As the signal of this third component is too weak to allow for meaningful fitting by itself, it was assumed to be characterized by the same lifetimes as those of the unmodified sample (the spline curve in Fig. 5c) and only the β parameter and the amplitude were fitted. The deduced spectral amplitudes are plotted in Fig. 5f. The reasonable agreement between the fitted β parameters in the modified (see Fig. S4 in the ESI†) and unmodified samples, respectively, justifies this approach.[58]

In order to get a quantitative comparison of the intensities of the blue bands of the unmodified and modified samples, integrals of the individual spectral components were calculated. Only the spectral region between 400 and 500 nm was considered to avoid the influence of the ultrafast component. Extreme care was taken to ensure that the samples were measured under the same excitation, collection and detection conditions, which enables a direct (relative) comparison of the measured PL intensities. The concentration of the SiNCs in the unmodified and modified suspensions were also roughly the same, which is evidenced by the transmittance spectra in Fig. S2 in the ESI:† the SiNCs cause scattering, resulting in the decreased transmittance in the 500–700 nm spectral region, which is the same in the two samples. Below 340 nm, the absorption of PAW sets in. Thus, the chosen excitation wavelength of 343 nm seems to ensure optimal conditions for the comparison of the two samples.

The red spectral component of the unmodified sample was also measured and spectrally integrated under the same conditions (not shown). The integrated intensities of the blue components are listed with respect to the integrated intensity of the red PL, see Fig. 5d. Since the quantum yield of the red component is known ($\approx 6\%$), rough estimates of the quantum yields of the blue components under the excitation conditions of this study can be made. They are on the order of 0.01% and are listed in Fig. 5d.

3.3 Computational study of the electronic states

In order to theoretically assess the influence of surface-present nitrogen on the SiNC states, we performed DFT calculations of the electronic states. In particular, the structure of a reasonably sized nanoparticle ($Si_{235}H$, corresponding to a diameter of about 2.0 nm) was relaxed, including two types of covalently bonded surfaces: (i) full passivation by –OH groups to simulate surface oxide and (ii) the replacement of some of the –OH groups with –O–NO_2 groups ($\approx 50\%$), emulating the surface-modified SiNCs. Importantly, the structure of the nanoparticle does not allow for full surface coverage with –O–NO_2 groups due to spatial constraints, suggesting that the partial surface chemistry change is a plausible scenario.

As is evident from the results of these calculations presented in Fig. 6, the surface-bonded nitrogen in the form of –O–NO_2 induces some changes in the density of states in the SiNC. First, low-density-of-states in-gap states close to the edges of both the valence and conduction bands appear. Secondly, disregarding these in-gap states, a small redshift in the bandgap energy (from 1.517 eV in the fully oxidized to 1.493 eV in the partly nitrated SiNC model) occurs. Thirdly, some degree of real-space localization at surface sites, mostly at O and N atoms, occurs for the states at the edge of the conduction band of the partly nitrated particle, as also evidenced by the 3D images in Fig. 6. However, if the radiative rates corresponding to these states are calculated, only an increase of several times is obtained when compared to the fully oxidized SiNC model.

The fact that nitrogen in the shell of the NC does not necessarily lead to an impurity state causing short PL decay is illustrated also in the calculations of the density of states of a partly –O–NH_2-terminated Si nanoparticle in Fig. S5 in the ESI,† where conduction-band states are much more influenced by the presence of nitrogen than the valence-band states.

4 Discussion

Both the results of our theoretical calculations and experiments confirm that, in the case of the SiNCs studied, the incorporation of non-negligible amounts of nitrogen in the form of nitrate–water complexes does not induce blue PL in the sample. The first-excited-state indirect-bandgap Γ_{15}–X transition is still observable, it is not dramatically affected and the presence of nitrogen in the shell does not give rise to localized in-gap "impurity-like" states.

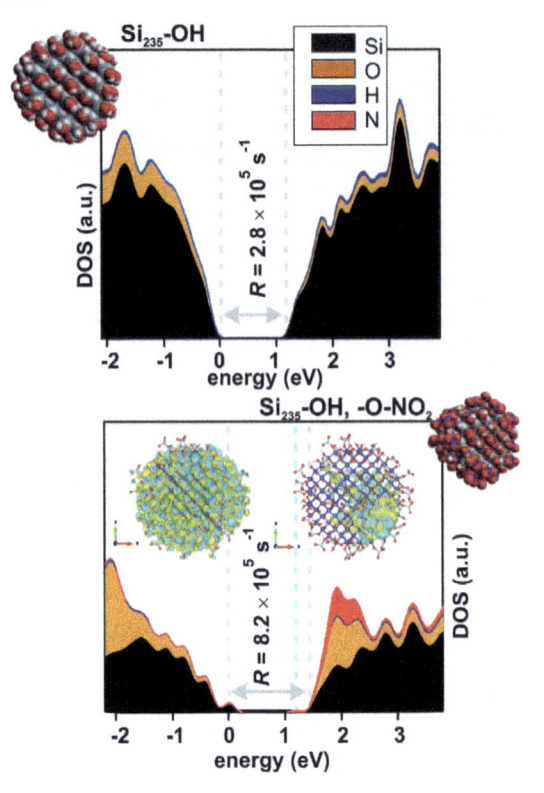

Fig. 6 Calculated real-space density of states as a function of energy for two Si nanoparticles with the same crystalline core (approx. 2.0 nm in diameter) but different surface passivations, showing real-space localization on the corresponding atoms. The insets in the nitrogen-containing Si nanoparticle show 3D real-space localizations of the two energy states as designated by the grey and light blue dashed lines on the left and right, respectively. R values are the calculated radiative rates of the first-excited-state transitions, shown by the grey dashed lines.

Focusing on the blue PL band as observed in the unmodified, oxidized SiNCs, it was made up by three components, two single-exponentials (exp) and one stretched-exponential (SE):

$$I_{PL}(t,\lambda) = \exp(\tau_1 = 40 \text{ ps}) + \exp(\tau_2 = 1.2 \text{ ns}) + SE(\tau_3(\lambda) \approx 10 \text{ ns}). \quad (2)$$

Clearly, the ultrafast component ($\tau_1 = 40$ ps) can be identified with the Raman signal of the solvent (water) since its spectrum is mostly situated in one peak (see Fig. 5b) (although some ultrafast signals can also be found in other parts of the spectrum). The difference between the excitation and emission wavelengths (343 and 384 nm, respectively) is ≈ 3000 cm^{-1}, implying the assignment of this peak to the typical vibrations of OH groups. Such Raman signal would be expected to shift under shorter-wavelength excitation of 315 nm to 348 nm, *i.e.* out of the collection spectral window, which we verified experimentally, see Fig. S4 in the ESI.†

The origin of the intermediate PL component exp($\tau_2 = 1.2$ ns), considering the spectral independence of its lifetime τ_2, most likely lies in the defect states of

SiO$_2$, as was suggested by previous studies.[34] The third component, then, is stretched-exponential in nature and despite being weak, its characteristics $\tau_3(\lambda)$ and $\beta_3(\lambda)$ were observed to depend on the emission wavelength (see Fig. 5c). Moreover, spectrally speaking, it seems to be a direct continuation of the "red" (stretched-exponentially decaying) PL band towards the shorter emission wavelengths. This trend was also confirmed when using a shorter excitation wavelength to avoid the influence of the Raman component, see Fig. 5b and c. Consequently, this third PL component can be assigned to the non-thermalized, direct Γ_{15}–Γ'_{25} transition, as discussed earlier.[32]

Interestingly, the presence of nitrates in the shells of the SiNCs dramatically decreases the PL intensity of the blue band in the modified SiNCs. This difference is quantified in Fig. 5d, where the integrated PL signals emitted between 400 and 500 nm are listed. Obviously, the observed blue PL band is relatively weak in all cases. The most important change, however, is the complete disappearance of the SiO$_2$-defect-related intermediate $\exp(\tau_2 = 1.2$ ns$)$ peak. The absence of this peak in the modified sample implies that the PAW treatment radically improves the quality of the surface oxide shell of the SiNCs.

The stretched-exponential component related to the non-thermalized direct Γ_{15}–Γ'_{25} emission also decreases in intensity by about five times. However, it is important to realize that the decrease in the intensity of this spectral component as a result of the change in the chemical composition of the surface shell of a NC does not automatically imply that this band originates in surface-related states, which might be the first intuitive conclusion. Being connected with non-thermalized (hot) carriers, this band is governed by both the radiative rates of the corresponding transition and the thermalization rates, which are generally very fast. Thus, it is plausible to deduce that the SiO$_2$-related defects enhanced the intensity of this non-thermalized emission in the unmodified sample by slowing down the thermalization rates. In the modified sample, the removal of these states clearly increased the thermalization rates and in turn decreased the intensity of the corresponding emission.

The most intriguing result here is clearly the completely different response of SiNCs to the exposure to a nitrogen- and oxygen-rich environment observed in this study (and previously also by Chen et al.[38]) and by Dasog et al.[1] Let us briefly review the existing evidence. Dasog et al.[1] prepare their SiNCs by three solution-based methods (thermally induced disproportionation of HSQ, the Na$_4$Si$_4$/NH$_4$Br reaction and the SiCl$_4$ reduction). In all cases, the surface was terminated with dodecene groups using thermal hydrosilylation. The Si core sizes were reported to be (3.5 $\pm$ 0.4), (6 $\pm$ 2) and (2.7 $\pm$ 0.6) nm, respectively, based on TEM analysis. Whereas the HSQ-based SiNCs exhibited slow red PL, the SiNC prepared by the other two methods emitted immediately fast blue PL, according to the authors because of trace amounts of nitrogen being incorporated as optically active impurities. Moreover, the exposure of the "red" emitting SiNCs (with H-terminated surfaces) caused a switch to the "blue" PL. Next, in the early work by Chen et al.,[38] the nitride-terminated SiNCs were prepared by plasma enhanced chemical vapor deposition using SiH$_4$ and N$_2$ as source gases with TEM-derived mean sized of 4.6 and 3.1 nm, exhibiting size-dependent slow PL (with no fast signal). A reference oxidized SiNC sample prepared by annealing of an amorphous Si layer in Ar at 1100 °C emitted slow red PL with hints of a fast bluer band. In this study, the exposure of electrochemically etched SiNCs, with mean core sizes of 2.5–3 nm, to nitrogen-rich PAW quenched the SiO_2-defect-related blue PL and dramatically decreased the core-related blue PL while not quenching the red PL.

Thus, it is very difficult, if not impossible, to formulate a hypothesis encompassing all of these reported results since they are contradictory. The differences between the samples seem to be neither the size (samples with comparable sizes behave differently) nor the surface (H-terminated SiNCs were exposed to nitrogen yielding both "red"[38] and "blue"[1] PL, even though during the preparation of the "red" SiNCs there was conceivably no oxygen present). We do not want to form unsubstantiated hypotheses here, which is why such an all-explaining hypothesis is beyond the scope of the current work and is a topic for further research. However, we want to highlight three observations which might help one find the solution to this problem:

(i) The SiNCs in this study are electrochemically etched and as such they are embedded in relatively thick layers of SiO_2. It is plausible to assume that these SiO_2 layers might make it more difficult for the nitrogen to get to the core of the SiNC and form a defect site inside it. Were this the case, the "red"-emitting SiNCs yielded by the plasma deposition in the presence of nitrogen[38] could be a result of the absence of oxygen, as suggested by Dasog *et al.*[1]

(ii) The "blue" PL as reported by Dasog *et al.*[1] is not by any means evocative of defect PL (it is very broad and excitation-wavelength dependent). Thus, one might hypothesize that nitrogen only quenches the PL of SiNCs and the source of the reported "blue" PL is a completely different material. Were this the case, nitrogen would have to bind differently to SiNCs in this study than in the other two discussed studies. Also, a thorough quantification of quenching of the "red" band would be helpful along with a more detailed analysis of the blue PL emitted by those samples.

(iii) There has also been an ongoing discussion about low- and high-temperature preparation methods yielding the "blue"- and "red"-emitting SiNCs,[2] respectively. Interestingly, in this regard, Dasog *et al.*[1] actually showed that "blue" PL was emitted by SiNCs prepared by a high-temperature method after their exposure to nitrogen, because HSQ disproportionation constitutes a high-temperature method due to the temperatures involved in the annealing step.

5 Conclusion

The influence of the exposure of oxidized SiNCs to a nitrogen- and oxygen-rich environment of plasma activated water was studied. The plasma-activated-water treatment resulted in the incorporation of nitrate/nitrite complexes onto the surface of the SiNCs and it affected the surface oxide layer, as evidenced by EDS and FTIR measurements. Primary focus was put on the blue PL emission band, which is present, but not dominant, in the unmodified oxidized SiNCs. Temporally and spectrally resolved PL measurements of this blue band analyzed in detail confirm that the blue PL emission is quenched after the intake of nitrogen into the SiNC. These results indicate that the exposure to a nitrogen-rich environment does not necessarily imply a switch to blue PL in all types of SiNCs, as was suggested by previous studies.

Conflicts of interest

There are no conflicts to declare.

Acknowledgements

The Czech Science Foundation funding, Grant No. 18-05552S (PG, TP, KK), the Operational Programme Research, Development and Education financed by European Structural and Investment Funds and the Czech Ministry of Education, Youth and Sports (Project No. SOLID21 CZ.02.1.01/0.0/0.0/16_019/0000760, KK), the CUCAM Centre of Excellence (OP VVV "Excellent Research Teams" project No. CZ.02.1.01/0.0/0.0/15_003/0000417, IM, IN) and NWO funding (FOM Project-ruimte No. 15PR3230, KDN) are gratefully acknowledged.

Notes and references

1 M. Dasog, Z. Yang, S. Regli, T. M. Atkins, A. Faramus, M. P. Singh, E. Muthuswamy, S. M. Kauzlarich, R. D. Tilley and J. G. C. Veinot, *ACS Nano*, 2013, **7**, 2676–2685.
2 B. Ghosh and N. Shirahata, *Sci. Technol. Adv. Mater.*, 2014, **15**, 014207.
3 B. F. P. McVey and R. D. Tilley, *Acc. Chem. Res.*, 2014, **47**, 3045–3051.
4 M. Dasog, J. Kehrle, B. Rieger and J. G. C. Veinot, *Angew. Chem., Int. Ed.*, 2016, **55**, 2322–2339.
5 K. Kůsová and K. Dohnalová, in *Silicon Nanomaterials Sourcebook: Low-Dimensional Structures, Quantum Dots, and Nanowires, Volume One*, ed. K. D. Sattler, Taylor&Francis Publisher, CRC Press, Boca Raton, Florida, USA, 2017, pp. 367–397.
6 M. V. Wolkin, J. Jorne, P. M. Fauchet, G. Allan and C. Delerue, *Phys. Rev. Lett.*, 1999, **82**, 197–200.
7 Z. Kang, Y. Liu, C. H. A. Tsang, D. D. D. Ma, X. Fan, N.-B. Wong and S.-T. Lee, *Adv. Mater.*, 2009, **21**, 661–664.
8 J. Choi, N. S. Wang and V. Reipa, *Langmuir*, 2007, **23**, 3388–3394.
9 C. M. Hessel, D. Reid, M. G. Panthani, M. R. Rasch, B. W. Goodfellow, J. Wei, H. Fujii, V. Akhavan and B. A. Korgel, *Chem. Mater.*, 2012, **24**, 393–401.
10 K. Kůsová, L. Ondič, E. Klimešová, K. Herynková, I. Pelant, S. Daniš, J. Valenta, M. Gallart, M. Ziegler, B. Hönerlage and P. Gilliot, *Appl. Phys. Lett.*, 2012, **101**, 143101.
11 J. R. Rodriguez Nunez, J. a. Kelly, E. J. Henderson, J. G. C. Veinot, J. R. Rodríguez Núñez, J. a. Kelly, E. J. Henderson and J. G. C. Veinot, *Chem. Mater.*, 2012, **24**, 346–352.
12 M. L. Mastronardi, F. Maier-Flaig, D. Faulkner, E. J. Henderson, C. Kübel, U. Lemmer and G. A. Ozin, *Nano Lett.*, 2012, **12**, 337–342.
13 S. P. Pujari, H. Driss, F. Bannani, B. van Lagen and H. Zuilhof, *Chem. Mater.*, 2018, **30**, 6503–6512.
14 F. Sangghaleh, I. Sychugov, Z. Yang, J. G. C. Veinot and J. Linnros, *ACS Nano*, 2015, **9**, 7097–7104.
15 M. Greben, P. Khoroshyy, X. Liu, X. Pi and J. Valenta, *J. Appl. Phys.*, 2017, **122**, 034304.
16 S. L. Brown, R. Krishnan, A. Elbaradei, J. Sivaguru, M. P. Sibi and E. K. Hobbie, *AIP Adv.*, 2017, **7**, 055314.
17 K. Dohnalová, L. Ondič, K. Kůsová, I. Pelant, J. L. Rehspringer and R.-R. Mafouana, *J. Appl. Phys.*, 2010, **107**, 053102.

18 K. Dohnalová, T. Gregorkiewicz and K. Kůsová, *J. Phys.: Condens. Matter*, 2014, **26**, 173201.
19 M. J. Sailor and E. J. Lee, *Adv. Mater.*, 1997, **9**, 783–793.
20 K. Dohnalová, K. Kůsová and I. Pelant, *Appl. Phys. Lett.*, 2009, **94**, 211903.
21 M. Dasog, G. B. D. los Reyes, L. V. Titova, F. A. Hegmann and J. G. C. Veinot, *ACS Nano*, 2014, **8**, 9636–9648.
22 T. Zhou, R. T. Anderson, H. Li, J. Bell, Y. Yang, B. P. Gorman, S. Pylypenko, M. T. Lusk and A. Sellinger, *Nano Lett.*, 2015, **15**, 3657–3663.
23 G. M. Carroll, R. Limpens and N. R. Neale, *Nano Lett.*, 2018, **18**, 3118–3124.
24 R. Anthony and U. Kortshagen, *Phys. Rev. B: Condens. Matter Mater. Phys.*, 2009, **80**, 115407.
25 J. M. Buriak, *Chem. Mater.*, 2014, **26**, 763–772.
26 D. Jurbergs, E. Rogojina, L. Mangolini and U. Kortshagen, *Appl. Phys. Lett.*, 2006, **88**, 233116–233118.
27 K. Dohnalová, K. Žídek, L. Ondič, K. Kůsová, O. Cibulka and I. Pelant, *J. Phys. D: Appl. Phys.*, 2009, **42**, 135102.
28 W. J. I. DeBenedetti, S.-K. Chiu, C. M. Radlinger, R. J. Ellison, B. A. Manhat, J. Z. Zhang, J. Shi and A. M. Goforth, *J. Phys. Chem. C*, 2015, **119**, 9595–9608.
29 L. Wang, Q. Li, H.-Y. Wang, J.-C. Huang, R. Zhang, Q.-D. Chen, H.-L. Xu, W. Han, Z.-Z. Shao and H.-B. Sun, *Light: Sci. Appl.*, 2015, **4**, e245.
30 J. Valenta, A. Fučíková, I. Pelant, K. Kůsová, K. Dohnalová, A. Aleknavičius, O. Cibulka, A. Fojtík and G. Kada, *New J. Phys.*, 2008, **10**, 073022.
31 W. D. A. M. de Boer, D. Timmerman, K. Dohnalova, I. N. Yassievich, H. Zhang, W. J. Buma and T. Gregorkiewicz, *Nat. Nanotechnol.*, 2010, **5**, 878–884.
32 L. Ondič, K. Kůsová, M. Ziegler, L. Fekete, V. Gärtnerová, V. Cháb, V. Holý, O. Cibulka, K. Herynková, M. Gallart, P. Gilliot, B. Hönerlage and I. Pelant, *Nanoscale*, 2014, **6**, 3837–3845.
33 A. A. Prokofiev, A. S. Moskalenko, I. N. Yassievich, W. D. A. M. de Boer, D. Timmerman, H. Zhang, W. J. Buma and T. Gregorkiewicz, *JETP Lett.*, 2009, **90**, 758–762.
34 B. Bruhn, B. J. Brenny, S. Dekker, İ. Doğan, P. Schall and K. Dohnalová, *Light: Sci. Appl.*, 2017, **6**, e17007.
35 K. Dohnalová, A. N. Poddubny, A. A. Prokofiev, W. D. A. M. de Boer, C. P. Umesh, J. M. J. Paulusse, H. Zuilhof and T. Gregorkiewicz, *Light: Sci. Appl.*, 2013, **2**, e47.
36 A. Fermi, M. Locritani, G. Di Carlo, M. Pizzotti, S. Caramori, Y. Yu, B. A. Korgel, G. Bergamini and P. Ceroni, *Faraday Discuss.*, 2015, **185**, 481–495.
37 G. Uchida, K. Yamamoto, M. Sato, Y. Kawashima, K. Nakahara, K. Kamataki, N. Itagaki, K. Koga and M. Shiratani, *Jpn. J. Appl. Phys.*, 2012, **51**, 01AD01.
38 H. Chen, J. H. Shin and P. M. Fauchet, *Appl. Phys. Lett.*, 2007, **91**, 173121.
39 K. Dohnalová, I. Pelant, K. Kůsová, P. Gilliot, M. Gallart, O. Crégut, J.-L. Rehspringer, B. Hönerlage, T. Ostatnický and S. Bakardjieva, *New J. Phys.*, 2008, **10**, 063014.
40 V. Zajac, H. Němec, C. Kadlec, K. Kůsová, I. Pelant and P. Kužel, *New J. Phys.*, 2014, **16**, 093013.
41 J. Khun, V. Scholtz, P. Hozák, P. Fitl and J. Julák, *Plasma Sources Sci. Technol.*, 2018, **27**, 065002.
42 J. Julák, A. Hujacová, V. Scholtz, J. Khun and K. Holada, *Plasma Phys. Rep.*, 2018, **44**, 125–136.

43 P. Hozák, V. Scholtz, J. Khun, D. Mertová, E. Vaňková and J. Julák, *Plasma Phys. Rep.*, 2018, **44**, 799–804.
44 K. Kůsová and T. Popelář, *J. Appl. Phys.*, 2019, **125**, 193103.
45 *Homepage of the CP2K open source package for molecular dynamics simulations*, http://www.cp2k.org.
46 J. P. Perdew, K. Burke and M. Ernzerhof, *Phys. Rev. Lett.*, 1996, **77**, 3865–3868.
47 S. Goedecker, M. Teter and J. Hutter, *Phys. Rev. B: Condens. Matter Mater. Phys.*, 1996, **54**, 1703–1710.
48 J. P. Perdew, *Int. J. Quantum Chem.*, 1985, **28**, 497–523.
49 G. Viera, M. Mikikian, E. Bertran, P. R. i. Cabarrocas and L. Boufendi, *J. Appl. Phys.*, 2002, **92**, 4684–4694.
50 C. T. Kirk, *Phys. Rev. B: Condens. Matter Mater. Phys.*, 1988, **38**, 1255–1273.
51 S. Gardelis, A. G. Nassiopoulou, M. Mahdouani, R. Bourguiga and S. Jaziri, *Phys. E*, 2009, **41**, 986–989.
52 D. Mawhinney, J. J. A. Glass and J. J. T. Yates, *J. Phys. Chem. B*, 1997, **101**, 1202–1206.
53 A. Burneau and C. Carterer, *Phys. Chem. Chem. Phys.*, 2000, **2**, 3217–3226.
54 K. Nakamoto, *Infrared and Raman Spectra of Inorganic and Coordination Compounds, Part B: Applications in Coordination, Organometallic, and Bioinorganic Chemistry*, John Wiley & Sons, Inc., New York, 1997.
55 S. G. Moussa, A. C. Stern, J. D. Raff, C. W. Dilbeck, D. J. Tobias and B. J. Finlayson-Pitts, *Phys. Chem. Chem. Phys.*, 2013, **15**, 448–458.
56 D. König, *AIP Adv.*, 2016, **6**, 085306.
57 The lifetime value of $\tau_2 = 1.2$ ns was obtained by performing a fit over all the wavelengths, computing a weighted mean of the fitted lifetimes and fixing the lifetime to this mean value in the next run.
58 As a result of the increased absorption of PAW under 340 nm and its consequent luminescence, see Fig. S6,† the PL of the modified sample under 315 nm excitation could not be accurately analyzed and separated into individual spectral components.

Faraday Discussions

PAPER

Tight-binding calculations of the optical properties of Si nanocrystals in a SiO$_2$ matrix

Mikhail O. Nestoklon,[a] Ivan D. Avdeev,[a] Alexey V. Belolipetskiy,[a] Ilya Sychugov,[b] Federico Pevere,[b] Jan Linnros[b] and Irina N. Yassievich[a]

Received 22nd September 2019, Accepted 25th October 2019
DOI: 10.1039/c9fd00090a

We develop an empirical tight binding approach for the modeling of the electronic states and optical properties of Si nanocrystals embedded in a SiO$_2$ matrix. To simulate the wide band gap SiO$_2$ matrix we use the virtual crystal approximation. The tight-binding parameters of the material with the diamond crystal lattice are fitted to the band structure of β-cristobalite. This model of the SiO$_2$ matrix allows us to reproduce the band structure of real Si nanocrystals embedded in a SiO$_2$ matrix. In this model, we compute the absorption spectra of the system. The calculations are in an excellent agreement with experimental data. We find that an important part of the high-energy absorption is defined by the spatially indirect, but direct in k-space transitions between holes inside the nanocrystal and electrons in the matrix.

1 Introduction

While silicon is most widely used in microelectronics materials, its use in optoelectronics is limited by the fundamental feature of an indirect band structure: the minima of the conduction band are near the edge of the Brillouin zone, and the top of the valence band is at the center of this zone. However, in nanocrystals (NCs), electrons and holes are quantized and no longer have a definite (quasi) momentum due to the Heisenberg uncertainty principle. As a result, the optical transitions are partially allowed. This has stimulated the development of technology, and experimental and theoretical work on the study of the optical phenomena in silicon nanoparticles. There are a large number of reviews and monographs devoted to the successful application of silicon nanocrystals in SiO$_2$ matrices in optoelectronics, photovoltaics and medicine.[1-4]

The simulation of Si NCs in SiO$_2$ in atomistic methods is complicated by the fact that SiO$_2$ is an amorphous material. There is a large set of different

[a] Ioffe Institute, Politekhnicheskaya 26, St.-Petersburg, 194021, Russia
[b] Department of Applied Physics, KTH-Royal Institute of Technology, Kista, 16440, Stockholm, Sweden

polymorphic modifications of crystalline SiO_2 with different band structures and the exact structure (possibly crystalline) of SiO_2 in the vicinity of the Si/SiO_2 interface is unknown. Recently, an empirical tight binding (ETB) approach[5] for the modeling of electronic states and energy levels of Si and SiGe nanocrystals embedded in a SiO_2 matrix was developed in Belolipetskiy et al.[6] The local density of states (LDOS) shows that excited electronic states with relatively small energy may strongly penetrate the SiO_2 matrix. The results of single-dot absorption spectroscopy, published by Sychugov et al.,[7] revealed the absorption states above the emission level. Thus, a way to directly compare experiment and theory has been provided which allows for the correction of the offset for electron and valence bands in SiO_2 matrices, too.

Ensembles of Si NCs typically exhibit a featureless absorption curve. In contrast, it was shown that individual nanoparticle spectra reveal subbands, visible as steps in the photoluminescence excitation experiment.[7] So for a better comparison with theory here we use data for single quantum dots, which are free from the ensemble inhomogeneities (see Sychugov et al.[8]).

In the present paper we compute the absorption cross-section of the NCs as a function of the incident light energy. By changing the conduction band offset between Si and SiO_2 we were able to obtain excellent agreement between the theoretical and experimental data on single dot absorption. The calculations show that the rise of the cross-section coefficient after ∼2.5 eV is associated with spatially indirect transitions between holes confined in the NC and electrons in the SiO_2 matrix. We show that the high energy feature in the absorption spectra around 3 eV is associated with strong mixing of the states inside the NC and in the matrix. The importance of the transitions in the matrix also makes the dielectric screening more important: the local field factor changes not only the amplitude of the transitions, but also the shape of the spectrum. In our calculations, we considered only the phononless transitions and did not take the processes of energy relaxation into account. We expect that with phonon assisted processes taken into account, the change of the absorption coefficient would be noticeable only for the transitions with energies close to the indirect gap.

2 Calculation methods

There are many possible approaches for the description of the electronic and optical properties of nanostructures. They may be separated into three "classes": (i) continuous empirical methods (effective mass and different versions of the $\mathbf{k} \cdot \mathbf{p}$ method[9,10]); (ii) density functional theory[2,11] (DFT) and (iii) empirical atomistic methods.[12–15] Each class has its advantages and disadvantages. Before the end of the 1990s only continuous empirical methods were acceptable for the calculations due to the limited computer power available. However, there was always a high demand for different atomistic methods because there is a large set of phenomena which have no parameter free unambiguous description.[16,17] They are mostly associated with symmetry reduction at the interfaces, both rotational[18,19] and translational.[20,21] Reduced symmetry leads to the mixing of states which cannot mix in bulk semiconductors, with different spin projection,[19] valley index[20] or both.[21,22]

A tremendous rise in computer speed at the beginning of the 21st century allowed for more complex, atomistic methods. However, even until now the use of

DFT calculations for semiconductors is quite limited. The standard DFT theory (with local LDA[23] and GGA[24] exchange–correlation potentials) introduces regular error which leads to significant (up to 1 eV) underestimation of the band gaps of typical semiconductors.[25] The use of hybrid potentials[25] or GW corrections[26] cures this problem, but at the cost of the dramatic increase of computational power needed for the calculations.

The compromise between accuracy and computational complexity may be resolved by using atomistic empirical methods which "fill the gap" between DFT calculations and continuous empirical methods. In the following text we will concentrate on the tight-binding method; similar arguments and concepts may be applied also to empirical pseudopotential method. The tight-binding method was first considered as one of the bases for *ab initio* calculations,[27] but very quickly it was realized that it does not give enough flexibility to allow for true *ab initio* calculations, being still too computationally expensive to be used as an empirical method. The first revival was in the mid 90s, when it was shown[5] that using just about 30 empirical parameters with a rather deterministic parametrisation scheme allows for a quantitatively accurate (within meV) description of the band structure of all bulk III–V materials. Moreover, it has been shown that the scheme gives transferable parameters which allow for accurate description of the heterointerfaces and deformation potentials. Recently, there was another big leap for the tight-binding method. A few groups have proposed schemes (conceptually equivalent, but different in technical details) which allow for the deterministic extraction of tight-binding parameters from DFT calculations.[28–30] This approach does not only allow for accurate description of the band structure. The resulting ETB scheme is the interpolation of the underlying DFT model with full correspondence between the ETB and DFT results. The most important consequence is the possibility to "reconstruct" the DFT wave functions from the empirical tight-binding calculations. As a result, it is possible to obtain results with the precision of the best DFT model for a computational cost comparable to that of regular empirical tight-binding calculations.

2.1 Tight-binding method

In the tight-binding method, the wave function of the quasiparticles is written in the basis of the orthogonal orbitals[31] ϕ_α, which are assumed to be localized near atoms.

$$\Phi(\mathbf{r}) = \sum_{n\alpha} C_{n\alpha} \phi_\alpha(\mathbf{r} - \mathbf{r}_n) \qquad (1)$$

Here $\mathbf{r}_n$ is the position of the n-th atom in the structure and α runs through the set of different orbitals at the atom. We use the standard empirical tight-binding theory in the sp^{3}d^{5}s* variant of Jancu et al.,[5] where index α runs through 20 orbitals (10 if spin is neglected) at each atom. In this basis, the Hamiltonian is a sparse matrix $H_{n\alpha,n'\beta}$ and the energies and wave functions of the states are found from the eigenproblem

$$\sum_{n'\beta} H_{n\alpha,n'\beta} C^j_{n'\beta} = E_j C^j_{n\alpha}. \qquad (2)$$

Modern iterative methods allow[32,33] a few solutions of the eigenproblem for large sparse matrices to be found very efficiently.

The matrix elements of the tight-binding Hamiltonian are

$$H_{n\alpha,n'\beta} = \begin{cases} E_\alpha^{t_n}, & n = n' \\ V_{\alpha\beta}^{t_n t_{n'}}(\mathbf{r}_{n'} - \mathbf{r}_n), & n \text{ and } n' \text{ are neighbors} \end{cases} \quad (3)$$

Here t_n is the chemical type of the n-th atom, $E_\alpha^{t_n}$ are diagonal energies and $V_{\alpha\beta}^{t_n t_{n'}}(\mathbf{r}_{n'} - \mathbf{r}_n)$ are transfer matrix elements which are constructed using the standard procedure of Slater and Koster[27] under the assumption that the basis functions ϕ_α transform under the action of the elements of the group O(3) (3D rotations plus inversion) as the atomic functions with the same orbital momentum,† the matrix elements $V_{\alpha\beta}^{tt'}(\mathbf{r}_{nn'})$ are decomposed into parts which also have definite angular momentum projection on the chemical bond vector $\mathbf{r}_{nn'} = \mathbf{r}_{n'} - \mathbf{r}_n$. This results in simple rules for constructing the matrix elements of the tight-binding Hamiltonian. As an example:

$$V_{p_x p_x}(\mathbf{r}_{nn'}) = V_{pp\sigma}(|\mathbf{r}_{nn'}|)\cos^2\Theta_x + V_{pp\pi}(|\mathbf{r}_{nn'}|)\sin^2\Theta_x \quad (4)$$

where Θ_x is the angle between the chemical bond vector $\mathbf{r}_{nn'}$ and the x axis. The parameters $V_{pp\sigma}$, $V_{pp\pi}$, etc. depend only on the chemical bond length and usually they are parametrized from the best fit of the bulk semiconductor band structure to the *ab initio* calculations and/or experimental data.

The spin–orbit splitting was added following the procedure of Chadi.[34] The idea is, similar to the $\mathbf{k}\cdot\mathbf{p}$ method, to put the spin–orbit splitting on the p-orbitals. In the spirit of the tight-binding method and based on the results of *ab initio* calculations, the spin–orbit splitting has only same-atom components. Then, using symmetry arguments, it is easy to derive the spin–orbit part of the tight-binding Hamiltonian:

$$H^{SO}_{np_\nu\xi, n'p_\eta\zeta} = -i\lambda^{SO}_{t_n}\delta_{nn'}\varepsilon_{\nu\eta\mu}\sigma^\mu_{\xi\zeta}. \quad (5)$$

Here $\delta_{nn'}$ is the Kronecker symbol and $\varepsilon_{\nu\eta\mu}$ is the Levi-Civita symbol. The parameter λ^{SO} depends only on the chemical type of the atom. For the quantitative analysis of the spin–orbit splitting in the tight-binding method refer to Jancu et al.[35]

For the Si atoms we used tight-binding parameters from Niquet et al.[36] Note that already the original parametrization of Jancu et al.[5] gives good quantitative agreement with the $\mathbf{k}\cdot\mathbf{p}$ parameters of bulk silicon such as the electron energies in the special points of the Brillouin zone, masses, *etc.* Also, there is no large improvement in the $\mathbf{k}\cdot\mathbf{p}$ parameters in the use of the ETB parameters of Niquet et al.[36] However, in the case of large quantum confinement energies we find that the two methods give noticeably different results.[37] This is related to the fact that in more recent parametrization[36] the energies of bands within the full Brillouin zone are included in the fitting procedure. As a result, when the quantum confinement is small and in quantitative agreement with the $\mathbf{k}\cdot\mathbf{p}$ method, two sets of parameters give very similar results. Meanwhile, when the quantum confinement is large, the set of parameters from Niquet et al.[36] is preferable over

---

† Note that in ETB the basis functions are never assumed to be real atomic functions.

the original Jancu et al.[5] parameters even if the strain is not included in the calculations.

2.2 Description of the SiO₂ matrix

The simulation of Si nanocrystals in SiO$_2$ by the tight-binding method is complicated by the fact that SiO$_2$ is an amorphous material. However, at the interface between SiO$_2$ and the NCs, there is a large band offset for both types of carriers. As a result, the electron and hole wave functions rapidly damp in the matrix. In this case, the role of disorder may be neglected and the most important factors are the general band structure of the material surrounding the nanocrystal and the boundary conditions between Si and SiO$_2$. This allows one to simulate SiO$_2$ as a virtual crystal.[6,37,38]

The virtual crystal approximation (VCA) is a standard model used for envelope function approaches for the description of alloys. When transferred to empirical atomistic methods, VCA is an ideal trade-off between atomistic resolution and computational complexity: instead of full atomistic calculation of an alloy, a tight-binding representation of a virtual binary material mimicking the alloy band structure may be used in heterostructure modeling. Despite the fact that in reality the random alloy may have a complicated strain distribution and cannot be easily mapped to a virtual crystal, the band structure of the alloy may be accurately described by the VCA even with the same set of ETB parameters.[39] Nestoklon et al.[38] proposed using the virtual crystal approximation to describe the amorphous SiO$_2$ and the results were proven to be more than acceptable for the description of low energy electron and hole states in Si and SiGe NCs.[6,37] In the current calculations, we extend this idea to the states which lie above the bottom of the SiO$_2$ conduction band.

To simulate the wide band gap SiO$_2$ matrix we construct a virtual crystal with a band structure close to the band structure of β-cristobalite near the band gap edges, following Belolipetskiy et al.[6] The β-cristobalite is the only polymorphous modification with a cubic lattice among SiO$_2$ crystals. We suppose that during the growth process, the SiO$_2$ matrix in the vicinity of Si NCs forms a structure similar to that of β-cristobalite.‡ The tight-binding parameters were fitted to the band structure of β-cristobalite calculated using density functional theory; see Belolipetskiy et al.[6] for the details. There is no strain from having NCs in an amorphous matrix and we set the lattice constant of the virtual crystal to match the lattice constant of bulk Si.

Fig. 1 demonstrates the energy position of the band edges of bulk silicon and the virtual crystal SiO$_2$. The virtual crystal "SiO$_2$" is a direct band gap material with a band gap equal to 7.38 eV matching the value in β-cristobalite; the extrema of the conduction and valence band lie in the Γ point. Bulk silicon is an indirect band gap semiconductor, with the 6 minima of the conduction band located between the Γ and X points of the Brillouin zone (the distance in k-space between the minima and the Γ point is 0.85 of the Γ–X distance). To consider the tunneling of electrons into the SiO$_2$ matrix, it is critical to reproduce exactly the

---

‡ Note that the validity of this assumption may depend on the technology used for the preparation of the NCs. If the NCs are formed at relatively low temperature, the structure of the SiO$_2$ near the NC surface may be different.

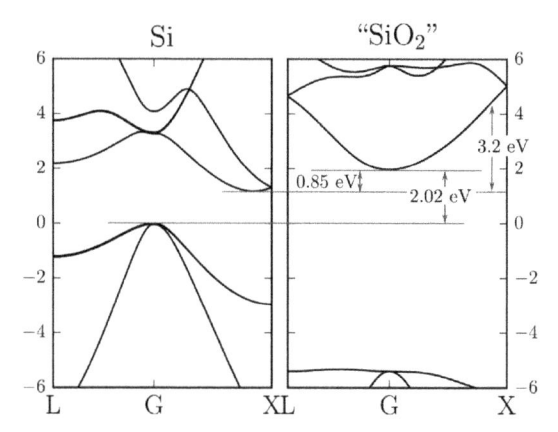

Fig. 1 Band structure of Si (left panel) and a virtual crystal which represents the SiO₂ wide band gap matrix (right panel).

energy positions of the edges of the conduction bands of SiO$_2$ in the X and Γ points. The distances from the top of the valence band at the point Γ to the points X and L of the conduction band for the virtual crystal are 10.42 eV and 10.02 eV, respectively, which are close to the values for β-cristobalite.

The important parameter which is hard to extract both from experimental data and from DFT calculations is the energy difference between the top of the valence band of bulk silicon and the top of the valence band in virtual crystal SiO$_2$. In a study by Belolipetskiy et al.[6] this parameter was set to 4.5 eV. However, we find that the best fit for the absorption spectra is at the value 5.36 eV (see below). It is more convenient to use not the energy difference between the valence band top, but the difference between conduction band minima in the two materials E_{CBO}. In the Belolipetskiy et al. study[6] the distance between the minimum of the conduction band in bulk Si to the bottom of the conduction band of VCA SiO$_2$ in the Γ point $E_{CBO} = 1.71$ eV. In the current work, we use the value $E_{CBO} = 0.85$ eV, with the position of the conduction band bottom in SiO$_2$ at 2.02 eV above the valence band top in Si which is set to "0". While the value E_{CBO} is much smaller than the widely accepted value for the difference between the bottom of the conduction bands of Si and SiO$_2$ of 3.2 eV,[40] we argue that the standard value should be associated not with the difference between the bottoms of the conduction bands (which are located at different points of the Brillouin zone in two materials), but to the difference in energy near the X point in both materials. With our definition, the difference between the bottom of the Si conduction band and the energy in the X point of SiO$_2$ is 3.89 eV and the difference between the energies at 0.85 of the Γ–X distance in the two materials is very close to 3.2 eV. The band structure of both materials is illustrated in Fig. 1.

3 Electron and hole states

For convenience, to compute electron and hole states in Si nanocrystals embedded in the SiO$_2$ matrix, we consider a cubic supercell of the virtual SiO$_2$ with a Si nanocrystal in the supercell center. In calculations, we use the periodic boundary conditions to discard the effects at the SiO$_2$ surface and choose the supercell to be large enough to neglect the tunneling between neighboring NCs. A NC with diameter D is constructed by placing the Si atoms inside a sphere with diameter $D + 0.5$ nm. An extra 0.5 nm are added to compensate for the interface effects, similar to the approach by Seino et al.[41] Most of the calculations presented below are done for a NC with diameter $D = 3$ nm, so with 1099 Si atoms in the 5 × 5 × 5 nm matrix of SiO_2, there are 4733 "atoms" of virtual crystal surrounding the NC.

Belolipetskiy et al.[6] demonstrated that the model of a VCA SiO_2 matrix reproduces the band gaps of Si NCs embedded in a SiO_2 matrix. However, the role of the E_{CBO} which we use here to tune the absorption was not studied. In Fig. 2 we show the effective band gap (Si band gap plus electron and hole energies of quantum confinement) of the Si NCs as a function of diameter for three values of E_{CBO}. The results show that the energy of the ground levels of the both electron and hole are not sensitive to this value.

In Fig. 3 we show the LDOS of some electron states. It was clearly seen that some of the states with energies above 2 eV are delocalized: they have almost no LDOS inside the NC. The ground state (Fig. 3a) and all states below the bottom of the conduction band of SiO_2 are localized inside the NC. For states with energies above the bottom of the SiO_2 conduction band, some of the states are located outside the NC (Fig. 3b). Note that as long as the bottom of the conduction band is formed by the states in different valleys, the states "inside" and "outside" the NC almost do not mix until the energy is large enough. The states localized in the NCs exponentially decay to the matrix (Fig. 3c) and the states localized in the matrix exponentially decay inside the NC. Analysis of the LDOS in k-space[38] confirms that the states localized in the matrix are formed by the electrons from the Γ valley while the states localized in the NC are mostly from the X valley. However, when the energy is large enough, a new type of state which has large LDOS both inside and outside the NC occurs (Fig. 3d).

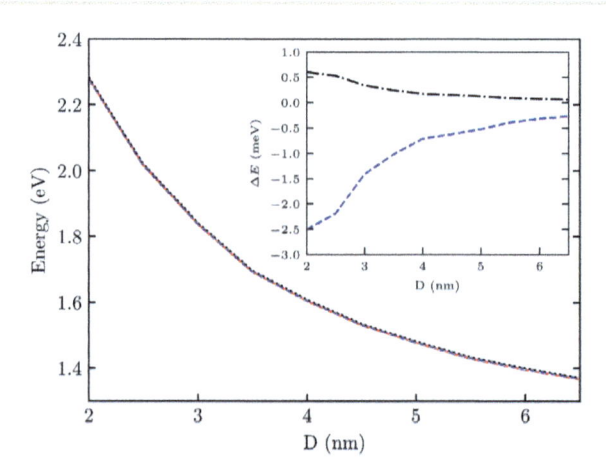

Fig. 2 Effective band gap of Si NCs as a function of NC diameter. The three lines show calculations with three different values of E_{CBO}: $E_{CBO} = 0.85$ eV (solid), $E_{CBO} = 1.01$ eV (dashed) and $E_{CBO} = 0.81$ eV (dotted). In the inset we show the difference between the effective band gaps of the Si NCs. The difference between NCs calculated using $E_{CBO} = 0.85$ eV and $E_{CBO} = 1.01$ eV is shown by the dashed line and the difference between $E_{CBO} = 0.85$ eV and $E_{CBO} = 0.81$ eV by the dashed-dotted line.

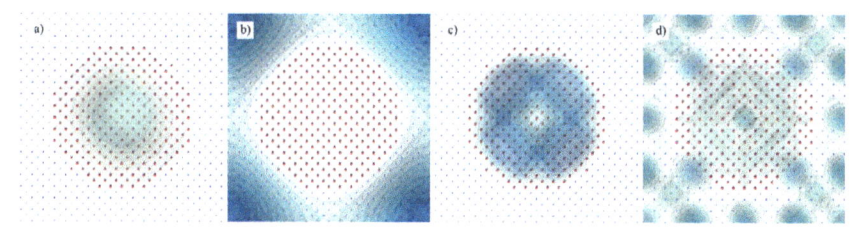

Fig. 3 Local density of states (LDOS) for few typical electron states in a Si NC embedded in the SiO$_2$ matrix. Panels (a)–(d) show the parallel projection of the 3D view[42] of atoms in the supercell used in the calculations and the level surfaces of LDOS to the (010) plane. We show the states with energies (a) 1.502 eV, (b) 2.044 eV, (c) 2.124 eV and (d) 2.836 eV. These states are (a) the ground electron state (b) the first state which is localized in the matrix (c) one of the states in the NC in the energy range where both electron states in the matrix and in the NC are allowed and (d) one of the first states with large LDOS both inside and outside the NC. The red dots show the positions of the Si atoms and the blue dots show the "atoms" of the virtual crystal.

Note that in our calculations we do not take exciton formation into account. In real structures, the states in the matrix will be weakly bound to the NC by the Coulomb interaction with the holes in the NC and localized within the vicinity of the NC surface at a distance of about an exciton Bohr radius a_B.

4 Absorption

Now we proceed to the analysis of the absorption coefficient as a function of the energy transition for the 3 nm Si NCs. Due to dielectric contrast between the SiO$_2$ matrix and the Si inside NC, the field inside the NC is smaller than the external field by the local field factor[43] $\mathbf{E}_\text{in} = \mathscr{F}\mathbf{E}_\text{out}$ which is given by:

$$\mathscr{F} = \frac{3n_\text{out}^2}{2n_\text{out}^2 + n_\text{in}^2}. \tag{6}$$

Here n_out is the refractive index of SiO$_2$ and n_in is the Si refractive index. For Si in SiO$_2$ $\mathscr{F} = 0.42$. We extend the absorption calculation procedure (for more details refer to Delerue and Lannoo[43]) to account for the transitions involving states both inside and outside nanocrystal. When the transition takes place both inside the NC and in the matrix, we should treat the matrix element of the product of velocity and electric field carefully. The transition amplitude is proportional to the matrix elements of the velocity operator multiplied by the electric field which may be written as:

$$\begin{aligned}\langle i|\mathbf{v}\cdot\mathbf{E}|j\rangle &= \int \psi_i^*(\mathbf{r})\mathbf{v}\cdot\mathbf{E}\,\psi_j(\mathbf{r}) \\ &= \int_{\mathbf{r}\in\text{NC}} \psi_i^*(\mathbf{r})\mathbf{v}\cdot\mathbf{E}_\text{in}\psi_j(\mathbf{r}) + \int_{\mathbf{r}\notin\text{NC}} \psi_i^*(\mathbf{r})\mathbf{v}\cdot\mathbf{E}_\text{out}\psi_j(\mathbf{r}) \\ &= \mathscr{F}\int_{\mathbf{r}\in\text{NC}} \psi_i^*(\mathbf{r})\mathbf{v}\cdot\mathbf{E}_\text{out}\psi_j(\mathbf{r}) + \int_{\mathbf{r}\notin\text{NC}} \psi_i^*(\mathbf{r})\mathbf{v}\cdot\mathbf{E}_\text{out}\psi_j(\mathbf{r}),\end{aligned} \tag{7}$$

where $\mathbf{v}$ is the velocity operator, $\mathbf{E}$ is the electric field of the light which is absorbed and $\psi_i(\mathbf{r})$ are the wave functions of the states. The procedure for absorption calculation[43,44] then may be extended through taking (7) into account. The absorption coefficient averaged over the polarizations is given by:

$$\sigma(\hbar\omega) = \frac{4\pi^2\hbar^2\alpha}{3n_{\text{out}}} \sum_{ij} \frac{\mu_{ij}}{E_i - E_j} \delta(\hbar\omega - E_i + E_j), \qquad (8)$$

where i and j enumerate states in the conduction and valence bands, E_i and E_j are the energies of these states and $\alpha \simeq 1/137$ is the fine structure constant. The local transition probability μ_{ij} is given by:

$$\mu_{ij} = \sum_{\lambda} |\mathcal{F}\langle i|v_\lambda|j\rangle_{\in \text{NC}} + \langle i|v_\lambda|j\rangle_{\notin \text{NC}}|^2. \qquad (9)$$

To compute the absorption in the ETB, we follow a standard procedure.[45] The velocity operator **v** can be defined as a commutator of the tight-binding Hamiltonian with the coordinate operator,

$$\mathbf{v} = \frac{i}{\hbar}[H, \mathbf{r}]. \qquad (10)$$

For the coordinate operator we use the diagonal approximation[46–48]

$$\mathbf{r} = \delta_{nn'}\delta_{\alpha\beta}\mathbf{r}_n \qquad (11)$$

where $\mathbf{r}_n$ is the coordinate of the n-th atom. This corresponds to neglecting the intraatomic matrix elements of the velocity operator for the optical transitions, which is reasonable for covalent semiconductors.[49]

Then, the velocity operator in the tight-binding basis is

$$\mathbf{v}_{nn'} = \frac{i}{\hbar}\mathbf{r}_{nn'} H_{n\alpha n'\beta}, \qquad (12)$$

where we define the vector of the chemical bond connecting neighbor atoms $\mathbf{r}_{nn'} = \mathbf{r}_{n'} - \mathbf{r}_n$.

As a result, in the tight-binding the local transition probability μ_{ij} is calculated as:

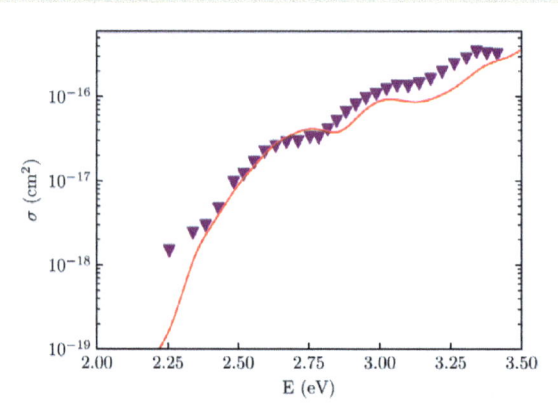

Fig. 4 Phononless absorption calculated in the tight-binding as a function of the incident photon energy. We also show the intensity of the absorption measured experimentally in symbols.

$$\mu_{ij} = \frac{i}{\hbar} \sum_{\lambda} |\sum_{n\alpha n'\beta} C^{i*}_{n\alpha} F_n r^{\lambda}_{nn'} H_{n\alpha n'\beta} C^{j}_{n'\beta}|^2, \qquad (13)$$

where we introduce the coordinate-dependent local field factor

$$F_n = \begin{cases} \mathcal{F}, & \mathbf{r}_n \in \mathrm{NC} \\ 1, & \mathbf{r}_n \notin \mathrm{NC} \end{cases} \qquad (14)$$

The above scheme (8–14) is rather standard, but in most cases the dependence of the local field factor on the coordinate is neglected. This is based on the assumption that the transitions occur only inside the NCs. In the case of Si NCs, the situation is rather non-standard if one wishes to take into account all the transitions. Transitions inside the NCs are strongly suppressed because they are indirect in k-space. The transitions between holes localized in the NCs and electrons in the matrix are direct in k-space, but indirect in real space, and as a result there are no solid qualitative arguments on their strength compared with that of the transitions between the ground states. We will discuss this in more detail in the next section.

5 Results

We used the scheme discussed in previous sections to compute the absorption coefficient for a NC with a 3 nm diameter. We found the best fit of the absorption spectrum to the experimental data for $E_{\mathrm{CBO}} = 0.85$ eV. The results are shown in Fig. 4 by the solid line. Without the consideration of phonon assisted transitions the absorption coefficient below ∼2.3 eV is negligible. We expect that with phonon assisted processes taken into account, the change of the absorption coefficient will be noticeable in this region.[50] For larger energies, the phononless transitions dominate.

For comparison, we also show the experimental absorption curve[7] for a single Si nanodot emitting at 1.86 eV. It was obtained from the photoluminescence excitation spectrum taken at room temperature. From Fig. 4 it can be seen that the overall shape of the calculated curve reproduces the experimental data surprisingly well. Note that the consideration of transitions with the electron states in the matrix is rather important, along with that of the local field factor (14) (see below).

Fig. 4 is calculated for a particular NC size and value of E_{CBO}. We also analyzed the dependence of the absorption spectrum on the NC size and the value of E_{CBO}. In Fig. 5 we show the spectra calculated for different NC sizes. The change in the effective band gap which is in qualitative agreement with Fig. 2 is clearly seen. However, the change in the high energy features with NC diameter is non-trivial. There was a discussion about the possibility of interpreting these features within the $\mathbf{k}\cdot\mathbf{p}$ model[51–53] without the definitive conclusion. We only comment on the amplitude of the absorption. One could expect that the absorption is higher for NCs with a smaller diameter as the optically indirect transitions are due to the quantum confinement which relaxes the momentum conservation law. However, we do not see a strong dependence in this diameter range and the amplitude of the optical transitions is more or less the same. Probably, this is due to the

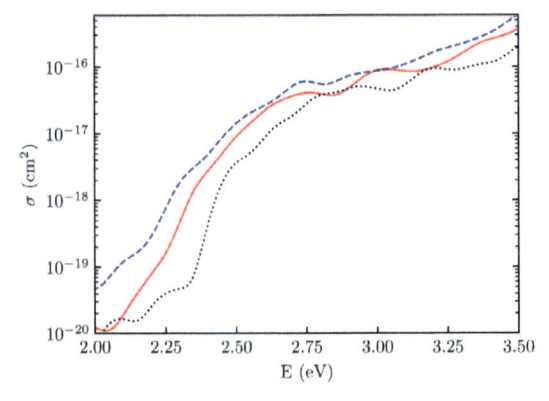

Fig. 5 Phononless absorption calculated in the tight-binding as a function of the incident photon energy for NCs with $D = 3$ nm (solid), 3.5 nm (dashed) and 2.5 nm (dotted).

compensation of the above mentioned effect by the larger density of states in larger NCs.

5.1 Role of the band offset and local field factor

We mentioned above that the shape of the absorption cross-section as a function of the energy may be tuned by changing the energy shift between the top of the valence bands of Si and the SiO_2 matrix E_{CBO}. As long as the change in this value did not change the effective band gap of the nanocrystal (see Fig. 2), we had the liberty to change it to fit the resulting absorption cross-section to the experimental data (see Fig. 4). To demonstrate the sensitivity of the absorption cross-section to E_{CBO}, in Fig. 6 we show the spectrum calculated for $E_{CBO} = 0.85$, 1.01 and 0.81 eV. The dependence of the spectrum on this parameter is rather interesting. As the band gap is almost unaffected by the variation of this parameter (see Fig. 2), the low energy transitions below ~2.25 eV have almost the same cross-section. However, the cross-section of the higher energy transitions is rather sensitive to this parameter. This reflects the fact that these transitions involve the

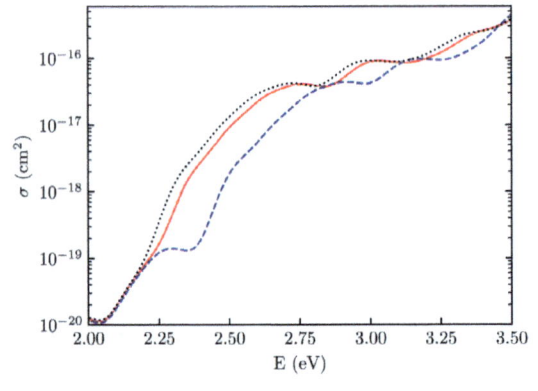

Fig. 6 Phononless absorption calculated in the tight-binding as a function of the incident photon energy for NCs with $E_{CBO} = 0.85$ eV (solid), 1.01 eV (dashed) and 0.81 eV (dotted).

electron states located in the matrix. As a result, the energies of these transitions (and the corresponding features in the absorption spectrum) shift together with E_{CBO}. One more conclusion may be drawn from the results in Fig. 6. The position of the higher energy peak in the vicinity of 3 eV is also "pinned" to the value of E_{CBO} even though the next feature of the band structure of either Si or SiO$_2$ is the position of the Γ valley in bulk Si which is located at significantly higher energy. From the LDOS of the wave functions with the electron energy around this peak position (see Fig. 3d) we conclude that this peak is associated with the electron states where electrons inside the NC and in the matrix are intermixed. As long as the minima of the conduction band and in the matrix are located in different points of the Brillouin zone, even if the states are located in the same energy range their mixing is highly suppressed by the momentum conservation. When the quantum confinement energy is large enough, the distribution of the electron in k-space is more and more spread through the Brillouin zone and the wave functions of the electrons and in the matrix start to overlap in k-space allowing for the direct admixture of electrons in the NC and in the matrix.

The role of the local field factor (14) is also rather pronounced. Note that in most cases, the local field factor is neglected or included as a prefactor, which is motivated by the fact that the optical transitions occur inside the NC. In our case, there are two main types of transitions: (i) direct in r-space and indirect in k-space transitions between holes and electrons both localized inside the NC and (ii) indirect in r-space and direct in k-space transitions between holes inside the NC and electrons in the matrix. The former are fully screened by the local field factor which gives the factor $\mathscr{F}^2$, while the latter are partially screened which results (roughly) in the factor $\mathscr{F}$. To show the role of the local field factor in Fig. 7 we show the absorption calculated (i) without a local field factor, (ii) with a constant local field factor and (iii) with a coordinate-dependent local field factor (14). The difference between the absorption calculated with constant and coordinate dependent local field factors shows the role of the optical transitions which involve the electron states in the matrix. For the optical transitions within the nanocrystal there is no difference. This can be shown in Fig. 7 for the transitions with energies below 2.25 eV. Starting from this energy, which coincides with the

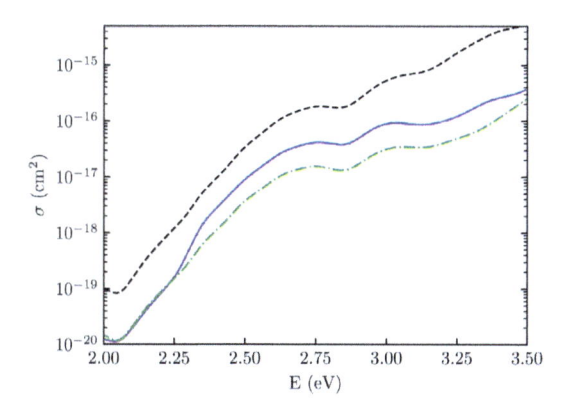

Fig. 7 Phononless absorption calculated with a coordinate-dependent (14) (solid) and constant (dash-dot) local field factor. We also show the "bare" absorption calculated without a local field factor (dashed).

bottom of the conduction band in the SiO$_2$ matrix we see a significant increase in the absorption coefficient. This is due to the fact that the local field factor significantly damps the amplitude of the absorption coefficient by dielectric screening. Note that the coordinate-dependent local field factor does not simply make the transitions with the electron states partially in the matrix more bright. It actually takes into account whether the overlap of electron and hole inside or outside the NC is involved in each particular optical transition.

6 Discussion

Fig. 4–7 show that in the calculations of the absorption of Si NCs in a SiO$_2$ matrix the role of the matrix is critically important. Indeed, in the analysis of the low energy transitions the role of the matrix may be neglected and this result is confirmed in our calculations. However, the position of the conduction band in the matrix means that the optical transitions with energies above ∼2.25–2.5 eV are rather sensitive to the matrix and accurate quantitative modeling of the matrix is absolutely necessary to describe the absorption spectra. Our findings show that the above mentioned discussion[51–53] of the nature of high energy transitions may be incomplete; the role of the matrix is not only in the boundary conditions for the electron states, but the optical transitions actually take place in the matrix. Two features in high energy absorption/photoluminescence may be associated with the states in the matrix. The first is the energy where electron states in the matrix are allowed. The second is the state where mixing of the states in the NC and in the matrix is allowed. Our findings show the importance of a proper model of the matrix for the accurate description of the optical properties of Si NCs.

While a virtual crystal approximation for the SiO$_2$ is rather effective for description of the absorption, our results show that it is highly desirable to consider a more accurate model of the glassy matrix in the atomistic method. We believe that our results are qualitatively correct, but for a quantitative analysis one should also consider the role of the disorder and defects in the matrix. As long as the optical transitions involving electron states in the matrix are important, our neglect of these effects which was motivated by the fact that the absorption is defined by the transitions in the NCs is no longer valid.

Another point which needs further analysis is the importance of the Coulomb interaction between the carriers. Its neglect with Si NCs is usually motivated by the small value of Coulomb interaction of the carriers inside the NC compared to the quantum quantization energy. For the electron states which are mostly localized in the matrix this assumption also should be reconsidered: the localization of the electrons in the matrix is mostly due to the Coulomb interaction with positively charged holes inside the NC. We mimic the effect of this interaction by considering a finite size matrix, but for quantitative analysis it is highly desirable to take the Coulomb interaction into consideration explicitly.

In the indirect transitions the role of the phonons is sometimes rather important. Our calculations were motivated by the fact that in the NCs, despite the indirect band structure of Si, phononless optical transitions are partially allowed. Still, the accurate consideration of the optical transitions with the participation of phonons may change our results quantitatively. We do not expect that phonons change the absorption spectra up to room temperature qualitatively, however. Note that for the calculation of the photoluminescence from the

Si NCs in a SiO$_2$ matrix, the phonon-assisted transitions and energy relaxation should be taken into account.

7 Conclusions

In conclusion, we have calculated the absorption spectra of Si NCs in a wide band gap SiO$_2$ matrix. We use the empirical tight-binding method. The SiO$_2$ matrix is taken into account as a virtual crystal with the band structure of β-cristobalite. We show that the electron states in the matrix are important for the description of the high-energy part of the absorption spectrum. Our calculations show that the high energy peak in the NC absorption is explained by large mixing of the electron states in the NC and in the matrix. The effect of the dielectric screening of the electric field inside the NC is also rather important and affects the shape of the absorption spectrum when the optical transitions involving the electron states in the matrix are taken into account.

Conflicts of interest

There are no conflicts to declare.

Acknowledgements

The work of MON and IDA was supported by the Russian Science Foundation (Grant No. 19-12-00051) and Foundation for the Advancement of Theoretical Physics and Mathematics "BASIS".

Notes and references

1 F. Priolo, T. Gregorkiewicz, M. Galli and T. F. Krauss, *Nat. Nanotechnol.*, 2014, **9**, 19.
2 S. Ossicini, L. Pavesi and F. Priolo, *Light Emitting Silicon for Microphotonics*, Springer, 1st edn, 2004, vol. 194.
3 K. Dohnalová, T. Gregorkiewicz and K. Kůsová, *J. Phys.: Condens. Matter*, 2014, **26**, 173201.
4 A. V. Gert, M. O. Nestoklon, A. A. Prokofiev and I. N. Yassievich, *Fiz. Tekh. Poluprovodn.*, 2017, **51**, 1325; *Semiconductors*, 2017, **51**, 1274.
5 J.-M. Jancu, R. Scholz, F. Beltram and F. Bassani, *Phys. Rev. B: Condens. Matter Mater. Phys.*, 1998, **57**, 6493.
6 A. V. Belolipetskiy, M. O. Nestoklon and I. N. Yassievich, *J. Phys.: Condens. Matter*, 2019, **31**, 385301.
7 I. Sychugov, F. Pevere, J.-W. Luo, A. Zunger and J. Linnros, *Phys. Rev. B*, 2016, **93**, 161413.
8 I. Sychugov, J. Valenta and J. Linnros, *Nanotechnology*, 2017, **28**, 072002.
9 T. Takagahara and K. Takeda, *Phys. Rev. B: Condens. Matter Mater. Phys.*, 1992, **46**, 15578.
10 A. S. Moskalenko, J. Berakdar, A. A. Prokofiev and I. N. Yassievich, *Phys. Rev. B: Condens. Matter Mater. Phys.*, 2007, **76**, 085427.
11 P. Hapala, K. c. v. Kůsová, I. Pelant and P. Jelínek, *Phys. Rev. B: Condens. Matter Mater. Phys.*, 2013, **87**, 195420.

12 A. Zunger, *Quantum theory of real materials*, Kluwer Academic Publisher, 1996, ch. 13, vol. 348, p. 173.
13 R. Benchamekh, M. Nestoklon, J.-M. Jancu and P. Voisin, *Semiconductor Modeling Techniques*, Springer, 2012, ch. 2, vol. 159, p. 19.
14 L.-W. Wang and A. Zunger, *Phys. Rev. Lett.*, 1994, **73**, 1039.
15 Y. M. Niquet, C. Delerue, G. Allan and M. Lannoo, *Phys. Rev. B: Condens. Matter Mater. Phys.*, 2000, **62**, 5109.
16 B. A. Foreman, *Phys. Rev. Lett.*, 1998, **81**, 425.
17 B. A. Foreman, *Phys. Rev. B: Condens. Matter Mater. Phys.*, 2007, **75**, 235331.
18 E. L. Ivchenko, A. Y. Kaminskii and I. L. Aleiner, *Zh. Eksp. Teor. Fiz.*, 1993, **104**, 3401; *J. Exp. Theor. Phys.*, 1993, **77**, 609.
19 E. L. Ivchenko, A. Y. Kaminski and U. Rössler, *Phys. Rev. B: Condens. Matter Mater. Phys.*, 1996, **54**, 5852.
20 T. B. Boykin, G. Klimeck, M. Friesen, S. N. Coppersmith, P. von Allmen, F. Oyafuso and S. Lee, *Phys. Rev. B: Condens. Matter Mater. Phys.*, 2004, **70**, 165325.
21 M. O. Nestoklon, L. E. Golub and E. L. Ivchenko, *Phys. Rev. B: Condens. Matter Mater. Phys.*, 2006, **73**, 235334.
22 M. O. Nestoklon, E. L. Ivchenko, J.-M. Jancu and P. Voisin, *Phys. Rev. B: Condens. Matter Mater. Phys.*, 2008, **77**, 155328.
23 J. P. Perdew and Y. Wang, *Phys. Rev. B: Condens. Matter Mater. Phys.*, 1992, **45**, 13244.
24 J. P. Perdew, K. Burke and M. Ernzerhof, *Phys. Rev. Lett.*, 1996, **77**, 3865.
25 J. Heyd, G. E. Scuseria and M. Ernzerhof, *J. Chem. Phys.*, 2006, **124**, 219906.
26 J. c. v. Klimeš, M. Kaltak and G. Kresse, *Phys. Rev. B: Condens. Matter. Phys.*, 2014, **90**, 075125.
27 J. C. Slater and G. F. Koster, *Phys. Rev.*, 1954, **94**, 1498.
28 Y. P. Tan, M. Povolotskyi, T. Kubis, T. B. Boykin and G. Klimeck, *Phys. Rev. B: Condens. Matter Mater. Phys.*, 2015, **92**, 085301.
29 D. Gresch, Q. Wu, G. W. Winkler, R. Häuselmann, M. Troyer and A. A. Soluyanov, *Phys. Rev. Mater.*, 2018, **2**, 103805.
30 J.-M. Lihm and C.-H. Park, *Phys. Rev. B*, 2019, **99**, 125117.
31 P.-O. Löwdin, *J. Chem. Phys.*, 1950, **18**, 365.
32 G. H. Golub and C. F. van Loan, *Matrix Computations*, Johns Hopkins University, Baltimore, 4th edn, 2013.
33 K. Wu and H. Simon, *SIAM J. Matrix Anal. Appl.*, 2000, **22**, 602.
34 D. J. Chadi, *Phys. Rev. B: Solid State*, 1977, **16**, 790.
35 J.-M. Jancu, R. Scholz, E. A. de Andrada e Silva and G. C. La Rocca, *Phys. Rev. B: Condens. Matter Mater. Phys.*, 2005, **72**, 193201.
36 Y. M. Niquet, D. Rideau, C. Tavernier, H. Jaouen and X. Blase, *Phys. Rev. B: Condens. Matter Mater. Phys.*, 2009, **79**, 245201.
37 A. V. Belolipetskiy, M. O. Nestoklon and I. N. Yassievich, *Fiz. Tekh. Poluprovodn.*, 2018, **52**, 1145; *Semiconductors*, 2018, **52**, 1264.
38 M. O. Nestoklon, A. N. Poddubny, P. Voisin and K. Dohnalova, *J. Phys. Chem. C*, 2016, **120**, 18901.
39 M. O. Nestoklon, R. Benchamekh and P. Voisin, *J. Phys.: Condens. Matter*, 2016, **28**, 305801.
40 S. M. Sze, *Physics of Semiconductor Devices*, John Wiley & Sons, New Jersey, 2nd edn, 1981.

41 K. Seino, F. Bechstedt and P. Kroll, *Phys. Rev. B: Condens. Matter Mater. Phys.*, 2010, **82**, 085320.
42 P. Ramachandran and G. Varoquaux, *Comput. Sci. Eng.*, 2011, **13**, 40–51.
43 C. Delerue and M. Lannoo, *Nanostructures. Theory and Modelling*, Springer Verlag, Berlin, Heidelberg, 2004.
44 A. N. Poddubny, A. A. Prokofiev and I. N. Yassievich, *Appl. Phys. Lett.*, 2010, **97**, 231116.
45 M. Graf and P. Vogl, *Phys. Rev. B: Condens. Matter Mater. Phys.*, 1995, **51**, 4940.
46 L. C. L. Y. Voon and L. R. Ram-Mohan, *Phys. Rev. B: Condens. Matter Mater. Phys.*, 1993, **47**, 15500.
47 E. Ivchenko and M. Nestoklon, *J. Exp. Theor. Phys.*, 2002, **94**, 644.
48 T. Sandu, *Phys. Rev. B: Condens. Matter Mater. Phys.*, 2005, **72**, 125105.
49 M. Cruz, M. R. Beltrán, C. Wang, J. Tagüeña-Martínez and Y. G. Rubo, *Phys. Rev. B: Condens. Matter Mater. Phys.*, 1999, **59**, 15381.
50 M. S. Hybertsen, *Phys. Rev. Lett.*, 1994, **72**, 1514.
51 W. D. A. M. de Boer, D. Timmerman, K. Dohnalova, I. N. Yassievich, H. Zhang, W. J. Buma and T. Gregorkiewicz, *Nat. Nanotechnol.*, 2010, **5**, 878.
52 J.-W. Luo, S.-S. Li, I. Sychugov, F. Pevere, J. Linnros and A. Zunger, *Nat. Nanotechnol.*, 2017, **12**, 930.
53 W. de Boer, D. Timmerman, I. Yassievich, A. Capretti and T. Gregorkiewicz, *Nat. Nanotechnol.*, 2017, **12**, 932.

Faraday Discussions

PAPER

Power-dependent photoluminescence decay kinetics of silicon nanocrystals under continuous and pulsed excitation

Michael Greben [ID]* and Jan Valenta [ID]

Received 1st October 2019, Accepted 1st November 2019
DOI: 10.1039/c9fd00100j

Power-dependent photoluminescence (PL) decay kinetics of silicon nanocrystals (Si NCs) in solid and liquid samples were studied under cw and pulsed excitation. The lifetime distribution and, consequently the measured PL kinetics are shown to depend on the excitation pulse duration until it is not sufficiently short (pulsed limit) or long (cw limit). These two excitation limits, however, are proven to excite different distributions of lifetime components and cannot be directly compared. We derive and experimentally confirm the equality of lifetimes averaged over amplitude and intensity for cw and pulsed excitation, accordingly. The absorption cross section (ACS) of Si NCs in solid and liquid samples is assessed and compared by two approaches under cw-excitation based on the treatment of power-modulated PL kinetics or PL amplitude saturation curves under low and moderate excitation powers, respectively. The discrepancy in extracted ACS values as well as the long-debated phenomena of incomplete PL saturation of matrix-embedded Si NCs is explained by a proposed model that is based on saturation of various components in an ensemble distribution at different excitation powers. The model finally allows us to explain the mystery of average decay lifetime dependence on excitation power in the non-linear power regime. By varying the excitation from cw to pulsed, we showed the reduction of average decay lifetime in the later case and attribute this to the increased relative PL contribution of fast lifetime components that results in at least one order of magnitude lower ACS values. Finally, exciting the solid sample with very high excitation powers, we detected a PL intensity decrease region that allowed us to extract the Auger lifetime which is about 170 ns.

1 Introduction

Time-resolved photoluminescence (PL) spectroscopy is one of the key and widely applied non-invasive methods used to study the excited-state relaxation in various systems. In spite of the great importance of the technique, there are very few reports describing proper data treatment.[1–4] In fact, the most common

Department of Chemical Physics and Optics, Faculty of Mathematics and Physics, Charles University, Ke Karlovu 3, 121 16 Prague 2, Czechia. E-mail: leibnits@gmail.com

approach is based on the mono-exponential decay model. However, these simplest kinetics are almost never observed in PL from a real sample formed by an ensemble of emitters. Most ensembles are inhomogeneous, possessing a continuous distribution of decay components. Interestingly, this results in a number of consequences (pitfalls) that are often overlooked in time-resolved PL experiments, leading to wrong estimations. For instance, three or four exponential models can adequately describe nearly any PL data with a continuous distribution of decay constants.[1] As a consequence, partial fitting components obtained from such an analysis usually have no physical relevance and could serve just as a tool to estimate the distribution width or average lifetime values. Another pitfall is that the distribution of decay constants is, in general, a function of the excitation pulse length.[2] In this work we show that lifetime distribution is pulse-independent only upon excitation with sufficiently short (pulsed excitation limit) or long (cw-excitation) pulses. To date, the profound analysis of these two special regimes with real experimental examples is still missing. In addition, a distribution of components principally complicates the practical usage of such models that were originally derived under the assumption of ensemble homogeneity for simplicity. Thus, the two-level model for absorption cross section (ACS) determination originally presented by Kovalev et al.[5,6] and then slightly modified in our recent papers[7,8] considers the single decay lifetime instead of the lifetime distribution. Here we demonstrate that proper lifetime averaging can be employed in order to make the model applicable to an inhomogeneously broadened ensemble of NCs. The model predicts that under moderate and high excitation powers, the PL variation on the excitation intensity should saturate. However, in solid samples with silicon nanocrystals (Si NCs) the expected saturation is never observed, even with extremely high excitation powers[9] (further increase may damage the sample). In this paper we demonstrate that this phenomena can be related to the distribution of components in the ensemble where each component saturates at different excitation powers. We measured the decay kinetics of Si NCs in solid and liquid samples under cw and pulsed excitation. The low, moderate and high power regimes were exploited for ACS determination using different approaches. After 30 years of research on nanocrystalline Si, our understanding of its non-linear optical properties in the strong-pumping regime is still poor since most studies are focused on maximizing PL quantum yield, and this is normally reached under weaker excitation. For instance, there is a large discrepancy in the literature on extracted values of Auger lifetime for Si NCs reported in the range from picoseconds[10] to hundreds of nanoseconds.[11] Investigation of the high power pulsed excitation leading to the decrease in PL with increasing power allowed us to extract the Auger recombination lifetime of 170 ns. Apparently, Auger lifetimes extracted from time-resolved measurements on the μs-scale (ref. 11 and this work) are several orders of magnitude larger in comparison with those extracted using ultra-fast techniques.[10,12] This is because the femtosecond/picosecond-pulsed excitation is far above the one-exciton-per-NC population. In order to obtain acceptable signal/noise ratio, pulse energy cannot be arbitrarily low. Then, one fs/ps-pulse generates several excitons per NC that can rapidly undergo Auger recombination. Under a microsecond pulse, we are close to the one exciton saturation and the first exciton usually has enough time to be thermalized and possibly get trapped.

2 Experimental details and data treatment

In this study solid and liquid Si-NC samples were investigated. The solid multilayer sample was fabricated by plasma-enhanced chemical vapour deposition (PECVD) as alternating layers (100×) of silicon-rich silicon oxide (SiO$_x$, $x = 1.1$) with 2.5 nm thickness with a stoichiometric SiO$_2$ barrier (3 nm thick) on a fused silica substrate. On top and below the multilayer stack, capping and buffer were deposited, respectively. Subsequently, the sample was annealed in a quartz tube furnace at 1150 °C for 1 h in high purity N$_2$ in order to form Si NCs and then the sample was passivated by annealing in H$_2$ at 500 °C for defect passivation. Further details on sample preparation can be found in related literature.[13]

As for the liquid sample, phenylpropyl-passivated Si NCs were synthesized by non-thermal plasma synthesis according to the procedure described in recent papers.[14,15] The sample was provided as a highly concentrated droplet on glass and was finally diluted in octane. The emission peak was at 765 nm, 1.54 eV under laser excitation of 405 nm (PL width was similar to dodecane-passivated Si NCs[16]). The mean diameter of the Si NCs was estimated to be 3.6 nm by the relation[14] that was experimentally derived by Liu *et al.* and the PL quantum yield $\eta = 49\%$ (measured in the setup based on an integrating sphere[17]) suggests that the Si NCs were of high quality.

The PL decay kinetics were measured with a 405 nm diode laser whose beam was modulated using a quartz acousto-optic cell. The edge switching time was about 100 ns. In this study we used mostly the square-shaped pulses with pulse durations of 4 μs or 400 μs denoted as pulsed and cw-excitation, accordingly (see discussion in the Section 3.1). The repetition rate was set to 500 Hz (*i.e.* the duty cycle was 20%) for cw-excitation of both the liquid and solid samples, in order to leave samples in the dark for 80% of the time, which can help to relax the generated heat under strong excitation. The duty cycle was 0.2% or 0.3% and the repetition rate 500 Hz or 800 Hz for the short-pulse excitation of solid or liquid samples, respectively. All power-dependent PL kinetics were measured at $\lambda_{em} = 740$ nm, which is close to PL peak of both samples. Because of the slow decay kinetics (μs-scale) there was no need to use the instrument response function deconvolution.

Finally, the PL decay kinetics of the solid sample were also measured with the 355 nm ns-pulsed laser in order to reach strong excitation power. This type of excitation is used only in Section 3.5. PL amplitudes were extracted from the power dependence of the corresponding PL spectra.

PL decays throughout the paper were modelled with a combination of mono-exponentials:

$$I(t) = \sum_{i=1}^{N} A_i e^{-t/\tau_i} \quad (1)$$

The intensity-averaged $\bar{\tau}$ and amplitude-averaged $\langle \tau \rangle$ lifetimes were calculated[1] as follows:

$$\bar{\tau} = \frac{\sum_{i=0}^{N} A_i \tau_i^2}{\sum_{j=0}^{N} A_j \tau_j} \quad (2)$$

$$\langle \tau \rangle = \frac{\sum_{i=0}^{N} A_i \tau_i}{\sum_{j=0}^{N} A_j} \quad (3)$$

For the liquid and solid samples it was necessary to use two and three exponentials to fit the onset kinetics, and three and four exponentials to fit the decay kinetics, respectively. A larger number of fitting components in the case of the solid sample was expected because decays are more non-exponential, possessing a broader distribution of decay constants compared with the liquid sample.[18]

3 Results and discussion

3.1 Amplitude and intensity averaged lifetimes

To illustrate the issue of average decay lifetimes, we consider a log-normal distribution of decay rates as an example:

$$H(k) = \frac{A}{k} e^{\frac{-\ln(k/\mu)^2}{2\sigma^2}} \quad (4)$$

where A is the amplitude, μ and σ parameters are related to the most frequent decay rate and the width of the lifetime distribution, respectively.

It can be easily shown that after transformation to lifetimes, this distribution is still log-normal. Indeed, considering $\tau = 1/k$, it can be shown as:

$$-\ln\left(\frac{k}{\mu}\right)^2 = \ln 1 - \ln(\tau\mu)^2 = -\ln\left(\frac{\tau}{1/\mu}\right)^2 \quad (5)$$

Finally, applying a proper correction factor,[1] we obtain:

$$f(\tau) \sim \frac{1}{\tau^2} H(k) = \frac{B}{\tau} e^{\frac{-\ln\left(\frac{\tau}{1/\mu}\right)^2}{2\sigma^2}} \quad (6)$$

where B stands for the new amplitude.

One may notice that the width of the $H(k)$ and $f(\tau)$ distributions is the same because σ is not changed. The fact that log-normal distribution preserves the log-normal law no matter whether it is expressed via lifetimes or rates allows for proper convergence of its tails unlike, e.g. in the Levy-stable distribution (strictly, the probability density function of stretched exponential decay).[1] We have to note that although log-normal distribution is not yet a standard in time-resolved PL data treatment, it often represents a decent approximation[1] of real decay rate distributions in Si NCs (Fig. 5) and therefore, was specifically chosen as an example for modelling. A set of log-normal distributions (Fig. 1) was simulated using eqn (6) where the $\mu = 0.01$ parameter was fixed and σ was varied from 0.05 to 1.05 with a step of 0.05 (see Fig. 1). The averaged lifetimes were calculated numerically (inset in Fig. 1) employing eqn (2) and (3) where the corresponding discretized distributions $f(\tau)$ were exploited as amplitudes of partial components. As the distribution becomes broader (σ increases), the difference between the

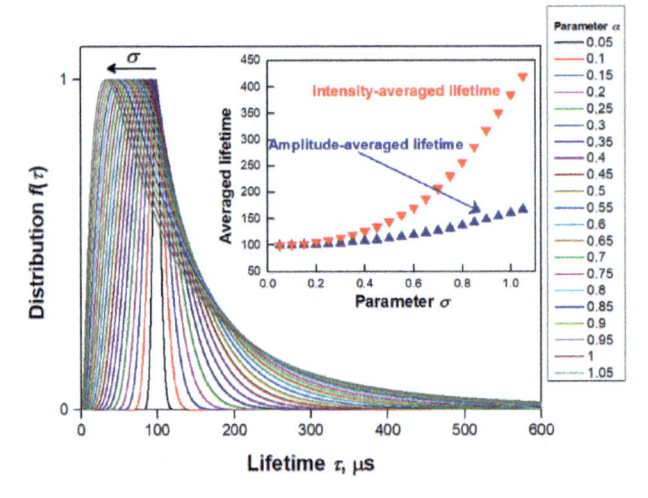

Fig. 1 Simulated log-normal distributions of lifetimes with the fixed parameter $\mu = 0.01$ and varied σ from 0.05 to 1.05 with a step of 0.05. The inset shows the numerically calculated intensity (eqn (2)) and the amplitude (eqn (3)) averaged lifetimes as functions of the σ parameter for corresponding distributions.

considered averaged lifetimes becomes more significant due to increasing long tails in the distribution.

3.2 Pulse duration (cw vs. pulsed excitation)

It is well known that depopulation kinetics of excited states directly depend on many experimental parameters, such as, e.g. excitation intensity and wavelength. However, in many studies, the excitation pulse duration is often assigned with marginal importance and therefore, sometimes is not even reported. We have already touched on the problem of pulse duration in our recent paper.[2] In this section we would like to elaborate further on this subject.

Unless the decay kinetics are mono-exponential, there is a distribution of decay components in a probed ensemble of emitters (time-varied decay rate is not considered here). Upon arrival of a long-enough pulse (labelled here as cw – continuous wave) with duration T, each component with lifetime τ_i will reach the cw-amplitude limit A_i^{cw} with different speeds:

$$A_i(T) = A_i^{cw}(1 - e^{-T/\tau_i}) \tag{7}$$

To illustrate the problem, we have simulated two components with eqn (7): fast (short) and slow (long) ones with corresponding lifetimes $\tau_{fast} = 20$ μs and $\tau_{slow} = 100$ μs as functions of the excitation pulse length (see Fig. 2). Considering the duration of the excitation pulse as $T = 50$ μs (yellow area) one may see that while the fast component is almost saturated (92% of A_1^{cw}), the slow component only reaches 39% of its maximum amplitude A_2^{cw}. Consequently, the fast component will contribute more to the detected PL decay signal when the excitation source is switched off in spite of the equality (by default) of the cw-amplitudes of the components ($A_1^{cw} = A_2^{cw} = 1$).

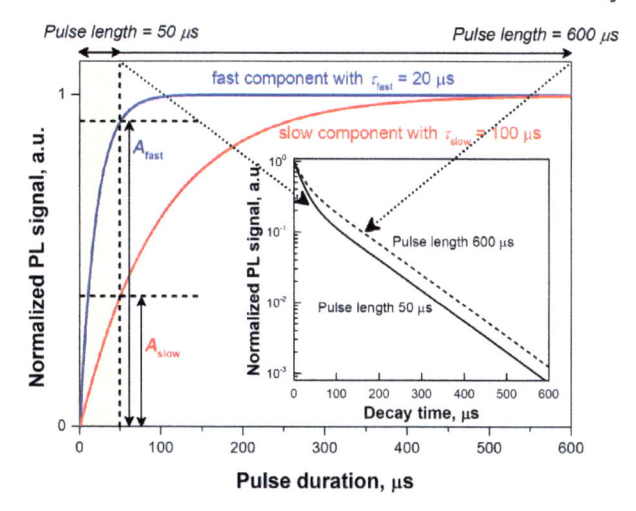

Fig. 2 Simulated (eqn (7)) PL signal onset of fast ($\tau_{fast} = 20$ μs, red line) and slow ($\tau_{slow} = 100$ μs, blue line) components with equal normalized cw-amplitudes ($A_1^{cw} = A_2^{cw} = 1$) as functions of the excitation pulse duration. The inset demonstrates simulated PL time traces (eqn (1)) for pulse durations of 50 μs (solid line) and 100 μs (dashed line).

As a result, decay curves under real cw-excitation (*e.g.* $T = 600$ μs) and shorter pulse excitation (*e.g.* $T = 50$ μs) are different (see the inset in Fig. 2). Here we assume that under low excitation the onset (rise) and decay lifetimes are the same. A similar effect is expected in real experiment where instead of just two components there can be a continuous distribution of components in an ensemble of emitters. Apparently, the extracted distribution of components will depend on whether we excite with short or long pulses. This raises the question: which pulse duration is better to correctly describe the real physical system?

To address this question, eqn (7) has to be considered where two pulse duration limits can be distinguished:

(1) Long pulse limit (cw-excitation);
In that limit we can assume $T \to \infty$ and from eqn (7) it follows $A_i(T) \to A_i^{cw}$;
(2) Short pulse limit (pulsed-excitation);
Here we consider $T \to 0$ that results in:

$$A_i^{pulsed}(T) \approx A_i^{cw}\left(\frac{T}{\tau_i} - \frac{1}{2}\left(\frac{T}{\tau_i}\right)^2 + \ldots\right) \approx A_i^{cw}\frac{T}{\tau_i} \quad (8)$$

In both cases the amplitude ratio of the ith and j^{th} components reaches a pulse-independent value, namely A_i/A_j for the long pulse limit and $A_i^{cw}\tau_j/A_j^{cw}\tau_i$ for the short pulse limit.[2] This means that the shape of the distribution of decay components is independent from the pulse duration for both limits. However, the absolute values of the amplitudes are different depending on the excitation type (cw or pulsed). Thus, in the short-pulse limit, the ultimate amplitudes, A_i^{cw}, from cw-excitation are weighted by the ratio of the lifetimes of the pulse and the corresponding decay component, respectively. Apparently, the longer the lifetime, τ_i, of a component is, the lower its contribution to the PL signal is, when keeping the amplitude A_i^{cw} constant.

Indeed, considering the amplitude ratio of the above simulated curves (see Fig. 2 and eqn (7)), for the short pulse limit it is equal to the ratio of onset lifetimes of the simulated components and for the long pulse limit it is equal to 1, both due to the equality of cw-amplitudes (see Fig. 3). Pulses with intermediate lengths, in between the two limits, must be avoided in experiments because in those cases, the distribution of components will depend on the excitation pulse duration and that makes the results severely dependent on the experimental conditions. In ref. 2 we estimated the values of the discussed limits quantitatively. Namely, $T_{short} \approx 0.1\tau_{PL}^{short}$ and $T_{long} \approx 3\tau_{PL}^{long}$, where τ_{PL}^{short} and τ_{PL}^{long} stay for the shortest and longest (significant) component lifetimes in the distribution, respectively. The selection of these significant lifetimes is subjective. In order to make very rough estimations, we took corresponding averaged lifetimes instead. Taking into account the shortest and longest intensity-averaged lifetimes obtained for the solid sample investigated in this paper (see next sections), the corresponding T_{short} and T_{long} limits can be estimated as 6.4 μs and 315 μs. Consequently, excitation by 4 μs short pulses and 400 μs long pulses used in this study fulfil the selected criteria for pulsed and cw excitations, respectively. In the case of the liquid sample, the addressed limits were 12 μs and 465 μs, respectively. Apparently, the 400 μs pulse is slightly below the long-pulse limit. However, we believe that the applied pulses still can be considered as quasi-pulsed and quasi-cw excitation without losing the main message.

One may notice that the addressed pulse limits result in different distributions of decay components (because of different amplitude ratios) and cannot be directly compared. The origin of this effect is simply related to the different onset speeds of the various components in a probed ensemble. In other words, different excitation types (pulse limits) result in excitation of probed components in different proportions (see inset in Fig. 2) of their amplitudes (contributions). Thus we face the question again: should we excite with pulses corresponding to the

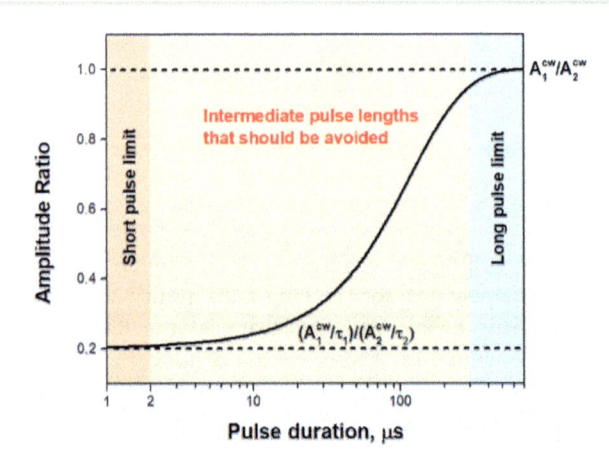

Fig. 3 Amplitude ratio for simulated fast (τ_{fast} = 20 μs) and slow (τ_{slow} = 100 μs) components ($A_1^{cw} = A_2^{cw} = 1$) upon varied pulse duration. The estimated short (orange area) and long (green area) pulse limits determine the region (yellow area) of intermediate pulses where the amplitude ratio significantly varies as a function of pulse duration.

short or long limit? Further, we will only use the terms of pulsed and cw-excitations that stay for short and long pulse limits, accordingly. We have to stress that only the cw-excitation probes the whole ensemble of components, leaving enough time for each of them to be excited. In contrast, with pulsed excitation, one preferably excites short components which readily contribute to the PL signal as opposed to suppressed long components (Fig. 2). We believe that one has to select cw or pulsed excitation according to the research focus and applications. For instance, if a fast response of the system is important, then pulsed excitation is preferable. In turn, if one wishes to excite the whole ensemble of components (emitters), the cw-excitation regime must be applied.

The issues demonstrated above on the simplest example of two components can easily be scaled up to an ensemble of emitters with an almost infinite number of components. In that case, instead of dealing with partial components (eqn (8)), average values of the addressed parameters should be considered. Thus, the intensity-averaged lifetime under cw-excitation can be calculated with eqn (2). Employing eqn (2) and (8), one can easily calculate the intensity-averaged lifetime for the pulsed excitation:

$$\overline{\tau}_{pulsed} = \frac{\sum_{i=0}^{N} A_i^{cw} \frac{T}{\tau_i} \tau_i^2}{\sum_{i=0}^{N} A_i^{cw} \frac{T}{\tau_i} \tau_i} = \frac{\sum_{i=0}^{N} A_i^{cw} \tau_i}{\sum_{i=0}^{N} A_i^{cw}} = \langle \tau_{cw} \rangle \qquad (9)$$

Eqn (9) shows that the intensity-averaged lifetime under pulsed excitation is equal to the amplitude-averaged lifetime under cw-excitation. This result is an important point of this section. In the literature, the decay lifetime is often calculated as amplitude-averaged under cw-excitation. Then, direct comparison with the intensity-averaged lifetimes under pulsed excitation can be substantiated. Again, the proper selection of lifetime averaging (intensity-averaged vs. amplitude-averaged) depends on the goal of the experiment.[1] Eqn (9) can be employed for testing whether the used excitation pulse durations belong to either of the two addressed limits (see the Section 3.5).

In order to validate eqn (9) experimentally, we considered PL decay kinetics of Si NCs measured using square pulses of a variable length.[2] Unlike in our previous work where the fitting was done with a double-exponential function, here we employ a combination of three exponents. This slightly improved the fits and consequently increased the precision in extracted values of averaged lifetimes. Intensity-averaged and amplitude-averaged lifetimes were calculated using eqn (2) and (3), respectively. It is clear (see Fig. 4) that the intensity-averaged lifetime $\overline{\tau}$ obtained under short pulses reaches the amplitude-averaged lifetime $\langle \tau \rangle$ under long pulses, i.e. $\lim_{T \to 0} \overline{\tau} \approx \lim_{T \to \infty} \langle \tau \rangle \approx 240$ μs.

In order to reveal the influence of the pulse duration on the distribution of decay constants, we applied unbiased fitting of the measured decay kinetics[2] by the maximum entropy method (using the open-source code[19,20]). The retrieved distributions of decay rates are presented in Fig. 5 where evidently the relative amplitudes of the short components (higher rates) increase upon decreasing pulse durations. Here the distributions were normalized at the peak to give an idea of the relative contributions of the short and long components.

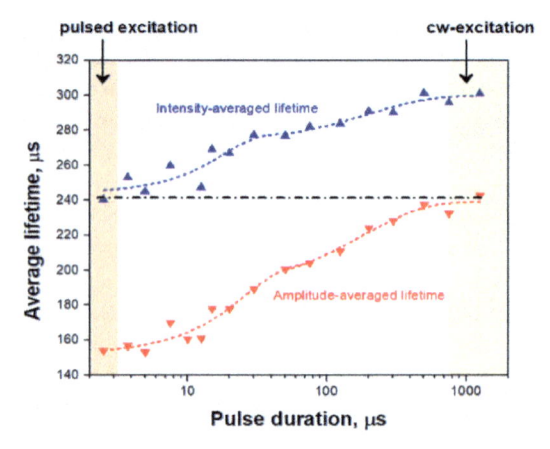

Fig. 4 Calculated intensity-averaged and amplitude-averaged lifetimes of the experimentally measured decay kinetics² following square excitation pulses with a variable duty cycle for Si NCs multilayers detected at 940 nm. Estimated short and long pulse limits are highlighted in orange and yellow colours, accordingly. The dashed lines serve as an eye guide.

3.3 Absorption cross-section under cw-excitation

In this section we estimate the absorption cross-section (ACS) of solid and liquid samples under 400 μs long pulses (cw-limit) with a 405 nm laser. There are two main approaches to estimate the ACS of a sample with emitters of unknown concentration (density). Both of them are based on a simple quasi-two-level model where only the ground-state, single-excited and double-excited states are considered and that results in a system of three differential equations.[8] A general solution of this system under certain assumptions can be presented[8] as the following:

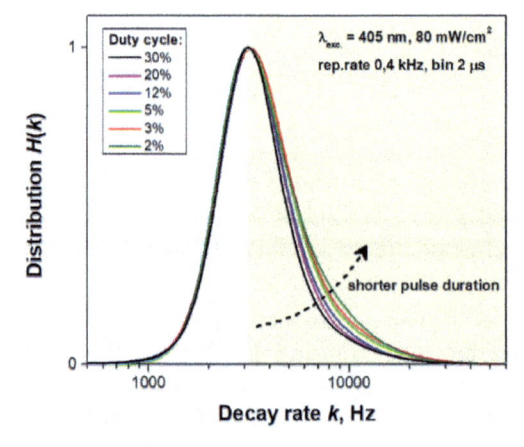

$$I_{PL}^{cw} = \frac{N_T I_{ex} \sigma / \tau_r}{1/\tau_{PL} + I_{ex}\sigma + I_{ex}^2 \sigma^2 \tau_A} \quad (10)$$

Fig. 5 Normalized distributions of rate constants recovered from corresponding time traces² under excitation with a variable duty cycle (pulse duration).

where I_{PL}^{cw} is the amplitude of the signal at time 0 (laser switched off), N_T is the total population (density) of emitters, σ describes the ACS, I_{ex} represents the excitation intensity and τ_r, τ_{PL}, and τ_A stand for radiative, total (contains radiative and non-radiative parts) and Auger lifetimes, respectively.

In spite of having the same starting model, each approach exploits different region of the excitation powers as described below.

3.3.1 The photoluminescence modulation technique.
It can be shown that under low excitation powers ($I_{ex} \ll 1/\tau_{PL} \ll 1/\tau_A$) eqn (10) is reduced to a linear function which serves as the cw-amplitude in the PL onset after switching on the constant excitation:[7]

$$I_{PL}(t) = N_T I_{ex} \sigma \frac{\tau_{PL}}{\tau_r} \times \left[1 - e^{-\left(\frac{1}{1/\tau_{PL}} + I_{ex}\sigma\right)t}\right] = I_{PL}^{cw}\left[1 - e^{-(t/\tau_{ON})}\right] \quad (11)$$

It follows from eqn (11) that the PL intensity exponentially approaches a maximum value, I_{PL}^{cw}, with characteristic onset time, τ_{ON}, which in turn is a function of the excitation intensity:

$$\frac{1}{\tau_{ON}} = \frac{1}{\tau_{PL}} + I_{ex}\sigma \quad (12)$$

Finally, the ACS can be directly calculated as a difference[8] in the slopes of the excitation power dependencies of the onset ($1/\tau_{ON}$) and decay ($1/\tau_{PL}$) rates. This approach, which is often called the PL modulation technique, seems to be easy and straightforward, but has a few pitfalls. Firstly, as assumed in eqn (11), the excitation powers must be sufficiently low to ensure that we stay in the PL linear regime, otherwise eqn (12) is not valid. Secondly, for very low powers, the onset and decay times should be equal $\tau_{ON} = \tau_{PL}$ and that results in large uncertainty of the extracted values. Consequently, there is usually a narrow region of excitation powers close to the PL saturation where the ACS can be reliably determined with this technique. Moreover, one may notice that in eqn (11) there is a mono-exponential function representing the solution of the system of differential equations describing the two-level model. In real experiments with ensembles of nanocrystals, mono-exponential decay is almost never observed. Staying within the same model, a way to circumvent this issue is to consider an average decay lifetime which will represent the decay (or onset) instead of a real distribution of lifetimes. The question is: what lifetime averaging should be used for ACS evaluation: intensity-averaging or amplitude-averaging? (Fig. 1) Let us discuss this problem.

A physically intensity-averaged lifetime means the average time during which the emitters remain in the excited state after the end of an excitation pulse. In contrast, an amplitude-averaged lifetime represents the mean lifetime value in the lifetime distribution.[1] We believe that in the original differential equations of the two-level model, what matters is the average time for which an emitter stays in the excited state, because this, by definition, represents the depopulation term in the equations. Therefore, intensity-averaging of lifetimes was used for the ACS calculations throughout this work.

To apply the described procedure, we measured the spectrally resolved kinetics (Fig. 6) of both the solid and liquid samples for different cw-excitation powers at the single wavelength of 740 nm. For each kinetic measurement, the PL amplitude

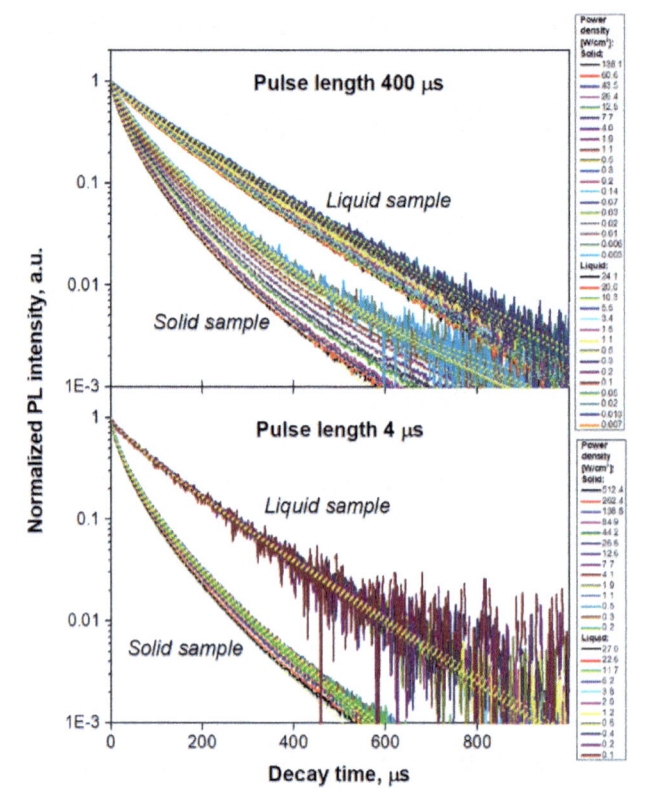

Fig. 6 Experimentally measured (λ_{em} = 740 nm) decay kinetics (solid lines) together with the corresponding fits (dashed yellow lines) by eqn (1) of the solid and liquid samples as a function of excitation power density under excitation by 400 μs (upper part) and 4 μs (lower part) pulses.

(peak) was determined with proper subtraction of the background signal (see Fig. 7a). In accordance with the model (eqn (11)), at low excitation powers, the PL amplitude increases linearly with excitation power. When approaching the saturation level, the PL starts to deviate from a linear dependence because the Auger recombination term starts to be significant. For each time trace, the PL onset and decay lifetimes were properly extracted from the fitting by a combination of several mono-exponentials (eqn (1)) and the corresponding intensity-averaged lifetimes were calculated with eqn (2). As predicted (eqn (12)), at low powers the onset and decay lifetimes approach similar values (Fig. 7b). Let us note that the PL lifetimes for the liquid sample were expectedly[9] longer than those measured for the solid sample (Fig. 6) and that resulted in slight truncation of the onset part of the kinetics of the liquid sample and subsequently, reduced the accuracy of the fits. Therefore, comparing the ACS estimations for the solid and liquid samples, the former one is more accurate. Employing eqn (12), the ACS values were calculated with dependence on applied power (Fig. 7c). The narrow region where the described model is valid is highlighted by a yellow color. The ACS was estimated to be $\sigma_1^{cw,L} \approx 2.3 \times 10^{-15}$ cm^2 and $\sigma_1^{cw,S} = 6.9 \times 10^{-15}$ cm^2 for liquid and solid samples, accordingly. So, comparing Si NCs embedded in a matrix and in a colloidal solution, the ACS of the former ones is roughly 3 times larger.

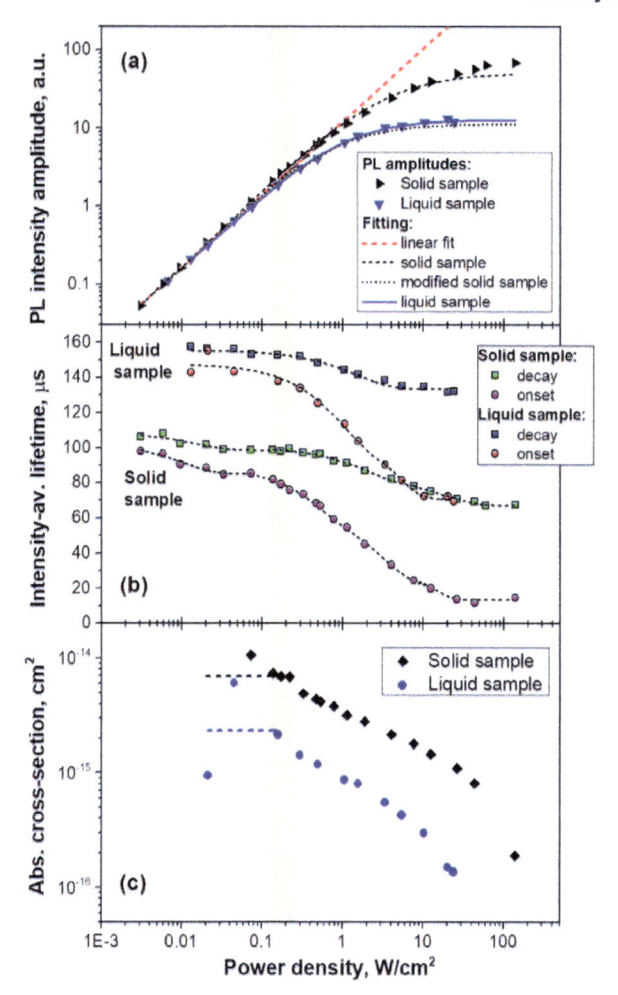

Fig. 7 (a) PL intensities of the solid (black triangles) and liquid (blue triangles) samples with dependence on excitation power density. The dashed black and solid blue lines represent the fits of the corresponding PL curves by eqn (13). The dotted black line represents the fitting model of the PL of the solid sample where the ACS value from the PL modulation technique is assumed. (b) The extracted onset (circles) and decay (squares) intensity-averaged lifetimes. The dashed lines serve as an eye guide. (c) The ACS obtained from the PL modulation technique for the solid (diamonds) and liquid (circles) samples with the restricted region (yellow stripe) of the trusted ACS determination.

3.3.2 Photoluminescence saturation curve fitting. Another approach to ACS determination utilizes Auger-related PL saturation phenomena and was successfully applied in several previous studies of Si NCs.[6,21,22] Here, unlike the above mentioned PL modulation technique, moderate excitation powers are a focus of interest. Indeed, assuming $I_{ex}\sigma \ll 1/\tau_A$, i.e. the Auger contribution is still relatively small, eqn (10) can be simplified as follows:

$$I_{PL}^{cw} = \frac{N_T I_{ex} \sigma / \tau_r}{1/\tau_{PL} + I_{ex}\sigma} \qquad (13)$$

In other words, at moderate excitation we can neglect the term $I_{ex}^2\sigma^2\tau_A$, but cannot ignore the contribution of the term $I_{ex}\sigma$ because the excitation power is moderately high.

From eqn (13) it follows, that under increasing excitation power, the PL firstly starts to respond non-linearly and then goes to saturation. Indeed, the PL curve of the liquid sample properly reaches saturation and can be well-fitted (Fig. 7a, blue solid line) by eqn (13). The extracted ACS is $\sigma_2^{cw,L} = 3.4 \times 10^{-15}$ cm^2 and is about 1.5 times larger than the roughly estimated value $\sigma_1^{cw,L}$ by the PL modulation technique. This difference is not large and can be attributed to the relatively low precision of the onset lifetimes for the liquid sample.

Unlike the liquid sample, the PL response curve of the solid sample does not seem to saturate properly and therefore, fitting by eqn (13) is not perfect (Fig. 7a, dashed black line). This problem was described before[7,8] and it was shown[9] that the PL curve does not saturate even for powers approaching 10^4 W cm^{-2}. De Jong and co-workers[23] attributed this problem to the laser-induced heating of the Si NCs embedded in the oxide matrix. The heating can modulate phonon population resulting in dramatic changes of both the lifetimes and the ACS[8] of the NCs. Indeed, we noticed[2] a slight difference between the cw-excitation onset curve and the amplitudes of PL signal under variation of the excitation duty cycle that can indicate possible sample heating. Indirectly, this idea can be supported by the current results of the liquid sample whose PL power dependence saturates properly and the effect of heating can be neglected in the colloidal solution. Additionally, the PL curves of a single Si NC excited by the 405 nm laser under cryogenic temperatures (10 to 35 K) were shown to saturate well.[11] However, the PL curves of an ensemble of matrix-embedded Si NCs do not tend to saturation[9] even under cryogenic temperatures (8 K). An alternative explanation is based on the assumption where radiative recombination from biexcitons[11] (two e–h pairs in one NC) is allowed and that generates additional photons and thus prevents the curve from saturation. However, it is not clear why PL from biexcitons is activated only in solid samples. Anyway, a rough estimation of the ACS from fitting (Fig. 7a, dashed black line) of the PL curve of the solid sample gives $\sigma_2^{cw,S} = 1.5 \times 10^{-15}$ cm^2 and that is about 4.5-times lower than $\sigma_1^{cw,S}$ extracted from the linear power regime. Our previous estimation[7] of the ACS from the PL curve fitting gave us a difference of about a 6 times from the ACS value obtained from PL modulation technique.

This discrepancy can be related to additional photons contributing to the signal above the expected saturation due to the possible origins listed above. These photons are not counted by the model (eqn (13)) and that results in underestimation of the ACS extracted from the PL curve fitting. If we enter the value $\sigma_1^{cw,S} = 6.9 \times 10^{-15}$ cm^2 (from the PL modulation technique) in the PL curve modelling of the solid sample, we obtain the PL curve (Fig. 7a, dotted black line) that almost coincides with the PL curve of the liquid sample.

Another explanation for the discussed discrepancy is that the model itself is wrong and the ACS values obtained by different methods are fundamentally different. Indeed, in the model it was assumed that the presence of an exciton in a NC does not change the ACS for the generation of the second one. However, in reality, it might occur that the cross section for the absorption of the second photon by a NC is lower. In such scenario ACS extracted from power-dependent PL curve fitting should be lower as compared with PL modulation technique similar to what we observed.

Although we do not exclude the possible explanations listed above, here we propose a simple and elegant way to explain the problem while staying within the starting model. To illustrate the idea, we again simulated two (short and long) components (Fig. 8a) with short $\tau_S = 20$ μs, and long $\tau_L = 100$ μs lifetimes and equal amplitudes employing eqn (13). The corresponding ACS values were assumed to be $\sigma_S = 3 \times 10^{-16}$ cm^2 and $\sigma_L = 3 \times 10^{-15}$ cm^2 (one order of magnitude difference) to mimic realistic values[7] where $\sigma_S < \sigma_L$. Both of the simulated PL curves (Fig. 8, cyan and green circles) properly saturate by definition. However, the saturation starts to take place at different power thresholds. This results in a power interval where the long component saturates, but the short

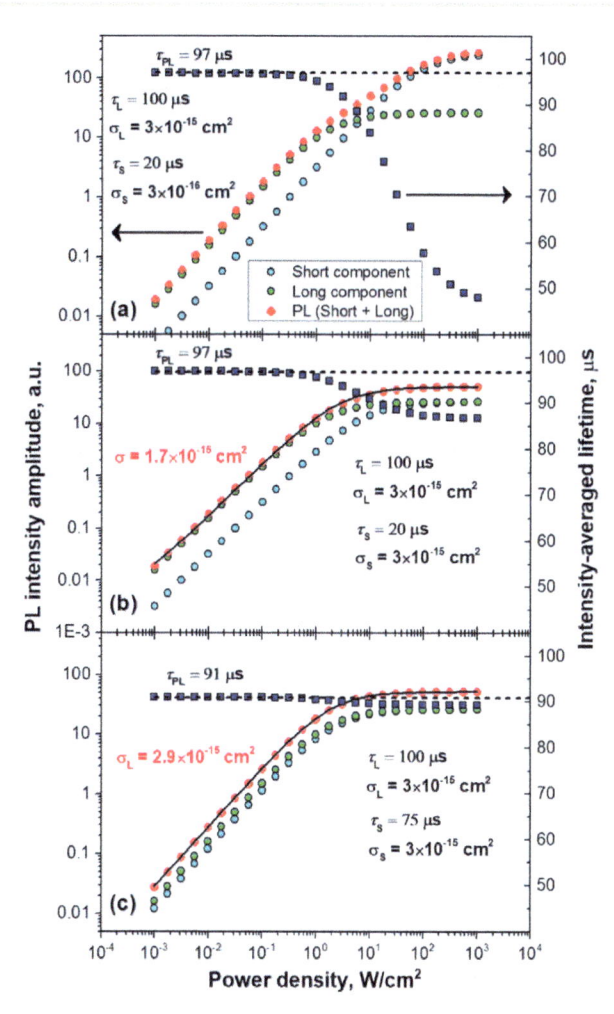

Fig. 8 Simulations of PL intensity (red circles) combined from short (cyan circles) and long (green circles) components as a function of excitation power density. The intensity-averaged lifetimes (squares) are presented on the right axis. In parts (a)–(c) ACS and partial lifetimes are varied. The black solid lines represent corresponding PL fits by eqn (13).

one still stays in the linear regime. Consequently, the contribution of the short component in the PL signal increases until it finally saturates as well. This situation is similar to the effect of short pulses illustrated in Fig. 2 where short components become more prominent in the PL signal when the pulse duration is decreasing. Notably, the resulting PL curve that is a simple sum of both components (Fig. 8a, red circles) is very close to the PL of the long component at low powers while it reaches the PL of the short component at high powers. An attempt to fit the simulated PL (red circles) with eqn (13) fails because the curve shape is complex and cannot be described by this simple equation. The proposed simple model gives an idea why the PL curve never reaches saturation.

With this model we can also simulate the power dependence of the average decay lifetime. Thus, intensity-averaged lifetimes were calculated using eqn (2) for every excitation power value where the PL curves of the partial components serve as the amplitudes (Fig. 8a). One can see that once the long component starts to saturate, the lifetime shortens with an increase of power density that is very similar to what we observe in the experiment with the solid sample (Fig. 7b or better Fig. 9b). In many papers, PL decay lifetime is erroneously assumed as

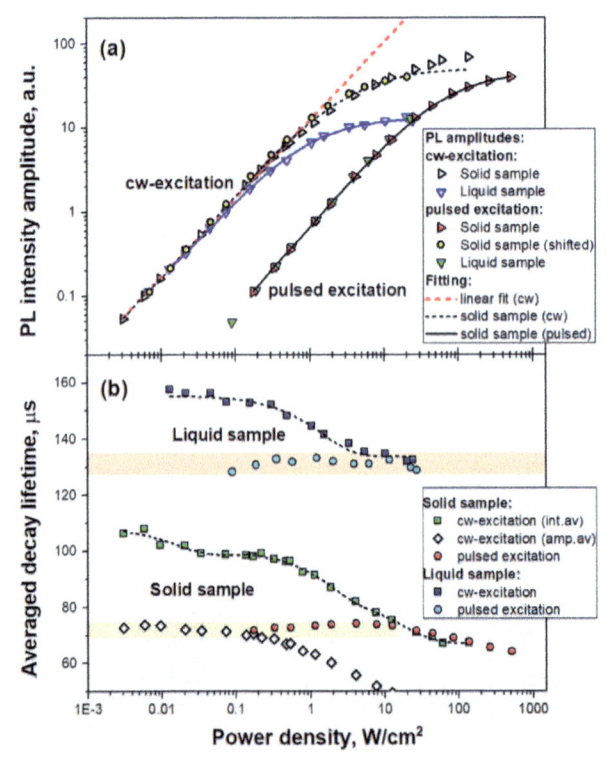

Fig. 9 (a) PL intensities of solid (horizontal triangles) and liquid (vertical triangles) samples under 400 μs (cw) and 4 μs (pulsed) excitation pulse durations as a function of excitation power density. The solid lines represent fits of the corresponding PL curves by eqn (13). The PL curve (yellow circles) of the solid sample under pulsed excitation was adjusted in power scale by reducing by a factor of 25 to overlap with the PL curve under cw-excitation. (b) The extracted intensity-averaged (solid) and amplitude-averaged (open diamonds) lifetimes of samples under cw (squares) and pulsed (circles) excitation.

power-independent. We reported[8] this problem back in 2017 and attributed this to saturation of non-radiative recombination decay paths resulting in an increase of the non-radiative lifetime. A similar lifetime decrease in porous Si with increasing excitation power was reported.[24,25] Here we demonstrated that a power-dependent lifetime decrease is nothing else but the increasing relative PL contribution of short components upon increasing excitation power.

Now we turn to the question of why the PL signal saturates for the colloidal sample in spite of the above described effect of the short components. Assuming that ACS and lifetimes do not vary much between components (Fig. 8c), then obviously the PL curve saturates well and the extracted value of the ACS from fitting of $\sigma = 2.9 \times 10^{-15}$ cm^2 is close to the simulated value $\sigma = 3 \times 10^{-15}$ cm^2. As a consequence of having similar values of partial lifetimes, the variation of the average decay lifetime with power is small (Fig. 8c). However, in reality, the PL lifetime of the liquid sample changes significantly (Fig. 9b) and that excludes the hypothesis of similar partial lifetimes. Alternatively, assuming similar values of the ACS for partial components, the simulated PL curve still goes to proper saturation under the variation of lifetimes and this is similar to a real experiment (Fig. 7b). It is important to note that in this case, the ACS obtained from PL curve fitting is almost two times lower than that entered in the model. Therefore, we can conclude that the discussed method of curve fitting is unreliable when cw-excitation is applied to a system of emitters with a significant distribution of optical parameters (ACS and/or lifetimes). Only in the case of a homogeneous ensemble of emitters are the extracted values of physical significance (see Fig. 7c). Taking this into account, that distribution of PL contribution components (lifetimes) in the liquid sample is narrower than in the solid one,[18] and one may assume, in addition to other possible reasons, that ACS variations between short and long components are not so significant in the liquid sample, resulting in the power dependence presented in Fig. 8b.

3.4 Absorption cross-section under pulsed excitation

In this section we consider PL properties of the same solid and liquid samples measured under similar conditions but the pulse duration is shortened from 400 μs (cw-excitation) to 4 μs (pulsed excitation). In spite of the similar conditions, the picture under pulsed excitation is different. The PL saturation of the solid sample starts at significantly higher powers (Fig. 9a, red triangles). This is expected as it was shown in Section 3.1, and excitation by short pulses results in a more pronounced relative contribution of short components. Comparing short and long components, the former ones saturate at higher excitation densities (Fig. 8a) simply because there is a shorter time available for the second photon be absorbed before the first exciton recombines. Consequently, more intense "bombarding" with photons (higher power density) is needed to make bi-exciton generation possible. We were not able to reach PL saturation of the liquid sample under pulsed (4 μs) excitation because the measurements had to be done in a cuvette (with objectives of longer working distances and smaller numerical apertures) and the power density that the liquid sample was exposed to could not reach the values applied to the solid sample (where a smaller excitation spot can be focused).

Before we continue further, it is important to consider lifetimes under pulsed excitation. In accordance with the model presented in Section 3.1, intensity-averaged lifetimes (Fig. 9b, green squares) under pulsed excitation (highlighted by the yellow stripe) in the linear power regime are equal to the amplitude-averaged lifetimes (Fig. 9b, open diamonds) under cw-excitation. This ensures that applied pulse durations properly stay in the pulse limits described above (see Fig. 3). Moreover, upon increase of the excitation power, decay lifetimes from cw and pulsed excitation shrink to similar values (highlighted by the orange stripe for the liquid sample). This suggests that the resulting effects of the pulse duration and non-linear powers are similar, leading to a matching reduction of lifetimes.

One may note that pulsed excitation lifetimes vary significantly less (Fig. 9b, red circles) compared to cw-excitation. This is related to the increased relative PL contribution (described above) by fast components (short lifetimes) activated under pulsed excitation that results in shorter average decay lifetimes (Fig. 9b, cyan and red circles). Consequently, the average lifetime will decrease less upon increasing excitation power as it is already significantly reduced. This is an important point, suggesting that under pulsed excitation, the effect of PL distortion by power-dependent components, described in the Subsection 3.3.2, can be suppressed.

If one were to adjust the PL curve of the solid sample under pulsed excitation by dividing the power densities by a factor of 25, this curve overlaps well with the corresponding PL curve obtained under cw-excitation (Fig. 9a, yellow circles). Although the shape of the fitting curve (Fig. 9a, dashed black line) is principally similar to the PL under cw-excitation scenario (Fig. 9a, solid black line) the actual position of the curve (Fig. 9a, red triangles) on the power scale makes the extracted ACS of $\sigma_2^{pulsed,S} = 9.7 \times 10^{-17}$ cm^2 significantly lower. This means that there is at least an order of magnitude difference in the ACS values even when compared with the underestimated value from the PL curve fitting (cw-excitation). The fact that the ACS is lower under pulsed excitation can be understood as follows. The relative contribution of the short components increases upon pulsed excitation, and in the case of the Si NCs this comes from the smaller NCs whose lifetimes are shorter[16] relative to the lifetimes of the larger ones. Taking into account that the ACS decreases[7] for smaller NC sizes, one may expect a resulting decrease of the ACS values with shortening of the pulse duration.

3.5 Probing Auger lifetime under pulsed excitation

In the previous section, the PL response and lifetimes under moderate excitation powers were considered. However, there was an intention to check higher power densities where eqn (13) is not valid and the general equation, eqn (10), should be used instead. Eqn (10) says that at very high power densities, the PL should start to decrease upon increasing excitation power. And indeed, this can be seen in the power-dependence of the PL amplitude in solid samples such as porous silicon[26] (Ar-ion laser, $\lambda_{ex} = 488$ nm, 514 nm) or similar multilayer samples[23] with Si NCs (Nd:YAG laser, $\lambda_{ex} = 355$ nm). In this study we applied the highest possible power densities from a 405 nm modulated diode laser (see previous section) but still did not reach powers sufficient to observe the discussed effect. In order to further increase the power density, it was necessary to apply a 355 nm pulsed laser (pulse

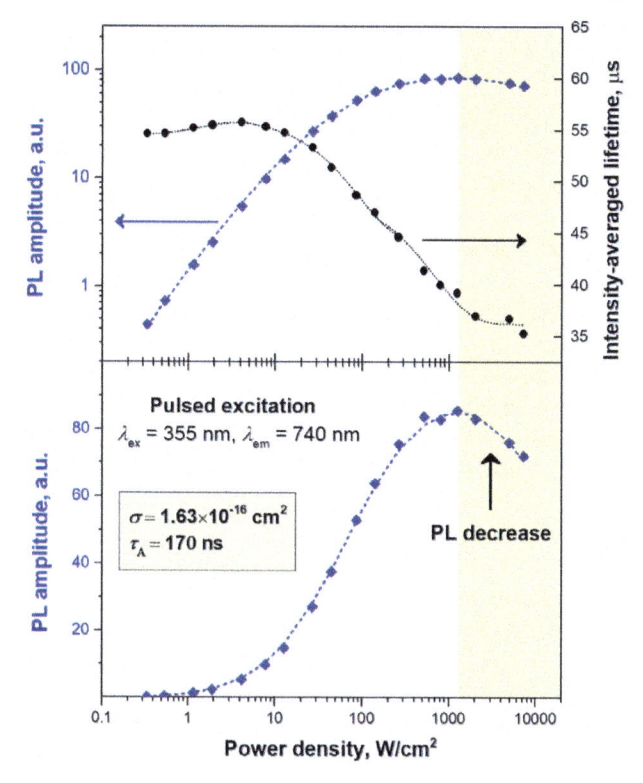

Fig. 10 PL intensity and its fit (dashed line) by eqn (10) as a function of excitation power density presented in the log–log (upper part) and semi-log (lower part) scales. The calculated intensity-averaged lifetimes are displayed on the right axis.

duration ∼ 5 ns). Similar to 405 nm excitation, a combination of 4 monoexponentials was applied for PL decay fittings. Fig. 10 demonstrates the power dependence of the extracted PL (from the PL spectra) and the lifetimes (from the decay kinetics). Indeed, one may notice a predicted PL decrease for high excitation powers where the lifetimes finally become independent from the power. This indicates that all contributing components (NCs) are in the saturation regime. The observed PL decrease is provided by the term $I_{ex}^2\sigma^2\tau_A$ that was considered as negligible for lower powers in previous sections. Finally, the fitting by eqn (10) gives the ACS of $\sigma = 1.6 \times 10^{-16}$ cm^2 that is about 1.5 times larger compared to that of 405 nm excitation. This is expected as the ACS is an increasing function of excitation photon energy.[7] The extracted Auger lifetime is 170 ns, and that is in good agreement[11] with values obtained for single Si NCs (30 to 300 ns).

4 Conclusions

In this work we studied the decay kinetics of Si NCs in solid and liquid samples under the variation of excitation power of long (cw) and short (pulsed) excitation. The excitation pulse duration was shown to play a major role in the PL response of ensembles of emitters. It was demonstrated that intermediate-long pulses affect the distribution of lifetime components and therefore should be avoided. This effect is a direct consequence of different excitation speeds of various components in the ensemble. The short (pulsed) and long (cw) pulse limits were demonstrated, resulting in different lifetime distributions that principally do not match and cannot be directly compared. Two averaged (by intensity and amplitude) lifetimes were distinguished and the discrepancy between them was demonstrated using the example of simulated log-normal distribution of varied width. We derived and experimentally confirmed that the amplitude-averaged lifetime under cw excitation is equal to the intensity-averaged lifetime under pulsed excitation. The ACS values of Si NCs in solid and liquid samples were assessed and compared by two independent techniques under cw-excitation. The ACS value of the liquid sample was estimated to be in the range from 2.3 to 3.4 × 10^{-15} cm^2. The ACS values of the solid sample were found to be 6.9 × 10^{-15} cm^2 from power-dependent PL kinetics and 1.5 × 10^{-15} cm^2 from power-dependent PL curve fitting. A simple model was proposed to explain the discrepancy in the obtained ACS values as well as the long-debated phenomenon of incomplete PL saturation of solid samples. The model suggests that power-dependent PL curve fitting from an ensemble of emitters (components) under cw-excitation can give incorrect results. The origin of this is related to the saturation of various components at different excitation powers. For the first time, we explicitly show that as a consequence, the average lifetime becomes a decreasing function of the excitation power in a non-linear power regime. However, under pulsed excitation, the average lifetime decrease is found to be significantly suppressed. We attribute this to the increased relative PL contribution of fast components under pulsed excitation that makes the average lifetime shorter in comparison with the cw-excitation scenario even at low excitation powers. The enhanced relative PL contribution of short components also results in an ACS that is at least one order of magnitude lower (9.7 × 10^{-17} cm^2) for the solid sample under pulsed excitation in comparison with that under cw-excitation. Finally, under very high powers of pulsed excitation, we were able to detect a PL decrease region related to an Auger term with the Auger lifetime as a parameter. This allowed us to estimate the Auger lifetime (without using ultra-fast techniques) which is equal to 170 ns and is consistent with data reported for single Si NCs. Apparently, the proposed approach can be extended to other nanocrystalline materials, besides Si, that reveal similar optical responses.

Conflicts of interest

There are no conflicts to declare.

Acknowledgements

The authors would like to thank Prof. X. Pi and coworkers from Zhejiang University, China for providing high quality colloidal samples and to Prof. M. Zacharias and coworkers from IMTEK, University of Freiburg, Germany for providing the PE-CVD grown multilayer sample. M. Greben would like to kindly acknowledge P. Steinbach for sharing his expertise and knowledge on the MEM code. This work was supported by the Charles University Research Centre program UNCE/SCI/010.

References

1. M. Greben, P. Khoroshyy, I. Sychugov and J. Valenta, *Appl. Spectrosc. Rev.*, 2019, **54**, 758–801.
2. M. Greben and J. Valenta, *Rev. Sci. Instrum.*, 2016, **87**, 126101.
3. M. N. Berberan-Santos, E. N. Bodunov and B. Valeur, *Chem. Phys.*, 2005, **315**, 171–182.
4. N. Boens and M. Van der Auweraer, *Photochem. Photobiol. Sci.*, 2014, **13**, 422–430.
5. D. Kovalev, H. Heckler, G. Polisski and F. Koch, *Phys. Status Solidi A*, 1999, **215**, 871–932.
6. D. Kovalev, J. Diener, H. Heckler, G. Polisski, N. Künzner and F. Koch, *Phys. Rev. B: Condens. Matter Mater. Phys.*, 2000, **61**, 4485–4487.
7. J. Valenta, M. Greben, Z. Remeš, S. Gutsch, D. Hiller and M. Zacharias, *Appl. Phys. Lett.*, 2016, **108**, 023102.
8. M. Greben, P. Khoroshyy, S. Gutsch, D. Hiller, M. Zacharias and J. Valenta, *Beilstein J. Nanotechnol.*, 2017, **8**, 2315–2323.
9. M. Greben, *Advanced spectroscopic characterization of quantum dot ensembles*, Charles University, 2018.
10. M. T. Trinh, R. Limpens and T. Gregorkiewicz, *J. Phys. Chem. C*, 2013, **117**, 5963–5968.
11. F. Pevere, I. Sychugov, F. Sangghaleh, A. Fucikova and J. Linnros, *J. Phys. Chem. C*, 2015, **119**, 7499–7505.
12. F. Trojánek, K. Neudert, M. Bittner and P. Malý, *Phys. Rev. B: Condens. Matter Mater. Phys.*, 2005, **72**, 1–6.
13. A. M. Hartel, D. Hiller, S. Gutsch, P. Löper, S. Estradé, F. Peiró, B. Garrido and M. Zacharias, *Thin Solid Films*, 2011, **520**, 121–125.
14. X. Liu, Y. Zhang, T. Yu, X. Qiao, R. Gresback, X. Pi and D. Yang, *Part. Part. Syst. Charact.*, 2016, **33**, 44–52.
15. X. Liu, S. Zhao, W. Gu, Y. Zhang, X. Qiao, Z. Ni, X. Pi and D. Yang, *ACS Appl. Mater. Interfaces*, 2018, **10**, 5959–5966.
16. M. Greben, P. Khoroshyy, X. Liu, X. Pi and J. Valenta, *J. Appl. Phys.*, 2017, **122**, 034304.
17. J. Valenta, *Nanosci. Methods*, 2014, **3**, 11–27.
18. M. Greben, P. Khoroshyy and J. Valenta, *Phys. Status Solidi A*, 2019, **2019**, 1900698.
19. P. J. Steinbach, R. Ionescu and C. R. Matthews, *Biophys. J.*, 2002, **82**, 2244–2255.
20. P. J. Steinbach, *Anal. Biochem.*, 2012, **427**, 102–105.
21. H. Rinnert, O. Jambois and M. Vergnat, *J. Appl. Phys.*, 2009, **106**, 023501.
22. C. Garcia, B. Garrido, P. Pellegrino, R. Ferre, J. A. Moreno, J. R. Morante, L. Pavesi and M. Cazzanelli, *Appl. Phys. Lett.*, 2003, **82**, 1595–1597.
23. E. M. de Jong, H. Rutjes, J. Valenta, M. T. Trinh, A. N. Poddubny, I. N. Yassievich, A. Capretti and T. Gregorkiewicz, *Light: Sci. Appl.*, 2018, **7**, 17133.
24. L. Pavesi, *J. Appl. Phys.*, 1996, **80**, 216–225.
25. A. L. Efros, M. Rosen, B. Averboukh, D. Kovalev, M. Ben-Chorin and F. Koch, *Phys. Rev. B: Condens. Matter Mater. Phys.*, 1997, **56**, 3875–3884.
26. M. Koós, I. Pócsik and É. B. Vázsonyi, *Appl. Phys. Lett.*, 1993, **62**, 1797–1799.

> # Faraday Discussions

DISCUSSIONS

Optical and electronic properties: from theory to experiments: general discussion

Mita Dasog, Anna Fucikova, Ankit Goyal, Michael Greben, Ali Reza Kamali, Kateřina Kůsová, Mikhail Nestoklon, Tomáš Popelář, Wei Sun and Ming Lee Tang

DOI: 10.1039/D0FD90005B

Ali Reza Kamali opened a general discussion of the paper by Mita Dasog: To prepare Si, SiO$_2$ particles were reacted with Mg powder at 650 °C for 0.5 h, followed by heating at 300 °C for 6 h. How does heating at 300 °C help, considering that the reactants are in the solid state at this low temperature?

Mita Dasog responded: Thank you, Ali, for your question. The magnesiothermic reduction is an exothermic reaction. Even though we heat it at 300 °C (which is lower than the melting point of Mg), the heat that is released during the first step is enough to keep the reaction going. We reported the details of this in a previous study.[1]

1 S. A. Martell, Y. Lai, E. Traver, J. MacInnis, D. D. Richards, S. MacQuarrie and M. Dasog, *ACS Appl. Nano Mater.*, 2019, 2, 5713–5719, DOI: 10.1021/acsanm.9b01207.

Ali Reza Kamali asked: The SEM morphology and the particle size in the initial SiO$_2$ and the Si product obtained by the magnesiothermic reduction of SiO$_2$ are almost the same (Fig. 1 in the article). Can the authors explain the reaction mechanism based on this observation?

Mita Dasog replied: The widely accepted mechanism for this reaction is that Mg diffuses into the silica structure and reduces it, which leads to the formation of MgO. The lattice unit of Si is smaller than that of SiO$_2$, which leads to shrinking of the local structure to form crystallites. The void space is occupied by the MgO byproduct. After washing with HCl acid, the MgO is removed. The space occupied by MgO now becomes a pore, leading to the formation of a porous structure. Therefore, the overall size of the nanoparticles remains the same. The change in lattice constants is compensated by the formation of pores.

Ali Reza Kamali queried: According to Fig. 3 in your paper, HF treatment of spherical Si particles caused a substantial morphological change. Were the HF-treated samples analysed by XRD, HRTEM or other characterization techniques to investigate whether there were any structural changes as well?

Mita Dasog answered: We examined the etched products using XRD and Raman spectroscopy. There was no change in the crystallinity of the nanoparticles after the etching process. However, prolonged exposure to air can lead to surface oxidation.

Anna Fucikova continued the discussion of the paper: This is a really nice article. I would just like to ask the authors how cost effective the production of hydrogen is compared to splitting water with electricity, since the release of H_2 is connected with dissolution of the Si matrix and therefore the material is consumed by H_2 production.

Mita Dasog replied: Thanks, Anna, for the question. The H_2 generation in our work is driven by light and Si nanoparticles are not consumed as seen in the stoichiometric studies. That said, their activity does decrease over time due to oxide build-up, which traps the carriers. The activity can be regenerated by treating the nanoparticles with HF to remove the oxide.

We have not personally performed cost calculations on our system, but technoeconomic analysis performed by other researchers predicts a production cost of \$2 kg^{-1} of H_2 for photocatalysis and \$5–6 kg^{-1} of H_2 using PV-electrolyzers (*i.e.* electricity). If we could prevent surface oxidation of the Si nanoparticles, it could be a cheaper alternative to make H_2.

Wei Sun remarked: Prof. Dasog has done brilliant work on using mesoporous Si to generate fuels. I just wonder if a more prolonged H_2 evolution experiment was ever done, say on the scale of days? Fig. 2 in the manuscript showed a linear curve within the first several hours, but I assume that similarly to CO_2 reduction, the H_2 evolution will exhibit a significantly decreased rate over time when Si–H is consumed by oxidation, and the curve will no longer be linear. I am just curious about what it would look like.

Mita Dasog responded: Thank you, Wei, for the question. Yes, we did perform H_2 evolution over an extended period of time, and yes, the activity does decrease significantly due to the oxide build-up. We are currently investigating the rate of oxidation over time and whether we can control that by changing the rate of hole transfer. So far, it looks like the photocatalytic activity dies after about 12 h.

Anna Fucikova opened a general discussion of the paper by Ming L. Tang: This is a really nice article. I would like to ask if the authors tested whether the nanoparticles keep their great optical properties upon drying, or if they studied them just in liquids.

Ming Lee Tang replied: Yes, the optical properties are intact upon drying. We sometimes dry these colloids under high vacuum in a glovebox to change solvent, and they are the same. Thanks for reading!

Ali Reza Kamali opened a general discussion of the paper by Kateřina Kůsová: In reference 17, the preparation of the "standard" sample is based on the electrochemical etching of boron-doped p-type Si wafers. In your work, SiNCs were fabricated using crystalline silicon wafers, but the type of silicon wafer and, consequently, the doping element(s), is not clear. I think the type of doping is rather important since it can potentially influence both the oxidation and etching behaviours, and to a higher extent the photoluminescence properties of SiNCs. Please can you elaborate on this?

Kateřina Kůsová responded: Usually, we use a boron-doped Si wafer and these wafers are also quite standard for other porous-Si-based NCs. Doping is generally essential in the electrochemical etching process, since it makes the wafer/electrode conductive and, moreover, it also provides the necessary holes. The doping of our Si wafers reaches 0.1 Ω cm, which corresponds to about 1×10^{17} B atoms cm^{-3} of bulk Si. Simple recalculation of this concentration by volume gives roughly 1 boron atom per 1000 NC, and this estimate is very likely unrealistically high, because boron atoms in the Si lattice would provide instability where the silicon lattice would crack during the etching process, and therefore pieces with boron atoms are less likely to produce nanocrystals.

That said, I would not think that the effect of boron atoms would be very high in our case. It should be pointed out that our case is significantly different from the heavily boron- and phosphorus-doped SiNCs studied by other groups.[1] We have been studying the luminescence properties of these porous-Si-based NCs for quite some time now, and we find the properties to be very similar to those of alkyl-capped SiNCs (although the QY is lower in our case).

1 H. Sugimoto, M. Fujii, K. Imakita, S. Hayashi and K. Akamatsu, *J. Phys. Chem. C*, 2013, **117**, 11850–11857, DOI: 10.1021/jp4027767.

Ali Reza Kamali commented: How does the electrochemical method of producing SiNCs used in your research work? In particular, what were the anode, cathode, electrolyte and the electrochemical parameters, such as voltage, current and time?

Kateřina Kůsová answered: The method we apply is described in ref. 17 of our paper.[1] We use the etched Si wafer (P-type: B-doped, (0.075–0.100) Ω cm, (100)-oriented) as the anode and place it on an aluminium substrate. Platinum is used as the top (negative) cathode immersed in the etching bath. The wafer is placed at the bottom of a Teflon chamber and we have a Teflon stirrer to homogenize the etching bath. For the etching bath, we mix 15 ml 50% HF with 37 ml 99% EtOH. The wafer area exposed to the etching bath is around 10 cm^2 and we etch with a constant current of 16 mA, setting the voltage to keep this current. Etching time is 2 h.

1 K. Dohnalová, L. Ondič, K. Kůsová, I. Pelant, J. L. Rehspringer and R.-R. Mafouana, *J. Appl. Phys.*, 2010, **107**, 053102, DOI: 10.1063/1.3289719.

Ankit Goyal remarked: You mentioned that magnetic stirring was used while the PL measurements were done, so did the SiNCs settle during the measurement?

Kateřina Kůsová replied: The original, unmodified sample (electrochemically etched SiNCs) settles a lot, which is why we choose to apply magnetic stirring. In the modified sample, the SiNCs disperse in water much better and settling is definitely not critical for the measurement. Nevertheless, we use magnetic stirring with nearly every liquid sample (even with dyes), because in our experience it can help prevent possible excitation-laser-induced changes in the sample.

Ankit Goyal asked: You mentioned that the size of the NCs was about 2.5–3 nm, which doesn't correspond to the HRTEM images provided in Fig. 1 in your paper. Could you please comment on this?

Kateřina Kůsová responded: You are certainly right in pointing this out. The problem here is that the electrochemical etching procedure produces a broad distribution of sizes (at least it does in our implementation). The size we cite in our paper is a result of previous studies (*e.g.* reference 39[1]) and we believe that this part of the size distribution is responsible for most of the observed luminescence, which is the reason why we consider the 2.5–3 nm SiNCs to be the most important ones in the luminescence studies. The HRTEM images in Fig. 1 in our article are clearly larger SiNCs from the tails of the size distribution, which are generally much easier to image. However, the presented SAED confirms that SiNCs with sizes around 3 nm are present (see *e.g.* ref. 2, below) and they make up a significant portion of the SiNCs in the sample.

1 K. Dohnalová, I. Pelant, K. Kůsová, P. Gilliot, M. Gallart, O. Crégut, J.-L. Rehspringer, B. Hönerlage, T. Ostatnický and S. Bakardjeva, *New J. Phys.*, 2008, **10**, 063014, DOI: 10.1088/1367-2630/10/6/063014.
2 M. L. Mastronardi, F. Hennrich, E. J. Henderson, F. Maier-Flaig, C. Blum, J. Reichenbach, U. Lemmer, C. Kübel, D. Wang, M. M. Kappes and G. A. Ozin, *J. Am. Chem. Soc.*, 2011, **133** 11928–11931, DOI: 10.1021/ja204865t.

Ankit Goyal queried: You used "plasma-activated water" treatment for the modifications, but the claims regarding the concentration of the N-species in PAW mentioned in the results and discussion are not supported by relevant characterization. Could you please comment on this?

Kateřina Kůsová answered: The presence of reactive nitrogen and oxygen species in plasma-activated water is a generally known fact.[1] During the modification procedure, the concentration of nitrogen-related species is routinely checked semiquantitatively using test strips for NO_2 and NO_3, but our knowledge is also based on previous experience with PAW (*e.g.* reference 43 in our paper[2]). The reason why all the details of the modification procedure are not listed in our *Faraday Discussions* paper is that we are currently working on a separate manuscript focused on the details of the modification procedure.

1 R. Thirumdas, A. Kothakota, U. Annapure, K. Siliveru, R. Blundell, R. Gatt and V. P. Valdramidis, *Trends Food Sci. Technol.*, 2018, **77**, 21–31, DOI: 10.1016/j.tifs.2018.05.007.
2 P. Hozák, V. Scholtz, J. Khun, D. Mertová, E. Vaňková and J. Julák, *Plasma Phys. Rep.*, 2018, **44**, 799–804, DOI: 10.1134/S1063780X18090040.

Ankit Goyal asked: In the EDS analysis you see several metallic impurities of Fe, Al, and Na which are not shown on the maps. The plasma process may be intense and could have caused the insertion of these impurities in SiNCs or at least at the surface. Could you please comment on this?

Kateřina Kůsová replied: There are several points we want to make here:

(1) Treatment with plasma-activated water is a relatively mild procedure.

(2) Doping: It is generally a difficult task to achieve doping of nanocrystals. Even though very small atoms such as lithium can get inside the silicon lattice,[1] it is not easy to achieve doping with larger atoms such as P or B[2,3] due to the massively increased ionization energies[4] and the self-purification effect.[5] This is why alternatives to conventional doping (dopant entering the lattice) are sought; such alternatives include energy-modulation doping[4] or surface doping,[6] *i.e.* the "dopant" atoms reside on the surface of the nanocrystal. Actually, even under relatively harsh conditions (annealing to 900–1200 °C), the "dopant" atoms tend to segregate at the surface of the SiNC rather than entering the Si lattice.[7] Keeping this in mind, I do not see a reason why atoms such as N, or even Na or Fe, should have a tendency to enter the Si lattice and stay there, especially when using such mild conditions as the ambient-condition non-thermal plasma treatment.

(3) We cannot definitively rule out the presence of these impurities on the surface of SiNCs, but we did not detect them using our characterization methods. Actually, the EDS measurements show that the concentration of these "impurities" decreases after modification, which suggests that they are washed out rather than incorporated into the SiNCs. They are originally present in the electrochemically etched SiNCs in the form of small compact pieces (fragments), which are not a good source for elemental incorporation.

(4) Doping by "impurities" usually manifests itself in the luminescence measurements by (1) the emergence of low-energy (sub-gap) states, *i.e.* the red-shift of luminescence, or (2) quenching of the material-related luminescence as a result of (a) the presence of extra free carriers and Auger recombination or (b) structural defects leading to trap states. (This is actually why it is difficult for me to imagine a scenario in which dopant levels would blue-shift the originally red luminescence to the blue spectral region.) The luminescence measurements presented in our paper focus on the blue (above the gap) emission, and as such cannot tell us anything about possible sub-gap states. Beyond the scope of our *Faraday Discussions* paper, we did not detect any emission which would be evocative of such dopant levels, but it is important to say that we did not study the modified SiNCs in the infrared region (our routine luminescence measurements extend to around 1050 nm).

(5) As far as I know, there is very limited literature on how such "exotic" surface-bonded species would affect an SiNC. Surface-bonded species can sometimes exhibit optical characteristics of "dopants" (see the "surface doping" with P and B;[7] it is highly probable that the Mn-, Ni-, Co-, and Cu-doped SiNCs shown in ref. 8 are the same case, even though the authors do not comment on the "position" of the dopants here) and lower the optical bandgap (see point (4)). On the other hand, judging from simulations, energetically higher states tend to be more surface-localized in real space, which implies that surface-bonded species could affect the above-the-gap non-thermalized emission studied in our paper. Here, we do not see any new signal in the modified sample. Rather, the incorporation of nitrogen into the shell of SiNCs seems to eliminate the signal from SiO_2 defect sites, which suggests that the procedure simply improves the quality of the surface oxide present on the SiNCs.

To sum up, we cannot unambiguously exclude the incorporation of these elements into or on the surface of SiNCs during the plasma-activated water treatment, but neither the chemical characterization nor luminescence measurements gave us any indication whatsoever that such impurities were present. Therefore, we expect their effect to be marginal at best, or none at all.

1 E. Klimešová, K. Kůsová, J. Vacík, V. Holý and I. Pelant, *J. Appl. Phys.*, 2012, **112**, 064322, DOI: 10.1063/1.4754518.
2 D. Hiller, J. López-Vidrier, S. Gutsch, M. Zacharias, M. Wahl, W. Bock, A. Brodyanski, M. Kopnarski, K. Nomoto, J. Valenta and D. König, *Sci. Rep.*, 2017, 7, 8337, DOI: 10.1038/s41598-017-08814-0.
3 D. Hiller, J. López-Vidrier, S. Gutsch, M. Zacharias, K. Nomoto and D. König, *Sci. Rep.*, 2017, **7**, 863, DOI: 10.1038/s41598-017-01001-1.
4 D. König, D. Hiller and S. Smith, *Phys. Rev. Appl.*, 2018, **10**, 054034, DOI: 10.1103/PhysRevApplied.10.054034.
5 G. M. Dalpian and J. R. Chelikowsky, *Phys. Rev. Lett.*, 2006, **96**, 226802, DOI: 10.1103/PhysRevLett.96.226802.
6 H. Sugimoto, M. Fujii, K. Imakita, S. Hayashi and K. Akamatsu, *J. Phys. Chem. C*, 2013, **117**, 11850–11857, DOI: 10.1021/jp4027767.
7 H. Sugimoto, M. Fujii and K. Imakita, *Nanoscale*, 2014, **6**, 12354–12359, DOI: 10.1039/c4nr03857f.
8 S. Chandra, Y. Masuda, N. Shirahata and F. M. Winnik, *Angew. Chem., Int. Ed.*, 2017, **56**, 6157–6160, DOI: 10.1002/anie.201700436.

Ankit Goyal remarked: From Fig. 4 in the paper, it doesn't seem that the blue band completely goes away; rather, its width becomes narrow, which could be due to N defects or metallic impurities in the Si bands. Could you please comment on this?

Kateřina Kůsová responded: You are completely correct, the blue band does not go away completely. As we describe in the paper, the blue band in the original unmodified sample is made up of three components, which based on their temporal and spectral characteristics are assigned to (1) ultrafast Raman signal, (2) intermediate SiO_2-defect-related PL, and (3) core-related non-thermalized direct recombination (equation 2). Out of these three, only (2) disappears in the modified sample, (1) is still present (since it mostly originates in the solvent), and (3) is diminished but still present. This means that we do not see any new signal in the modified sample that is not present in the original sample. I would guess that what you refer to as the "narrow" signal is actually the ultrafast component. The ultrafast signal is a Raman signal, as seen from (a) its spectral position and (b) the fact that its position shifts with the changing excitation wavelength. This ultrafast component has the same spectral profile in both samples, as shown in Fig. 1 in the discussion. Thus, the difference between the spectral profiles of the unmodified and modified samples is a result of the disappearance of component (2), which reveals the spectral shape of component (1). Even though "spectral narrowing" is seemingly present, it does not occur.

Thank you for this question, because I can see now how the presentation I used could be misleading, and thank you for going through our paper so thoroughly.

Tomáš Popelář opened a general discussion of the paper by Mikhail O. Nestoklon: You calculate that the excitation of Si NCs in the SiO_2 matrix of higher energy photons occurs with assistance from the SiO_2 matrix – the transition is indirect in *r*-space and direct in *k*-space. Yet the recombination then happens between the states of lower energy, presumably between the electron and hole both inside the NC. What is the mechanism of the transition from absorption taking place in the matrix to recombination in the nanocrystal?

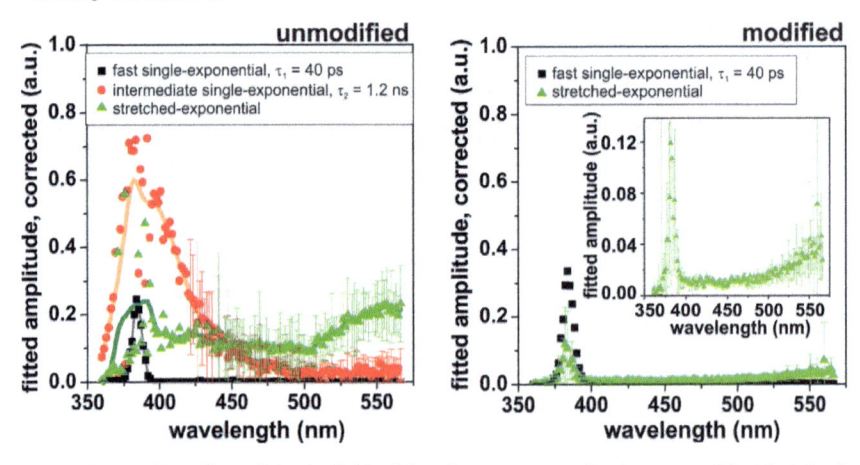

Fig. 1 Spectral profiles of the individual fitted components in the unmodified (original) sample and the modified sample.

Mikhail Nestoklon answered: To compute the photoluminescence one has to take into account the phonon-assisted relaxation of the photogenerated hot carriers and phonon-assisted optical transitions for the states with lower energy. Some estimations of these processes may be found in the literature, *e.g.* ref. 1–3. To accurately consider the energy relaxation and phonon-assisted optical transitions, our model of electronic states and optical transitions should be supplemented with (1) the calculations of phonon states, (2) interactions between the phonon and electron systems, and (3) dynamics of the phonon system. Technically, the virtual crystal approximation we use for the SiO_2 matrix can hardly be used to model the phonons in amorphous silica. We plan to develop a more accurate model of amorphous SiO_2 and perform phonon calculations in the future, but this is quite a challenging task.

1 A. N. Poddubny, A. A. Prokofiev and I. N. Yassievich, *Appl. Phys. Lett.*, 2010, **97**, 231116.
2 A. S. Moskalenko, J. Berakdar, A. N. Poddubny, A. A. Prokofiev, I. N. Yassievich and S. V. Goupalov, *Phys. Rev. B*, 2012, **85**, 085432.
3 A. A. Prokofiev, A. N. Poddubny and I. N. Yassievich, *Phys. Rev. B*, 2014, **89**, 125409.

Kateřina Kůsová opened a general discussion of the paper by Michael Greben: The PL decays are analysed using a sum of exponentials. As the authors say themselves, a combination of 3 or 4 exponentials can describe nearly any PL decay data, which means that using this description the system is overparameterized. How does this overparameterization influence the result of the fit? Is the result sensitive to small changes in the fitted data? Statistical techniques such as factor analysis can be used to determine the number of free variables of a problem; have you considered applying this approach? Also, throughout the paper, only the average lifetime is used to characterize the data. Do you think that the shape of the decay curve has no relevance?

Michael Greben replied: The fact that almost any decay curve can be described by a sum of 3 (or 4) exponentials does not mean that the system is overparameterized. We used the minimum number of exponentials to obtain good quality fitting (with careful control of residuals). In fact, fitting is not necessarily needed in the case of excellent data quality. Indeed, average decay lifetimes can be calculated by employing just experimentally measured decay curves (integral method):[1]

$$\bar{\tau} = \frac{\int_0^\infty tI(t)\mathrm{d}t}{\int_0^\infty I(t)\mathrm{d}t}$$

As an example, we calculated the intensity-averaged lifetimes (see Fig. 2 in the discussion) of a solid sample under cw-excitation by both fitting and integral methods. One can note that the results are very close, with a slight deviation (note the log-scale of both axes) for high powers. This is related to the relatively long time-step (2 μs) of the experimental curves and, therefore, the limited number of curve points due to shortening of the average decay lifetime. As result, the integral method gives slightly overestimated values. This nuance was the only motivation to employ any fitting.

In order to estimate uncertainties, we generated 1000 decay curves by the addition of random Gaussian noise to the initial four-exponential fit. For the

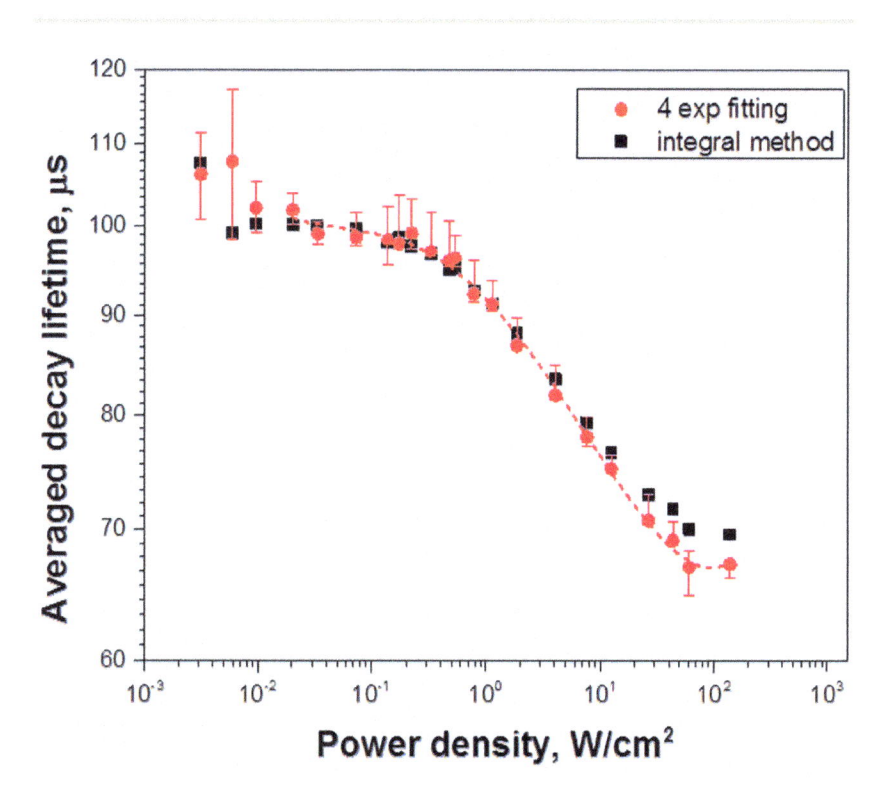

Fig. 2 Intensity averaged decay lifetimes of solid sample under cw-excitation as a function of excitation power density calculated from four-exponential fitting (red circles) and employing experimental curves (integral method).

standard deviation of pseudo-random values of noise, we used residuals from the initial fit. Then all the generated kinetics were fitted by four-exponential functions and the corresponding lifetime values were collected in a histogram with subsequent estimation of the lower and upper limits (5% and 95% of all counts, respectively). One may note that despite many (nine) fitting parameters, the estimated errors are low and do not invalidate any conclusions presented in the paper. Therefore, we did not consider statistical techniques to determine the number of free variables in this case.

Finally, we believe that the decay curve shape has major importance and we recently analyzed it in detail elsewhere.[2] Here we employed mainly averaged lifetime values (though one can see the normalized lifetime distributions in Figure 5 in our article) due to the focus of this study and limited space in the paper.

1 M. Greben, P. Khoroshyy, I. Sychugov and J. Valenta, *Appl. Spectrosc. Rev.*, 2019, **54**, 758–801.
2 M. Greben, P. Khoroshyy and J. Valenta, *Phys. Status Solidi A*, 2020, **217**, 1900698.

Tomáš Popelář remarked: You show that pulsed excitation preferentially shows the short components of PL decay in the measurement. I assume that these short components are more or less connected to smaller NCs in the ensemble which have lower ACS than larger NCs with long PL decay. Shouldn't these two facts compensate for each other? We saturate the short components more, but excite fewer NCs which do have these short components, so should they be weighted less in the PL decay?

Michael Greben responded: Well, in fact we excite fewer NCs with shorter lifetimes due to their lower absorption cross-section independently of pulse duration (this probability parameter is considered to be pulse-independent). Therefore, there is no compensation of any parameters when pulse duration is varied. This can easily be shown mathematically as follows.

The excitation of components in an ensemble is a statistical process where the contribution of partial components (radiative and decay lifetimes are τ_r^i and τ_{PL}^i, respectively) to the PL signal under low excitation intensity and pulse duration T is given by equation 11 in the article:

$$I_{PL}^i(T) = N_T^i I_{ex} \sigma_i \frac{\tau_{PL}^i}{\tau^i}(1 - e^{-T/\tau_{PL}^i})$$

Here, N_T^i is a population of partial components, I_{ex} represents the excitation intensity, and σ_i describes the cross-section for the absorption of the given components. Let's consider two limits of pulse duration. Under low-power cw-excitation ($T \to \infty$) the amplitude of the ith component is:

$$I_{PL}^i(T_{cw}) = N_T^i I_{ex} \sigma_i \frac{\tau_{PL}^i}{\tau^i},$$

while under pulsed excitation ($T \to 0$) it can be expressed as (see equation 8 in the paper):

$$I^i_{PL}(T_P) = N^i_T I_{ex} \sigma_i \frac{\tau^i_{PL}}{\tau^i_r} \frac{T_P}{\tau^i_{PL}} = N^i_T I_{ex} \sigma_i \frac{T_P}{\tau^i_r}.$$

Now let's find the amplitude ratio of short (N^S_T, σ_S, τ^S_r, τ^S_{PL}) and long (N^L_T, σ_L, τ^L_r, τ^L_{PL}) components under cw or pulsed excitation:

$$\frac{I^S_{PL}(T_{cw})}{I^L_{PL}(T_{cw})} = \frac{N^S_T}{N^L_T} \frac{\sigma_S}{\sigma_L} \frac{\tau^L_r}{\tau^S_r} \frac{\tau^S_{PL}}{\tau^L_{PL}}; \quad \frac{I^S_{PL}(T_P)}{I^L_{PL}(T_P)} = \frac{N^S_T}{N^L_T} \frac{\sigma_S}{\sigma_L} \frac{\tau^L_r}{\tau^S_r}$$

From the two above equations it follows:

$$\frac{I^S_{PL}(T_P)}{I^L_{PL}(T_P)} = \frac{I^S_{PL}(T_{cw})}{I^L_{PL}(T_{cw})} \frac{\tau^L_{PL}}{\tau^S_{PL}}$$

If fact, we obtain the equation that is presented in the article (a direct consequence of equation 8 in the paper). This equation unambiguously shows that regardless of the different parameters such as absorption cross-section or population (density) of the partial components, the relative contribution of short components under pulsed excitation ($I^S_{PL}(T_P)/I^L_{PL}(T_P)$) increases in comparison with cw-excitation ($I^S_{PL}(T_{cw})/I^L_{PL}(T_{cw})$) because $\tau^L_{PL} > \tau^S_{PL}$. In relation to the question, the short and long components mentioned above can represent the contributions of small and large SiNCs, respectively.

Conflicts of interest

There are no conflicts to declare.

Faraday Discussions

PAPER

Luminescent silicon nanoparticles for distinctive tracking of cellular targeting and trafficking

Gi-Heon Kim,[†a] Goun Lee,[†bc] Myoung-Hee Kang,[a] Minjong Kim,[d] Yusung Jin,[be] Sungjun Beck,[bf] Jihyun Cheon,[bg] Junyeong Sung[bg] and Jinmyoung Joo [*ah]

Received 25th November 2019, Accepted 12th December 2019
DOI: 10.1039/c9fd00124g

Developing therapeutic nanoparticles that actively target disease cells or tissues by exploiting the binding specificity of receptors presented on the cell surface has extensively opened up biomedical applications for drug delivery and imaging. An ideal nanoparticle for biomedical applications is required to report confirmation of relevant targeting and the ultimate fate in a physiological environment for further verification, *e.g.* to adapt dosage or predict response. Herein, we demonstrate tracking of silicon nanoparticles through intrinsic photoluminescence (PL) during the course of cellular targeting and uptake. Time-resolved analysis of PL characteristics in cellular microenvironments provides dynamic information on the physiological conditions where the silicon nanoparticles are exposed. In particular, the PL lifetime of the silicon nanoparticles is in the order of microseconds, which is significantly longer than the nanosecond lifetimes exhibited by fluorescent molecules naturally presented in cells, thus allowing discrimination of the nanoparticles from the cellular background autofluorescence in time-gated imaging. The PL lifetime is a physically intensive

---

[a]*Department of Biomedical Engineering, School of Life Sciences, Ulsan National Institute of Science and Technology (UNIST), Ulsan 44919, Republic of Korea. E-mail: jjoo@unist.ac.kr*

[b]*Biomedical Engineering Research Center, Asan Institute for Life Sciences, Asan Medical Center, Seoul 05505, Republic of Korea*

[c]*Department of Life Sciences, Pohang University of Science and Technology (POSTECH), Pohang 37673, Republic of Korea*

[d]*Department of Biological Science, School of Life Sciences, Ulsan National Institute of Science and Technology (UNIST), Ulsan 44919, Republic of Korea*

[e]*Department of Convergence Medicine, University of Ulsan College of Medicine, Seoul 05505, Republic of Korea*

[f]*Skaggs School of Pharmacy and Pharmaceutical Sciences, University of California, San Diego, La Jolla, CA 92093, USA*

[g]*Department of Chemical Engineering, Pohang University of Science and Technology (POSTECH), Pohang 37673, Republic of Korea*

[h]*Department of Chemical Engineering, School of Energy and Chemical Engineering, Ulsan National Institute of Science and Technology (UNIST), Ulsan 44919, Republic of Korea*

† These authors have contributed equally to this work.

property that reports the inherent characteristics of the nanoparticles regardless of surrounding noise. Furthermore, we investigate a unique means to inform the lifespan of the biodegradable silicon nanoparticles responsive to local microenvironment in the course of endocytosis. A multivalent strategy of nanoparticles for enhanced cell targeting is also demonstrated with complementary analysis of time-resolved PL emission imaging and fluorescence correlation spectroscopy. The result presents the promising potential of the photoluminescent silicon nanoparticles toward advanced cell targeting systems that simultaneously enable tracking of cellular trafficking and tissue microenvironment monitoring.

Introduction

Since the discovery of strong visible photoluminescence (PL) at room temperature in porous silicon prepared by anodic etching in aqueous hydrofluoric acid, much attention has been turned to the fabrication and characterization of nanostructured silicon materials.[1] In particular, visible light emission from nanostructured silicon has stimulated tremendous interest over past decades owing to its potential applications in optoelectronic and biomedical devices.[2-5] Indeed, the idea that silicon is able to emit luminescence has produced great attention to the synthesis and characterization of both porous silicon and silicon nanostructures (*e.g.*, nanoparticles, nanowires, *etc.*).[6-12] The intrinsic PL characteristics are attributed to quantum confinement effects which occur when the size of the crystalline silicon becomes smaller than the 5 nm exciton diameter of bulk silicon.[13] Through the quantum confinement in nanocrystalline silicon, the indirect band gap of silicon becomes brightly photoluminescent, displaying colors that can be tuned across the visible spectrum by controlling the size of the silicon nanocrystals.

Colloidal silicon nanoparticles, a simplified formulation of nanocrystalline silicon, have attracted increasing interest in a broad range of fields due to their elemental advantages including high abundance, low cytotoxicity, favorable biocompatibility, and high feasibility with robust surface chemistry.[14-20] One such application is luminescent labeling, where quantum dots have advantages for *in vitro* and *in vivo* imaging compared to conventional fluorescent molecular dyes.[21] Desirable properties such as biocompatibility, biodegradability and photoluminescence have led to a variety of applications in biomedical research. As an attractive alternative, silicon nanoparticles have become a promising probe as many conventional quantum dots contain toxic heavy metal elements such as cadmium. In this regard, it has been extensively shown that silicon nanoparticles can be used for cell imaging, single-molecule tracking, and *in vivo* labeling with expanding potential for medical probes.[22-25] Furthermore, silicon nanoparticles have the noteworthy feature that they are readily degraded in the biological environment to form nontoxic silicic acid, providing a pathway for their safe excretion from the body.[26-28]

Despite the potential application in biomedical imaging, there have been certain issues rendering silicon nanoparticles less attractive than conventional quantum dots such as cadmium selenide. First of all, the indirect band gap of silicon gives rise to relatively strong thermal broadening of the PL spectra, and this is a particularly important issue for biomedical labeling, in which emission spectra must be sufficiently isolated to permit color discrimination across multiple channels. In addition, the indirect bandgap possessed by crystalline silicon inhibits light emission from electron–hole pair recombination, resulting in inefficient emission (lower quantum efficiency) due to conserved momentum accompanied by phonons. These optical properties of nanocrystalline silicon have been considerably improved by enhancing the overlap of electron and hole wave functions in quantum-confined silicon, which leads to faster recombination and reduced nonradiative rate.[29–33] Moreover, the indirect band structure of silicon gives rise to PL lifetimes in the range of hundreds of nanoseconds to microseconds rather than the typical nanosecond fluorescence lifetimes of molecular dyes or quantum dots, which opens up a way to demonstrate time-gated imaging that eliminates short-lived tissue autofluorescence *in vivo*.[34] Polydispersity of nanocrystalline silicon size is another factor affecting spectral broadening. In particular, the wide emission spectra of porous silicon nanoparticles are attributed to the large range in crystalline silicon size in the skeletal silicon matrix. Efforts toward narrowing the size distribution of silicon fluorophores has resulted in narrower emission spectra, however, they have essentially remained wide because of the ensemble photoluminescence from the nanostructured network of crystalline silicon.[35] In addition, the formation of uniform pore wall thickness, where quantum confinement occurs, in the course of electrochemical etching is hardly achievable due to certain limitations in the pore forming mechanism such as hydrogen evolution and rapid reaction kinetics.[36] In the past decade, chemical synthetic methods for colloidal silicon nanoparticles have been explored for size sorting and purification to produce a narrower size distribution. Typically, these methods include fabrication of a solid-state matrix which embeds the silicon nanocrystals, followed by release of the individual nanoparticles into a solution.[37–40] Direct synthetic methods such as laser ablation and microwave-assisted reaction have also been proposed.[41–43] Since the improvement of silicon nanoparticle PL efficiency is a major issue in the field, a number of studies have reported not only synthetic methods for well-defined quantum domains but also the effect of surface chemistry on the quantum yield and PL bandwidth.[44–50] For instance, rational variation in the surface moieties showed effective tunability of PL emission across the entire visible spectral region without changing the nanoparticle size.[51]

Herein, we present the utility of photoluminescent silicon nanoparticles for tracking cellular uptake and microenvironment monitoring. Harnessing the intrinsic PL of silicon nanoparticles, time-resolved spectroscopic analysis reports the physiological status of the silicon nanoparticles depending on the microenvironment. When it comes to optical properties, silicon nanoparticles are still far behind direct band gap quantum dots (*e.g.*, lower quantum efficiency), and it remains crucial to seek effective ways to improve their optical properties. However, the microsecond-long luminescence lifetime, another indication of the indirect band gap nature of the optical transition in silicon nanoparticles, provides feasible information for tracking cellular microenvironment. Since the PL lifetime is substantially longer than the nanosecond lifetimes typically exhibited by autofluorescence from cells and tissues, emission from the silicon nanoparticles is readily discriminated in the time-resolved PL spectrum.[52] Furthermore, combinatorial analysis of PL intensity and lifetime allows both tracking and assessment of the biodegradation state of the silicon nanoparticles.

Besides this, we demonstrate the use of silicon nanoparticles obtained as side products during the course of ultrasonic fragmentation of porous silicon films. Although study of these silicon nanocrystals is typically neglected, we suggest the reuse of these byproducts as a promising imaging probe. In particular, the PL characteristics of the byproduct silicon nanoparticles show higher quantum efficiency and narrower PL emission spectra compared to the ensemble PL of porous silicon. The silicon nanoparticles have enhanced capability for surface passivation, and the dense oxide layer efficiently prevents non-radiative recombination of the excited electrons upon light irradiation. Along with the improved PL characteristics, the silicon nanoparticles show significant modulation capabilities for cellular targeting and microenvironment monitoring.

Experimental

Preparation of silicon nanoparticles

Silicon nanoparticles were prepared by electrochemical etching of crystalline silicon wafer, followed by ultrasonic fracture as described previously. In brief, boron-doped p-type silicon wafer (polished on ⟨100⟩ face) was anodically etched in an electrolyte consisting of 3 : 1 (v/v) 48% aqueous hydrofluidic acid : ethanol under current control. The etching current waveform consisted of two current level profiles (50 mA cm^{-2} for 1.6 s; 400 mA cm^{-2} for 0.3 s), repeated for 200 cycles. The mesoporous film was then removed from the silicon substrate by applying a current pulse of 4.0 mA cm^{-2} for 250 s in 1 : 30 (v/v) 48% aqueous hydrofluidic acid : ethanol electrolyte. The freestanding porous silicon films were fragmented by ultrasonication overnight in water. The resulting nanoparticles were separated by centrifugation in 21 000 rcf for 15 min. Porous silicon nanoparticles (PSiNPs) and smaller-sized silicon nanoparticles (SiNPs) were collected from the pellet and supernatant, respectively (Fig. 1a). Both the PSiNPs and SiNPs were dispersed in deionized water to grow a thin layer of silicon oxide on the surface for activation of photoluminescence as described elsewhere.[53] SiNPs were then conjugated with RPARPAR peptide by using maleimide–PEG–succinimidyl carboxy methyl ester (MAL–PEG–SCM, MW: 5000) as a linker. A controlled density of the RPARPAR peptide on the SiNPs was obtained by mixing the conjugation linkers MAL–PEG–SCM and mPEG–SCM with the designated ratio.

Characterization of silicon nanoparticles

Transmission electron microscopy (TEM) images were obtained on a JEOL-1200 EX II operating at 120 kV. Dynamic light scattering (DLS, Zetasizer ZS90, Malvern Instruments) was used to determine the hydrodynamic size of the nanoparticles. Time-resolved spectral analyses were conducted with a home-built system as described previously.[52] The imaging system was equipped with an intensified CCD camera (iSTAR-sCMOS-18F-73, Oxford Instruments), spectrograph (Kymera 193i, Oxford Instruments) and a light emitting diode (LED, λ_{ex}: 365 nm, Ocean Optics) was used as the excitation source at a repetition rate of 10 Hz, which was externally synchronized and triggered by a function generator (Keithley 3390). Andor Solid software was used to acquire the PL spectral and optical images as well as to analyze the data. The integrated PL intensity was obtained in the wavelength range 550–900 nm.

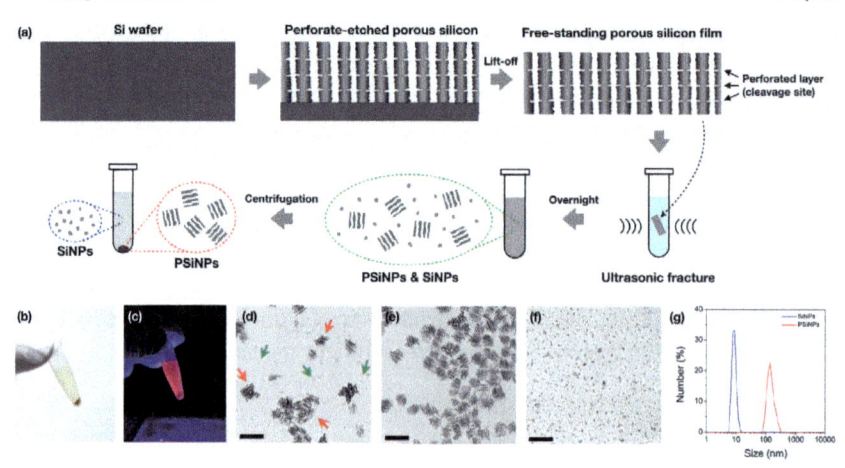

Fig. 1 (a) Schematic illustration depicting the silicon nanoparticle preparation. (b and c) Photographs of the silicon nanoparticles under (b) ambient light and (c) UV irradiation (λ: 365 nm) after centrifugal separation. Note that the supernatant contains smaller-sized SiNPs while the PSiNPs were spun down as a pellet. (d–f) Transmission electron microscopy images of the silicon nanoparticles (d) before and (e and f) after centrifugation. Prior to centrifugal separation, both SiNPs (green arrows) and PSiNPs (red arrows) were obtained through the ultrasonic fragmentation process (d), and they were then completely separated by centrifugation: (e) re-suspended PSiNP pellet and (f) SiNPs collected in the supernatant. Scale bar: 200 nm. (g) Size distribution of the silicon nanoparticles after centrifugal separation measured by dynamic light scattering.

In vitro cell targeting and biodegradation tracking

SiNPs were suspended in phosphate-buffered saline (PBS) and maintained at 37 °C with mild shaking (300 rpm) in an incubator. Dynamic PL spectral characteristics were measured using the above imaging system with an LED excitation source. For the cell targeting experiments, SW-480 cells were seeded and cultured in glass-bottom dishes (Nunc, ThermoFisher Scientific). Cells were mounted and observed through fluorescence microscopic imaging with a time-gated system. In parallel, cells were also analyzed with a confocal laser scanning microscope (LSM880 equipped with fluorescence correlation spectroscopy, Carl Zeiss).

Results and discussion

Silicon nanoparticles were prepared by electrochemical perforation etching of highly doped p-type single crystalline silicon wafers in ethanolic hydrofluoric acid electrolyte, followed by lift-off and ultrasonic fracture, as reported in the literature (Fig. 1a).[54] The perforation etching yields porous silicon nanoparticles (PSiNPs) with a well-defined mesoporous structure (pore size ∼10 nm) and narrow size distribution (Fig. 1b).[12] Interestingly, smaller-sized silicon nanoparticles (SiNPs) were also obtained as a side product during the course of ultrasonic fragmentation of the porous silicon film, and were separated from the PSiNPs by centrifugation (Fig. 1c). Although these SiNPs, the crystalline silicon fragments, are typically subjected to removal after centrifugation and were not intended for further study, we purified and chemically functionalized these SiNPs to utilize them as an imaging probe in the present study. It should be noted that various synthetic methods to produce highly crystalline SiNPs typically involve microwave irradiation or heat treatment (>1000 °C).[43–45,51,55] These synthetic approaches can be used to easily produce high-quality SiNPs, which enables a broad range of applications for functionalized SiNPs. However, the present preparation route also provides facile production of highly crystalline SiNPs as part of a top-down fabrication strategy, which is an extra advantage as it does not require an additional heating process.

Both PSiNPs and SiNPs were then stored in deionized water to form a native oxide layer, which leads to photoluminescence (PL) activation attributed to passivation of the nonradiative surface defects. The passivating silicon oxide layer grows on the hydrogen-terminated silicon surface through nucleophilic attack, and allows the replacement of dangling bonds and other surface states which are responsible for the nonradiative recombination of charge carriers in nanocrystalline silicon.[56] However, the nanocrystalline silicon embedded in the silicon oxide matrix of PSiNPs has a broad size distribution of quantum-confined domains, which thus leads to a wide range of PL emission in the near-infrared region (500–950 nm), while the individual SiNPs show a relatively narrower PL emission peak (Fig. 2a). Moreover, the SiNPs showed a higher PL quantum yield (∼29%) than the PSiNPs (∼9%). The PL quantum yield of the ensemble of silicon nanocrystals embedded in PSiNPs have generally been reported to be in the few percent range.[24] Improvement of the optical properties has been achieved through careful passivation of the silicon surface, and an ensemble PL quantum yield for PSiNPs as high as 23% was reported by controlled oxidation with sodium borate.[53] However, electrochemically etched porous silicon consists of a network of silicon nanocrystals, which have generally lower optical performance and ensemble PL quantum yield due to the generation of nonradiative defects at the initially hydrogen-terminated surface upon the current flowing and oxidation process. In contrast, SiNPs, the unintended byproduct of PSiNP preparation, have favorable accessibility for water molecules for surface passivation, and thus show an improved quantum yield of up to 29%. The enhanced degree of surface oxidation which promotes sufficiently passivated nonradiative defects, enabled by

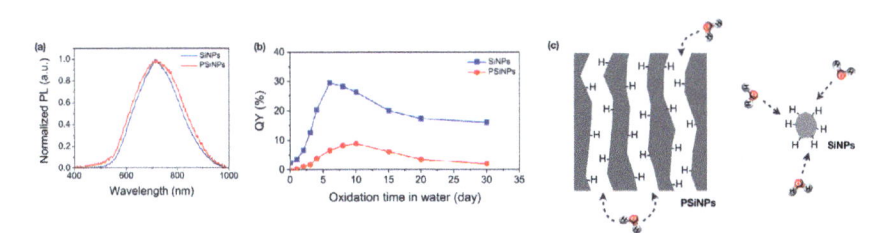

Fig. 2 (a) Normalized PL spectra of SiNPs (blue) and PSiNPs (red) showing the narrower spectral emission bandwidth of the SiNPs. (b) PL quantum yield (QY) of SiNPs (blue) and PSiNPs (red) measured as a function of oxidation time in deionized water at the same concentration (4 ml ml^{-1}). (c) Schematic illustrations of PSiNP and SiNP oxidation in water. Note that the hydrogen-terminated PSiNPs obtained from the etching process have a hydrophobic surface, possibly resulting in less infiltration of water molecules, while the SiNPs are highly accessible to water for sufficient oxide layer formation and defect passivation.

nucleophilic attack on SiNPs being more accessible than on PSiNPs, results in rapid PL activation and a higher quantum yield for a longer period of time (Fig. 2b and c). The results can be interpreted by the significantly lowered interfacial defect density due to the favorable features for surface passivation in the SiNPs, which is responsible for the higher quality oxide layer, isolating the silicon nanocrystals from the defects and dangling bonds.

The intrinsic PL diminishes during the course of biodegradation in a physiological environment (phosphate-buffered saline (PBS), pH 7.4, 37 °C), as shown in Fig. 3a. The monotonic blue shift of the PL peak wavelength along with the biodegradation is indicative of the shrinking in size of the quantum-confined crystalline silicon domain. In addition, the PL intensity is observed to continuously decrease due to gradual dissolution of the SiNPs, in accordance with the decreasing number of silicon fluorophores and increasing contribution of non-radiative recombination such as surface defects formed by silicon dissolution. Due to the PL characteristics reporting the biodegradation status, we can discern the course of biodegradation by tracking the continuous PL intensity decrease. However, PL intensity-based information is often incorrect and causes misleading results due to endogenous autofluorescence caused by inherent cellular components such as collagen, tryptophan, and nicotinamide found in all living cells. The tissue autofluorescence overlapping with the PL emissive spectral range of the imaging probes hampers accurate analysis. Moreover, continuous cellular uptake of the nanoparticles and biodegradation cause concerns about conflicting interpretations, as they show opposite contributions on the PL intensity accumulation.

Alternatively, the PL decay lifetime can be easily differentiated from the complicated interference raised by the surrounding tissue autofluorescence as it is a physically intensive property, which enables discernment of the intrinsic status of the SiNPs with cellular environment. In this regard, we next investigated the dynamic PL characteristics in time-resolved analysis. It should be noted that longer PL emission wavelengths (lower emission energy) exhibit a slower decay rate compared to that of shorter PL emission wavelengths (higher emission

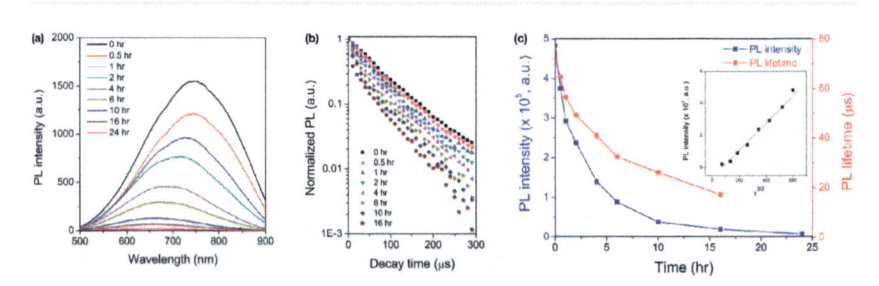

Fig. 3 (a) Steady-state PL emission spectra (λ: 365 nm) of the SiNPs during the course of biodegradation in PBS (pH 7.4, 37 °C). (b) Corresponding time-resolved PL decay curve. The PL spectra are integrated over the wavelength range 500–900 nm and normalized for comparison in the course of biodegradation. (c) Representative PL intensity and decay lifetime as a function of biodegradation time. Inset: correlation between the PL intensity (*I*) and lifetime (τ): ($I \propto \tau^{3/2}$). Note that the PL intensity decreases as a result of the reduced number of silicon fluorophores in the given volume (r^3) of nanoparticle through degradation, while the PL lifetime is determined by the area (r^2) of the quantum-confined domain (*r*: size of SiNPs).

energy), in accordance with the quantum-confinement model of silicon.[35,52] In addition, since the PL spectrum showed blue shift in the continuous biodegradation process, the PL decay lifetime of integrated PL intensity tends to shorten correspondingly (Fig. 3b and c). This is a significant PL feature of the SiNPs which is worthy of attention for cellular behavior tracking, because the PL lifetime is readily acquired and differentiated from surrounding noise. On the other hand, the PL intensity-based tracking of nanoparticles often needs to be carefully moderated due to the autofluorescence artifacts raised by endogenous fluorophores of cellular components, by using particular emission filters to reduce the background noise. Although both the steady-state PL intensity under continuous excitation and the time-resolved PL signature with pulsed excitation show strong correlation and decreasing profiles related to the biodegradation of SiNPs during PBS incubation (Fig. 3c), cellular uptake and tissue accumulation should be considered as a more complex system involved with the simultaneous process of increasing (uptake of nanoparticles) and decreasing (biodegradation of nanoparticles) factors. The PL lifetime is, however, a physically intensive property solely accompanying the biodegradation status of SiNPs, and has a novel feature that informs the inherent properties of the nanoparticle itself with the cells.

To further demonstrate the feasibility of long-lived PL emission for cellular tracking, the SiNPs were then conjugated with a tumor-targeting peptide, RPARPAR. The RPARPAR peptide has been known to bind specifically to neuropilin-1 (NRP-1), a cell-surface receptor overexpressed onto tumor cells.[57] NRP-1 plays an essential role in angiogenesis, regulation of vascular permeability, and the development of the nervous system. When incubated with NRP-1 positive colorectal cancer cell line, SW-480 cells, significant uptake of the SiNPs was observed (Fig. 4a). It should be noted that endogenous cellular components were also detected in the same excitation/emission filter set of SiNPs, thus making it difficult to clearly distinguish the nanoparticles from the background noise in conventional continuous wave imaging (CWI) acquired under light irradiation. However, time-gated imaging (TGI) is able to completely eliminate short-lived autofluorescence, and presents only PL signals originating from SiNPs while suppressing unwanted fluorescence arising from cellular components (Fig. 4a). Therefore, we can clearly confirm that the SiNPs tend to be evenly distributed at the cytosol in the early state of incubation (~2 h), and then gradually accumulate at the perinuclear region in later stages.

By tracking the PL signal during the course of cellular uptake, increasing PL intensity in the cytoplasmic area was obtained upon extended exposure time up to 2 h, corresponding to continuous intracellular uptake of SiNPs. However, it should be noted that a gradual decrease in PL intensity followed for the cells incubated with SiNPs for a longer period of time (Fig. 4b). Although cellular uptake of nanoparticles is supposed to increase as time progresses, the decrease in PL intensity might result from the reduced PL intensity emitted by each SiNP due to biodegradation as noted above (Fig. 3c). Therefore, tracking the nanoparticles and their associated cellular uptake with a PL intensity-based study raises concerns about conflicting interpretations. This should be considered for biodegradable imaging probes as they inform the cellular behavior through emission intensity at desired optical filter sets. There are two conflicting changes in the characteristics of PL intensity: (1) the SiNPs continuously accumulate in cells through receptor-mediated endocytosis (the increasing factor), while (2) the

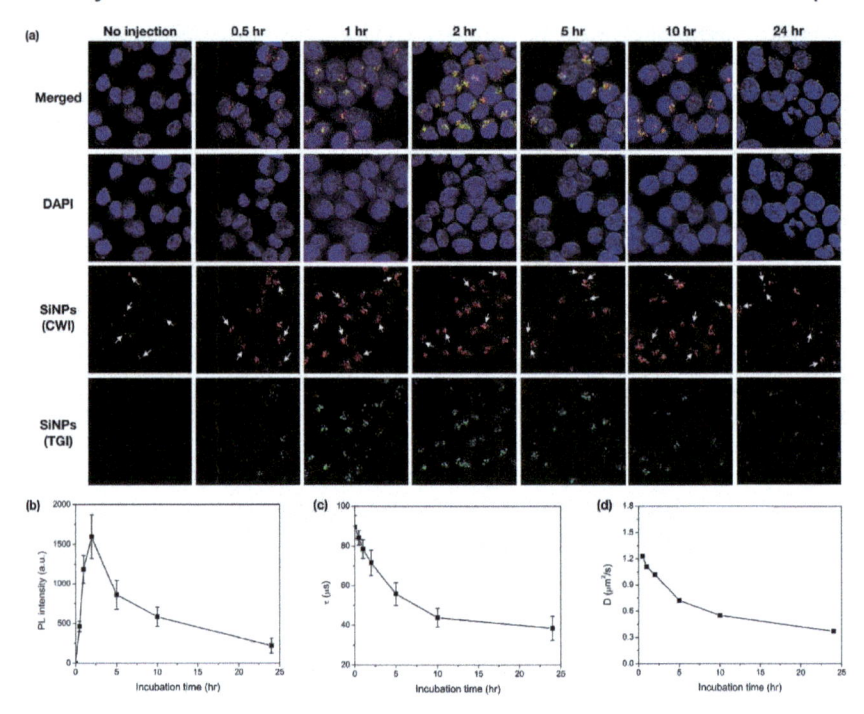

Fig. 4 (a) Cellular trafficking of SiNPs via receptor-medicated endocytosis. Note that cellular component-induced autofluorescence (white arrows) obtained under continuous wave imaging (CWI, red) is removed in time-gated imaging (TGI, green). (b) PL intensity and (c) decay lifetime detected from the intercellular regions, obtained from TGI images as a function of incubation time. (d) Mean diffusion coefficient (D) of SiNPs in the cytosol measured using fluorescence correlation spectroscopy (FCS) at various time points of endocytosis. The decreasing trend of D indicates trapped SiNPs in endosomes and lysosomes rather than free diffusion in cytosol.

SiNPs simultaneously degrade to be less photoluminescent emitters (the decreasing factor). However, the PL decay lifetime showed a monotonic decrease with incubation time (Fig. 4c), resulting from the continuous biodegradation of SiNPs in the cellular uptake process. By using the PL lifetime as an optical signature, one can estimate the present state of the SiNPs in terms of biodegradation, providing a comprehensive platform when combined with the intensity-based analysis.

Intracellular trafficking of the SiNPs over the extended period of endocytosis also shows a gradual formation of clusters accompanied with slower diffusional movement (Fig. 4d). Fluorescence correlation spectroscopy (FCS) analysis revealed that a significant proportion of the SiNPs initially show fast diffusion of individual nanoparticles, followed by cluster formation in the perinuclear region of cytosol at later stages of endocytosis. This process is also presumably related to the intracellular clearance pathway of SiNPs over time associated with autophagy as well. Correspondingly, relatively slow biodegradation of the SiNPs in the course of intracellular trafficking compared to *in vitro* incubation (PBS, pH 7.4, 37 °C in Fig. 3c) is responsible for the lower local pH at endosomes and lysosomes. The SiNPs are vulnerable to biological media owing to continuous oxidation and hydrolysis reaction, which leads to complete degradation under aqueous conditions, and it has been widely accepted that the degradation kinetics of silica nanoparticles is slower in an acidic environment.[58–60] The physiological environment in PBS and in culture media is different, and particularly the lower pH of endosomes (~6.0) and lysosomes (4.0–5.0) inside the cells correspondingly suggest that the SiNPs degrade slowly once they enter the endosomes *via* endocytosis.

We then investigated the validity of the SiNPs to examine the concept of cell-targeted multivalent strategies for developing theranostic systems.[61] A key challenge in diagnostics and therapeutics is the ability to specifically target cells and tissues. Targeting strategies typically rely on both (1) identifying a suitable marker such as overexpressed receptors and (2) choosing a ligand that strongly and specifically binds to the receptor.[62] Since the cellular uptake of multivalent nanoparticles depends on strong surface adhesion over a period of time to achieve endocytosis,[63–66] the multivalent mechanism plays a critical role in the cell specificity of nanoparticle adhesion to the tumor-associated receptors. To verify the use of multivalent nanoparticles decorated with a homing peptide, RPARPAR, for efficient colorectal cancer targeting in our system, cellular uptake of SiNPs which have different degree of multivalency (peptide density on the surface) has been conducted. These SiNPs are bound to and taken up into the SW-480 cells with varied populations which is related to the peptide density (Fig. 5a). With the

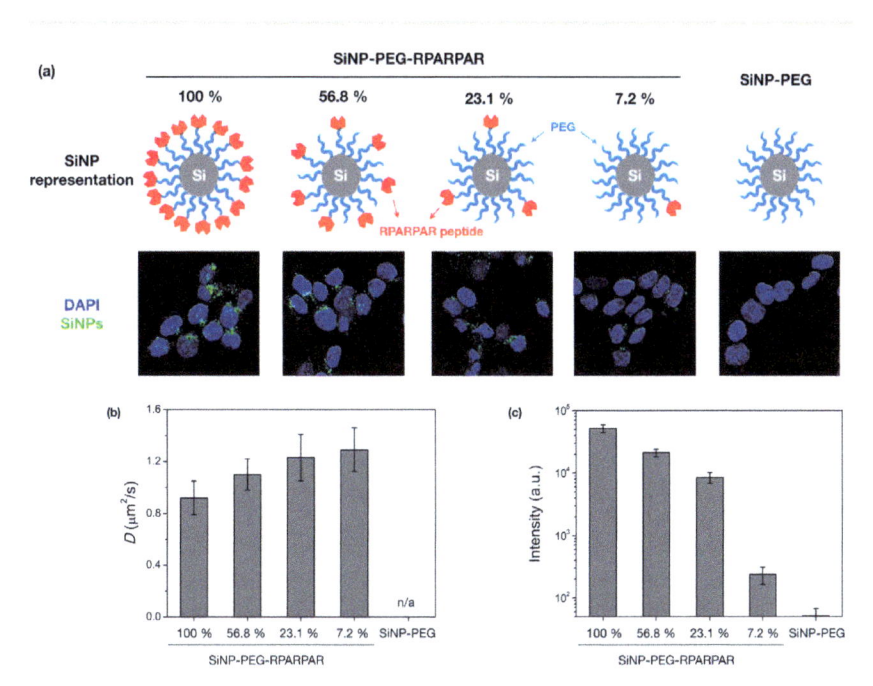

Fig. 5 (a) Schematic representation of SiNPs decorated with differing coverage of RPARPAR peptide on the surface, and the corresponding fluorescence images (for DAPI staining, blue) and time-gated images (for SiNPs, green) showing cellular uptake. (b) Mean diffusion coefficient of SiNPs detected through fluorescence correlation spectroscopy. (c) Mean PL intensity detected from the unit area in cells as a function of the peptide coverage showing enhanced cellular targeting capability upon a higher degree of multivalency.

injection concentration being constant, each SiNP shows a substantially different efficacy of binding and cellular uptake at the same period of incubation time, resulting in a strong relationship with the degree of multivalency. A gradual shift to the lower diffusion coefficient (D) in the later stages of endocytosis indicates decelerating diffusional motion, possibly due to the increased population of SiNPs trapped in either endosomes or lysosomes (Fig. 4d). Similarly, the lower mean diffusion coefficient (D) of the SiNPs with a higher degree of multivalency indicates rapid uptake and cellular trafficking of SiNPs for higher peptide density conjugates (Fig. 5b). The results suggest that the SiNPs with a lower degree of multivalency are also bound to the NRP-1 receptors on the cell surface but dissociate faster than when bound to a higher density of RPARPAR peptides. Thus, the decrease in peptide density on the SiNPs reduces not only the quantity of the nanoparticles adsorbed to the target cells but also the overall strength of the nanoparticle-cell surface interaction (Fig. 5c). Since the SW-480 cells have a high surface density of overexpressed NRP-1, the RPARPAR-targeted SiNPs can engage numerous NRP-1 receptors simultaneously to provide enhanced interactions over the lower multivalent SiNPs. We believe these findings are supportive of the mechanism of specific cell targeting being derived by certain multivalent nanoparticles adhering tightly and therefore more selectively to the cell surface with a higher degree of targeting motif decoration. Our study combined with time-gated imaging and time-resolved PL emission analysis on the cellular trafficking of SiNPs provides a unique platform that reports cellular binding kinetics determined by the multivalency of nanoparticles with negligible artifacts caused by endogenous autofluorescence background noise.

Conclusion

Photoluminescent silicon nanoparticles have great potential in the field of bioimaging. In particular, long-lived PL emission allows complete elimination of the substantial tissue autofluorescence background, resulting in clear distinguishment of the imaging probe itself without the possible concerns of misleading results caused by endogenous fluorophores in cellular components. Feasible tracking of the PL emission lifetime over the short-lived autofluorescence background further enables facile reporting of the status of biodegradation upon cellular uptake. Along with two key features of intrinsic PL-based analysis, we have verified the targeted binding and uptake of SiNPs as a receptor-mediated endocytosis, where the nanoparticles showed reduced biodegradation kinetics affected by continuous changes in the local pH of endosomes and lysosomes. In addition, enhanced binding affinity and uptake efficiency of the nanoparticles through multivalent interactions with the target cells were clearly proven with complementary analysis of fluorescence correlation spectroscopy. The unique PL signature of the SiNPs will open up the potential to analyze the *in vivo* behavior during circulation and targeting as a novel theranostic nanomedicine. Moreover, the PL properties can be adopted to report microenvironmental changes in the local region of tissue by controlling the surface chemistry or enzymatic reactions.

Conflict of interest

The authors declare no competing financial interests.

Acknowledgements

This work was supported by the Basic Science Research Program through the National Research Foundation of Korea funded by Ministry of Education (2017R1D1A1B03035525), Settlement Research Fund (1.190040.01) of UNIST, and POSCO Science Fellowship of the POSCO TJ Park Foundation.

References

1 L. T. Canham, *Appl. Phys. Lett.*, 1990, **57**, 1046–1048.
2 D. Timmerman, I. Izeddin, P. Stallinga, I. N. Yassievich and T. Gregorkiewicz, *Nat. Photonics*, 2008, **2**, 105–109.
3 D. Kovalev and M. Fujii, *Adv. Mater.*, 2005, **17**, 2531–2544.
4 C. M. Hessel, M. R. Rasch, J. L. Hueso, B. W. Goodfellow, V. A. Akhavan, P. Puvanakrishnan, J. W. Tunnel and B. A. Korgel, *Small*, 2010, **6**, 2026–2034.
5 M. L. Mastronardi, E. J. Henderson, D. P. Puzzo and G. A. Ozin, *Adv. Mater.*, 2012, **24**, 5890–5898.
6 M. C. Beard, K. P. Knutsen, P. R. Yu, J. M. Luther, Q. Song, W. K. Metzger, R. J. Ellingson and A. J. Nozik, *Nano Lett.*, 2007, **7**, 2506–2512.
7 S. Sato and M. T. Swihart, *Chem. Mater.*, 2006, **18**, 4083–4088.
8 J. D. Holmes, K. J. Ziegler, R. C. Doty, L. E. Pell, K. P. Johnston and B. A. Korgel, *J. Am. Chem. Soc.*, 2001, **123**, 3743–3748.
9 D. Mariotti and R. M. Sankaran, *J. Phys. D: Appl. Phys.*, 2010, **43**, 323001.
10 R. M. Sankaran, D. Holunga, R. C. Flagan and K. P. Giapis, *Nano Lett.*, 2005, **5**, 537–541.
11 J. Joo, T. Defforge, A. Loni, D. Kim, Z. Y. Li, M. J. Sailor, G. Gautier and L. T. Canham, *Appl. Phys. Lett.*, 2016, **108**, 153111.
12 Z. Qin, J. Joo, L. Gu and M. J. Sailor, *Part. Part. Syst. Charact.*, 2014, **31**, 252–256.
13 A. G. Cullis and L. T. Canham, *Nature*, 1991, **353**, 335–338.
14 M. A. Green, J. H. Zhao, A. H. Wang, P. J. Reece and M. Gal, *Nature*, 2001, **412**, 805–808.
15 L. Pavesi, L. Dal Negro, C. Mazzoleni, G. Franzo and F. Priolo, *Nature*, 2000, **408**, 440–444.
16 Z. F. Ding, B. M. Quinn, S. K. Haram, L. E. Pell, B. A. Korgel and A. J. Bard, *Science*, 2002, **296**, 1293–1297.
17 A. Shiohara, S. Hanada, S. Prabakar, K. Fujioka, T. H. Lim, K. Yamamoto, P. T. Northcote and R. D. Tilley, *J. Am. Chem. Soc.*, 2010, **132**, 248–253.
18 Q. Li, Y. He, J. Chang, L. Wang, H. Z. Chen, Y. W. Tan, H. Y. Wang and Z. Z. Shao, *J. Am. Chem. Soc.*, 2013, **135**, 14924–14927.
19 S. Hussain, J. Joo, J. Kang, B. Kim, G. B. Braun, Z.-G. She, D. Kim, A. P. Mann, T. Mölder, T. Teesalu, S. Carnazza, S. Guglielmino, M. J. Sailor and E. Ruoslahti, *Nat. Biomed. Eng.*, 2018, **2**, 95–103.
20 T. Gong, Y. J. Li, B. F. Lei, X. J. Zhang, Y. L. Liu and H. R. Zhang, *J. Mater. Chem. C*, 2019, **7**, 5962–5969.
21 U. Resch-Genger, M. Grabolle, S. Cavaliere-Jaricot, R. Nitschke and T. Nann, *Nat. Methods*, 2008, **5**, 763–775.
22 F. Erogbogbo, K. T. Yong, I. Roy, G. X. Xu, P. N. Prasad and M. T. Swihart, *ACS Nano*, 2008, **2**, 873–878.

23 H. Nishimura, K. Ritchie, R. S. Kasai, N. Morone, H. Sugimura, K. Tanaka, I. Sase, A. Yoshimura, Y. Nakano, T. K. Fujiwara and A. Kusumi, *J. Cell Biol.*, 2013, **202**, 967–983.
24 J. H. Park, L. Gu, G. von Maltzahn, E. Ruoslahti, S. N. Bhatia and M. J. Sailor, *Nat. Mater.*, 2009, **8**, 331–336.
25 A. P. Mann, P. Scodeller, S. Hussain, J. Joo, E. Kwon, G. B. Braun, T. Mölder, Z.-G. She, V. R. Kotamraju, B. Ranscht, S. Krajewski, T. Teesalu, S. Bhatia, M. J. Sailor and E. Ruoslahti, *Nat. Commun.*, 2016, **7**, 11980.
26 E. C. Wu, J. S. Andrew, A. Buyanin, J. M. Kinsella and M. J. Sailor, *Chem. Commun.*, 2011, **47**, 5699–5701.
27 V. Chirvony, A. Chyrvonaya, J. Ovejero, E. Matveeva, B. Goller, D. Kovalev, A. Huygens and P. de Witte, *Adv. Mater.*, 2007, **19**, 2967–2972.
28 L. Gu, D. J. Hall, Z. T. Qin, E. Anglin, J. Joo, D. J. Mooney, S. B. Howell and M. J. Sailor, *Nat. Commun.*, 2013, **4**, 2326.
29 L. E. Brus, P. F. Szajowski, W. L. Wilson, T. D. Harris, S. Schuppler and P. H. Citrin, *J. Am. Chem. Soc.*, 1995, **117**, 2915–2922.
30 C. Delerue, G. Allan and M. Lannoo, *Phys. Rev. B: Condens. Matter Mater. Phys.*, 2001, **64**, 193402.
31 R. Anthony and U. Kortshagen, *Phys. Rev. B: Condens. Matter Mater. Phys.*, 2009, **80**, 115407.
32 W. D. A. M. de Boer, D. Timmerman, K. Dohnalova, I. N. Yassievich, H. Zhang, W. J. Buma and T. Gregorkiewicz, *Nat. Nanotechnol.*, 2010, **5**, 878–884.
33 D. Timmerman, J. Valenta, K. Dohnalova, W. D. A. M. de Boer and T. Gregorkiewicz, *Nat. Nanotechnol.*, 2011, **6**, 710–713.
34 J. Joo, X. Y. Liu, V. R. Kotamraju, E. Ruoslahti, Y. Nam and M. J. Sailor, *ACS Nano*, 2015, **9**, 6233–6241.
35 W. L. Wilson, P. F. Szajowski and L. E. Brus, *Science*, 1993, **262**, 1242–1244.
36 M. J. Sailor and E. C. Wu, *Adv. Funct. Mater.*, 2009, **19**, 3195–3208.
37 C. M. Hessel, E. J. Henderson and J. G. C. Veinot, *Chem. Mater.*, 2006, **18**, 6139–6146.
38 A. Fojtik and A. Henglein, *J. Phys. Chem. B*, 2006, **110**, 1994–1998.
39 H. Sugimoto, M. Fujii, K. Imakita, S. Hayashi and K. Akamatsu, *J. Phys. Chem. C*, 2012, **116**, 17969–17974.
40 L. Mangolini, E. Thimsen and U. Kortshagen, *Nano Lett.*, 2005, **5**, 655–659.
41 T. M. Atkins, A. Y. Louie and S. M. Kauzlarich, *Nanotechnology*, 2012, **23**, 294006.
42 N. Shirahata, M. R. Linford, S. Furumi, L. Pei, Y. Sakka, R. J. Gates and M. C. Asplund, *Chem. Commun.*, 2009, 4684–4686, DOI: 10.1039/b905777c.
43 S. P. Pujari, H. Driss, F. Bannani, B. van Lagen and H. Zuilhof, *Chem. Mater.*, 2018, **30**, 6503–6512.
44 C. M. Hessel, D. Reid, M. G. Panthani, M. R. Rasch, B. W. Goodfellow, J. W. Wei, H. Fujii, V. Akhavan and B. A. Korgel, *Chem. Mater.*, 2012, **24**, 393–401.
45 S. M. Liu, Y. Yang, S. Sato and K. Kimura, *Chem. Mater.*, 2006, **18**, 637–642.
46 Q. Li, T. Y. Luo, M. Zhou, H. Abroshan, J. C. Huang, H. J. Kim, N. L. Rosi, Z. Z. Shao and R. C. Jin, *ACS Nano*, 2016, **10**, 8385–8393.
47 F. J. Hua, F. Erogbogbo, M. T. Swihart and E. Ruckenstein, *Langmuir*, 2006, **22**, 4363–4370.

48 W. Y. So, Q. Li, C. M. Legaspi, B. Redler, K. M. Koe, R. C. Jin and L. A. Peteanu, *ACS Nano*, 2018, **12**, 7232–7238.
49 B. V. Oliinyk, D. Korytko, V. Lysenko and S. Alekseev, *Chem. Mater.*, 2019, **31**, 7167–7172.
50 A. Inoue, H. Sugimoto and M. Fujii, *J. Phys. Chem. C*, 2017, **121**, 11609–11615.
51 M. Dasog, G. B. De los Reyes, L. V. Titova, F. A. Hegmann and J. G. C. Veinot, *ACS Nano*, 2014, **8**, 9636–9648.
52 Y. S. Jin, D. Kim, H. Roh, S. Kim, S. Hussain, J. Y. Kang, C. G. Pack, J. K. Kim, S. J. Myung, E. Ruoslahti, M. J. Sailor, S. C. Kim and J. Joo, *Adv. Mater.*, 2018, **30**, 1802878.
53 J. Joo, J. F. Cruz, S. Vijayakumar, J. Grondek and M. J. Sailor, *Adv. Funct. Mater.*, 2014, **24**, 5688–5694.
54 J. Joo, E. J. Kwon, J. Y. Kang, M. Skalak, E. J. Anglin, A. P. Mann, E. Ruoslahti, S. N. Bhatia and M. J. Sailor, *Nanoscale Horiz.*, 2016, **1**, 407–414.
55 M. Fujii, H. Sugimoto and S. Kano, *Chem. Commun.*, 2018, **54**, 4375–4389.
56 A. Sa'ar, *J. Nanophotonics*, 2009, **3**, 032501.
57 T. Teesalu, K. N. Sugahara, V. R. Kotamraju and E. Ruoslahti, *Proc. Natl. Acad. Sci. U. S. A.*, 2009, **106**, 16157–16162.
58 S. A. Yang, S. Choi, S. M. Jeon and J. Yu, *Sci. Rep.*, 2018, **8**, 185.
59 H. Meng, W. X. Mai, H. Y. Zhang, M. Xue, T. Xia, S. J. Lin, X. Wang, Y. Zhao, Z. X. Ji, J. I. Zink and A. E. Nel, *ACS Nano*, 2013, **7**, 994–1005.
60 J. L. Townson, Y. S. Lin, J. O. Agola, E. C. Carnes, H. S. Leong, J. D. Lewis, C. L. Haynes and C. J. Brinker, *J. Am. Chem. Soc.*, 2013, **135**, 16030–16033.
61 M. E. Davis, Z. Chen and D. M. Shin, *Nat. Rev. Drug Discovery*, 2008, **7**, 771–782.
62 C. B. Carlson, P. Mowery, R. M. Owen, E. C. Dykhuizen and L. L. Kiessling, *ACS Chem. Biol.*, 2007, **2**, 119–127.
63 T. Curk, J. Dobnikar and D. Frenkel, *Proc. Natl. Acad. Sci. U. S. A.*, 2017, **114**, 7210–7215.
64 J. E. Silpe, M. Sumit, T. P. Thomas, B. H. Huang, A. Kotlyar, M. A. van Dongen, M. M. B. Holl, B. G. Orr and S. K. Choi, *ACS Chem. Biol.*, 2013, **8**, 2063–2071.
65 T. P. Thomas, B. H. Huang, S. K. Choi, J. E. Silpe, A. Kotlyar, A. M. Desai, H. Zong, J. Gam, M. Joice and J. R. Baker, *Mol. Pharm.*, 2012, **9**, 2669–2676.
66 M. Mammen, S. K. Choi and G. M. Whitesides, *Angew. Chem., Int. Ed.*, 1998, **37**, 2755–2794.

Faraday Discussions

PAPER

The effects of drying technique and surface pre-treatment on the cytotoxicity and dissolution rate of luminescent porous silicon quantum dots in model fluids and living cells†

Maxim B. Gongalsky, ‡[a] Uliana A. Tsurikova,‡[a] Catherine J. Storey,[b] Yana V. Evstratova,[c] Andrew A. Kudryavtsev,[ac] Leigh T. Canham *[b] and Liubov A. Osminkina *[ad]

Received 9th October 2019, Accepted 31st October 2019
DOI: 10.1039/c9fd00107g

Tailoring of the biodegradation of photoluminescent silicon quantum dots (Si QDs) is important for their future applications in diagnostics and therapy. Here, the effect of drying and surface pretreatment on the dissolution rate of Si QDs in model liquids and living cells was studied *in vitro* using a combination of photoluminescence and Raman micro-spectroscopy. Porous silicon particles were obtained by mechanical milling of electrochemically etched mesoporous silicon films, and consist of interlinked silicon nanocrystals (QDs) and pores. The samples were subjected to super-critical drying with CO_2 solvent (SCD) or air drying (AD) and then annealed at 600 °C for 16 hours in 1% oxygen to obtain nano-sized Si QDs. The obtained samples were characterized by a core–shell structure with a crystalline silicon core and a SiO_2 layer on the surface. The sizes of the crystalline silicon cores, calculated from Raman scattering spectra, were about 4.5 nm for the initial AD-SiQDs, and about 2 nm for the initial SCD-SiQDs. Both the AD-Si QDs and the SCD-Si QDs exhibited visible photoluminescence (PL) properties due to quantum confinement effects. The dissolution of the nanocrystals was evaluated through their PL quenching, as well as by the presence of a low-frequency shift, broadening, and a decrease in the intensity of the Raman signal. The stability of the AD-Si QDs and the complete dissolution of the SCD-Si QDs during 24 hours of incubation

[a]*Department of Physics, Lomonosov Moscow State University, Leninskie Gory 1, 119991 Moscow, Russian Federation. E-mail: osminkina@physics.msu.ru*
[b]*Nanoscale Physics, Chemistry, and Engineering Research Laboratory, University of Birmingham, Birmingham B15 2TT, UK. E-mail: l.t.canham@bham.ac.uk*
[c]*Institute of Theoretical and Experimental Biophysics, Russian Academy of Science, 142290 Pushchino, Moscow Region, Russian Federation*
[d]*Institute for Biological Instrumentation of Russian Academy of Sciences, 142290 Pushchino, Moscow Region, Russian Federation*

† Electronic supplementary information (ESI) available. See DOI: 10.1039/c9fd00107g
‡ Equally contributing authors.

with cells have been demonstrated. This might explain the apparent lower cytotoxicity observed for SCD-Si QDs.

Introduction

Porous silicon (PSi) is a nanomaterial of great interest since the discovery of its visible photoluminescence (PL).[1] The origin of PL is quantum confinement of excitons in silicon nanocrystals or silicon quantum dots (Si QDs). The quantum yield of PL can exceed 80% for individual Si QDs[2] and 30% for Si QD ensembles.[3] Furthermore, PSi fractured into nanoparticles (PSi NPs) can carry various drugs due to its developed pore network.[4] The combination of PL and drug delivery in the same nanoparticles opens up the potential for PSi NPs to be theranostics agents.[5] Moreover, the therapeutic effect of PSi NPs can be enhanced by efficient sensitizing of external fields, *e.g.* light[6] and ultrasound.[7]

On the one hand, PSi based nanomaterials demonstrating complete biodegradability is desirable for biomedical applications.[8,9] The mechanism of biodegradation is hydrolytic dissolution of silicon resulting in formation of silicic acid $Si(OH)_4$.[8] The products of dissolution and small particles are easily excreted from an organism *via* renal pathways.[10] On the other hand, the use of biodegradability requires precise tailoring of both degradation and *in vivo* clearance times, which are dependent on a variety of factors.[11] Some of them correspond to the intrinsic morphology of the nanocontainers such as the size of the nanoparticles, the porosity, and the size of the pores.[12] Other factors involve variation of the surface composition achieved by surface treatment such as thermal[13] or chemical[14,15] oxidation of Si QDs or coating with biopolymers[16] or silica.[13] The combination of these factors facilitates adjustment of the biodegradation time for PSi NPs in the range of 2 hours to several weeks. However, the degradation time may also be affected by variation of the local environment, *e.g.* the concentration of reactive oxygen species.[17] Surface modification can alter the biodistribution of nanoparticles[9] and change their toxicity profiles.[18]

Intense development of multifunctional nanoparticles requires powerful characterization methods for Si QD biodegradation which are universally applicable *in vitro* and *in vivo*. Unlike many commonly used indirect methods, Raman spectroscopy provides convenient monitoring of Si QD biodegradation including phase analysis and estimation of their diameter *in vitro*.[19] This approach can also provide information on the drug release from silicon nanocontainers.[20] The combination of Raman scattering with PL of Si QDs could potentially be a perfect framework for the study of biodegradation, since Raman spectra give detailed information on the morphology while PL monitors localization of Si QDs in cell cultures and tissues. However, the relationship between the two types of spectral information has not been properly investigated yet.

Recently an increase in both the PL quantum yield (QY) and the specific surface area of micron-sized porous silicon particles (PSi Ps) through supercritical drying (SCD) has been demonstrated.[3,21] SCD prevents the small pores from collapsing and potentially increases the drug payload due to higher levels of porosity being achievable. Thus, an improved material for theranostics, combining drug delivery and luminescence visualization, has been proposed. However, the effect of supercritical drying on the biodegradation profile and modification of the structural properties of PSi Ps and Si QDs has not been studied systematically to date.

Therefore, the aim of this study is to report the structural and biological features of silicon particles modified by supercritical drying compared to conventional air drying (AD), to measure the typical time for biodegradation in model fluids and in cell cultures, and to combine information obtained by Raman and PL spectroscopy for the same nanostructures. The results of the study can be used for improvement of PSi based drug delivery systems and multifunctional particles.

Results and discussion

Fig. 1A shows a schematic representation of a freshly prepared silicon quantum dot (as-Si QD) and its core–shell structure after air drying (AD-Si QD) or supercritical drying (SCD-Si QD), milling, and then annealing at 600 °C for 16 hours in 1% oxygen.

After their manufacture, the average hydrodynamic size of the porous silicon particles (PSi Ps) was 45 μm. After additional grinding, the PSi Ps had a broad size distribution from nano to micro sized. According to the TEM micrographs presented in Fig. 1B and D, both types of particles are irregular in shape due to the top-down fabrication process and consist of small Si QDs and pores. The crystallinity of the particles was confirmed by electron diffraction measurements obtained in "transmission" geometry (see Fig. 1C and E). It is noticeable that in the diffraction pattern of the SCD-Si QDs there is a greater number of bright spots lying on the diffraction rings, compared to on the AD-Si QDs ones. This indicates the improved polycrystallinity of the nanocrystal network.

The Si QD size distribution inside the Ps is shown in ESI Fig. S1† and was calculated from the dark-field TEM images. The mean diameter of AD- or SCD-Si QDs was found to be 4 nm and 2.5 nm, respectively.

Nitrogen gas adsorption/desorption analysis was used to determine the specific surface area, pore volume and pore diameters of the obtained samples (see Table 1).

The large values of specific surface area for SCD- in comparison to those for AD-Si QDs indicate a less pronounced pore collapse for these samples during the drying process.

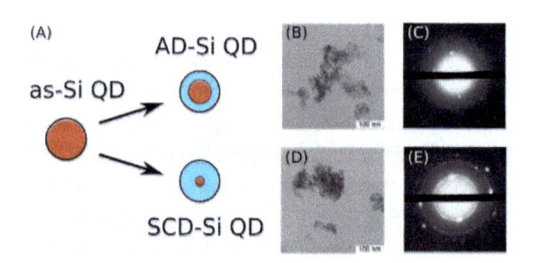

Fig. 1 (A) Schematic representation of a freshly prepared silicon quantum dot (as-Si QD) and its core–shell structure after air drying (AD-Si QD) or supercritical drying (SCD-Si QD), followed by milling and annealing at 600 °C for 16 hours in 1% oxygen. (B) TEM image of AD-Si QDs. (C) The electron diffraction pattern from an AD-Si QD. (D) TEM image of SCD-Si QDs. (E) The electron diffraction pattern from a SCD-Si QD.

Table 1 Nitrogen gas adsorption analysis of AD- and SCD-thermally passivated Si QDs

Sample type	AD-Si QDs	SCD-Si QDs
Specific surface area, m^2 g^{-1}	224	365
Pore volume, cm^3 g^{-1}	0.91	1.762
Pore diameter, nm	15.3	17.2

Fourier transform infrared (FTIR) transmittance spectra of the AD and SCD-Si QDs are shown in Fig. 2.

The IR spectra of both samples show the presence of various absorption bands which correspond to δSi–O–Si deformation vibration at 484 cm^{-1}, to Si$_x$O$_y$ vibration at 800 cm^{-1}, to νSi–O–Si vibration of the transversal optic (TO) modes at 1060 cm^{-1} and longitudinal optic (LO) phonons at 1200 cm^{-1}.[22] Both types of structure therefore have similar surface chemistry. According to the TEM and FTIR results, both AD- and SCD-Si QDs have a core–shell structure, where the core is a Si QD and the shell is a SiO$_2$ layer.

The changes in the photoluminescence (PL) spectra during storage of the samples in a dialysis bag in PBS at room temperature (RT) are shown in Fig. 3.

Both samples are characterized by broad PL spectra with only moderate quantum yield and a maximum at 800 nm, which are well explained by the radiative recombination of excitons confined in Si QDs with an average size of about 2–6 nm.[1] Therefore, the exciton emission energy and the position of the maximum in the PL spectra were determined by the diameters of the PL Si QDs (D_{PLQDs}) with the empirical equation:[23]

$$D_{PLQDs} = \left(\frac{3.73}{\Delta E_g}\right)^{0.72} \quad (1)$$

where ΔE_g is the increase in the effective band gap enlarged due to quantum confinement in comparison to the bulk silicon band gap E_{g0} = 1.12 eV. So, the predominant size of the luminescent Si QDs, calculated from the PL spectra peak position in Fig. 3, was about 4.7 nm.

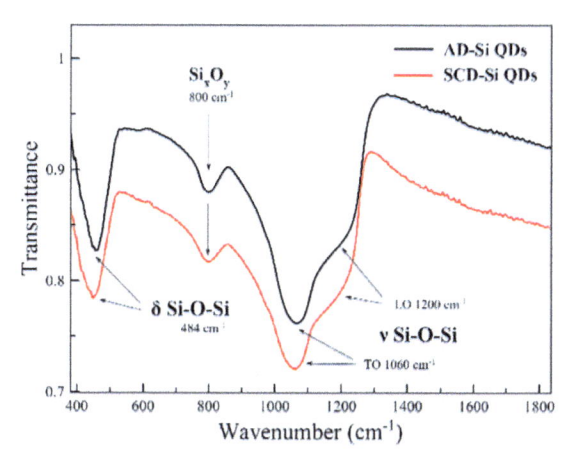

Fig. 2 Infrared transmittance spectra of the AD-Si QDs (black) and SCD-Si QDs (red).

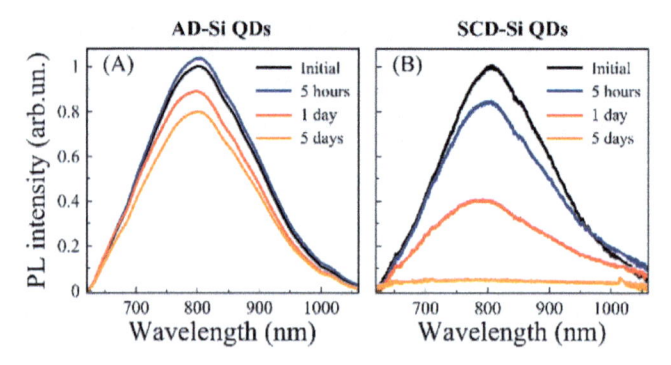

Fig. 3 Changes in the photoluminescence spectra of the AD-Si QDs (A) and SCD-Si QDs (B) during storage in dialysis bag in PBS at RT.

During the dissolution process, the PL of AD-Si QDs increased very slightly during the first 5 hours of incubation which could be associated with passivation of the silicon dangling bonds, and was then slightly quenched with further incubation of the samples for 5 days. This indicates the partial stability of the PL properties of AD-Si QDs. Meanwhile, the PL intensity of SCD-Si QDs dramatically dropped and blue shifted after 1 day of incubation, and then disappeared after 5 days. This is attributed to decreasing SCD-Si QD size and their complete dissolution in PBS.

Fig. 4 shows the Raman spectra of the initial samples of AD-Si QDs (A) and SCD-Si QDs (B), and the changes in the Raman spectra during storage of the samples in a dialysis bag in PBS at RT (C and D).

It is well known that the Raman band at about 520 cm^{-1} is explained by scattering by longitudinal optical (LO) phonons in the Si lattice.[24] The low frequency shift and broadening in the Raman signal are explained by the phonon confinement model, related to the change of spectral efficiency of the Raman scattering on optical phonons in Si QDs. So, the Raman peak intensity is proportional to the concentration of silicon, whereas the Raman peak position is governed by the diameter of the crystalline silicon core of the Si QDs (D_{RSQDs}):

$$D_{RSQDs} = 0.543 \left(\frac{52.3}{\Delta\omega}\right)^{0.63} \quad (2)$$

where $\Delta\omega$ is a Raman band shift from the bulk c-Si position of 520.5 cm^{-1}.[25] The average diameter of the crystalline silicon core, calculated from eqn (2), is about 4–5 nm for the initial AD-SiQDs, and about 2 nm for the initial SCD-SiQDs. The different sizes of the QDs calculated from the PL and Raman spectra/TEM can be explained by the higher quantum yield of photoluminescence for QDs with larger sizes than for small ones, and by the fact that many QDs are not luminescent.[26,27] Thus, we can assume that the determination of the sizes of the QDs from their Raman spectra is more accurate for the total ensemble.

Deconvolution of the Raman signal by Gaussians demonstrate the presence of amorphous silicon (a-Si) bands at 480 cm^{-1} and 328 cm^{-1},[28,29] and also of an amorphous SiO$_2$ band at 420 cm^{-1}.[29-31] According to the results obtained, the amorphous components are greater for AD-Si QDs than for SCD-Si QDs, which is

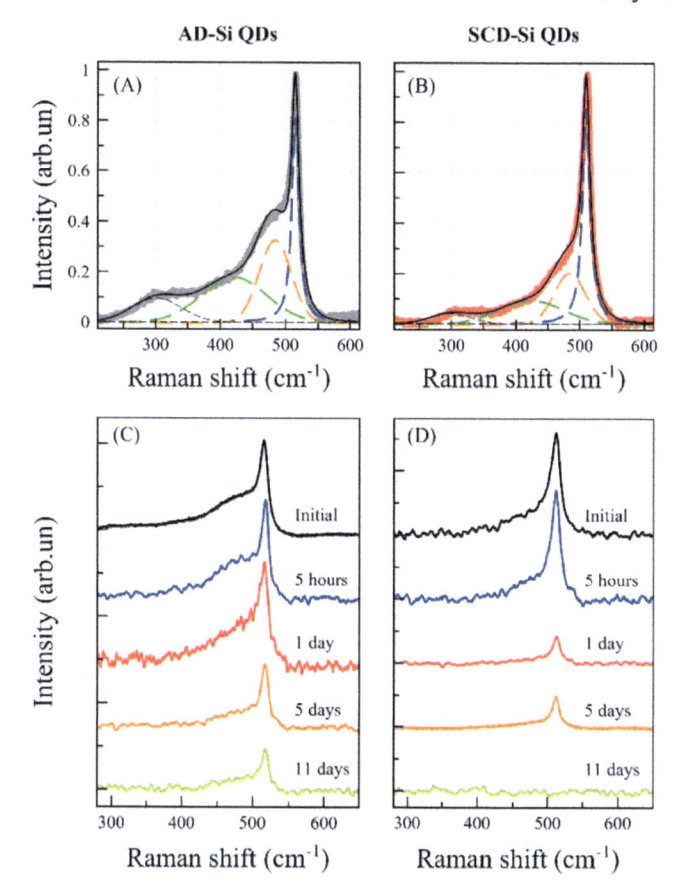

Fig. 4 Raman spectra of AD-Si QDs (A) and SCD-Si QDs (B); changes in the Raman spectra of the AD-Si QDs (C) and SCD-Si QDs (D) during storage in dialysis bag in PBS at RT.

a consequence of the capillary force induced damage in AD-Si QDs that is avoided in SCD-Si QDs.

Fig. 4C and D show the changes in the Raman spectra from AD- and SCD-Si QDs during their dissolution in PBS at 20 °C. A schematic representation of the dissolution of the Si QDs obtained on the basis of these Raman measurements is shown in Fig. 5.

From the Raman spectra shown in Fig. 4, it can be concluded that the AD-Si QDs are quite stable and have a slow dissolution rate under the conditions used. This is indicated by a rather high signal intensity, a two-fold drop of which is observed only after 11 days of incubation. The growth of the shoulder of the amorphous band at 480 cm^{-1} indicates the occurrence of the process of additional amorphization of the AD sample upon dissolution (see Fig. 5).

The SCD-Si QDs were characterized by accelerated dissolution. After 5 days, the Raman signal intensity decreased threefold, and after 11 days it was not possible to register the signal, which indicates complete dissolution of the samples. Note that no amorphization of the nanocrystals was observed. This can be explained by the initial small size of the nanocrystalline silicon core in the SCD-Si QDs (see Fig. 5).

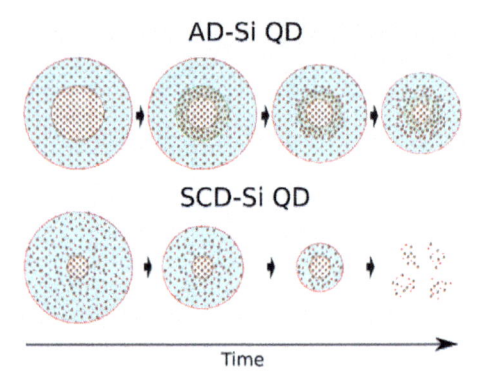

Fig. 5 A schematic representation of the dissolution process of the AD-Si QDs and the SCD-Si QDs. The nanocrystalline Si core is shown in brown, the SiO₂ layer in blue, and the amorphous layer in green.

Fig. 6 shows the dependence of the viability of 3T3 NIH cells after 24 hours of incubation with different concentrations of AD- and SCD-Si QDs.

In order to better understand the effect of the concentration of mesoporous silicon particles on the process of cell viability, the obtained experimental curves in the region for particle concentrations of more than 10 µg mL^{-1} in the well were extrapolated by the logistic curve:

$$V = \frac{A}{1 + \left(\dfrac{C}{C_0}\right)^p} \quad (3)$$

where V is the cell viability, A, p are adjustable parameters, C is the QD concentration and C_0 is the critical QD concentration. According to the parameters obtained from eqn (3), the 50% critical concentration of QDs reached 480 µg mL^{-1} for AD-Si QDs, and 1200 µg mL^{-1} for SCD-Si QDs. This might suggest a higher toxicity for AD-Si QDs than SCD-Si QDs but may instead be due to significant

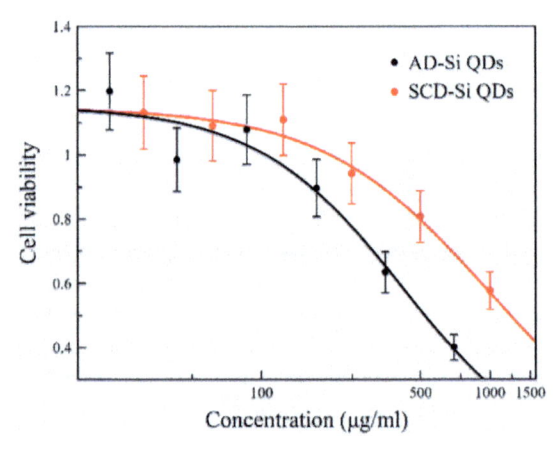

Fig. 6 *In vitro* cytotoxicity of AD- and SCD-Si QDs to 3T3 NIH cells after 24 hours of incubation.

dissolution of the latter occurring within the incubation timescale, which can protect the cells against the formation of toxic biodegradation products on the surface of the particles.

The presence of the PL properties of the obtained QDs allows them to be used as a luminescent labels for cell imaging. Fig. 7A and B show confocal luminescence images of 3T3 NIH cells after 9 hours of incubation with the QDs. The effective photoluminescence of both AD- and SCD-PSi Ps with localization on the cell membranes and cytoplasm is clearly visible here.

From Fig. 7B and C it can be clearly seen that after 24 hours of incubation, the AD-Si QDs still continue to luminesce, while the SCD-Si QDs are not visible. To understand whether PL quenching is associated with complete dissolution of the SCD-Si QDs or only with the appearance of the centers of non-radiative exciton recombination during incubation, Raman microspectroscopy of the samples inside the cells was also carried out.

The Raman images of 3T3 NIH cells incubated with AD- or SCD-Si QDs for 9 and 24 hours, and the corresponding Raman spectra of the samples, are represented in Fig. 8.

Note, that the Raman images in Fig. 8 were obtained from Raman spectra of the cells and Si QDs. The green color in Fig. 8A–D corresponds to the cell body and reflects the protein composition of the cells. The red-orange colour corresponds to the silicon particles. A Raman spectrum of the cells shows such typical characteristics peaks as an amide-I band at 1632 cm^{-1} and a CH$_2$-deformation band at 1452 cm^{-1} (see ESI Fig. S2†).

Raman spectra of AD- and SCD-Si QDs, measured directly inside the cells, are presented in Fig. 8E and F. For the AD-Si QDs, the size of the nanocrystalline core, calculated according to eqn (2), was 5 nm after 9 hours and changed to 2.8 nm after 24 hours of incubation with the cells. Along with the low-frequency shift of the maximum of the Raman signal, a decrease in the signal intensity and an increase in

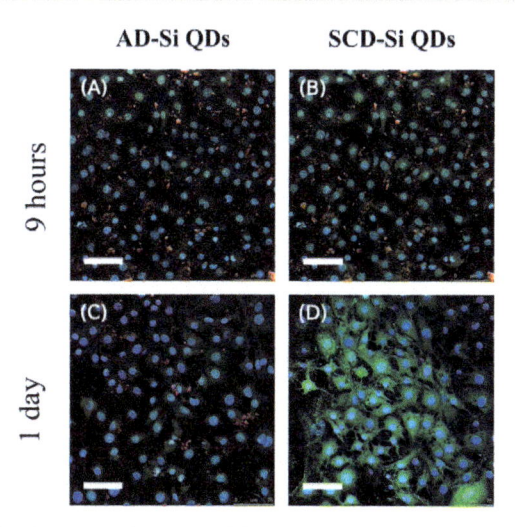

Fig. 7 Confocal luminescence images of 3T3 NIH cells incubated with 0.2 mg mL^{-1} AD-Si QDs and SCD-Si QDs for (A and B) 9 hours, and (C and D) 24 hours. The scale bar is 75 μm.

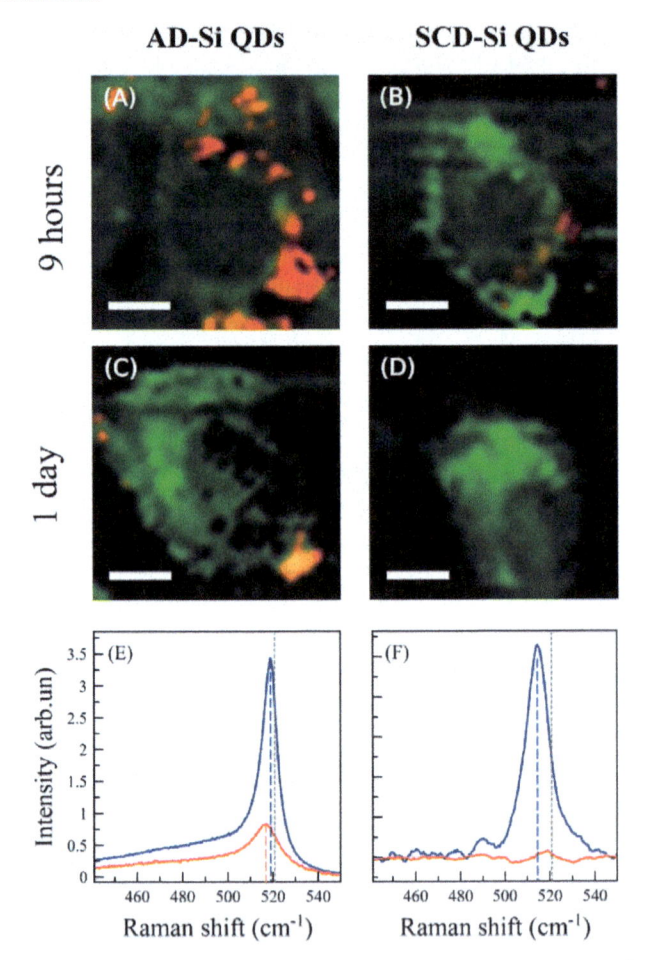

Fig. 8 Confocal Raman images of 3T3 NIH cells incubated with 0.2 mg mL^{-1} AD-Si QDs or SCD-Si QDs for (A and B) 9 hours, and (C and D) 24 hours. The scale bar is 10 μm. Corresponding Raman spectra of AD-Si QDs (E) and SCD-Si QDs (F) after 9 and 24 hours of incubation with the cells.

the amorphous phase were observed, which indicates the dissolution of these QDs. For the SCD-Si QDs, the size of the nanocrystalline core was 2 nm after 9 hours, and then after 24 hours of incubation with the cells, the Raman signal had completely disappeared. The results indicate the significant impact of the drying technique and surface pretreatment on the dissolution rate of luminescent Si QDs.

Experimental

Formation of particles

The porous silicon (PSi) layers were manufactured by electrochemical etching of a p$^+$ boron wafer with a resistivity of 0.005–0.02 ohm cm at a current density of about 100 mA cm^{-2} for 60 minutes in hydrofluoric acid (30 wt%) with methanol additive. To separate the layer from the substrate, an electropolishing process was used. Then the films were soaked in methanol to rinse away any residual HF solution and air dried on a hotplate at around 60 °C for the subsequent air drying (AD) process, or stored in ethanol for the subsequent supercritical drying (SCD). For AD, the PSi films were milled at 300 rpm for 1 minute in a Fritsch Pulverisette 6 planetary ball miller, and then placed in a Carbolite HTR 11/75 rotary furnace at 600–700 °C for 16 hours in 1% oxygen (with 99% nitrogen). For SCD, the PSi films were filled with liquid CO_2, followed by a standardized sequential procedure with a Quorum K850 critical point drier, reported previously,[3] and then oxidised under the same conditions as for the AD process. Finally, both the samples were additionally dried at 120 °C for 24 hours in a vacuum oven. Then the samples were stored in air at RT. Before the experiment, the obtained particle powders were ground in an agate mortar for 2–5 minutes.

Characterization of particles

Structural analysis of the samples was carried out using transmission electron microscopy (TEM, LEO912 AB OMEGA). The samples of Ps for the TEM studies were prepared by deposition of the particle powder on standard carbon-coated gold TEM-grids. TEM images were processed with Image J Software to obtain the QD size distribution. To determine the hydrodynamic size of the particles a Malvern Instruments Mastersizer 2000 was used. The pore sizes and pore volumes were determined with N_2 adsorption (Micromeritics Tristar 3000). The pore area was found from the adsorption branch using BET theory (Brunauer, Emmett, and Teller), and the pore size was calculated from the desorption branch using BJH theory (Barrett, Joyner, and Halenda). Surface chemistry was studied with a Fourier-transform infrared (FTIR) spectrometer (Bruker IFS 66v/S). Before recording the FTIR spectra, the suspensions were dried in air on monocrystalline silicon wafers. The FTIR measurements were carried out at room temperature in vacuum under a residual pressure of 10^{-3} Torr.

Dissolution in model liquids

PSi Ps were introduced into phosphate buffered saline (PBS, pH 7.2) to achieve a concentration of 1 mg mL^{-1}, then 4 mL of nanoparticle suspension was placed into a dialysis bag (4.5 kDa pore diameter) and then into 1 litre of PBS at room temperature.

To measure the PL or Raman spectra, 0.1 mL of the suspension was dried on a metal plate after 5 hours, 1 day, 5 days and 11 days of incubation. The PL spectra of the samples were measured under excitation with a CW Ar-ion laser at 364 nm (power of 10 mW, spot diameter of 1 mm). The PL signal was detected using a grating monochromator (MS750, SOLAR TII) equipped with a CCD array. Raman spectra were measured using a Confotec™ MR350 confocal Raman microscope with laser excitation at 633 nm and a weak power of 1 mW to protect the samples from overheating.

Cytotoxicity studies

The studies were performed using mouse fibroblast 3T3NIH cells (obtained from the bank of cultures at the Institute of Cytology RAS). All cells were grown in DMEM (Gibco) medium, containing 10% inactivated bovine serum (Thermo Fisher Scientific, Gibco), in the presence of antibiotics (penicillin, streptomycin,

100 U mL^{-1}: Thermo Fisher Scientific, Gibco) and 5% CO_2 at 37 °C. After 24 h, the culture medium was replaced with a fresh one for treatment with different concentrations of AD- or SCD-PSi Ps. The number of living cells was determined on a Partec PAS III flow cytometer.[7] The excitation wavelength was 488 nm. Fluorescence was recorded on the FL-3 channel. At the same time, light scattering of the cells in the FSC and SSC channels was recorded. All cells in the phases of the G1, S, G2M cell cycle were considered alive. The final number of living cells was compared with their final number in the control group grown without particles.

Luminescence imaging

For imaging the concentration of particles of 200 μg mL^{-1} was used.

Confocal luminescence imaging experiments were performed using a Leica optical microscope with excitation at 405 nm and 488 nm. An oil-immersion objective of 63×/NA 1.40 was used for imaging. Particles with a concentration of 0.2 mg mL^{-1} were incubated with 3N3NIH cells for 9 or 24 hours before the experiments. Cells were then stained by adding 3 μg mL^{-1} of Calcein-AM, Sigma for the cytoplasm and 5 μg mL^{-1} of bisbenzimide H 33342 (Hoechst, Calbiochem) for the nucleus.

Raman imaging and spectral data analysis

Micro-Raman spectroscopy data were acquired using the confocal Raman microscope Confotec™ MR350, with laser excitation at the 633 nm wavelength and 40 mW power. Particles with the concentration of 0.2 mg mL^{-1} were incubated with 3T3NIH cells for 9 or 24 hours before the experiments. The 3T3NIH cells were grown and then fixed on special CaF_2 glass aiming to prevent background scattering.[19] Afterwards the glasses with the cells were placed in standard 10 cm Petri dishes filled with distilled water. This prevents the cells from overheating during the measurements due to high laser power and enables longer imaging of the cell clusters. For this purpose, a 63×/NA 1.0 water dipping objective Zeiss was used to focus the laser on the cell. The obtained images of the cell and the intracellular uptake of particles were generated by applying a CCD camera (Andor) operated at −60 °C. Raman images were obtained by Gaussian deconvolution of the initial spectra. Baseline correction was applied prior to calculations. The intensities of the Raman bands 510–520 cm^{-1} and 1452 cm^{-1} corresponding to the Si QD band and the cell components, respectively, were used for color representation.

Conclusions

Here we have shown that the drying process and surface pre-treatment strongly affect the dissolution rate of luminescent silicon QDs in model fluids and living cells.

Initial porous silicon particles were obtained by mechanical milling of electrochemically etched mesoporous silicon films, and consisted of silicon nanocrystallite networks and pores. The samples were then subjected to super-critical drying with CO_2 solvent (SD) or air drying (AD), and then annealed at 600 °C for 16 hours in 1% oxygen to obtain the nano-sized Si QDs.

The obtained samples were characterized by a core–shell structure and had a crystalline silicon core with a SiO$_2$ layer on the surface. Nitrogen gas adsorption/desorption analysis determined larger values for specific surface area, pore volume and pore diameter for SCD- than for AD-Si QDs, which indicates a less pronounced pore collapse for SCD samples during the drying process. IR transmittance spectra show the presence of various absorption bands which correspond to different Si–O–Si adsorption modes for both the samples.

Cell viability measurements indicate a higher apparent toxicity for AD-Si QDs than for SCD-Si QDs which may be associated with the quick dissolution of the latter, which can protect the cells against the formation of toxic biodegradation products on the surface of the particles.

The AD- and SCD-SiQDs were characterized by broad PL spectra. The PL of AD-Si QDs was quite stable during their incubation in PBS. Meanwhile, the PL of the SCD-Si QDs dramatically dropped and blue shifted after 1 day of incubation, and then disappeared after 5 days. The same situation was observed after incubation of the 3T3 NIH cells with the QDs. After 24 hours of incubation, the AD-Si QDs still continued to luminesce while the SCD-Si QDs were not visible.

To understand whether the PL quenching was associated with complete dissolution of SCD-Si QDs or only with the appearance of the centers of non-radiative exciton recombination during incubation, Raman microspectroscopy of the samples in PBS and inside the cells was also carried out. It was shown that the size of the nanocrystalline core of an AD-Si QD was 5 nm after 9 hours and changed to 2.8 nm after 24 hours of incubation with the cells. In accordance with the proposed model, a low-frequency shift of the maximum of the Raman signal, a decrease in the signal intensity and an increase in the amorphous phase were observed during the dissolution of the AD-Si QDs. For the SCD-Si QDs, the dissolution proceeded without amorphization of the nanocrystals, which is associated with their initially small size. So, for the SCD-Si QDs the size of the nanocrystalline core was 2 nm after 9 hours, and then after 24 hours of incubation with the cells, the Raman signal completely disappeared. Thus, the data obtained using the two methods of photoluminescence and Raman microspectroscopy are in good agreement. Given that the surface chemistries are similar for the two types of material, we attribute the faster dissolution of SCD-Si to a combination of two morphological factors: larger pore volume and pore diameters which promotes faster fluid penetration (better "wetting"); larger internal surface and smaller nanocrystals which promotes faster dissolution to non-luminescent structures.

So, for the first time this combination of the complementary methods of photoluminescence and Raman microspectroscopy was used for investigation of the dissolution rate of Si QDs. The obtained results indicate the significant impact of drying and surface pretreatment on the dissolution rate of luminescent Si QDs, which is important for their use in theranostics combining drug delivery and luminescence imaging.

Conflicts of interest

There are no conflicts to declare.

Acknowledgements

This work was supported by the Russian Science Foundation (Grant No. 19-72-10131).

Notes and references

1. L. T. Canham, *Appl. Phys. Lett.*, 1990, **57**, 1046–1048.
2. D. Jurbergs, E. Rogojina, L. Mangolini and U. Kortshagen, *Appl. Phys. Lett.*, 2006, **88**, 233116.
3. J. Joo, T. Defforge, A. Loni, D. Kim, Z. Y. Li, M. J. Sailor, G. Gautier and L. T. Canham, *Appl. Phys. Lett.*, 2016, **108**, 153111.
4. E. J. Anglin, L. Cheng, W. R. Freeman and M. J. Sailor, *Adv. Drug Delivery Rev.*, 2008, **60**, 1266–1277.
5. T. Kumeria, S. J. P. McInnes, S. Maher and A. Santos, *Expert Opin. Drug Delivery*, 2017, **14**, 1407–1422.
6. L. A. Osminkina, M. B. Gongalsky, A. V. Motuzuk, V. Y. Timoshenko and A. A. Kudryavtsev, *Appl. Phys. B: Lasers Opt.*, 2011, **105**, 665–668.
7. L. A. Osminkina, A. L. Nikolaev, A. P. Sviridov, N. V. Andronova, K. P. Tamarov, M. B. Gongalsky, A. A. Kudryavtsev, H. M. Treshalina and V. Y. Timoshenko, *Microporous Mesoporous Mater.*, 2015, **210**, 169–175.
8. L. T. Canham, *Nanotechnology*, 2007, **18**, 185704.
9. J.-H. Park, L. Gu, G. von Maltzahn, E. Ruoslahti, S. N. Bhatia and M. J. Sailor, *Nat. Mater.*, 2009, **8**, 331–336.
10. J. Liu, M. Yu, C. Zhou and J. Zheng, *Mater. Today*, 2013, **16**, 477–486.
11. J. G. Croissant, Y. Fatieiev and N. M. Khashab, *Adv. Mater.*, 2017, **29**, 1604634.
12. J. O. Martinez, C. Chiappini, A. Ziemys, A. M. Faust, M. Kojic, X. Liu, M. Ferrari and E. Tasciotti, *Biomaterials*, 2013, **34**, 8469–8477.
13. N. K. Hon, Z. Shaposhnik, E. D. Diebold, F. Tamanoi and B. Jalali, *J. Biomed. Mater. Res., Part A*, 2012, **100A**, 3416–3421.
14. J. Joo, J. F. Cruz, S. Vijayakumar, J. Grondek and M. J. Sailor, *Adv. Funct. Mater.*, 2014, **24**, 5688–5694.
15. M. B. Gongalsky, J. V. Kargina, J. F. Cruz, J. F. Sanchez-Royo, V. S. Chirvony, L. A. Osminkina and M. J. Sailor, *Front. Chem.*, 2019, **7**, 165.
16. B. Godin, J. Gu, R. E. Serda, R. Bhavane, E. Tasciotti, C. Chiappini, X. Liu, T. Tanaka, P. Decuzzi and M. Ferrari, *J. Biomed. Mater. Res., Part A*, 2010, **94**, 1236–1243.
17. A. Tzur-Balter, Z. Shatsberg, M. Beckerman, E. Segal and N. Artzi, *Nat. Commun.*, 2015, **6**, 6208.
18. S. Bhattacharjee, I. M. C. M. Rietjens, M. P. Singh, T. M. Atkins, T. K. Purkait, Z. Xu, S. Regli, A. Shukaliak, R. J. Clark, B. S. Mitchell, G. M. Alink, A. T. M. Marcelis, M. J. Fink, J. G. C. Veinot, S. M. Kauzlarich and H. Zuilhof, *Nanoscale*, 2013, **5**, 4870.
19. E. Tolstik, L. A. Osminkina, C. Matthäus, M. Burkhardt, K. E. Tsurikov, U. A. Natashina, V. Y. Timoshenko, R. Heintzmann, J. Popp and V. Sivakov, *Nanomed. Nanotechnol. Biol. Med.*, 2016, **12**, 1931–1940.
20. P. Maximchik, K. Tamarov, E. V. Sheval, E. Tolstik, T. Kirchberger-Tolstik, Z. Yang, V. Sivakov, B. Zhivotovsky and L. A. Osminkina, *ACS Biomater. Sci. Eng.*, 2019, **5**(11), 6063–6071.

21 A. Loni, L. T. Canham, T. Defforge and G. Gautier, *ECS J. Solid State Sci. Technol.*, 2015, **4**, P289–P292.
22 M. J. Sailor, *Porous Silicon In Practice*, Wiley, 2012.
23 G. Ledoux, O. Guillois, D. Porterat, C. Reynaud, F. Huisken, B. Kohn and V. Paillard, *Phys. Rev. B: Condens. Matter Mater. Phys.*, 2000, **62**(23), 15942.
24 I. H. Campbell and P. M. Fauchet, *Solid State Commun.*, 1986, **58**(10), P739–P741.
25 V. Paillard, P. Puech, M. A. Laguna, R. Carles, B. Kohn and F. Huisken, *J. Appl. Phys.*, 1999, **86**, P1921–P1924.
26 Y. Yu, G. Fan, A. Fermi, R. Mazzaro, V. Morandi, P. Ceroni, D. M. Smilgies and B. A. Korgel, *J. Phys. Chem. C*, 2017, **121**, 23240–23248.
27 X. Liu, Y. Zhang, T. Yu, X. Qiao, R. Gresback, X. Pi and D. Yang, *Part. Part. Syst. Charact.*, 2016, **33**, 44–52.
28 Z. Iqbal and S. Veprek, *J. Phys. C: Solid State Phys.*, 1982, **15**(2), P377.
29 Y. Zhu, *et al.*, *Nucl. Instrum. Methods Phys. Res.*, 2015, **365**, P123–P127.
30 F. Ruiz, J. R. Martınez and J. González-Hernández, *J. Mol. Struct.*, 2002, **641**(2-3), P243–P250.
31 M. Chligui, G. Guimbretiere, A. Canizares, G. Matzen, Y. Vaills and P. Simon, *New features in the Raman spectrum of silica: Key-points in the improvement on structure knowledge*, 2010, hal-00520823.

Faraday Discussions

PAPER

Synthesis and characterisation of isothiocyanate functionalised silicon nanoparticles and their uptake in cultured colonic cells†

Yimin Chao, [ID] *[a] Ashley I. Marsh,[a] Mehrnaz Behray,[a] Feng Guan,[ab] Anders Engdahl,[c] Yueyang Chao,[d] Qi Wang[e] and Yongping Bao[e]

Received 18th September 2019, Accepted 14th January 2020

DOI: 10.1039/c9fd00087a

The functionalisation of silicon nanoparticles with a terminal thiocyanate group, producing isothiocyanate-capped silicon nanoparticles (ITC-capped SiNPs) has been successfully attained. The procedure for the synthesis is a two-step process that occurs via thermally induced hydrosilylation of hydrogen terminated silicon nanoparticles (H-SiNPs) and further reaction with potassium thiocyanate (KSCN). The synthesis was confirmed by Fourier transform infrared (FTIR) spectroscopy and X-Ray photoelectron spectroscopy (XPS). At the same time, the internalisation and the cytotoxicity of the ITC-capped SiNPs *in vitro* were assessed in two cell lines: Caco-2, human colorectal cancer cells and CCD-841, human colon "normal" cells. The results showed that above concentrations of 15 μg ml^{-1}, the cell viability of both cell lines was depleted significantly when treated with ITC SiNPs, particularly over a 48 hour period, to approximately 20% cell viability at the highest treatment concentration (70 μg ml^{-1}). Flow cytometry was employed to determine cellular uptake in Caco-2 cells treated with ITC SiNPs. It was observed that at lower SiNP concentrations, uptake efficiency was significantly improved for time periods under 12 hours; overall it was noted that cellular uptake was positively dependent on the period of incubation and the temperature of incubation. As such, it was concluded that the mechanism of uptake of ITC SiNPs was through endocytosis. Synchrotron FTIR spectroscopy, by means of line spectral analysis and IR imaging, provided further evidence to suggest the internalisation of ITC SiNPs displays a strong localisation, with an affinity for the nucleus of treated Caco-2 cells.

[a]*School of Chemistry, University of East Anglia, Norwich NR4 7TJ, UK. E-mail: Y.chao@uea.ac.uk*
[b]*MAX IV Laboratory, Lund University, P. O. Box 118, 22100 Lund, Sweden*
[c]*School of Pharmacy, Heilongjiang University of Chinese Medicine, Haerbin 150040, China*
[d]*Birmingham Heartlands Hospital, Bordesley Green E, Birmingham B9 5SS, UK*
[e]*Norwich Medical School, University of East Anglia, Norwich NR4 7TJ, UK*
† Electronic supplementary information (ESI) available. See DOI: 10.1039/c9fd00087a

1. Introduction

In the US alone it was predicted that there would be a total of 1 735 350 new cases of cancer diagnosed and up to 609 640 fatalities owing to cancer in the year of 2018.[1] Both figures were higher than those in 2017, with an increase of 46 570 and 8720 respectively. Whilst survival rates of cancer are projected to double by 2030 in the UK, the number of cases in the UK alone has risen by 500 000 over the years of 2010–2015 and is projected to increase from 2.5 million to 4 million by 2040.[2] Despite the increase in survival rate however, some of the principles behind some the most successful and frequently employed anti-cancer treatments are proving to be almost as harmful to patients as cancer itself. Traditional anti-cancer treatments and therapies, such as chemotherapy and radiotherapy, despite their success, can lead to the onset of a number of physical and psychological side effects such as long term cardiovascular disease, cognitive and neurological deficits and bone density reductions.[3-5]

Consequently, a lot of time and money is being invested in the development of nanoscale therapeutic agents that can be employed to provide a safer, time efficient, more direct route to identification, diagnosis and treatment of cancer as well as a whole host of other diseases. The functionalisation of silicon quantum dots with ligands that exhibit anti-cancer properties holds several advantages *versus* traditional therapies. One functional group that has been extensively proven to inhibit carcinogens is the isothiocyanate group (ITC).[6,7]

ITCs have been identified as a highly effective chemopreventive group of molecules that are associated with a number of anti-cancer processes, namely the prevention of the propagation of carcinogen activating phase I enzymes, the induction of arrests in cell cycles, leading to apoptosis of genetically mutated cells, and the induction of phase II enzymes that act as carcinogen detoxifiers.[8,9]

Phase II enzymes such as quinone reductase (QR) and glutathione *S*-transferase (GST) have been identified as potent detoxifying enzymes that can reduce the onset of chemical carcinogenesis, in effect preventing the initiation of malignant and invasive cell behaviour.[10] It has been well documented that a number of naturally occurring isothiocyanates, found in cruciferous vegetables, are able to not only prevent the initiation of carcinogenesis but further inhibit the progression of tumorigenesis by increasing the tissue levels of these detoxifying enzymes in a number of rat tissues.[8,11]

This type of interaction does not rely on the binding of ITCs to specific enzymes and thus occurs as an indirect interaction, however the ability for ITC compounds to perform such a role has a reliance on direct interactions and binding mechanisms to well identified targets. A number of studies have focused on the effects of plant derived isothiocyanates and their effect on carcinogenesis.[12,13] Herein, we present an investigation utilising silicon nanoparticles (SiNPs) functionalised to contain a terminal isothiocyanate group in order to more fully develop and utilize it.

The use of silicon nanoparticles as observable carriers for chemopreventive molecules holds a number of advantages over traditional fluorophore and/or heavy metal nanoparticle systems. Silicon nanoparticles offer a material with a broad excitation spectrum, which is tuneable by size, owing to the quantum confinement effect.[14] Traditional dyes such as fluorescein or rhodamine cannot provide such favourable characteristics in comparison. Therefore, from a biomedical standpoint, the main advantage of this effect is in their use in multi-coloured quantum dot probes. The inherent photoinduced bleaching effects observed with traditional dye methods can be overcome by utilising silicon nanoparticles that provide a much more photostable alternative for 3D imaging.[15] Photoinduced cytotoxicity is an issue that has been reported when utilising traditional dyes and has resulted in the formation of free radicals and singlet oxygen.[16]

Moreover, group II–VI heavy metal core nanoparticles provide size tuneable optical properties and a high quantum yield much in the same way that silicon quantum dots are able to.[17] However upon exposure to aqueous environments or irradiation with UV light, cadmium based quantum dots have demonstrated core material leakage, leading to the requirement of preventative measures being put in place to inhibit cytotoxicity.[18–20]

In this investigation, hydrogen terminated SiNPs (H-SiNPs) were prepared by the electrochemical etching of porous silicon. These H-SiNPs were reacted with allyl bromide in a hydrosilylation reaction and the resulting bromine terminated nanoparticles (Br-SiNPs) refluxed with potassium thiocyanate to produce ITC terminated silicon nanoparticles (ITC-SiNPs). In addition, the resultant ITC-SiNPs were examined in terms of their cytotoxicity, cellular uptake efficiency and cellular internalisation in colorectal cancer cells and primary cells.

2. Results and discussion

2.1 Chemical properties

As confirmation that the silicon nanoparticle surface was successfully functionalized with the isothiocyanate ligand, FTIR measurements in transmission mode were taken. The obtained spectra (Fig. 1) were measured using nanoparticle samples dissolved in distilled toluene.

The peak visible at about 2080 cm^{-1} is a clear indicator of the presence of the Si–H band. The observed features at 1259 cm^{-1} at 1454 cm^{-1} can be ascribed the symmetric bending modes and vibrational scissoring modes of Si–CH$_2$

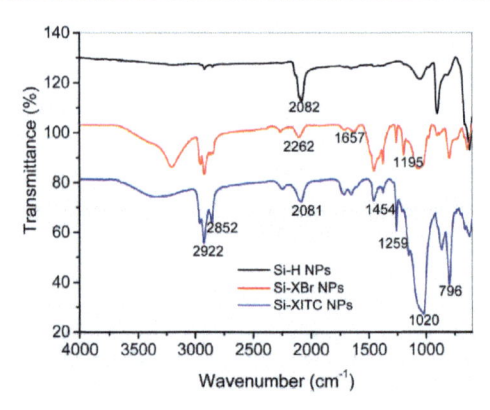

Fig. 1 FTIR spectra of H-terminated SiNPs, allyl bromide capped SiNPs, and ITC-capped SiNPs.

Table 1 FTIR spectral features of ITC and allyl bromide functionalised SiNPs

SiNPs	Si–C	Si–O	C=S	–CH$_2$	–CH$_2$–	C=N	Si–H	=C=N–	–CH$_2$–	–CH$_2$–
ITC-capped		1020	1195	1259	1454	1657	~2080	2262	2852	2922
Allyl bromide capped	~800	1020	1195	1259	1454	1657	~2080	2262	2852	2922
H-Terminated	~800	1020					~2080			

respectively, whilst the dominant feature at 1020 cm^{-1} is indicative of modes of silicon surface oxidation and characteristic of Si–O bonds.[21,22] The polarity of this bond in particular is what extenuates it's observed transmittance.[23] All major data are listed in Table 1.

The surface properties of the functionalized nanoparticles were investigated utilising X-ray photoelectron spectroscopy. The full survey spectrum of the ITC SiNPs can be found in Fig. 2a, whilst high resolution spectra representing the N 1s, C 1s and Si 2p orbitals are displayed in Fig. 2b–d. The spectrum displayed in Fig. 2b corresponds to the high resolution N 1s spectrum. It was fitted with three components and a Shirley background. The three observed peaks correspond to N=C=S at 397.5 eV, N–C at 398.7 eV and N–O at 399.3 eV. The C 1s peak observed in Fig. 2c was subjected to four components being fitted and a Shirley background. The four peaks observed correspond to C–Si, C–C, C–N and C–O bonding at 283.7, 284.1, 285.0 and 285.4 eV respectively. Fig. 2d shows the high resolution spectrum of Si 2p, which was fitted with three components and a Shirley background. The three observed peaks appear at 102.8, 101.4 and 100.8 eV. The first peak is believed to arise from Si–O bonding. The peak at 101.4 eV is representative of Si–C bonding and the third peak at 100.8 eV is taken to be characteristic of Si–Si bonding. According to the surface chemistry data obtained from XPS, the surface

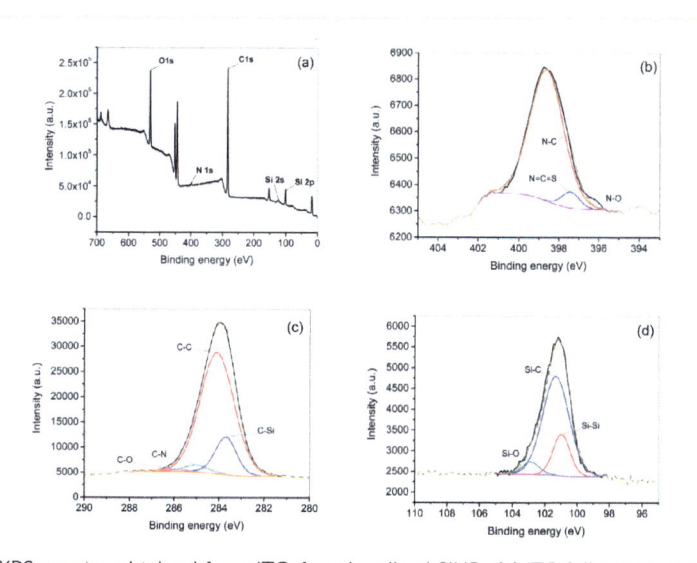

Fig. 2 XPS spectra obtained from ITC-functionalised SiNPs (a) ITC full survey spectrum, (b) N 1s, (c) C 1s, and (d) Si 2p. Dotted lines correspond to the fitting obtained for the spectrum components.

of the SiNPs is fully capped with the thiocyanate group. The presence of the ITC functionality has also been confirmed by energy-dispersive X-ray spectroscopy (EDX) mapping of the particles (Fig. S1†). The disappearance of the Br signal in the EDX of the ICT SiNPs suggested that all of the surface Br moieties have been converted to ITC groups (Fig. S1†).

2.2 Optical properties

Fig. 3a displays the emission and absorption spectra of ITC SiNPs in distilled toluene, measured at room temperature. The solid line represents the photoluminescence emission spectrum of the ITC SiNPs that were subject to an excitation wavelength of 360 nm. The maximum observed emission was centred at approximately 430 nm, characteristic of blue light emission. In the UV-vis spectra obtained (Fig. 3) there was a clear increase in the absorbance detected when decreasing the excitation wavelength from approximately 500 nm, and this is a characteristic feature of the absorption corresponding to indirect bandgap semiconductors such as silicon.

The fluorescence quantum yield was measured and calculated according to the method outlaid by Williams *et al.*, with 0.1 M H_2SO_4 as the reference for measurements.[24] As detailed in ESI Fig. S2 and S3,† it was observed that for ITC SiNPs, at an excitation wavelength of 360 nm, the photoluminescence quantum yield (PLQY) was measured to be about 13%, which complies with previous investigations that cite PLQY values of up to 25%.[25–27]

2.3 Particle size

To determine the mean size and polydispersity of the ITC-capped silicon nanoparticles, TEM images and XRD and DLS spectra were obtained. Fig. 4 displays an obtained TEM image of a sample of ITC SiNPs with data that displays the range of sizes obtained from these images. From 100 random nanoparticles it was

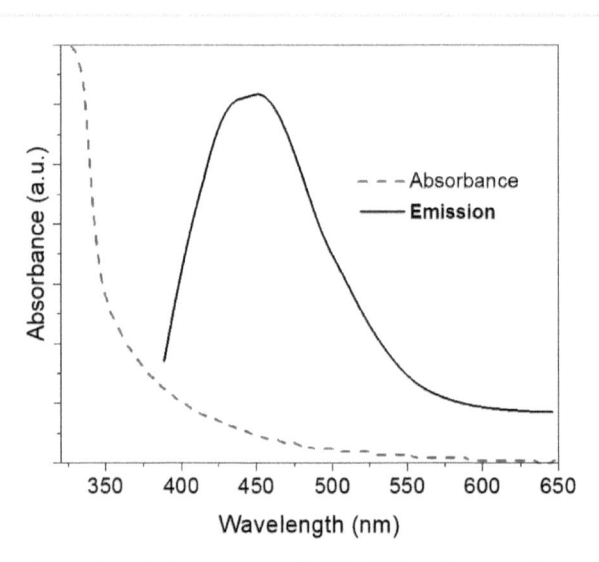

Fig. 3 Absorption and emission spectra of ITC SiNPs with excitation wavelength of 360 nm.

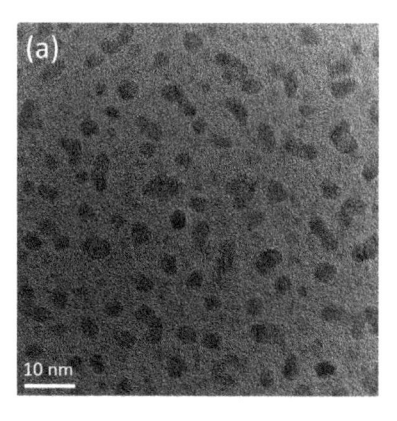

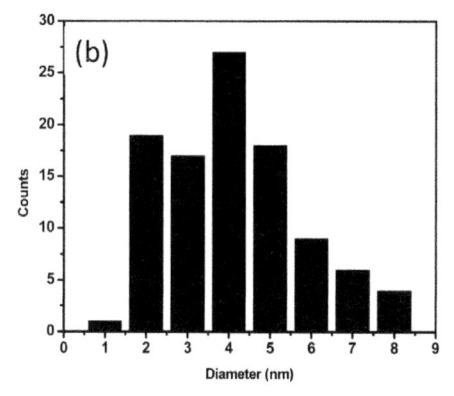

Fig. 4 TEM image of ITC-functionalized SiNPs drop cast onto a copper grid with a scale bar 10 nm (a) and (b) corresponding histogram of sizes obtained.

observed that the sample ITC SiNPs had a mean diameter of 4.01 nm. The observed polydispersity of these values can be attributed to the lattice spacing of the silicon crystals. Results obtained by DLS and the value for the mean diameter calculated utilising XRD, concur with this value and these results can be found in the ESI (Fig. S4 and S5†).

Deviations in the calculated values of diameter are likely due to the differing individual methods used for size calculation. It is highly likely that due to the low electron density of the ITC chain, only the silicon core of our ITC SiNPs is measurable by TEM. Meanwhile, values obtained by DLS can be taken only as an estimate of the hydrodynamic radius, that is the hypothetical size value of a solid sphere that diffuses in the same manner as our nanoparticles, as calculated by the Stokes–Einstein relation. Values obtained by utilising XRD spectra and the Scherrer equation can similarly be regarded as no more than an estimate.

2.4 Cellular internalisation

2.4.1 Internalisation of ITC SiNPs observed by confocal microspectroscopy.
In this experiment, Caco-2 colonic cancer cells were treated with a solution of ITC SiNPs with a concentration of 50 µg ml^{-1}. Cells were incubated with the ITC SiNPs for 6 and 12 hours at a temperature of 37 °C. The series of images in Fig. 5a show

the fluorescence exhibited under confocal microscopy due to (1) phalloidin, used to stain actin, in red (2) ITC SiNPs which in this case is null due to these being images of control cells, in green and (3) a combined image of both the phalloidin and ITC SiNP fluorescence. In Fig. 5b2 the cellular internalisation of ITC SiNPs after 6 hours of incubation is displayed. What was noted when comparing Fig. 5b2 and c2, is that the internalization and localization of ITC SiNPs was clearly time-dependent, given that the image obtained after 12 hours, Fig. 5c3, exhibits much more intense ITC SiNP fluorescence and clearly a higher number of ITC SiNPs were observed.

2.4.2 **Internalisation of ITC SiNPs observed by synchrotron FTIR.** Both Caco-2 and CCD cell lines were treated with a 50 μg ml^{-1} ITC SiNP solution for 6 hours prior to point spectroscopy measurements being taken. Each measurement consisted of a minimum of 8 points selected across the cell for IR identification. Along with this, a visual image and protein specific IR image were taken by using the integrated absorbance of the amide I region between 1600–1700 cm^{-1}, a critical band which arises due to the stretching of C=O contained in the peptide bonds of the proteins within the cell. This gives an indication of protein distribution when treated as shown in Fig. 6a and S6,† and untreated, Fig. 6b and S7.†

Comparing the IR images obtained for the treated and untreated Caco-2 cells, the first striking difference was the relative uniformity of protein concentration

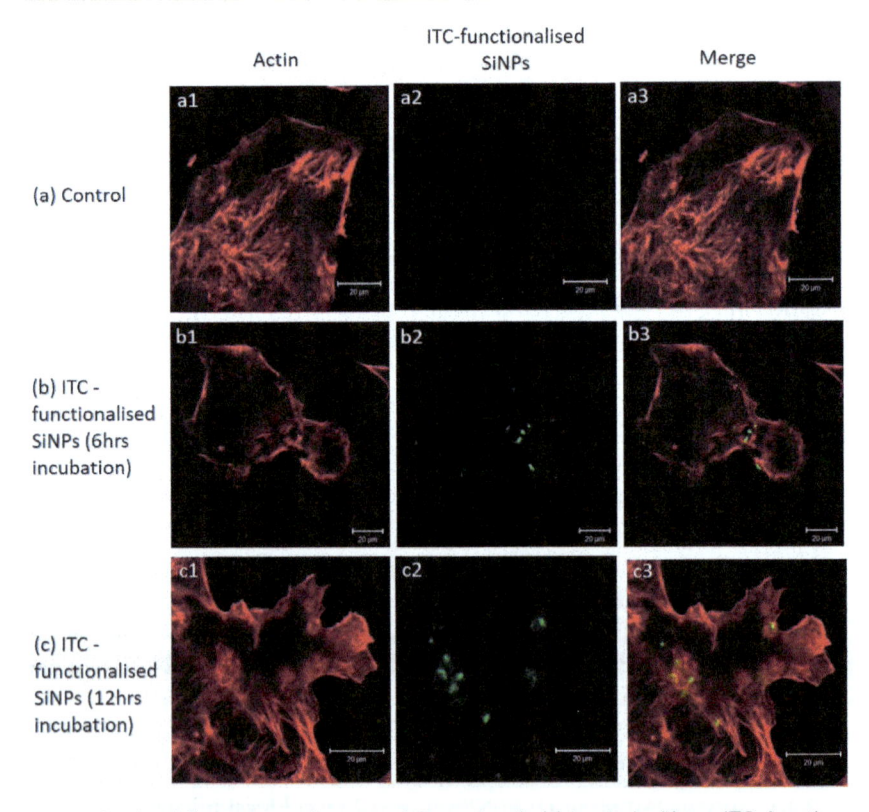

Fig. 5 Confocal fluorescence images of Caco-2 cells (a) treated with no ITC-functionalised SiNPs (control) and Caco-2 cells incubated with ITC-functionalised SiNPs for (b) 6 hours and (c) 12 hours. Red fluorescence corresponds to phalloidin staining of actin and green corresponds to ITC-functionalized SiNPs.

throughout large parts of the untreated cell compared to what appears to be nucleus localisation in those treated with ITC SiNPs. The line spectra confirm the observations taken from the IR images, as can be seen when highlighting the aforementioned amide I region between 1600–1700 cm^{-1}. It was noted that the intensity of the peak in this region remained the same throughout the untreated cell, whereas it was clear that converging upon the nucleus in the untreated Caco-2 cell, that the peak intensity increased, with the greatest absorbance observed within the nucleus. A number of features that correspond to the vibrations of DNA and phospholipid can be observed when comparing treated and untreated cells. At the nucleus in particular, it was seen that the vibration around 1078 cm^{-1}, corresponding to phosphodiester groups in DNA, changed in both cell lines,

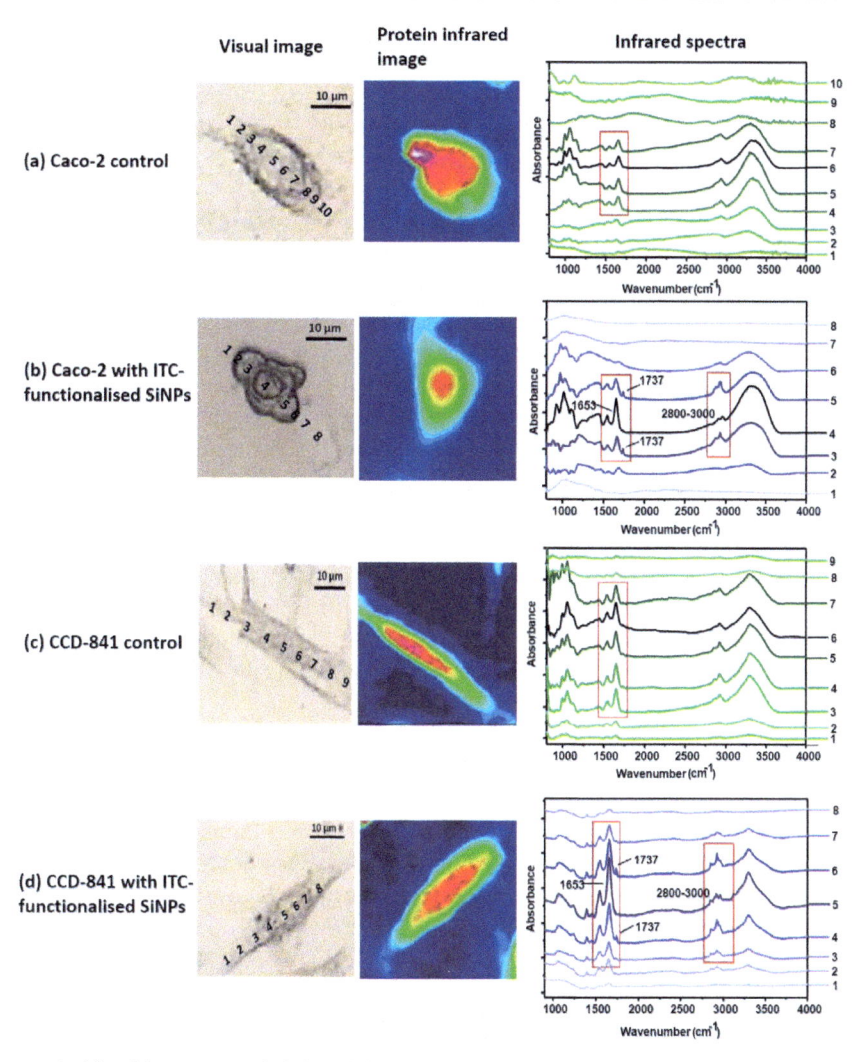

Fig. 6 Visual image, protein infrared image and corresponding infrared spectra acquired along the line of a Caco-2 control cell (a), a Caco-2 cell treated with ITC-functionalised SiNPs (b), a CCD-841 control cell (c), and a CCD-841 cell treated with ITC-functionalised SiNPs (d).

indicating changes in the DNA environment and a shift in DNA distribution across the cell.

Differences in the phospholipid content within the cellular membrane gave rise to a new peak at 1737 cm^{-1} which corresponds to C=O stretching, and an increase in the C–H absorption observed between 2800–3050 cm^{-1}. These features indicate an increase in the phospholipid content in this area of the cell. Changes in the cell membrane content have been described previously in work using a number of inducing molecules, including ITC, citing apoptosis as a consequence of these cellular alterations.[28,29] Further literature concurs that isothiocyanate has been proven to induce apoptosis.[30,31]

The observed concentration of proteins in the nucleus after treatment with ITC concurs with several publications which note the translocation of Nrf2 to the nucleus following transfection with extracellular kinases which occurred as a result of the chemical stress caused by the introduction of ITC.[7,32,33]

A similar outcome was observed when comparing the IR spectra obtained for the treated and untreated CCD-841 cells, in Fig. 6c and d. The same fluctuation in absorbance in the amide I band was observed moving from the space outside the cell, through the cytoplasm and nucleus.

2.5 Cell viability by MTT assay

In order to obtain a measure for the cell viability and cytotoxicity of cells treated with ITC functionalised SiNPs, an MTT assay was performed on both Caco-2 and CCD-841 cell lines, treated with ITC SiNPs. As a control, amine capped SiNPs have been established as a non-cytotoxic biological substance.[34] The assay was replicated over time periods of 24 and 48 hours to investigate the potential longer term effects of treatment with ITC SiNPs, as indicated in Fig. 7.

After an incubation period of 24 hours, at concentrations of 15 μg ml^{-1} and below, for both cell lines, it was observed that both nanoparticle systems performed with relative parity with regards to the viability of the cells treated. From herein, at concentrations above 15 μg ml^{-1}, it was noted that the cell viability of both lines when treated with ITC SiNPs suffered an immediate reduction of approximately 20%, in stark contrast to the control cells which from this point, maintained relatively stable viability values of 90% and 80% in Caco-2 and CCD-841 lines respectively.

The viability data obtained after 48 hours of incubation for both cell lines showed a much more dramatic and clearer reductive trend in cell viability with increasing concentration of ITC SiNPs. This effect again becomes particularly apparent at concentrations above 15 μg ml^{-1}. Below 15 μg ml^{-1} however, there was no observed statistical difference in the cell viability of ITC treated CCD-841 cells after 48 hours, the same can be said for Caco-2 cells treated with ITC at low concentrations of 5 and 10 μg ml^{-1}.

The cytotoxic nature of the ITC SiNPs is highlighted in the steady reduction of cell viability from 80% at 10 μg ml^{-1} to just above 20% at the highest concentration; their effect on healthy CCD cells however was more dramatic, resulting in an immediate loss of approximately 40% viability at 20 μg ml^{-1} and a steady decline from 40% to 20% cell viability thereafter up to the maximum concentration.

As expected, after 48 hours of incubation, control cells treated with amine capped SiNPs produced similar results in both healthy and unhealthy cell lines,

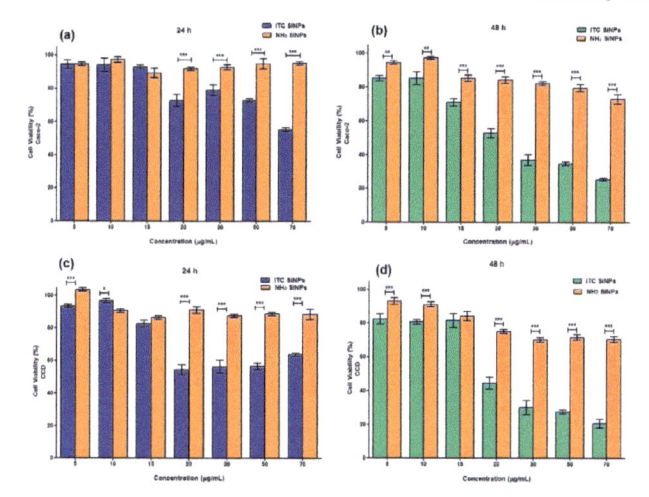

Fig. 7 Cell viability results of Caco-2 cells treated with ITC SiNPs after (a) 24 hours incubation and (b) 48 hours incubation time, and CCD-841 cells treated with ITC-functionalised SiNPs after (c) 24 hours incubation and (d) 48 hours incubation, with NH_2-functionalised SiNPs as a control. Statistical significance was determined by two-way-ANOVA followed by a Bonferroni post-test (***$P < 0.001$, **$P < 0.01$, *$P < 0.05$). Results are expressed as mean ± SEM ($n = 3$).

and overall displayed a significant reduction in the cell viability from lower to higher concentrations, resulting in the lowest cell viability of 70–80% in both lines.

By varying both time and concentration of cells incubated with ITC SiNPs, this study provided further evidence for the relatively low cytotoxicity of amine capped silicon nanoparticles and the large time and concentration dependence on the reduction of cell death in healthy and cancerous cell lines treated with ITC SiNPs. Whilst this is the desired effect for cancerous cells treated with this ITC system, it is worth noting that particularly with increased incubation; the observed cytotoxicity in healthy CCD cells provides no more benefit than the already well-established traditional chemopreventive treatments.

2.6 Cellular uptake and uptake efficiency

The efficiency of cellular uptake, which takes the fluorescence of cells treated with ITC SiNPs relative to unstained control cells into account, was measured at both concentrations at four incubation periods between 6 and 48 hours as detailed in Fig. 8a.

The relative fluorescence of cells treated with 20 μg ml^{-1} ITC SiNPs steadily rose from 8 to 20% as incubation increased, which contrasts greatly to those treated with the higher concentration (70 μg ml^{-1}) of ITC SiNPs. The median fluorescence observed remained virtually unchanged between 6 and 12 hours of incubation, meaning, at 12 hours, that at both concentrations of SiNP, the observed fluorescence was close to parity. At this point it can be said that the uptake efficiency of cell treated with the lower concentration of SiNPs was higher than that of those treated at a higher dose.

After 24 hours of incubation however, cells treated with 70 μg ml^{-1} displayed far greater fluorescence in comparison to those treated at a lower concentration.

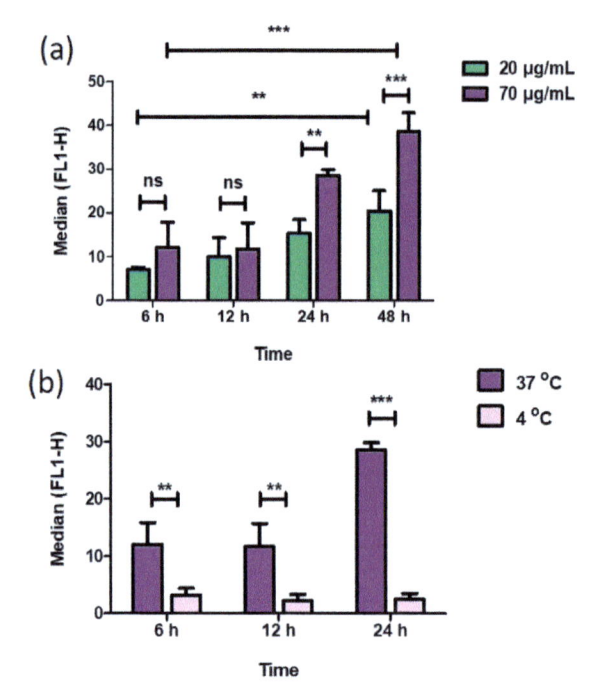

Fig. 8 Flow cytometry results, displaying (a) cellular uptake as the fluorescence median of cells treated with ITC-functionalized SiNPs vs. time at SiNP concentrations of 20 and 70 μg ml^{-1}. (b) Cellular uptake as the fluorescence median of cells treated with ITC-functionalized SiNPs vs. time at incubation temperatures of 37 °C and 4 °C.

As it was the case that the median fluorescence observed was higher in those cells treated with a higher concentration of ITC SiNPs, and given that the general trend observed with increasing incubation time was one of increased fluorescence, it can be stated with confidence that the cellular uptake of ITC SiNPs displays a dependence on both time and concentration. However, to gain more of an understanding of the mechanism of ITC uptake within the treated cells, it was important to identify whether their cellular uptake exhibited a relationship with the temperature of incubation. It is well documented that along with sample viscosity and material size, that the rate and uniformity at which molecules are transported and diffused across the cell membrane is highly dependent on temperature.[23]

Endocytosis is generally accepted as being the most common uptake mechanism for nanoparticles of sizes above 5 nm, and uptake efficiency by this mechanism has been largely linked to the physical and chemical parameters of the materials uptaken.[35]

Caco-2 cells were treated with the highest concentration (70 μg ml^{-1}) of ITC SiNPs and incubated over three periods of time between 6 and 24 hours, at 4 and 37 °C, as detailed in Fig. 8b.

It was found that cellular uptake, particularly in the period of 12 to 24 hours, displayed a vast increase at 37 °C whilst at the lower temperature, the value remained virtually unchanged overtime. What this indicated is that the mechanism dictating Caco-2 cell membrane penetration by ITC SiNPs depends as much on environmental factors as nanoparticle concentration.

3. Experimental methods

3.1 Synthesis procedure of ITC-capped SiNPs

A layer of porous silicon with terminal hydrogen was synthesized by the galvanostatic anodisation of boron-doped porous silicon p-Si (100) chip, 1.3 × 1.3 cm, 1–10 Ω cm^{-1} resistivity (PI-KEM Ltd, Tamworth, UK) in a 1 : 1 (v/v) solution of ethanol and 48% aq. HF, in a modification to the method established by Chao et al.[36] The resulting hydrogen terminated SiNPs were reacted with allyl bromide (2 ml) under reflux for 18 hours in distilled toluene. The result was a yellow-orange solution which when irradiated with UV light exhibited orange-red visible photoluminescence. The solution was dried under reduced pressure at 60 °C resulting in approximately 30 mg allyl bromide terminated SiNPs. The resulting nanoparticles were redispersed in distilled EtOH (10 ml), and potassium isothiocyanate (KSCN) was added in a 0.6 : 1 ratio of KSCN : allyl bromide terminated SiNPs and the mixture left to reflux for 18 hours. The solution was allowed to cool before being dried under reduced pressure. The product obtained was dissolved in 3 × 10 ml diethyl ether and washed with 3 × 12 ml deionised water. The yellow, translucent organic layer was dried and filtered, and the solvent was removed from the filtrate by drying under vacuum, resulting in an average of 10 mg ITC-capped SiNPs. The process principles of the above two steps are explained in Scheme 1.

3.2 Chemical properties analysis

FTIR measurements were carried out using a PerkinElmer ATR-FTIR spectrometer and conducted on solid hydrogen, allyl bromide and isothiocyanate terminated silicon nanoparticle samples. Spectra were obtained after 16 scans following background correction.

XPS measurements were conducted using a K-Alpha XPS Instrument, NEXUS, Newcastle, UK. The solid sample was pressed into 1 × 1 cm indium foil substrate

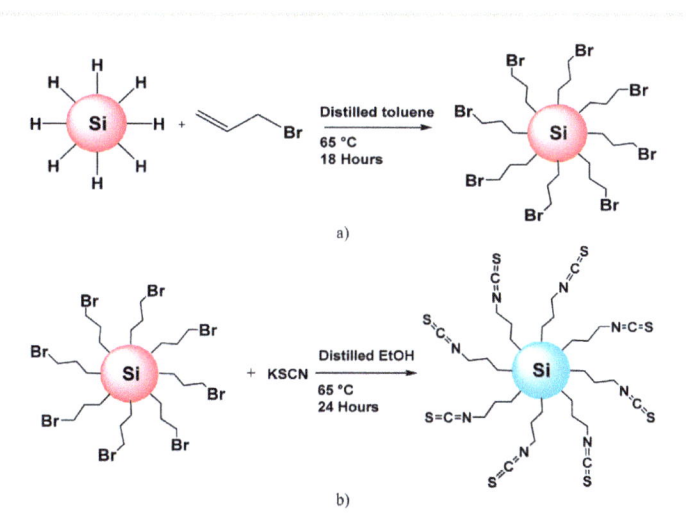

Scheme 1 Schematic representation of the two-step synthesis of ITC SiNPs. (a) The thermally induced hydrosilylation of hydrogen terminated SiNPs with allyl bromide, and (b) the subsequent reaction between allyl bromide functionalised SiNPs and KSCN.

(0.5 mm thickness, 99.999% trace metals basis, Sigma-Aldrich). The foil was placed in a N_2 load-lock for drying before being placed under ultra-high vacuum conditions, with pressure under 5×10^{-9} mbar. All obtained spectra were acquired at room temperature and normal emission using Al Kα radiation. All values of binding energy (BE) were referenced to the In $3d_{5/2}$ peak obtained from the measurement of the indium foil in direct contact with the ITC SiNPs, which was observed at a BE of 444 eV.

3.3 Optical properties

All optical measurements were conducted with samples of ITC SiNPs dispersed in distilled toluene. Quinine hemisulfate salt monohydrate diluted in 0.1 M H_2SO_4 was used to obtain reference UV-vis and PL spectra.

UV-vis absorption (UV-vis) spectra of ITC-terminated SiNPs were obtained using a PerkinElmer 35 UV-vis double beam spectrometer with the sample in a quartz cuvette (10 mm × 10 mm). The scan range was 200–800 nm with a scan speed of 200 nm min^{-1}. Prior to each measurement, a baseline scan of distilled toluene was recorded so that it could be subtracted from the subsequent sample spectrum. Photoluminescence emission (PL) spectra were recorded using a PerkinElmer LS55 spectrometer with the sample in a quartz cuvette (10 mm × 10 mm). The spectrometer scan range was set to 370–800 nm, with a fixed excitation of 360 nm, excitation slit width of 12.5 nm and emission slit of 2.5 nm. All emission spectra were corrected using the solvent spectrum as the background.

The photoluminescence quantum yield (QY) of the ITC SiNPs was obtained using quinine sulfate, with a known QY of 54.6% in 0.5 M H_2SO_4, as a standard. The absorbance and emission spectra obtained for the ITC SiNPs and the quinine sulfate show a gradient of the plotted integrated intensity of the ITC SiNP (in distilled toluene) and quinine sulfate (in 0.1 M H_2SO_4) solutions with recorded absorbances of between 0.1 and 0.01. Utilising the calculated gradients of these plots and the known quantities of solvent refractive indices and QY of quinine sulfate, a value of photoluminescence quantum yield for the ITC SiNPs could be generated using the following equation:

$$Q = Q_R \left(\frac{\text{grad}_S}{\text{grad}_R}\right)\left(\frac{\eta^2}{\eta_R^2}\right)$$

where Q is the quantum yield, grad refers to the gradient obtained from a plot of integrated fluorescence emission intensity vs. absorbance for the sample, and η is symbolic of the refractive index. The subscript S and R represent the sample and reference respectively.

3.4 Particle size measurements

Dynamic Light Scattering (DLS) was used to measure the hydrodynamic radius and polydispersity index of ITC-capped SiNPs dispersed in distilled toluene. Spectra were obtained with a Zetasizer Nano ZS (Malvern Instruments Ltd, Worcestershire, UK) and a triplet of measurements was taken for each sample solution. Measurements were taken at 20 °C with a 30 second equilibration preceding each set of measurements.

X-Ray Powder Diffraction (XRD) measurements were undertaken utilising an ARL XTRA Powder Diffractometer (Thermo Scientific, USA) in order to obtain the crystallinity and the mean nanoparticle size of the ITC SiNPs. Applying the Scherrer equation to the obtained spectra it is possible to calculate an estimate for the mean size of a powder sample τ:

$$\tau = \frac{k\lambda}{d\cos\theta}$$

where k is a shape factor, typically taken as 0.9, λ is the wavelength of the incident X-ray, d is the full width half maximum (FWHM) of the peak in radians, also denoted as $\Delta(2\theta)$ and θ is the Bragg angle in degrees.

Transmission electron microscopy (TEM) studies were performed using a JEOL, JEM 2100 microscope running a LaB$_6$ (lanthanum hexaboride crystal) emitter and an applied voltage of 200 kV. TEM samples were prepared by drop-casting a dilute suspension of the sample dispersed in distilled ethanol onto a 200-mesh carbon-coated copper grid. The grids were dried before the measurements. TEM micrographs were taken at different spots of the grid.

3.5 Cell culture

Two cell lines, Caco-2 and CCD-841, human colorectal adenocarcinoma and human normal colon epithelial cells respectively, were sub-cultured in RPMI with 10% fetal bovine serum, 2 mM L-glutamine, 100 μg ml^{-1} streptomycin and 100 μg ml^{-1} penicillin. Incubation and maintenance occurred at a temperature of 37 °C under a 5% CO$_2$ atmosphere. Both cell lines exhibited negative results for mycoplasmosis.

3.6 Cellular internalisation

Initially, Caco-2 cells were seeded in a 12-well plate, covered by glass slips. After 24 hours, the cells were incubated for 6 and 12 hours respectively with 50 μg ml^{-1} ITC SiNPs. Following this incubation period, cells were washed and fixed with 4% (v/v) paraformaldehyde (Sigma-Aldrich) in phosphate buffered solution (PBS), at room temperature for 20 minutes.

Actin present within the cells was stained using 6.6 μM Texas Red-X Phalloidin (Life Technologies). Laser Scanning Confocal Microscopy (LSCM) was conducted using a Zeiss LSM510 META confocal microscope, utilising a 63× oil immersion objective lens. The excitation wavelengths of the beams exciting the ITC SiNPs and actin were 488 nm and 543 nm respectively.

3.7 Cytotoxicity

Both cell lines were subjected to standard MTT assays to determine the cytotoxicity of the ITC SiNPs. Each line was seeded into a 96 well plate at a density of 4.5 × 10^{-5} cells per well and left for 24 hours. After this, the media used in seeding was removed and replaced with varying concentrations of ITC SiNP solution. The ITC SiNP suspensions were concentrated with values of 5, 10, 15, 20, 30 50 and 70 μg ml^{-1}. Each concentration and time grouping were replicated a minimum of three times and results normalised against an untreated control. The absorbance of the resultant wells was measured using a BMG Labtech Polar Star Optima at a wavelength of 570 nm with results expressed as mean values ± standard deviation.

3.8 Flow cytometry

Both cell lines were seeded in 24 well plates at a seeding density of 5×10^4 cells per well. Both were treated with ITC SiNPs at concentrations of 20 and 70 µg ml^{-1} for time periods of 6, 12, 24 and 48 hours. Following trypsinisation, the cells were washed three times with PBS and centrifuged for 5 minutes at 1200 rpm between washes. Cells were fixed with 4% (v/v) paraformaldehyde in fetal bovine serum for 20 minutes and washed with PBS. Cells were resuspended in a buffer of 5% (v/v) fetal bovine serum in PBS prior to flow cytometry measurements. 10 000 gated cells were examined for cellular uptake by observing ITC SiNP fluorescence using the FL1 channel detector and a BD FACSCalibur (BD Bioscience). Each concentration and time pairing was performed three times on different SiNP samples. Data was analyzed with CellQuest Pro® software (BD Biosciences) and values presented as median ± standard deviation.

3.9 Synchrotron FTIR

High resolution synchrotron FTIR microscopy was performed at the Max IV Laboratory facility and measurements were taken using a Bruker IFS66V FTIR spectrometer linked to a Hyperion 3000 IR microscope utilising synchrotron radiation as the illumination source. Operating in transmission mode with a 10×10 µm^2 aperture, 15× objective lens, a single condenser and a 100×100 µm^2 mercury cadmium telluride (MCT) detector, the video linked microscope achieved a magnification of 215× for the purpose of identifying cells. Data were obtained and analysed utilising OPUS 7 (Bruker) software. Spectra with a strong signal to noise ratio were maintained for analysis, 100 per substrate, and were acquired through the culmination of 256 scans at spectral resolution of 4 cm^{-1}.

Both cell lines were seeded on cadmium fluoride IR substrates, within a 12 well plate, at a seeding density of 5×10^4 cells per well per substrate. Following 24 hours of incubation at 37 °C, the media used during seeding was removed and the cells were washed three times with PBS. To each well, a 50 µg ml^{-1} solution of ITC SiNPs was added and the plate was incubated for 6 hours. Following the removal of the growth media, cells were washed with PBS on three occasions and 0.9% NaCl, before being fixed with 4% PFA in PBS for 30 minutes. Following this fixing period, the resultant cell plates were washed with distilled water, dried at room temperature and refrigerated prior to analysis.

FTIR images were obtained again using the Bruker IFS66V and Hyperion 3000 equipment, with a standard global acting as the thermal light source and a 128×128 focal plane array detector (Bruker). Measurements were undertaken with both instruments under a nitrogen atmosphere to reduce variations in the local atmosphere. The measurement conditions for the capture of spectral images were set at 128 scans and a resolution of 8 cm^{-1}. The final images were compiled from a selection of spectral areas characteristic of protein constituents, taking their integrated intensities and adding colour according to their distribution.

3.10 Statistical analysis

Statistical methods throughout this study are represented as mean ± SEM taken over a minimum of three independent experiments. Statistical significance was measured by two-way-ANOVA followed by a Bonferroni post-test using GraphPad Prism version 5.03 for Windows, GraphPad Software, San Diego California USA.

4. Conclusions

Isothiocyanate-functionalized silicon nanoparticles, their synthesis and the measurement of their optical, chemical and biomedical properties and behaviour were described in this report.

From the MTT assay, it was concluded that above concentrations of 15 μg ml^{-1}, the ITC SiNPs inflicted drastic cytotoxicity on Caco-2 and CCD-841 cells after 48 hours. This outcome is one of mixed interpretations given that cytotoxicity in Caco-2 cells is clearly favourable in the detoxification and/or promotion of apoptosis in carcinogenic cells, however this highlights that specificity towards cancerous cells is an area in which this research can be developed.

Through confocal microscopy and high resolution synchrotron FTIR, a better understanding as to how and where ITC SiNPs distribute upon cellular internalization was gained. The infrared images and synchrotron IR spectra obtained provided strong evidence for the specific concentration of ITC SiNPs within the nucleus of Caco-2 cells.

With analysis of confocal images and flow cytometry cellular uptake data, it was clear that internalisation of ITC SiNPs within Caco-2 cells was highly time and temperature dependent. From 12 to 48 hours, the cellular uptake increased steadily at variable concentration and it was evident that at low temperatures (4 °C) that uptake was hindered in the period monitored up to 24 hours.

The outcomes of this project provide succinct evidence for the versatility of ITC SiNPs as a naturally photoluminescent, easily monitored nano-system for the delivery of anti-cancer materials. A lot of encouragement can be taken from the results outlined above that provide a solid base from which future work on these materials can be drawn.

Conflicts of interest

There are no conflicts to declare.

Acknowledgements

X-ray photoelectron spectra were obtained at the National EPSRC XPS Users' Service (NEXUS) at Newcastle University, an EPSRC Mid-Range Facility.

References

1 R. L. Siegel, K. D. Miller and A. Jemal, *Ca-Cancer J. Clin.*, 2018, **68**, 7–30.
2 J. Maddams, M. Utley and H. Moller, *Br. J. Cancer*, 2012, **107**, 1195–1202.
3 A. S. Ahmad, N. Ormiston-Smith and P. D. Sasieni, *Br. J. Cancer*, 2015, **112**, 943–947.
4 V. T. DeVita, T. S. Lawrence, S. A. Rosenberg, R. A. DePinho and R. A. Weinberg, *Cancer: Principles and Practice of Oncology*, Wolters Kluwer Health, 2012.
5 F. P. Downie, H. G. Mar Fan, N. Houede-Tchen, Q. Yi and I. F. Tannock, *Psychooncology*, 2006, **15**, 921–930.

6 X. Wu, Q. H. Zhou and K. Xu, *Acta Pharmacol. Sin.*, 2009, **30**, 501–512.
7 Y. S. Keum, W. S. Jeong and A. N. Kong, *Mutat. Res.*, 2004, **555**, 191–202.
8 M. Rex and M. M. Christine, *J. Agric. Food Chem.*, 2004, **52**, 1867–1871.
9 J. W. Fahey, Y. S. Zhang and P. Talalay, *Proc. Natl. Acad. Sci. U. S. A.*, 1997, **94**, 10367–10372.
10 P. Talalay, J. W. Fahey, W. D. Holtzclaw, T. Prestera and Y. S. Zhang, *Toxicol. Lett.*, 1995, **82–83**, 173–179.
11 R. Munday and C. M. Munday, *Nutr. Cancer*, 2002, **44**, 52–59.
12 A. Basu and S. Haldar, *Int. J. Oncol.*, 2008, **33**, 657–663.
13 A. Pawlik, M. Wala, A. Hac, A. Felczykowska and A. Herman-Antosiewicz, *Phytomedicine*, 2017, **29**, 1–10.
14 J. W. Fahey, Y. S. Zhang and P. Talalay, *Proc. Natl. Acad. Sci. U. S. A.*, 1997, **94**, 10367–10372.
15 L. L. Song, E. J. Hennink, I. T. Young and H. J. Tanke, *Biophys. J.*, 1995, **68**, 2588–2600.
16 Y. Zhao, Y. Ye, X. Zhou, J. Chen, Y. Jin, A. Hanson, J. X. Zhao and M. Wu, *Theranostics*, 2014, **4**, 445–459.
17 M. G. B. and D. J. Norris, *Am. Phys. Soc.*, 1996, **53**, 16338–16346.
18 X. Gao, Y. Cui, R. M. Levenson, L. W. Chung and S. Nie, *Nat. Biotechnol.*, 2004, **22**, 969–976.
19 N. Lewinski, V. Colvin and R. Drezek, *Small*, 2008, **4**, 26–49.
20 A. M. Derfus, W. C. W. Chan and S. N. Bhatia, *Nano Lett.*, 2004, **4**, 11–18.
21 J. Wang, D.-X. Ye, G.-H. Liang, J. Chang, J.-L. Kong and J.-Y. Chen, *J. Mater. Chem. B*, 2014, **2**, 4338–4345.
22 L. H. Lie, M. Duerdin, E. M. Tuite, A. Houlton and B. R. Horrocks, *J. Electroanal. Chem.*, 2002, **538–539**, 183–190.
23 B. Shokri, M. A. Firouzjah and S. I. Hosseini, *Proc. 19th Int. Symp. Plasma Chem. Soc.*, 2009.
24 A. T. R. Williams, S. A. Winfield and J. N. Miller, *Analyst*, 1983, **108**, 1067–1071.
25 M. Behray, C. A. Webster, S. Pereira, P. Ghosh, S. Krishnamurthy, W. T. Al-Jamal and Y. Chao, *ACS Appl. Mater. Interfaces*, 2016, **8**, 8908–8917.
26 J. Wang, S. Sun, F. Peng, L. Cao and L. Sun, *Chem. Commun.*, 2011, **47**, 4941–4943.
27 Y. Zhong, F. Peng, F. Bao, S. Wang, X. Ji, L. Yang, Y. Su, S. T. Lee and Y. He, *J. Am. Chem. Soc.*, 2013, **135**, 8350–8356.
28 P. Caravan, J. J. Ellison, T. J. McMurry and R. B. Lauffer, *Chem. Rev.*, 1999, **99**, 2293–2352.
29 S. Jongmin, A. R. Md, K. M. Kyeong, I. G. Ho, L. J. Hee and L. I. Su, *Angew. Chem., Int. Ed.*, 2009, **48**, 321–324.
30 A. M. Smith, H. Duan, A. M. Mohs and S. Nie, *Adv. Drug Delivery Rev.*, 2008, **60**, 1226–1240.
31 D. He, D. Wang, X. Shi, W. Quan, R. Xiong, C.-y. Yu and H. Huang, *RSC Adv.*, 2017, **7**, 12374–12381.
32 L. J. Kan-Zhi Liu, S. M. Kelsey, *et al.*, *Apoptosis*, 2001, **6**, 269–278.
33 J. H. Ahire, Q. Wang, P. R. Coxon, G. Malhotra, R. Brydson, R. Chen and Y. Chao, *ACS Appl. Mater. Interfaces*, 2012, **4**, 3285–3292.
34 X. Jia, N. Li, W. Zhang, X. Zhang, K. Lapsley, G. Huang, J. Blumberg, G. Ma and J. Chen, *Nutr. Cancer*, 2006, **54**, 179–183.

35 C. Xu, X. Yuan, Z. Pan, G. Shen, J. H. Kim, S. Yu, T. O. Khor, W. Li, J. Ma and A. N. Kong, *Mol. Cancer Ther.*, 2006, **5**, 1918–1926.
36 N. H. Alsharif, C. E. Berger, S. S. Varanasi, Y. Chao, B. R. Horrocks and H. K. Datta, *Small*, 2009, **5**, 221–228.

Faraday Discussions

PAPER

Shedding light on the aqueous synthesis of silicon nanoparticles by reduction of silanes with citrates†

John L. Z. Ddungu,[ab] Simone Silvestrini, [a] Alessandra Tassoni [a] and Luisa De Cola [*ab]

Received 4th December 2019, Accepted 2nd January 2020
DOI: 10.1039/c9fd00127a

The synthesis of silicon nanoparticles in water has recently attracted a lot of attention. However, many scientists have expressed concerns on the nanomaterials obtained. We decided to explore two different routes to obtain silicon nanoparticles starting from a silane precursor. We report our findings regarding the preparation of nanomaterials under microwave irradiation and hydrothermal conditions starting from aqueous mixtures of (3-aminopropyl) triethoxysilane and sodium citrate. The microwave process, in particular, has been reported to yield silicon quantum dots bearing a surface layer of 3-aminopropyl moieties, allowing for the preparation of luminescent substrates amenable to biological-friendly amide chemistry. However, rigorous experimental design and thorough characterization of the products definitely rules out the formation of silicon nanoparticles. By highlighting the main issues linked to the proper characterization of these materials, we prove that the nanoparticles produced under both microwave and hydrothermal conditions, are a mixture of silica and carbon quantum dots.

Introduction

Silicon nanoparticles (Si NPs) have been proposed for a wide range of applications, owing to their chemical, optical and electronic properties.[1] These nanomaterials can be produced in different ways, such as by the reduction of silicon tetrachloride,[2] the laser ablation of silicon wafers[3] or their electrochemical etching.[4] The typical products bear surface Si–H groups that can undergo further functionalization via metal-catalyzed hydrosilylation,[5] to yield particles with diverse surface groups. Ω-Aminoalkyl chains, for example, are very popular in the

[a] Laboratoire de Chimie et des Biomatériaux Supramoléculaires, Institut de Science et d'Ingénierie Supramoléculaires (ISIS), CNRS, 8 Allée Gaspard Monge, Strasbourg, 67000, France. E-mail: decola@unistra.fr; Fax: +33 368855242; Tel: +33 368855220
[b] Institut für Nanotechnologie (INT), Karlsruher Institut für Technologie (KIT) Campus North, Hermann-von-Helmholtz-Platz 1, Eggenstein-Leopoldshafen, 76344, Germany
† Electronic supplementary information (ESI) available. See DOI: 10.1039/c9fd00127a

development of new solutions for theranostic applications, since they give surface access to biologically-friendly amide chemistry.[6]

A particularly attractive method for the preparation of Si NPs bearing 3-aminopropyl chains on their surfaces has been reported by Y. He and coworkers.[7] This method makes use of 3-aminopropyl trimethoxysilane (APTMS) as a silicon source, to ensure the presence of amino groups on the surface of the resulting highly crystalline NPs. The conversion of the silane into nanoparticles is carried out in water, using microwave (MW) irradiation to drive the reaction and trisodium citrate as a reducing agent. In their paper, He and co-workers state that reduction of APTMS by citrate yields the formation of nuclei that grow to yield fully-fledged nanocrystals via Ostwald ripening processes. The paper however does not focus on the reduction mechanism, or the fate of the excess 3-aminopropyl chains, that must be eliminated in the Ostwald ripening process in order to decrease the specific surface of the nanoparticles while providing an all-silicon core. The material cited on the topic deals with MW growth (rather than nucleation) of CdTe-type quantum dots where citrate is not involved and, no reaction mechanisms are described. Finally, it should be noted that an APTMS-to-citrate molar ratio of 10 : 1 is reported by the authors for the reaction, while the same reducing agent is typically employed in large excess in the preparation of metal nanoparticles.

Citrates are indeed well-known reducing agents for the preparation of metal nanoparticles, often doubling as capping agents to suppress excessive particle growth.[8] The mechanism of reduction by citrates has been studied in detail and, in the case of tetrachloroauric salts, goes through a ligand substitution by carboxylate, that attacks the metal and displaces a chloride anion, yielding the transition state that can evolve towards the products.[9] In the case of APTMS, it is hard to envisage a mechanism where methanoate is displaced by a better leaving group such as a carboxylate. On the other hand, this step may be possible by prior hydrolysis of the alkoxysilane by a hydroxyl group, followed by its substitution. In addition, the Si–C bond is much less polarized than Si–O, with the 3-aminopropyl chain serving as an even worse leaving group than methanoate, so its reduction by citrate may be more problematic. This is an important aspect since, by considering spherical monocrystalline silicon nanoparticles with radii between 2.0 and 2.5 nm, we can evaluate the number of surface Si atoms as roughly 50–60% of the total number of Si atoms in the particle, meaning that about half of the Si–C bonds in the starting APTES molecules need to be substituted by Si–Si ones, if a crystalline particle is to be formed.

The characterization of the chemical nature of the nanoparticles reported includes high resolution TEM, X-ray powder diffractometry, FT-IR and energy-dispersive X-ray spectroscopy. The latter technique does highlight the presence of silicon atoms, but cannot discriminate between redox states and prove the presence of Si^0 in the sample. Moreover, the FT-IR spectrum displays a strong C–O vibration that is not discussed, but hardly fits in the chemical scenario proposed by the authors. Finally, the XRD pattern shows two reflections typical of the diamond structure of crystalline silicon, the (111) and (220), with the (311) being mentioned but not discernible from the background noise. According to the Scherrer formula,[10] the FWHM – or $\Delta(2\theta)$ – for the diffraction peaks recorded for a 2.2 nm isotropic crystalline silicon nanoparticle, should be around 3.7° for the (111) reflection and 3.9° for the (220) one, considering an ideal instrument that does not contribute to further line broadening, operating with monochromatic Cu Kα radiation. Many authors have shown that the Scherrer formula can be used to calculate the mean crystallite size for Si NPs, with results closely matching the particles sizes measured by high-resolution TEM.[11,12] While the quality of the data reported by He and co-workers does not allow for a quantitative measurement of the FWHM of the diffraction peak, it is undoubtedly much smaller than 3°, apparently contradicting the DLS and HR-TEM data reported in the same article.

Very recently, while we were compiling our data, Sergei Alekseev et al.[13] published a communication in which they also doubt the formation of Si NPs using the simple water based methodology. Their analysis, mostly based on the photophysical properties of the obtained nanoparticles and on the critical interpretation of the previously reported data, indicated that the particles obtained are not Si NPs.

In this paper, we intend to deepen such analysis and to report our findings on the MW-assisted preparation of nanomaterials starting from aqueous mixtures of APTMS and sodium citrate, as well as a hydrothermal (HT) variant, where the mixture is thermally treated in a conventional oven for an extended period of time.

Upon approaching He's protocol, we made two observations: (i) that the conditions of the synthesis, namely the presence of water and an electrolyte, along with the final basic pH (*vide infra*), may very well prompt a hydrolysis-condensation process on the alkoxysilanes (as in a Stöber-like process,[14] or the synthesis of oligomeric silsesquioxanes[15]), thus yielding silica (SiO_2) rather than silicon (Si^0); (ii) that both the hydrothermal treatment of citrate and its microwave-assisted pyrolysis have been reported to yield large macromolecules, graphene oxide or carbon quantum dots with peculiar fluorescence properties and, in some cases, quantum yields exceeding 20%.[16–19]

Other authors reported the preparation of carbon quantum dots from mixtures of 3-aminopropyl trialkoxysilanes and citric acid[20] under hydrothermal conditions, labelling the products as Si-doped CQDs, but final proof of the inclusion of silicon into the carbon core of the quantum dots was never presented. Therefore, we designed a set of experiments to characterize the materials prepared, in order to elucidate their chemical nature. Syntheses were carried out by mixing the starting materials (APTMS and trisodium citrate) in water and providing either MW irradiation or prolonged thermal treatment. Mock reactions, lacking the reducing agent or the silicon source were also performed to evaluate the roles of the reagents. The chemical structure of the samples was assayed by means of infrared and X-ray photoelectron spectroscopies, while their morphologies and crystallinity were evaluated using electron microscopy and X-ray powder diffraction. Finally, UV-vis spectroscopy and emission measurements complete the characterization of the materials.

Results and discussion

Synthesis and characterisation of the nanoparticles

Aqueous mixtures of trisodium citrate and APTMS were treated either in a microwave reactor, as reported by He and coworkers[7] (MW-series samples), or under hydrothermal conditions,[6] in a static oven within tightly sealed Teflon containers (HT-series samples). Table 1 summarizes the composition of the

Table 1 Samples and experimental conditions

Sample name	[Citrate] (M)	APTMS (M)	Heating	Time (min)
MW-CS	0.126	1.11	MW	15
MW-C	0.126	0	MW	15
HT-CS	0.126	1.11	HT	300
HT-C	0.126	0	HT	300

starting mixtures for the different tests, with –C samples being prepared without the addition of the silicon source in the reaction medium and –CS samples containing both reagents. All combinations gave crude reaction mixtures with a basic pH, between 8.0 and 9.5, and samples MW-CS, HT-CS and HT-C displayed intense blue emission upon UV irradiation at 365 nm. Sample MW-C was only weakly luminescent after 15 min MW treatment, but prolonging the irradiation time led to crude samples with increasing emission.

The complete characterization of these samples was carried out after purification of the products by dialysis against water through a 1 kDa membrane (corresponding to a pore diameter below 1.5 nm).

Fig. 1, compares the FT-IR spectra (collected in ATR configuration) of trisodium citrate and those of the products of its treatment in water, MW-C and HT-C, highlighting the dominance of the symmetric and asymmetric stretching of the carbonyl groups at 1586 and 1396 cm^{-1} respectively (for the starting material). Both signals, the former in particular, are slightly shifted to lower wavenumbers for the treated samples, together with a general broadening of the weaker peaks.

The trend observed for the signals of the carboxyl group is confirmed for MW-CD and HT-CS. To better comprehend the features brought about in the FT-IR spectra by the addition of APTMS into the reaction mixture, we compare the data (summed up in Table 2) with those recorded for a pure sample of the trifluoromethane-sulfonic acid salt of octa(3-aminopropyl)silsesquioxane (POSS, green trace), prepared as reported elsewhere.[21] All the features of the FT-IR traces, aside from those described for the carboxylates, can indeed be tracked back to the green spectrum. N–H bends and C–N stretches appear as shoulders for the stronger carbonyl peak, while the signal for the Si–C is more isolated and clear. In polyhedral silsesquioxanes, the asymmetric (1114 cm^{-1}) and symmetric (1030 cm^{-1}) stretches of the Si–O–Si units are very sharp due to the well-defined, symmetric geometry of the molecular system.[22] The same two peaks typically collapse into one single, broad band for amorphous silica. Interestingly, symmetric and asymmetric stretches are clearly recognizable in the trace of HT-CS, while the asymmetric stretching appears as a mere shoulder for MW-CS, indicating a more disordered structure for the latter sample.

High resolution XPS scans for the Si 2p and C 1s lines of MW-CS and HT-CS are reported in Fig. 2 (survey scans can be found in the ESI†). Si 2p lines did not show well-defined spin–orbit components, therefore Si $2p_{1/2}$ and Si $2p_{3/2}$ contributions have not been separated in the analysis. Clear separation of these components ($\Delta = 0.63$ eV) is typical for bulk silicon only. No signal for elemental silicon was observed, expected at 99.4 eV for bulk silicon or as high as 99.9 eV for small NPs.[11,12] The elemental peak for silicon has previously been reported to

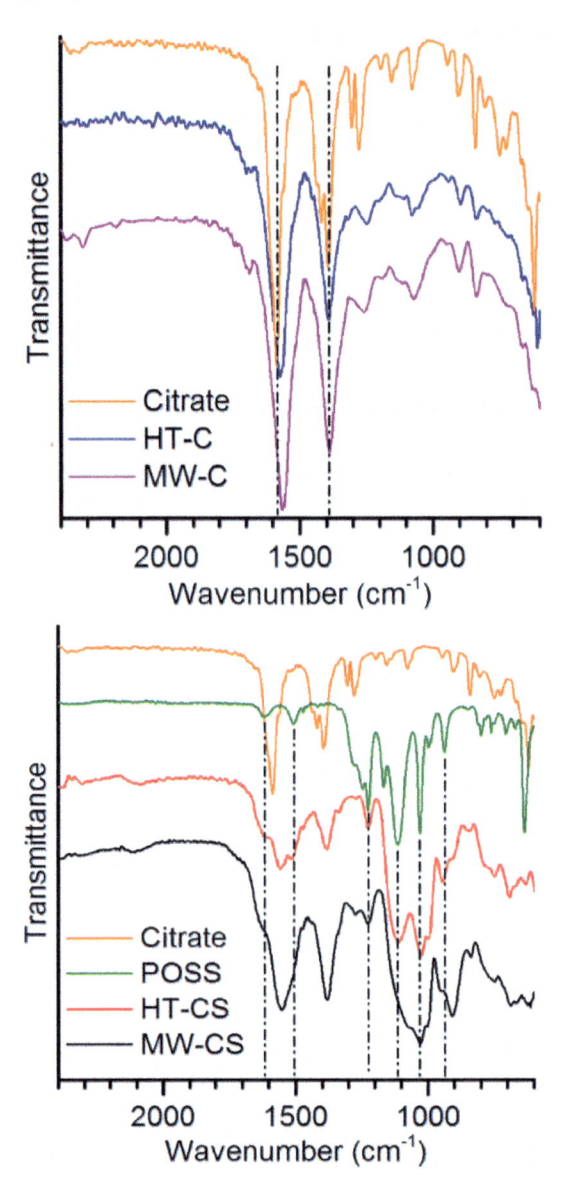

Fig. 1 (top) FT-IR spectra of the samples MW-C, HT-C. Trisodium citrate alone is reported for comparison. (bottom) Comparison of FT-IR spectra of samples MW-CS and HT-CS with trisodium citrate and octa-amino T8-polyhedral oligomeric silsesquioxane (POSS).

progressively disappear as the diameter of Si NPs decreases since most of the silicon atoms are on the surface, but can always be appreciated for particles above 1 nm in diameter.[11] In this case, the samples display very similar profiles, with two components centered in the 101.5–103.0 eV region, compatible with the presence of silicon species bearing organic groups such as alkyl chains.

A third, weaker peak at higher binding energy, usually recorded for SiO_2, was also observed. Elemental compositions, in particular the silicon-to-nitrogen

Table 2 Peak assignments for the samples, in comparison with pure references. (sh) denotes shoulders

Signal	Wavenumbers (cm^{-1})			
	MW-CS	HT-CS	POSS	Citrate
Carbonyl (sym)	1548	1559	—	1586
Carbonyl (asym)	1379	1382	—	1396
N–H bending	1625 (sh)	1619 (sh)	1616	—
C–N	1506 (sh)	1515	1508	—
Si–C	1225	1223	1225	—
Si–O–Si (ring-asym)	1117 (sh)	1110	1114	—
Si–O–Si (ring-sym)	1028	1026	1030	—

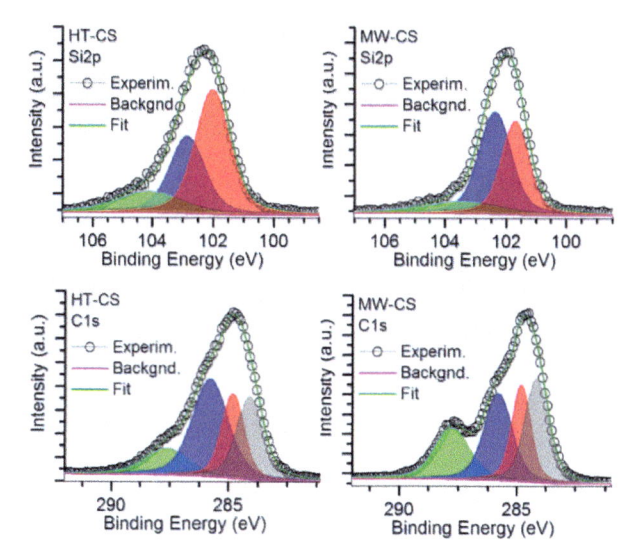

Fig. 2 XPS high resolution scans of the Si 2p (top) and C 1s (bottom) lines of samples HT-CS (left) and MW-CS (right).

atomic ratios are in both cases close to 1 (see Table 3), suggesting that the 3-aminopropyl chains of the silane starting material have survived the treatment. Interestingly, deconvolution of the C 1s lines shows a component with BE close to 284.0 eV, indicating the presence of sp^2-hybridized carbon double bonds, typically observed in carbon quantum dots.

Transmission electron microscopy imaging of the samples was performed to assay the morphologies of the various materials and the presence of crystalline domains. This kind of data is typically reported by most authors to provide visual evidence of the presence of nanoparticles and can further support other analytical techniques such as X-ray powder diffraction. As shown in Fig. 3, recorded for sample HT-CS, the samples are made of particles displaying slightly irregular round shapes. The size distributions recorded for HT-CS and MW-CS by direct measurement of 80 particles are reported in Fig. 4.

We wish to highlight that upon high resolution imaging, we observed a fast phase transition in the particles subjected to the electron beam, that went from

Faraday Discussions Paper

Table 3 Atomic composition of the samples probed by XPS

	MW-CS	HT-CS
Si	12.7%	10.1%
O	26.3%	31.0%
N	10.5%	9.1%
C	50.5%	49.8%

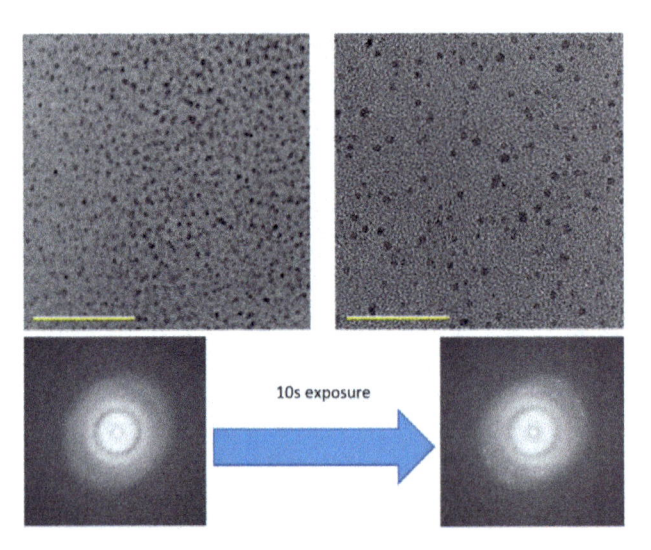

Fig. 3 (top) Representative TEM images for sample HT-CS (left) and MW-CS (right) (scale bars = 50 nm). (bottom) Observed evolution of the live FFT for one particle over time.

amorphous to crystalline as shown by the Fourier transforms shown at the bottom of Fig. 3. This is likely due to the heating provided by the electron beam, combined with the high vacuum conditions typical of the analysis chamber. We conclude that, for this kind of sample, crystallinity should be probed by means of X-ray diffraction, rather than selected area diffraction experiments on TEM equipment, that should only confirm X-ray diffraction findings.

Indeed, X-ray powder diffraction traces (Fig. 5) show amorphous phases with no discernible peaks. Crystalline Si NPs should display three diffraction peaks characteristic of (111), (220) and (311) reflections of the diamond structure, with a full width at half maximum compatible with the mean crystallite size according to the Scherrer formula.[10] Both diffractograms are characterized by two broad peaks. The one at *ca.* 22° can be ascribed to silica while the other at *ca.* 11° must result from a second amorphous phase[23] – most probably amorphous carbon.

Fig. 6 shows the normalized emission spectra recorded for MW-CS and HT-CS upon excitation at 340 nm, comparing them with the emission spectra of sample HT-C and MW-C, prepared without a silicon source. There is good superimposition of the normalized spectra for all the samples, with a slight shift to lower wavelengths for MW-C and a somewhat broader emission spectrum for HT-CS. The similarities between the spectra for the samples prepared with and without

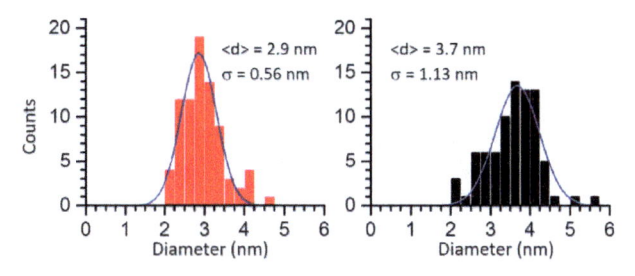

Fig. 4 Size distributions of the particles observed by TEM (histograms) and a Gaussian size distribution function (blue line), for samples HT-CS (left) and MW-CS (right).

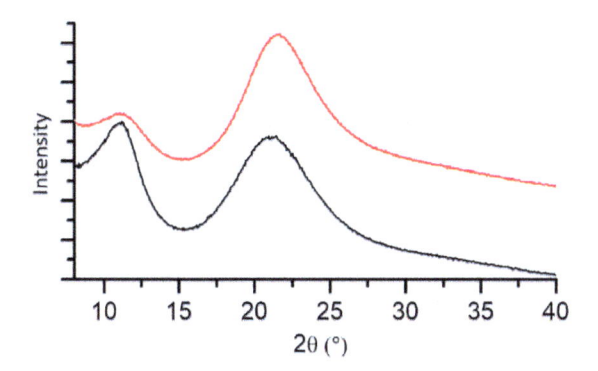

Fig. 5 X-Ray powder diffractograms for samples MW-CS (black) and HT-CS (red).

a silicon source suggest that the emission may be due to the presence of carbon-based species, rather than silicon nanoparticles. Emission spectra recorded at different excitation wavelengths are reported in the ESI,† while Fig. 7 highlights the shift of the emission maxima as a function of the excitation wavelength. This is a well-known feature for carbon quantum dots and the limited shift under UV irradiation below 400 nm has been described as a consequence of the passivation of the carbon surface.[16] MW-C was the only sample in the series that displayed a pronounced shift of the emission maxima below the 400 nm threshold, but we deemed it necessary to assign a large error bar to the data recorded using long excitation wavelengths (380 and 400 nm). Such measurements were marred by the high intensity of the Raman band of the solvent relative to the signal given by the sample, due to the low concentration of MW-C resulting from the scarce yield of the process.

To further prove that the photoluminescence recorded for MW-CS is not ascribed to the presence of silicon quantum dots, aliquots of the sample were treated with either hydrofluoric or hydrochloric acids. The former was employed to dissolve any silicon or silica particles in suspension, the latter to provide an acidic pH without compromising the suspended materials. Fig. 8 shows the photoluminescence emission spectra (only the spectra recorded at 340 nm are shown, for clarity) of the acid-treated aliquots and compares them to the pristine material, to which water was added to account for dilution effects. While the

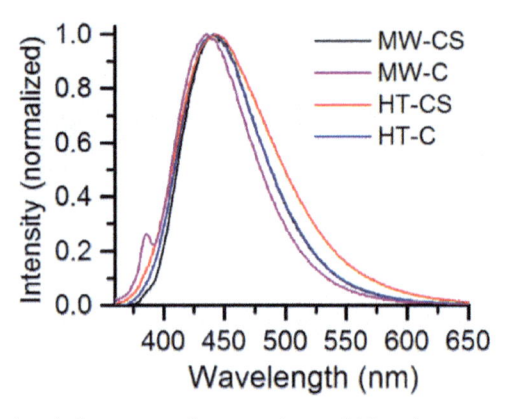

Fig. 6 Normalized emission spectra (in water, λ_{exc} = 340 nm).

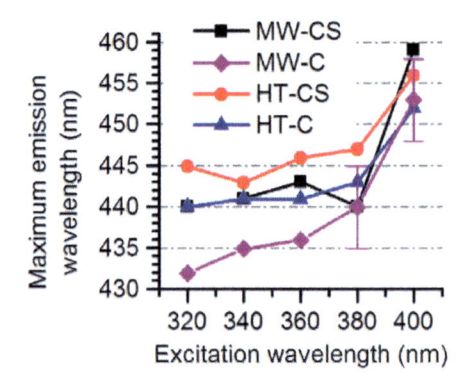

Fig. 7 Position of the maxima in the PL emission spectra upon excitation at different wavelengths.

signal recorded for the HF-treated sample is lower in intensity than the pristine one, this quenching effect is caused by the acidic pH, as proven by the spectra recorded when employing hydrochloric acid. This behavior was previously

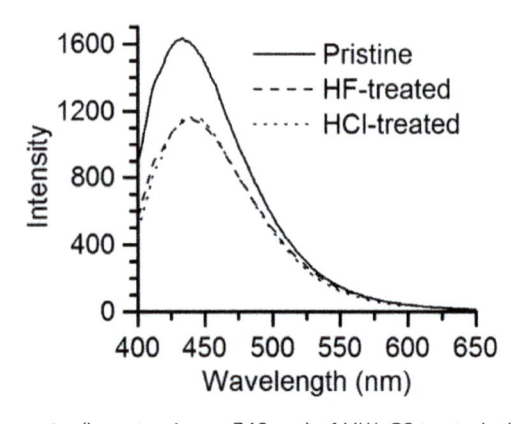

Fig. 8 Emission spectra (in water, λ_{exc} = 340 nm) of MW-CS treated with different acids.

reported for carbon quantum dots[24] and further suggests their presence instead of the emissive silicon dots.

Conclusions

In this paper, we critically analyzed the process proposed by Y. He and coworkers for the preparation of silicon nanoparticles with 3-aminopropyl surface modification.[7] Comparison of ATR-FTIR spectra with those of the reagents and pure reference compounds, and high resolution XPS analyses and powder X-ray diffraction data suggest that the nanoparticles produced by heating mixtures of sodium citrate and (3-aminopropyl)triethoxysilane in water are not crystalline Si^0. Rather, they are silica nanoparticles with 3-aminopropyl-decorated surfaces mixed with the thermal degradation products of citrate (carbon quantum dots) that are well-known to display photoluminescence properties very similar to those we observed for our samples.

Our report highlights a number of critical issues that material scientists should be aware of when approaching the preparation of silicon but also any other nanoparticles. In particular, we prove that high resolution transmission electron microscopy should not be used alone to assay the crystallinity of nanoparticles, since (i) the material may evolve under the electron beam and (ii) both interplanar distances and mean crystallite sizes should be related to powder X-ray diffraction data. Moreover, we give strong examples for the reason why photoluminescence properties should not be considered as proof of the presence of silicon nanoparticles. A proper characterization must be performed to assess the real nature and structure of nanomaterials or any material.

Experimental methods and materials

Materials

(3-Aminopropyl)trimethoxysilane (97%), trisodium citrate dihydrate (≥99.0%) and technical grade hydrofluoric acid (48%) were purchased from Sigma-Aldrich. Hydrochloric acid (36.5 to 38.0% wt) was purchased from Fisher Scientific. All chemicals were used without additional purification. All solutions were prepared using DI water (Milli-Q system by Millipore) as the solvent. A Memmert 200 oven (Memmert GbmH) was used for hydrothermal treatments, while microwave processing was carried out with a Discover model CEM focused microwave synthesis system.

Optical measurements were performed at room temperature under ambient air conditions. UV-vis absorption spectra were recorded with a Shimadzu UV-3600 UV-vis-NIR spectrophotometer. Photoluminescence was assayed using a Horiba Jobin-Yvon FluoroLog-3 spectrofluorimeter. TEM samples were prepared by dispersing aqueous suspensions of the samples onto holey-carbon-on-copper grids with the excess solvent evaporated. The TEM/HRTEM overview images were recorded using a Jeol 2100F electron microscope equipped with Cs-corrected condenser, imaging filter, biprism and 2 CCD cameras, operated at 200 kV. For FTIR measurements, samples were lyophilised and the resulting solids were ground with an agate mortar and laid on the ATR crystal. IR spectra were recorded on a Shimadzu IRAffinity-1 FTIR spectrometer in ATR-FTIR configuration and 64 scans were cumulated at a resolution of 4 cm^{-1}. XPS measurements were performed using a Thermo Scientific K-Alpha X-Ray Photoelectron Spectrometer using monochromatic Al Kα radiation (1486.6 eV). High resolution scans were performed with a 50 eV analyser pass energy and 0.1 eV step size. Binding energies of each element were referenced to the carbon C 1s peak at 284.8 eV. Powder X-ray diffraction (XRD) spectra were recorded using a Bruker D2 PHASER diffractometer operating in θ–2θ mode with Cu Kα (8.04 keV) radiation passed through a graphite monochromator, scanning between 8 and 40° with a 0.05° step angle and a 100 s dwell time to suppress noise.

Synthesis of the nanoparticles

These procedures are taken and adapted from the previous report by He and coworkers.[7] A precursor solution was prepared by adding 10 mL of (3-aminopropyl)trimethoxysilane to 40 mL of an Argon-saturated aqueous solution dispersed with 1.66 g of trisodium citrate dihydrate. The mixture was stirred for 10 min. The solution was transferred into the appropriate reaction vessels (pressure-sealable Teflon vessels for the hydrothermal assisted synthesis or sealable glass tubes for the microwave assisted synthesis). Teflon vessels containing the reaction solution were heated at 160 °C for 5 hours, while glass tubes containing the same solution were irradiated in a microwave reactor for 15 minutes with the temperature set to 160 °C. In both cases, samples were removed when the temperature cooled to <30 °C naturally and were then neutralized to pH 7 with conc. HCl. Residual reagents were removed by dialysis (MWCO = 1 kDa) using Spectra/Por 7 standard regenerated cellulose pre-treated dialysis tubing.

Conflicts of interest

There are no conflicts to declare.

Abbreviations

Si NP	Silicon nanoparticles
APTMS	(3-Aminopropyl)tri-methoxysilane
MW	Microwave
(HR-)TEM	(High resolution-) transmission electron microscopy
HT	Hydrothermal
POSS	Octa-amino T8-polyhedral oligomeric silsesquioxane

Acknowledgements

We are thankful for financial support from the Helmholtz Association (funding through the Helmholtz Virtual Institute NanoTracking, Agreement Number VH-VI-421) and from L' Institut Universitaire de France, IUF. SS acknowledges the financial support from the European Union's Horizon 2020 research and innovation programme under the Marie Skłodowska-Curie grant agreement No. 707944. JD acknowledges the financial support from the French-German University (DFH-UFA).

References

1 Z. Kang, Y. Liu and S.-T. Lee, *Nanoscale*, 2011, **3**, 777–791.

2 J. Zou, R. K. Baldwin, K. A. Pettigrew and S. M. Kauzlarich, *Nano Lett.*, 2004, **4**, 1181–1186.
3 T. Yoshida, S. Takeyama, Y. Yamada and K. Mutoh, *Appl. Phys. Lett.*, 1996, **68**, 1772–1774.
4 J. L. Heinrich, C. L. Curtis, G. M. Credo, M. J. Sailor and K. L. Kavanagh, *Science*, 1992, **255**, 66–68.
5 A. B. Sieval, A. L. Demirel, J. W. M. Nissink, M. R. Linford, J. H. van der Maas, W. H. de Jeu, H. Zuilhof and E. J. R. Sudhölter, *Langmuir*, 1998, **14**, 1759–1768.
6 N. Licciardello, S. Hunoldt, R. Bergmann, G. Singh, C. Mamat, A. Faramus, J. L. Z. Ddungu, S. Silvestrini, M. Maggini, L. De Cola and H. Stephan, *Nanoscale*, 2018, **10**, 9880–9891.
7 Y. Zhong, F. Peng, F. Bao, S. Wang, X. Ji, L. Yang, Y. Su, S.-T. Lee and Y. He, *J. Am. Chem. Soc.*, 2013, **135**, 8350–8356.
8 J. Turkevich, P. C. Stevenson and J. Hillier, *Discuss. Faraday Soc.*, 1951, **11**, 55–75.
9 I. Ojea-Jiménez and J. M. Campanera, *J. Phys. Chem. C*, 2012, **116**, 23682–23691.
10 A. L. Patterson, *Phys. Rev.*, 1939, **56**, 978–982.
11 M. L. Mastronardi, F. Maier-Flaig, D. Faulkner, E. J. Henderson, C. Kübel, U. Lemmer and G. A. Ozin, *Nano Lett.*, 2012, **12**, 337–342.
12 K. Sato, T. Izumi, M. Iwase, Y. Show, H. Morisaki, T. Yaguchi and T. Kamino, *Appl. Surf. Sci.*, 2003, **216**, 376–381.
13 B. V. Oliinyk, D. Korytko, V. Lysenko and S. Alekseev, *Chem. Mater.*, 2019, **31**, 7167–7172.
14 W. Stöber, A. Fink and E. Bohn, *J. Colloid Interface Sci.*, 1968, **26**, 62–69.
15 D. B. Cordes, P. D. Lickiss and F. Rataboul, *Chem. Rev.*, 2010, **110**, 2081–2173.
16 X. Zhai, P. Zhang, C. Liu, T. Bai, W. Li, L. Dai and W. Liu, *Chem. Commun.*, 2012, **48**, 7955–7957.
17 Y. Guo, Z. Wang, H. Shao and X. Jiang, *Carbon*, 2013, **52**, 583–589.
18 M. Xu, G. He, Z. Li, F. He, F. Gao, Y. Su, L. Zhang, Z. Yang and Y. Zhang, *Nanoscale*, 2014, **6**, 10307–10315.
19 W. H. Green, K. P. Le, J. Grey, T. T. Au and M. J. Sailor, *Science*, 1997, **276**, 1826.
20 J. Chen, W. Liu, L.-H. Mao, Y.-J. Yin, C.-F. Wang and S. Chen, *J. Mater. Sci.*, 2014, **49**, 7391–7398.
21 M. Janeta, Ł. John, J. Ejfler and S. Szafert, *Chem.–Eur. J.*, 2014, **20**, 15966–15974.
22 E. S. Park, H. W. Ro, C. V. Nguyen, R. L. Jaffe and D. Y. Yoon, *Chem. Mater.*, 2008, **20**, 1548–1554.
23 P. Riello, M. Munarin, S. Silvestrini, E. Moretti and L. Storaro, *J. Appl. Crystallogr.*, 2008, **41**, 985–990.
24 N. Dhenadhayalan, K.-C. Lin, R. Suresh and P. Ramamurthy, *J. Phys. Chem. C*, 2016, **120**, 1252–1261.

Faraday Discussions

PAPER

Ultrasmall silicon nanoparticles as a promising platform for multimodal imaging†

Garima Singh,[a] John L. Z. Ddungu,[bc] Nadia Licciardello,[‡a] Ralf Bergmann,[a] Luisa De Cola[*bc] and Holger Stephan[*a]

Received 25th September 2019, Accepted 11th October 2019
DOI: 10.1039/c9fd00091g

Bimodal systems for nuclear and optical imaging are currently being intensively investigated due to their comparable detection sensitivity and the complementary information they provide. In this perspective, we have implemented both modalities on biocompatible ultrasmall silicon nanoparticles (Si NPs). Such nanoparticles are particularly interesting since they are highly biocompatible, have covalent surface functionalization and demonstrate very fast body clearance. We prepared monodisperse citrate-stabilized Si NPs (2.4 ± 0.5 nm) with more than 40 accessible terminal amino groups per particle and, for the first time, simultaneously, a near-infrared dye (IR800-CW) and a radiolabel (^{64}Cu-NOTA = 1,4,7-triazacyclononane-1,4,7-triacetic acid) have been covalently linked to the surface of such Si NPs. The obtained nanomaterials have been fully characterized using HR-TEM, XPS, UV-Vis and FT-IR spectroscopy. These dual-labelled particles do not exhibit any cytotoxicity *in vitro*. *In vivo* studies employing both positron emission tomography (PET) and optical imaging (OI) techniques revealed rapid renal clearance of dual-labelled Si NPs from mice.

Introduction

Molecular imaging is a non-invasive method that provides reliable information about anatomy, as well as physiological and pathophysiological processes in

---

[a]*Institute of Radiopharmaceutical Cancer Research, Helmholtz-Zentrum Dresden-Rossendorf, Bautzner Landstraße 400, Dresden, D-01328, Germany. E-mail: h.stephan@hzdr.de; Fax: +49 3512603232; Tel: +49 3512603091*

[b]*Laboratoire de Chimie et des Biomatériaux Supramoléculaires, Institut de Science et d'Ingénierie Supramoléculaires (ISIS), 8 allée Gaspard Monge, Strasbourg, 67000, France. E-mail: decola@unistra.fr; Fax: +33 368855242; Tel: +33 368855220*

[c]*Institut für Nanotechnologie (INT), Karlsruher Institut für Technologie (KIT) Campus North, Hermann-von-Helmholtz-Platz 1, Eggenstein-Leopoldshafen, 76344, Germany*

† Electronic supplementary information (ESI) available. See DOI: 10.1039/c9fd00091g

‡ Current address: International Iberian Nanotechnology Laboratory, Avenida Mestre José Veiga s/n, 4715-330 Braga, Portugal.

living systems. Various imaging modalities such as computed tomography (CT), magnetic resonance imaging (MRI), optical imaging (OI), single-photon emission computed tomography (SPECT) and positron emission tomography (PET) are available for this purpose.[1-7] Each imaging modality has its own unique strengths and intrinsic limitations, mainly regarding spatial/depth resolution and sensitivity, making the achievement of precise and reliable information at the disease site difficult. In order to circumvent the limitations of single imaging techniques and exploit their advantages synergistically, multimodal molecular imaging has recently gained importance. Of particular interest is the combination of nuclear and optical methods. In this respect, the use of PET, due to its high specificity/sensitivity (fM to pM range), the possibility of quantifying the data, and near infrared fluorescence imaging, which enables high resolution, relatively deep penetration (cm range) and the use of fluorescence-guided surgery is a winning combination.

Implementation of both modalities on the same system requires multifunctional molecules or nanomaterials. Due to their diverse modification possibilities, nanoscale materials have become significantly more important in recent years for imaging applications.[8-16] This is due to the fact that it is possible to anchor a large number of similar or different modalities to a probe and thus increase the sensitivity to follow *in vitro* and *in vivo* processes in more detail. In addition, biological vector molecules can be introduced simultaneously for pharmaceutical targeting. For medical imaging however, very small materials that reach their target quickly and are excreted renally are desirable. The size considered optimal for such application is less than 6 nm, and particles with such small dimension are referred to as ultrasmall nanoparticles. Ultrasmall renally clearable nanoparticles possess enormous potential as cancer imaging agents,[17-20] and in this respect, biocompatible silicon nanoparticles (Si NPs) are highly attractive. Their covalent surface functionalization allows the introduction of different labels for *in vivo* imaging.[21,22] As far as biomedical applications are concerned, to date, Si NPs have been applied in bioimaging[23-27] and in real-time immunofluorescence imaging.[28-33] For instance, to target cells in a specific way, Si NPs have been successfully coupled with single-stranded DNA,[34-36] folic acid,[37] sugars,[38,39] peptides[29] and antibodies,[31,33] and several *in vitro* studies have been performed on their cellular uptake[40-42] and cytotoxicity.[43-45] *In vitro* investigations are mostly based on the intrinsic photoluminescence of the Si NPs. Through targeted surface modification, the luminescence properties can be tailored and, in particular, the quantum yield significantly increased.[46-50]

In contrast to *in vitro* studies, there are only a few *in vivo* studies with Si NPs so far using magnetic resonance and fluorescence imaging.[37,51-55] For the first time, biodistribution and pharmacokinetic properties of radiolabelled Si NPs were reported by Kauzlarich *et al.*[56] Here, dextran-coated, manganese-doped, Si NPs were functionalized with a 1,4,7,10-tetraazacyclododecane-1,4,7-triacetic acid (DO3A) derivative for labelling with ^{64}Cu, enabling PET studies. Owing to their relatively large hydrodynamic diameter (HD) (15.1 ± 7.6 nm), the particles have considerable liver uptake. Recently, we were able to show that ultrasmall ^{64}Cu-labelled Si NPs (size below 5 nm) are rapidly cleared from the body.[57] Despite their very small size, multiple and different functionalities can be grafted onto the nanoparticle surface.

Based on our previous studies, we have selected citrate-stabilized Si NPs for further functionalization. The concurrent use of the bifunctional chelating agent NOTA (1,4,7-triazacyclononane-triacetic acid) and a near-infrared dye (IR800-CW) allows detailed *in vivo* studies using PET and optical imaging in small animals. Herein, we describe the synthesis of these dual-labelled particles, together with absorption, emission and structural properties. Subsequent assessment of nanotoxicity, as well as radiolabelling of Si NPs using the PET radionuclide ^{64}Cu is also reported in detail.

Results and discussion

Synthesis and characterisation of silicon-based nanoparticles

Water-dispersible amine-terminated silicon-based nanoparticles (Si NPs) were synthesised under hydrothermal conditions following our previously reported method.[57] A general synthetic route is depicted in Scheme 1. A broad characterisation of the Si NPs covering size, chemical composition and photoluminescence was carried out using a number of analytical techniques to show that the particles were completely comparable with our previously published work.[57]

Observing the Si NPs *via* Transmission Electron Microscopy (TEM) (Fig. 1A) revealed a system of low polydispersity with an average size of 2.4 ± 0.5 nm. This smaller size of Si NPs compared to the size those achieved in our previous study[57] was achieved through optimisation of the synthetic procedure. Increasing the mixing time of the reagents from a total of 30 min to 60 min before heating in the oven afforded a much more effective dispersion of the reagents during the hydrothermal process, creating more nuclei from which Si NPs could grow. A similar hypothesis has been reported for other species of silicon-based nanoparticles.[58]

The pristine amine-terminated Si NPs exhibit a near neutral surface charge due to the formation of a stable citrate shell confirmed by DLS measurements (zeta potential ζ −5.3 ± 2.8 mV). More information on the surface structure of the Si NPs was elucidated through FT-IR spectroscopy (Fig. 1B). Notable are the bands occurring at 1557 cm^{-1} and 1385 cm^{-1} for symmetric and asymmetric vibrational modes of C=O bonds respectively, as well as a band at 1224 cm^{-1} originating from stretching of C–O bonds. These are assumed to be due to the presence of residual citric acid in the system as a result of interactions with the protonated

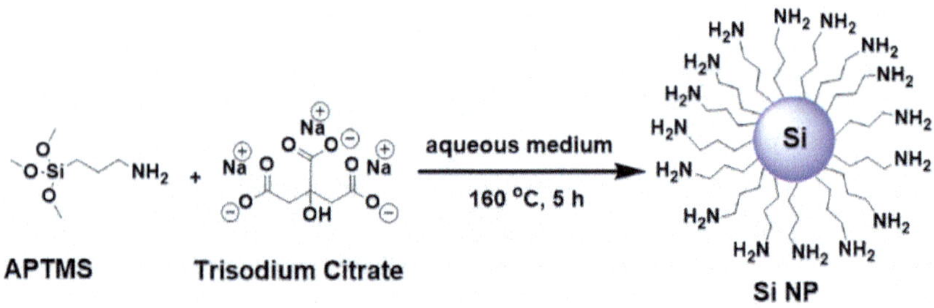

Scheme 1 Reaction scheme for the synthesis of amine-terminated Si NPs through a hydrothermal method [APTMS = (3-aminopropyl)trimethoxysilane].

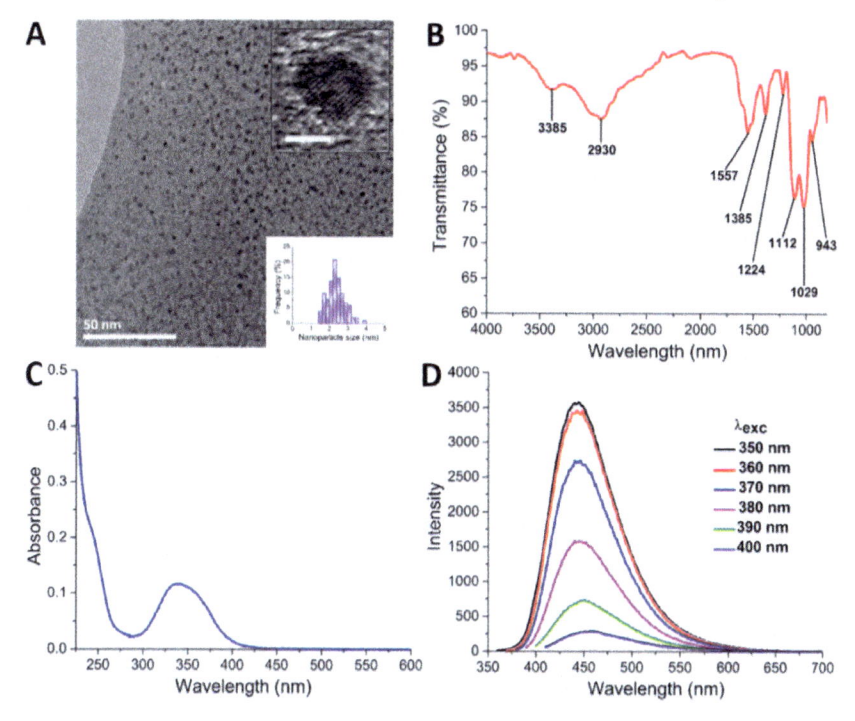

Fig. 1 Characterisation of Si NPs: (A) TEM and HR TEM (inset scale bar = 2 nm) images with associated size distribution, (B) FT-IR spectrum, (C) UV-Vis absorption spectrum, and (D) emission spectra.

amines on the Si NPs, confirming the expected structure, as reported in our previous work.[57]

This presence of citric acid can be further identified by X-ray photoelectron spectroscopy (XPS). Deconvolution of the high resolution carbon scan (ESI Fig. S1/Table S1†) shows a component with a binding energy of 287.6 eV, indicating the presence of oxygen bound carbon with sp^2 hybridisation. This is in addition to a major component with a binding energy of 285.8 eV, corresponding to oxygen bound carbon with sp^3 hybridisation. Taking the presence of both binding environments into account, it can be assumed that citric acid forms a "shell" of sorts around the Si NP through interactions with the surface amine groups as illustrated in Fig. 2. This was indeed already suggested in our previous work through NMR studies.[57] The UV-Vis absorption spectrum of the Si NPs was acquired in order to verify the similarity of the new batch to that of our previous work[57] and to compare to the literature.[33] As shown in Fig. 1C, the spectrum appears almost identical to those reported previously, with an absorption band observed at 340 nm. The emission spectra of the Si NPs are shown in Fig. 1D. An emission maximum is centred at about 450 nm for each excitation wavelength from 350 to 400 nm, matching the stable photoluminescence characteristics in the blue region as reported in the literature.[33,57] The emission quantum yield for these NPs is 15% in water. The surface amine groups were quantified and the average concentration was estimated to be 4.16 µmol NH_2 per mg Si NPs which

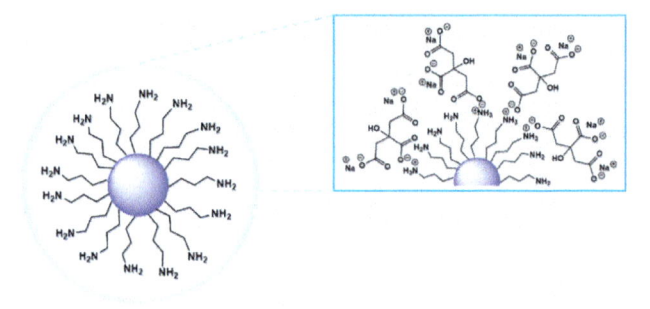

Fig. 2 Schematic representation of a Si-based nanoparticle surrounded by a "shell" of citric acid.

was used to calculate a value of approximately 43 amine groups per particle (ESI Scheme S1/Fig. S2/Table S2/eqn (S1) and (S2)†).

Synthesis and characterisation of Si NPs conjugates NOTA-Si NPs, IR800-Si NPs and NOTA-IR800-Si NPs

The NOTA-Si NPs, as reported,[57] were prepared by reacting amine-terminated Si NPs to NOTA-Bn-SCN (0.1 µmol per mg Si NPs) and were purified by dialysis (MWCO: 1 kDa) against water (Scheme 2A). The reaction of the Si NPs with NHS-activated IR800 dye (0.5 µmol per mg Si NPs), on the other hand, produced IR800-Si NPs (Scheme 2B). The near-infrared dye CW800 was employed for this modification because its emission wavelength (λ_{em} 789 nm) does not overlap with the autofluorescence of biological tissues and thus, provides the necessary penetration depth for *in vivo* studies.[59,60] The IR800-Si NPs were purified by dialysis (MWCO: 3.5–5 kDa), followed by size exclusion chromatography (SEC) using Sephadex G-25. Dual-labelled NOTA-IR800-Si NPs were synthesized by a sequential addition method where activated IR800 dye (0.5 µmol per mg Si NPs) and NOTA-Bn-SCN (0.1 µmol per mg Si NPs) were added sequentially in small amounts to an aqueous Si NP

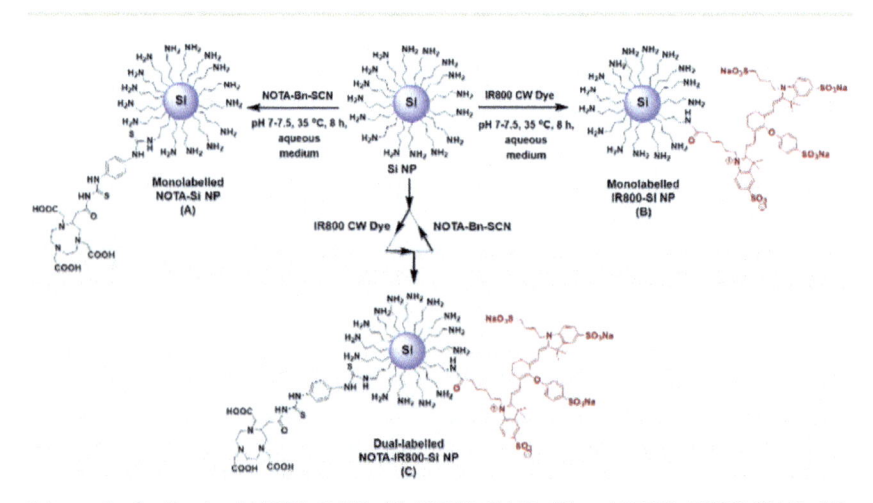

Scheme 2 Synthesis of NOTA-Si NPs (A), IR800-Si NPs (B) and NOTA-IR800-Si NPs (C).

solution over a period of 1 h (Scheme 2C). It is worth mentioning that simultaneous addition of NOTA and IR800 led to non-uniform surface modifications of Si NPs, which could not be applied for dual imaging. The crude reaction mixture was also purified by dialysis (MWCO: 3.5–5 kDa) and SEC. The purity of the IR800-Si NPs and NOTA-IR800-Si NPs was verified by fluorescence (reverse phase thin layer chromatography RP-TLC) and radiolabelling (Radio-instant thin layer chromatography Radio-iTLC) studies (ESI Fig. S3†).

The zeta potentials of both the unimodal (NOTA-Si NPs and IR800-Si NPs: −5.4 ± 4.1 mV) and the bimodal particles (NOTA-IR800-Si NPs: −5.1 mV ± 3.5 mV) were in the same range as that of the pristine particles.

The hydrophilicities of the [^{64}Cu]Cu-NOTA-Si NPs and the [^{64}Cu]Cu-NOTA-IR800-Si NPs were determined by 1-octanol/water distribution experiments to predict the behaviour of the conjugates under physiological conditions. A log $D_{7.4}$ value of −4.12 ± 0.04 was obtained for the [^{64}Cu]Cu-NOTA-Si NPs and a value of −3.84 ± 0.10 was obtained for the [^{64}Cu]Cu-NOTA-IR800-Si NPs. The slight increase in the distribution ratio of the bimodal Si NPs is attributed to the presence of the additional IR800 dye. The negative distribution coefficients indicate the hydrophilic character of the Si NPs, which is a prerequisite for the rapid clearance of the particles via the renal pathway.

Radiometric and UV-Vis titrations were performed to quantify the amounts of NOTA and IR800 dye molecules present on the surface of the modified Si NPs. Demanding only nanograms of the sample, the assays effectively derive the quantified values along with the purity of the conjugates.

The radiometric assay employed a mixture of radioactive ([^{64}Cu]CuCl$_2$) and non-radioactive copper (CuSO$_4$)[61] at different concentrations to determine the amount of NOTA on the surface of the Si NPs. The values were calculated to be ∼0.090 μmol NOTA per mg NOTA-Si NPs and ∼0.091 μmol NOTA per mg NOTA-IR800-Si NPs (ESI Fig. S4†).

The amount of conjugated IR800 dye was determined using UV-Vis analysis based on the Beer–Lambert law. The results reveal values of ∼0.013 μmol IR800 per mg IR800-Si NPs and ∼0.010 μmol IR800 per mg NOTA-IR800-Si NPs. The slight difference in the quantities for unimodal and bimodal Si NPs is due to the additional NOTA present on the surface of NOTA-IR800-Si NPs (ESI Fig. S5†).

The synthetic routes applied lead to Si NPs with defined amounts of NOTA and IR800 molecules, allowing for the simultaneous application of PET and OI after appropriate radiolabelling of the particles. A summary of the physicochemical properties of the mono and dual-labelled Si NPs is presented in Table 1.

Table 1 Summary of characteristic features of NOTA-Si NPs, IR800-Si NPs and NOTA-IR800-Si NPs

	NOTA-Si NPs	IR800-Si NPs	NOTA-IR800-Si NPs
IR800 conjugation (per mg Si NP conjugate)	—	0.013 μmol	0.010 μmol
NOTA conjugation (per mg Si NP conjugate)	0.090 μmol	—	0.091 μmol
Distribution coefficient (log $D_{7.4}$)	−4.12 ± 0.04	—	−3.84 ± 0.10

Absorption and emission properties of the IR800-Si NPs and the NOTA-IR800-Si NPs

The maximum absorbance for the IR800-Si NPs was observed at λ_{abs} 611 nm, which portrays a huge hypsochromic shift, compared to the absorbance of the IR800 dye (λ_{abs} 775 nm) (Fig. 3A). The probable reason for this massive shift and the profile broadening of the absorbance peak could be due to the formation of H-aggregates, as observed for other cyanine dyes in the presence of protonated amines, or their interaction with the surface.[62,63] Nonetheless, the absorbance of the IR800-Si NPs remains in the NIR region (600–900 nm) and thus, these nanoparticles are applicable for *in vivo* trials as potential imaging agents. The emission band of the IR800-Si NPs is observed at λ_{em} 753 nm (λ_{exc} 611 nm) which also shows a small blue shift compared to that of the free dye (IR800 dye λ_{em} 792 nm, λ_{exc} 775 nm).

The absorbance maxima for dual-labelled NOTA-IR800-Si NPs is also observed at λ_{abs} 611 nm with a typical cyanine shoulder at 470 nm, which confirms the conjugation of the IR800 dye to the Si NPs. The maximum emission, on the other hand, for NOTA-IR800-Si NPs is observed at λ_{em} 748 nm (λ_{exc} 611 nm). Comparative UV-Vis and fluorescence spectra of the IR800-Si NPs and the NOTA-IR800-Si NPs are shown in Fig. 3. The photoluminescence quantum yield (PLQY) for the IR800-Si NPs (22.5%) is reported to be higher compared to that of the dual-labelled NOTA-IR800-NOTA (15.7%) and the IR800 dye (14.2%).[64] Table 2 summarizes the photophysical properties of the Si NPs and the associated derivatives.

Radiolabelling studies of NOTA-Si NPs and NOTA-IR800-Si NPs

The radiolabelling experiments were carried out according to our reported protocol for ^{64}Cu-labelling.[57] In brief, NOTA-Si NPs and NOTA-IR800-Si NPs were labelled with [^{64}Cu]CuCl$_2$ radionuclide at pH 6 (MES/NaOH buffer) at room temperature for 30 min. [^{64}Cu]Cu-NOTA-Si NPs and [^{64}Cu]Cu-NOTA-IR800-Si NPs were obtained in >90% radiochemical yields (RCY) with a specific activity of ~3 MBq and ~2 MBq per 100 μg Si NP conjugates determined by Radio-iTLC. Variation in the parameters, such as increasing the reaction time and slight

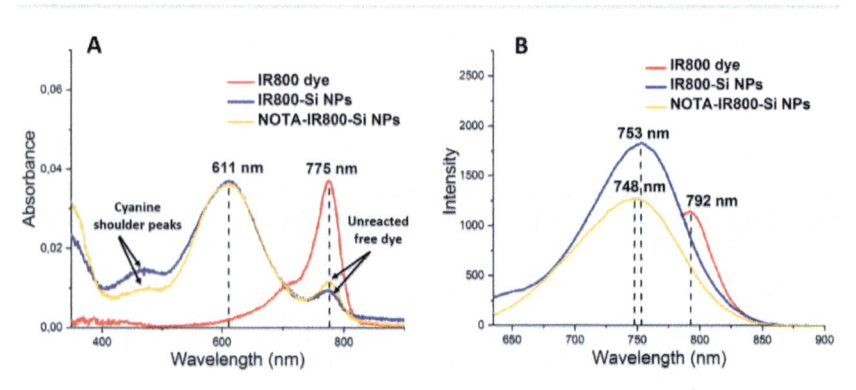

Fig. 3 Comparative absorption (A) and emission (B) spectra of IR800-Si NPs, NOTA-IR800-Si NPs and IR800 dye in water (IR800-Si NPs and NOTA-IR800-Si NPs λ_{exc} 611 nm; IR800 dye λ_{exc} 775 nm).

Table 2 Comparative analysis of the photophysical properties of the mono-labelled vs. the dual-labelled Si NPs

	IR800 dye	Si NPs	IR800-Si NPs	NOTA-IR800-Si NPs
Max. absorbance λ_{abs}	775 nm	355 nm	612 nm	611 nm
Max. fluorescence λ_{em}	792 nm	456 nm	754 nm	748 nm
λ_{exc}	775 nm	360 nm	611 nm	611 nm
PLQY	14.2%	15%	22.5%	15.7%

elevation of the temperature did not notably affect the radiolabelling of the Si NP conjugates. The radiolabelled Si NPs were purified by SEC to obtain a ~98% radiochemical purity. Challenge experiments were performed using a 10 000 fold excess of EDTA for 24 h with no indication of transchelation.

Nanotoxicity of NOTA-Si NPs, IR800-Si NPs and NOTA-IR800-Si NPs

For nano-objects to be developed for pre-clinical and clinical imaging trials, it is of the utmost importance to evaluate their integrity and biocompatibility. Our previous studies with citrate-stabilized Si NPs equipped with NOTA revealed that these particles have a very low cytotoxicity, if any, in a human embryonic kidney (HEK 293) cell line.[57] This is in accordance with the results obtained with similar silicon particles in other cell lines, *i.e.*, HeLa, HepG2, Caco-2 and CCD-841.[33,65,66]

The current renally clearable particles, NOTA-Si NPs, IR800-Si NPs and NOTA-IR800-Si NPs were tested in HEK293 kidney cells utilising MTS ([3-(4,5-dimethylthiazol-2-yl)-5-(3-carboxymethoxyphenyl)-2-(4-sulfophenyl)-2*H*-tetrazolium]), LDH (lactate dehydrogenase) and CyQuant® assays to assess their *in vitro* behaviours. For the cell viability and cytotoxicity measurements, the HEK293 cells were incubated with 1, 10, 50 and 100 μg mL^{-1} Si NP conjugates and analyzed after 24, 48 and 72 hours. The MTS assay determines the number of viable cells. The LDH cytotoxicity assay investigates the amount of released lactate dehydrogenase from damaged cells and thus provides information on whether the cell membrane is intact. The results obtained using the MTS and LDH assays are summarized in the ESI (Fig. S6 and S7).† Negligible effects of the NOTA-Si NPs, the IR800-Si NPs and the NOTA-IR800-Si NPs on cell metabolism, viability and integrity were observed for HEK293 cells. In addition, we performed the cell proliferation assay CyQuant®. This fluorescence-based assay detects the viable cellular DNA. As shown in Fig. 4, cell viability was not affected for all three kinds of Si NP at all concentrations, even after 72 h of incubation.

The results clearly show that the investigated particles do not have any cytotoxic effects. This is an important prerequisite for further *in vivo* investigations with regard to the biodistribution and pharmacokinetic properties.

Fluorescence imaging of IR800-Si NPs in healthy mice

For the biodistribution studies, ~100 μg of the IR800-Si NPs were administered intravenously in healthy mice (*n* = 4) which were analyzed at different time points

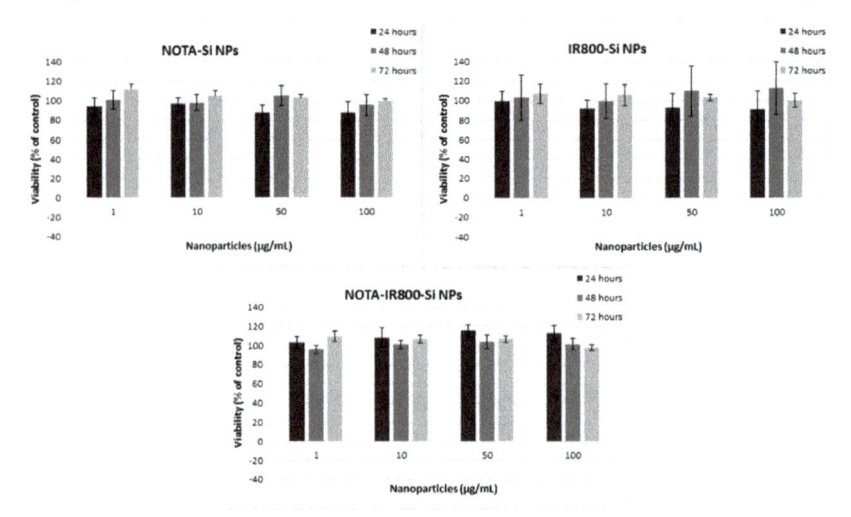

Fig. 4 Effect of NOTA-Si NPs, IR800-Si NPs and NOTA-IR800-Si NPs on the viability of HEK293 cells. The cells were exposed to 1, 10, 50 and 100 μg mL^{-1} Si NP dispersions for 24, 48 and 72 h respectively in Dulbecco's Modified Eagle Medium (DMEM) supplemented with 10% fetal bovine (FBS) and viability was determined using CyQuant® assay.

(1 min, 10 min and 2 h) using whole body fluorescence scanners to detect the fluorescence intensity in each organ. Exemplary images of the mice at different time points are displayed in Fig. 5A. The fluorescence image 1 min p.i. projects a uniform distribution of the IR800-Si NPs all over the body *via* blood circulation. Within 10 min the particles start to pass through the kidneys, towards the bladder and are eventually eliminated *via* urine. The later time point (2 hours p.i.) shows that the majority of the IR800-Si NPs had already been excreted from the body with traces of fluorescence observed in the kidneys, visualized only with high intensity projections. In order to provide information on the *in vivo* stability of the IR800-Si NPs, urine from the mice was collected and analyzed

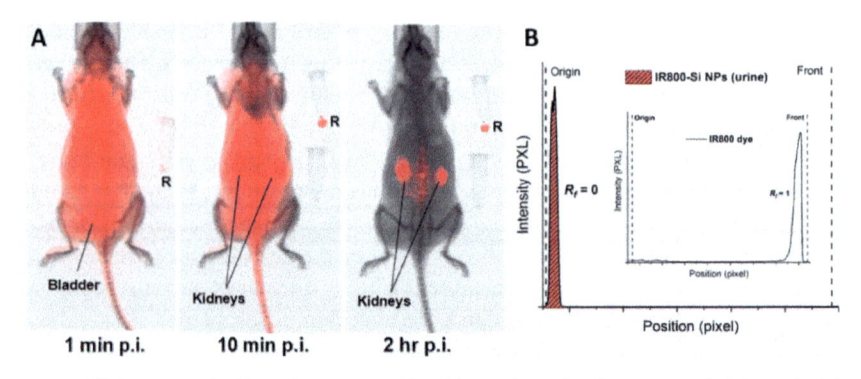

Fig. 5 (A) *In vivo* optical imaging scans of healthy nu/nu mice intravenously injected with IR800-Si NPs after 1 min, 10 min and 2 h p.i. (R: injected IR800-Si NPs as reference). (B) RP-TLC analysis of the urine from the experimental healthy mice 1 h p.i. IR800-Si NPs (IR800 dye inset); TLC system: RP-TLC plates, solvent system: methanol.

(1 h p.i.). The RP-TLC displayed a single peak at the origin corresponding to the IR800-Si NPs ($R_f = 0$), and no free/detached fluorescence (IR800 dye, $R_f = 1$) was observed at the solvent front (Fig. 5B). This concludes that the nanoparticle conjugates were rapidly eliminated from the *in vivo* system *via* the renal pathway and confirms the integrity of the IR800-Si NPs in the presence of complex biological milieu.

Biodistribution of [^{64}Cu]Cu-NOTA-IR800-Si NPs in healthy mice

To study the biodistribution of dual labelled [^{64}Cu]Cu-NOTA-IR800-Si NPs, the nanoparticles were intravenously injected into non-anesthetised healthy nu/nu mice ($n = 5$). Selected organs and tissues of the mice were excised 5 min, 1 h and 24 h p.i. for activity measurements. The amount of activity in the complete organs was calculated as % injected dose (% ID) and the activity concentrations in the organs and tissues were normalized to the body weight (SUV = activity concentration in the tissue/injected activity × body weight). These values were evaluated to define the biodistribution and mode of elimination of the dual mode Si NPs. The biodistribution studies showed the elimination of 30% of the [^{64}Cu]Cu-NOTA-IR800-Si NPs *via* urine from the kidneys 5 min p.i., which spiked to 75% at 1 h and 100% after 24 h (Fig. 6A and B). Minor amounts of activity (<5% ID) were detected 1 hour p.i. in the liver, intestine and other organs/tissues. The normalized SUV values give a better overview of the biodistribution of the Si NPs with negligible activity in the whole body, particularly in the kidneys and the liver after 24 h (Fig. 6C). The results confirmed the extremely fast and complete renal clearance of these hydrophilic Si NPs from the *in vivo* system.

The largest amount of activity was found in the urine. Therefore, the urine from the experimental mice was collected 5 min and 1 h p.i. and examined by Radio-iTLC (Fig. 7). The TLC chromatograms of the injected Si NPs and the urine samples correspond to the bimodal Si NPs ($R_f = 0$) and hence, confirm the *in vivo* integrity of the Si NPs.

PET imaging of dual-labelled [^{64}Cu]Cu-NOTA-IR800-Si NPs in NMRI nu/nu mice

In addition to the biodistribution studies, dynamic PET was carried out. [^{64}Cu]Cu-NOTA-IR800-Si NPs were injected intravenously (~10 MBq each) into mice ($n = 4$)

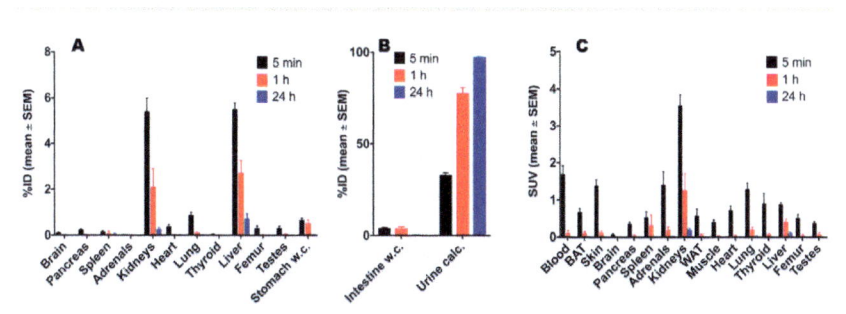

Fig. 6 Biodistribution of [^{64}Cu]Cu-NOTA-IR800-Si NPs in NMRI nu/nu mice at 5 min, 1 h and 24 h p.i. in extracted organs and tissues. (A) % ID in selected organs, (B) eliminated activity in the intestine and urine (w.c.: with content), (C) activity concentrations (SUV) of selected organs and tissues. (BAT brown adipose tissue, WAT white adipose tissue).

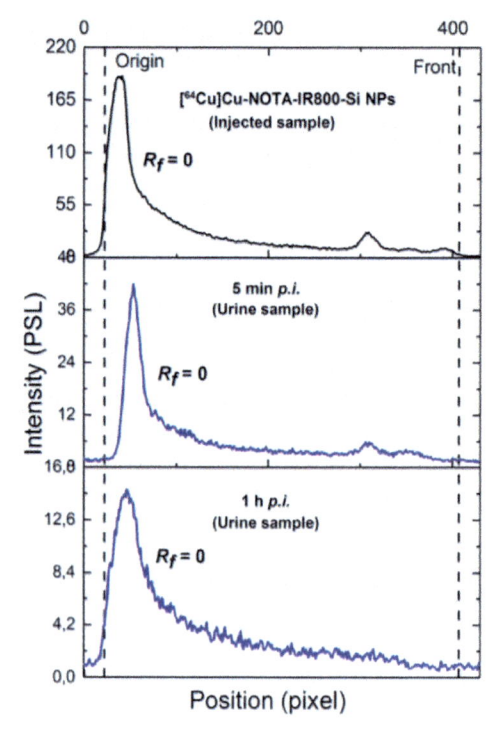

Fig. 7 Radio-iTLC of the injected and urine samples of [^{64}Cu]Cu-NOTA-IR800-Si NPs injected mice (5 min and 1 h p.i.). TLC system: Radio-iTLC plates, 2 M NH$_4$OAc : MeOH (1 : 1, v/v), [^{64}Cu]Cu-NOTA-IR800-Si NPs $R_f = 0$.

and whole body PET scans were recorded over 1 h. Exemplary figures of mid-frame time points at 5 min, 20 min, 30 min and 60 min are given in Fig. 8 and show fast clearance of the radiolabelled Si NPs from the kidneys into the urinary bladder.

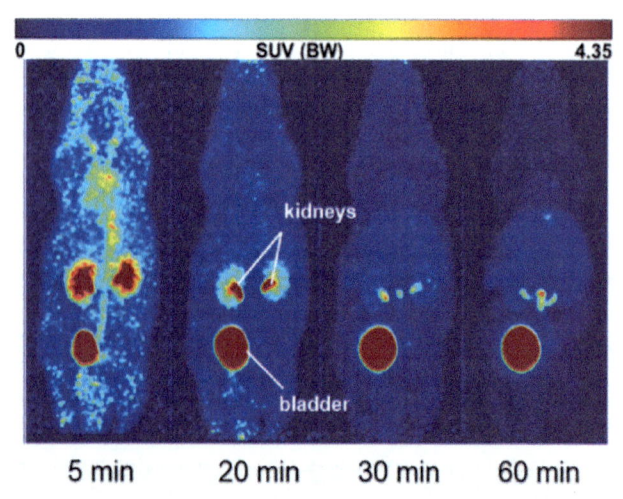

Fig. 8 *In vivo* PET imaging of [^{64}Cu]Cu-NOTA-IR800-Si NPs 5, 20, 30, 60 min p.i.

Within the first 5–10 minutes, the bimodal Si NPs distribute uniformly in the whole body and then start eliminating *via* the kidneys. The particles show no signs of sticking in the renal medulla and easily pass through the glomerular pores within 1 h. The activity was temporarily retained only in the kidney pelvis from where it was excreted into the bladder.

Comparative nuclear and optical imaging of [^{64}Cu]Cu-NOTA-IR800-Si NPs in healthy mice

The [^{64}Cu]Cu-NOTA-IR800-Si NPs were not only examined by PET by also by fluorescence imaging (Fig. 9). For this purpose, one mouse was sacrificed 1 h after injection of the NPs. PET/CT and planar X-ray 2D measurements of the mouse were recorded in prone and supine positions for the detection of ^{64}Cu-activity and the fluorescence signal distribution in the body. All three imaging modalities revealed that within 1 h, the majority of the dual-labelled Si NPs remained in the kidneys and large amounts were excreted into the bladder. The higher kidney retention for this mouse in comparison to the previous PET study (see Fig. 8) can be attributed to a biological variation of this mouse, *i.e.*, different reaction to anaesthesia and hydration of the animal. However, in accordance with the other experiments, the activity and fluorescence images allow for the detection of only traces of activity and fluorescence in other organs. The PET study shows small amounts in the gall bladder. This is another indication of the high metabolic stability of the dual-labelled NPs.

The complementary biocompatibility and rapid renal clearance of the NOTA-IR800-Si NPs explained by the biodistribution, PET and optical imaging, confirms the bimodality of the Si NP system. The results also displayed excellent pharmacokinetics and the extraordinary biocompatibility of the system, which is a prerequisite for any nanoparticle platform adapted for *in vivo* studies. The most acceptable reason for this behaviour is the minimum required surface modifications performed on the Si NP surface to achieve the imaging goals without hampering the size and surface charges of the particles.

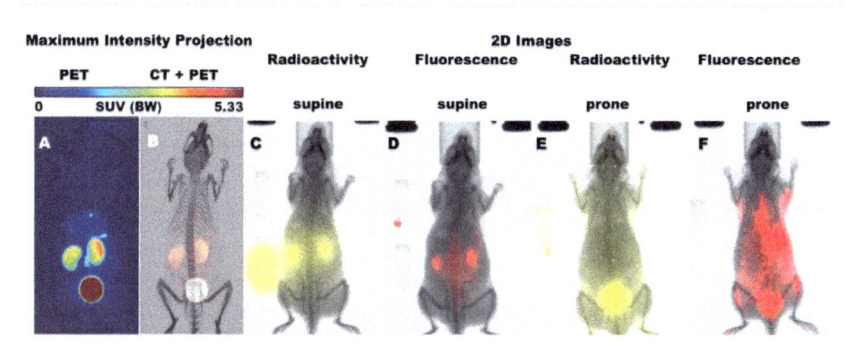

Fig. 9 Biodistribution of [^{64}Cu]Cu-NOTA-IR800-Si NPs 1 h after single intravenous injection in a gas anesthetized NMRI nu/nu mouse. (A) Maximum intensity projection (MIP) of PET, (B) an overlay of MIP with an X-ray tomogram, (C) a planar 2D view of the activity distribution in the mouse in the supine position to the camera, (D) a spectral unmixed fluorescence image, (E) activity distribution in the prone position, (F) fluorescence in the prone position (injected sample reference in Eppendorf tubes).

Conclusions

In summary, ultrasmall amine-terminated Si NPs, prepared by hydrothermal synthesis, were equipped with the near-infrared dye IR800 and the bifunctional chelating agent NOTA to obtain a dual imaging probe. The subsequent *in vivo* investigations based on PET and OI show that these multifunctional particles are cleared very quickly *via* the kidneys and that the particles remain intact after excretion. The minimal cytotoxicity, if any, and *in vivo* integrity, as well as the favourable pharmacokinetic properties, underline the suitability of the NOTA-IR800-Si NPs for *in vivo* applications. Overall, these particles thus represent a promising platform for targeted multimodal imaging. Despite the very small size, several small targeting vector molecules such as peptides can be grafted on the nanoparticle surface. However, it must be ensured that the size and charge of the particles are changed only minimally in order to maintain rapid circulation in the bloodstream and full clearance.

Experimental methods and materials

All chemicals were purchased from commercial suppliers and used as received without further purification. *S*-2-(4-Isothiocyanatobenzyl)-1,4,7-triazacyclononane-1,4,7-triacetic acid (SCN-Bn-NOTA; #B-605) was purchased from Macrocyclics and the infrared dye 800CW from Li-COR Biosciences. A Direct-Q 3 UV water purification system from Millipore (Merck KGaA) was used for producing deionized water. The resistance of the deionized water was 18.2 MΩ cm. Dialysis tubes were purchased from Serva (MWCO: 1 kDa and 3.5–5 kDa; Spectra/Por® 7 dialysis tubing); Sephadex G-25 desalting columns were purchased from GE healthcare. High pressure resistant homemade Teflon vessels (similar to those reported by Calzaferri *et al.*[67]) were used for hydrothermal syntheses throughout this work.

Analysis, characterisation and purification techniques

TEM analysis was performed using a Jeol 2100F electron microscope operated at 200 kV. Samples were prepared by dispersing the aqueous suspensions of the Si NPs onto holey-carbon-on-copper grids (Quantifoil, GmbH) with the excess solvent evaporated. Hydrodynamic size and zeta potential measurements were performed using dynamic light scattering (DLS) using a Zetasizer NanoZS (Malvern Instruments, UK) with an angle detection of 173° at 25 °C with a beam wavelength of 632.8 nm. Particle size determination, as well as zeta potential measurements, was conducted at a concentration of 2 mg mL^{-1} in 10 mM phosphate buffer saline (pH 7.4) in disposable plastic cuvettes with a square aperture (Malvern Instruments Ltd.). Prior to each experiment, the solutions were filtered through 0.02 μm or 0.2 μm syringe filters (GE Healthcare, Germany) and equilibrated for 2 min. The stated value is a mean of three independent measurements. Zeta potential measurements were performed in a universal dip cell (Malvern Instruments Ltd.). The stated values are a mean of three independent measurements using the Smoluchowski model. Data were evaluated with Malvern Zetasizer Software version 7.11. Ultraviolet-Visible absorption spectra for Si NPs were recorded on a Shimadzu UV-3600 double-beam UV-VIS-NIR spectrophotometer. Samples were measured using quartz cuvettes with a path length of 1 cm and volume of 3.5 mL. Absorption spectra recordings for the IR800-Si NPs and the NOTA-IR800-Si NPs were conducted using a Specord® S50 from Analytik Jena (Germany). The samples were measured using quartz micro cuvettes ($d = 1$ cm) with volume 400–500 µL with a wavelength range from 190–1100 nm. The standard calibration curves and the molar extinction coefficient were calculated using the Beer–Lambert equation: $A = \varepsilon \times l \times c$ where A is the absorbance at λ_{max}, ε denotes the extinction coefficient, c denotes the concentration and l denotes the path length of the cuvette (1 cm). For quantum yield experiments, the absorbance was kept between 0.01–0.1. The measurements were set relative to a blank solution (reference). Data were processed using WinASPECT® software. For the synthesised Si NPs obtained after purification, steady-state emission spectra were recorded using a HORIBA Jobin-Yvon IBH FL-322 Fluorolog 3 spectrometer equipped with a 450 W xenon arc lamp as the excitation source, double-grating excitation and emission monochromators (dispersion of 2.1 nm mm^{-1}; 1200 grooves per mm), and a TBX-04 single-photon-counting device as the detector. Emission and excitation spectra were corrected for source intensity (lamp and grating) and emission spectral response (detector and grating) by standard correction curves. Fluorescence emission measurements and quantum yield experiments for the IR800-Si NPs and the NOTA-IR800-Si NPs were done using an LS 55 (230 V) luminescence spectrometer from PerkinElmer equipped with a xenon flash lamp source for excitation. Samples were measured in transparent disposable cuvettes with a volume of 2 mL with a wavelength range of 200–900 nm and a slit width of 10 nm. FL UV WinLab® software was used for data analysis and acquisition. Fourier transform infrared (FTIR) spectra were recorded using a Shimadzu IRAffinity-1 spectrometer used in attenuated total reflectance (ATR) mode. Si NP samples were prepared by lyophilization. The resulting solids were ground with an agate mortar and were laid on the ATR crystal. Spectra were cumulated from 64 scans at a resolution of 4 cm^{-1}. Dialysis was performed against water with regenerated cellulose membranes with ultra-pure cellulose ester Float-A-Lyzer membranes from Spectrum Labs, Germany with various molecular weight cut-off (MWCO: 1 kDa, 3.5–5 kDa). Size exclusion chromatography was performed using a PD-10 Desalting G25 column (Sephadex, G25 medium). The eluents used were Millipore water or 0.9% NaCl (wt/v) as indicated in the experimental section. Thin layer chromatography was conducted on reverse stationary phase silica gel using methanol as the mobile phase in a vertical chamber for elution. A UV lamp (254 nm) and an Amersham Typhoon 5 Scanner (300–900 nm) were used for analysis of the fluorescently labelled nanoparticles.

Synthesis of monomodal Si NP-based imaging agents

Si NPs. This procedure was taken and adapted from a previous report by He and coworkers.[28] 20 mL deionised H$_2$O was added to a 50 mL 2-neck flask. Argon gas was bubbled through the solution with stirring for 30 min. Citric acid trisodium salt (0.83 g, 3.75 mmol) was added and stirring and bubbling of the mixture was continued for a further 30 min. 3-Aminopropyltrimethoxysilane (APTMS, 5 mL, 28.65 µmol) was then added with continued stirring and bubbling and shortly after, the mixture was transferred to a Teflon pressure sealed vessel. Bubbling with argon gas was maintained until the vessel was sealed and following this, the vessel was placed into a preheated oven at 160 °C for 5 h. After cooling to room temperature, the vessel was opened and the solution inside was neutralised with dropwise addition of conc. HCl. Finally, the crude product was purified by dialysis (MWCO: 1 kDa) against water for 48 h (water was changed every 2 h). The final product was a colourless dispersion of Si NPs in water.

NOTA-Si NPs. NOTA-Si NPs were prepared by dispersing 1 mg of amine-terminated Si NPs in deionized water (150 μL) in a 1.5 mL low protein binding Eppendorf-tube. The pH of the dispersion was set between 7–7.5 using 0.01 M NaOH solution. NOTA-Bn-SCN (0.1 μmol) was dissolved in water (100 μL) and was added to the Si NPs dispersion. The reaction was shaken on a thermomixer at 750 rpm for 16 h at 25 °C. The NOTA-Si NPs were purified by dialyzing against water (MWCO: 1 kDa). A defined aliquot of the sample was freeze-dried to obtain the exact weight of the particles.

IR800-Si NPs. To amine-terminated Si NPs (1 mg) dispersed in water (150 μL) at pH 7–7.5 (0.01 M NaOH), NHS-IR 800 (0.05 μmol) dissolved in water (100 μL) was added. The reaction was performed in a 1.5 mL low protein binding Eppendorf-tube in the dark. The reaction mixture was shaken using a thermomixer at 750 rpm at 35 °C. The green solution started to turn blue within 15 min. Upon completion of the reaction, the product was purified by dialysis (MWCO: 3.5–5 kDa) against Millipore water. The IR800-Si NPs were analysed using fluorescence measurements obtained *via* RP-TLC. The sample was further purified by SEC (Sephadex G-25) with 0.9% (wt/v) NaCl as the eluent, to obtain pure IR800-Si NPs. A known volume of the dispersion was lyophilized to determine the concentration of the nanoconjugate.

NOTA-IR800-Si NPs. Si NPs (1 mg) dispersed in water (500 μL) were neutralised to pH 7–7.5 (0.01 M NaOH). To this solution, 20 μL NHS-IR800 (0.05 μmol in 100 μL water) and 20 μL NOTA-Bn-SCN (0.1 μmol in 100 μL water) were added in 5 cycles over a period of 1 h. The reaction was run for 8 h at 35 °C. NOTA-IR800-Si NPs were purified by dialyzing the solution against water (MWCO: 3.5–5 kDa) followed by SEC. The concentration of the purified NP conjugate was obtained by freeze-drying a known volume of the sample.

Determination of dye concentration

The IR 800 dye concentration for the IR800-Si NPs and the NOTA-IR800-Si NPs was determined by UV-Vis titration. A stock solution of a known concentration of IR 800 dye was prepared in Millipore water. 5 samples with increasing concentrations were prepared from the stock. The absorbance of each dye sample was measured, keeping the value between 0.01–0.1 AU. A graph was plotted with this data, (absorbance *vs.* concentration) which led to a straight line passing through the origin. The absorbance of a sample with a known amount of Si NP conjugate was measured and plotted on the same graph. The Beer–Lambert equation was used for the determination of the amount of dye present per mg of Si NP conjugate.

Relative quantum yield determination

The photoluminescence quantum yields of the IR800 dye coupled to unimodal and bimodal Si NP conjugates were determined using free IR800 dye as a reference. The QY of free NHS-IR 800 dye, according to previously published data, is 14.2%. The UV-Vis absorbances of the water dispersed Si NP conjugates and free IR800 dye were measured. To minimize re-absorption effects, the absorbance signals were always kept between 0.01–0.10 AU. The maximum absorbance (~615 nm) was used as the excitation wavelength for the fluorescence emission measurements. A slit width of 5 mm was kept constant for all the measurements. The resulting fluorescence signal was integrated, and the quantum yield was calculated.

Radiolabelling of Si NP conjugates (NOTA-Si NPs/NOTA-IR800-Si NPs)

The production of ^{64}Cu was performed using a Cyclone®18/9 (Helmholtz-Zentrum Dresden-Rossendorf). For the ^{64}Ni(p,n)^{64}Cu nuclear reaction, 15 MeV protons with a beam current of 12 µA for 150 min were used. The yields of the nuclear reactions ^{64}Ni(p,n)^{64}Cu were 3.6–5.2 GBq [at the end of bombardment (EOB)] with molar activities of 150–250 GBq µmol^{-1} Cu diluted in HCl (10 mM).[68] An aqueous solution of [^{64}Cu]CuCl$_2$ (10 MBq) was added to ~0.1 mg (1 mg mL^{-1} stock solution) of NOTA-Si NPs and NOTA-IR800-Si NPs dispersed in water. 100 mM MES buffer at pH 6 was used to make the final volume up to 400 µL. The reaction mixtures were shaken at room temperature for 30 min, which resulted in RCY ~80–90%. The radiolabelling was monitored using Radio-iTLC.

Radiochemical analysis of the yields and kinetics of the experiments was carried out using different stationary phase and eluent systems. iTLC plates were purchased from Agilent Technologies. The air-dried TLC strips were exposed to a high-resolution phosphor imaging plate (GE Healthcare). The exposed plates were scanned with an Amersham Typhoon 5 Scanner (GE Healthcare) or read using a radioluminography laser scanner BAS-1800II (Raytest). Aida Image Analyzer version 4.0 software was used for data analysis and the conversion of data to ASCII before graph-plotting in Origin.

Radio-SEC was performed using PD-10 Desalting G25 columns (Sephadex, G25 medium) to purify the traces of free [^{64}Cu]CuCl$_2$. The ^{64}Cu-complexes were purified using water or 0.9% NaCl solution as the mobile phase. The radiochemical yields for [^{64}Cu]Cu-NOTA-Si NPs were determined to be 90–95% whereas for [^{64}Cu]Cu-NOTA-IR800-Si NPs they were 80–95%.

Radiometric titration to determine the amount of NOTA on the Si NPs

Radiometric titrations were conducted using a mixture of radioactive ^{64}Cu(II) and non-radioactive Cu(II) for the determination of the amount of NOTA conjugated to the Si NPs. To 10 µL aliquots of the NOTA-Si NPs and the NOTA-IR800-Si NPs (0.2 mM stock solution in 100 mM MES buffer, pH 5.5), known amounts of a mixture of [^{64}Cu]CuCl$_2$ and CuSO$_4$ with a final concentration of 0.4 mM (stock solution) were added. The solutions were stirred for 1 h at room temperature on a thermomixer at 750 rpm. After complete complexation (confirmed using Radio-iTLC), an 8-fold excess of EDTA was added to each vial of solution to remove any non-specifically bound Cu(II) ions. The resulting solutions were stirred at 37 °C for 40 min at 750 rpm. The RCY was determined via Radio-iTLC and also using Radio-SEC (Sephadex G-25). The amount of chelator per mg of Si NP conjugate was calculated using the percentage of Cu(II) ions bound to the conjugate according to the equation:

$$\text{Moles of chelator conjugated} = \text{moles(II)}[\%\ ^{64}\text{Cu(II)}(R_\text{f} < 0.2)/\%\ ^{64}\text{Cu(total)}]$$

The amount of NOTA conjugated to the Si NPs was calculated.

In vitro assessment of nanotoxicity

All biocompatibility assays were conducted as per the protocol specified in the ISO norm 10993-5:2009. All cell culture reagents were purchased from Biochrom AG and Sigma-Aldrich unless otherwise stated. The human embryonic kidney cell lines HEK293 (DMSZ number ACC 305) were cultured in DMEM + 10% fetal calf serum (FCS) and incubated in a humidified atmosphere of 95% air/5% CO_2 at 37 °C. The cell number and the viability of the cells were determined using a CASY cell counter (Roche Diagnostics) following the manufacturer's protocol. The cell lines were regularly tested and were confirmed to be mycoplasma negative using a LookOut mycoplasma PCR detection kit (Sigma-Aldrich). For the cell viability and LDH assays, the cells were treated with penicillin/streptomycin to avoid any bacterial growth during the experiments.

The in vitro assessment was performed using a MTS (3-(4,5-dimethylthiazol-2-yl)-5-(3-carboxymethoxyphenyl)-2-(4-sulfophenyl)-2*H*-tetrazolium) calorimetric assay and the CyQuant® Direct Cell Proliferation Assay. The MTS assay is based on the conversion of the tetrazolium salt to a formazan dye, presumably by NADPH and NADH produced by dehydrogenase enzymes in metabolically-active cells. The formation of the dye in the presence of the NPs was monitored by optical absorbance measurements at 490 nm, and the viability was finally compared with an untreated cellular control. In addition to that, the membrane integrity was measured using an LDH assay which quantifies lactate dehydrogenase (LDH), an enzyme that is released upon cell lysis. The CyQuant® assay is a fluorescence-based proliferation and cytotoxicity assay and therefore can be used to assess cell growth, cell viability, or compound toxicity. For the evaluation of cytotoxicity, the non-cancerous cell line, HEK 293 Human epithelial kidney, was seeded in sterile 96-well microtiter plates (Greiner Bio-One) at a density of 1.8×10^4 cells per well in DMEM media and allowed to grow for 24 h prior to the addition of the selected samples. CyQuant® is a direct fluorescence-based proliferation and cytotoxicity assay consisting of a fluorescent nucleic acid stain that permeates live cells and concentrates in the nuclei. The second component is a suppressor dye that cannot permeate live cells and suppresses green fluorescence. 100 μL of CyQuant® dye was added to the wells after 24 h of incubation. The fluorescence was measured after 60 min of incubation. After 24 h incubation, increasing concentrations of NOTA-Si NP, IR800-Si NP and NOTA-IR800-Si NP conjugates were added to the cells in triplicate. After exposure for 72 h, 20 μL of MTS (Promega) was added to the wells and incubated for 40 min. Optical densities at 492 nm were measured with a microplate reader (TECAN sunrise). The viability of cells was expressed as a percentage of viable cells grown in the absence of the NPs.

Additionally, to measure the membrane integration, an LDH assay was used on the same cell lines. For this, 2.1×10^4 cells per well were seeded in a 96-well microtiter plate. After 24 h incubation, the stock solutions of the conjugates were added and incubated for 72 h. 10 μL of LDH lysis buffer (cytoscan LDH from G-Biosciences) was added to the cells. The optical density was measured at 492 nm. The percentage of LDH release was expressed as the percentage of LDH release from cells grown in the absence of the NPs. The data were analysed by normalizing the absorbance from each well containing the samples to the absorbance values of the cells without any samples. As a positive control, a triplicate set of cells were treated with the supplied lysis buffer and the percentage of LDH release was expressed relative to this control. As a negative control, bovine serum albumin was used for a triplicate set of cells.

Animal experiments

In vivo **small animal PET.** All animal experiments were carried out according to the guidelines of the German Regulations for Animal Welfare and the protocols approved by the local Ethical Committee for Animal Experiments (AZ 24-9168.21-4/2004-1, AZ 24-9168.11-4/2012-1). Animals received standard food and tap water *ad libitum* and were anesthetized prior to animal experiments using 10% desflurane in 30% oxygen/air. Male and female NMRI nu/nu mice (RjOrl: NMRI-Foxn1nu/Foxn1nu) were received from Janvier Labs (Saint-Berthevin Cedex, France). Small animal PET imaging was performed using a microPET® P4 scanner (Siemens Medical Solutions, Knoxville) or a NanoScanPET/CT (Mediso, Budapest). Animals were positioned on a heated bed to maintain body temperature at 37 °C. Mice received 10–30 MBq (50–100 µg) of the Si NP conjugates in 0.5 mL and 0.2 mL E-153, respectively, intravenously over 0.5 min into a lateral tail vein. The activity of the injection solution was measured using a dose calibrator (Isomed 2000, Dresden, Germany) cross-calibrated to the PET scanners. Attenuation correction of the PET data was performed by means of a 10 min transmission scan using a rotating point source of ^{57}Co (microPET®) or by whole body CT (NanoScan®PET/CT). Data acquisition and reconstruction were performed as described elsewhere. Three-dimensional regions of interest (ROI) were determined within masks around different organs by thresholding PET data within these masks. ROI time activity curves (TAC) were derived from subsequent data analysis. Data were calculated as standard uptake values (SUV, SUV = activity concentration in tissue [Bq g^{-1}] × body weight [g]/injected dose [Bq]) at a certain time point p.i. The ROI data and TAC were further analysed using R (R is available as Free Software under the terms of the Free Software Foundation's GNU General Public License in source code form) and specially developed program packages (Jörg van den Hoff, Helmholtz-Zentrum Dresden-Rossendorf, Dresden, Germany). The data were calculated in standard uptake values (SUV, in units of grams per milliliter), defined as the tracer concentration at a certain time point normalized to injected dose per unit of body weight. The SUV was used for better comparisons across animals of different sizes and weights and with other species. The graphs were calculated with GraphPad Prism version 5.00 for Windows (GraphPad Software, San Diego California USA, http://www.graphpad.com).

Biodistribution studies. For biodistribution experiments, NMRI nu/nu mice were intravenously injected with ∼0.5 mL 0.5–10 MBq (10–100 µg) of ^{64}Cu-labelled [^{64}Cu]Cu-NOTA-Si NPs and [^{64}Cu]Cu-NOTA-IR800-Si NPs. Animals were sacrificed at different time points. Organs and tissues of interest were excised, weighed, and the radioactivity was determined using a Wizard™3″ gamma counter. Activity in selected organs and tissues was expressed as % injected dose (% ID), % ID/g tissue, or SUV. The obtained blood volume amounted to 400–500 µL. Afterwards, urine samples (200–400 µL) were obtained by puncture of the bladder.

Optical imaging. For optical imaging experiments, anesthetized healthy NMRI nu/nu mice were intravenously injected with 20 µl IR800-Si NPs and NOTA-IR800-

Si NPs (~80 μg each) respectively. Fluorescence was measured using a small animal In-vivo-Imaging System FX Pro (Carestream Health, USA) in fluorescence mode (Ex 600 nm, Em 753 nm; reference image for autofluorescence: Ex 430 nm, Em 535 nm). Fluorescence pictures were merged with the X-ray acquisitions. Afterwards, animals were sacrificed, organs and tissues of interest were excised, and the fluorescence was measured again.

Conflicts of interest

There are no conflicts to declare.

Acknowledgements

We thank Utta Herzog and Karin Landrock for their excellent technical assistance. Financial support by the Helmholtz Association (funding through the Helmholtz Virtual Institute NanoTracking, Agreement Number VH-VI-421) is gratefully acknowledged.

References

1 S. Achilefu, *Chem. Rev.*, 2010, **110**, 2575.
2 P. J. Cassidy and G. K. Radda, *J. R. Soc., Interface*, 2005, **2**, 133.
3 J. V. Frangioni, *J. Clin. Oncol.*, 2008, **26**, 4012.
4 M. L. James and S. S. Gambhir, *Physiol. Rev.*, 2012, **92**, 897.
5 G. Singh, M. D. Gott, H. J. Pietzsch and H. Stephan, *Nuklearmedizin*, 2016, **55**, 41.
6 R. Weissleder, *Science*, 2006, **312**, 1168.
7 A. Louie, *Chem. Rev.*, 2010, **110**, 3146.
8 B. R. Smith and S. S. Gambhir, *Chem. Rev.*, 2017, **117**, 901.
9 G. Chen, I. Roy, C. Yang and P. N. Prasad, *Chem. Rev.*, 2016, **116**, 2826.
10 B. Pelaz, S. Jaber, D. J. de Aberasturi, V. Wulf, T. Aida, J. M. de la Fuente, J. Feldmann, H. E. Gaub, L. Josephson, C. R. Kagan, N. A. Kotov, L. M. Liz-Marzán, H. Mattoussi, P. Mulvaney, C. B. Murray, A. L. Rogach, P. S. Weiss, I. Willner and W. J. Parak, *ACS Nano*, 2012, **6**, 8468.
11 K. L. Viola, J. Sbarboro, R. Sureka, M. De, M. A. Bicca, J. Wang, S. Vasavada, S. Satpathy, S. Wu, H. Joshi, P. T. Velasco, K. MacRenaris, E. A. Waters, C. Lu, J. Phan, P. Lacor, P. Prasad, V. P. Dravid and W. L. Klein, *Nat. Nanotechnol.*, 2015, **10**, 91.
12 S. Marchesan and M. Prato, *ACS Med. Chem. Lett.*, 2013, **4**, 147.
13 O. S. Wolfbeis, *Chem. Soc. Rev.*, 2015, **44**, 4743.
14 B. Pelaz, C. Alexiou, R. A. Alvarez-Puebla, F. Alves, A. M. Andrews, S. Ashraf, L. P. Balogh, L. Ballerini, A. Bestetti, C. Brendel, S. Bosi, M. Carril, W. C. W. Chan, C. Chen, X. Chen, X. Chen, Z. Cheng, D. Cui, J. Du, C. Dullin, A. Escudero, N. Feliu, M. Gao, M. George, Y. Gogotsi, A. Grünweller, Z. Gu, N. J. Halas, N. Hampp, R. K. Hartmann, M. C. Hersam, P. Hunziker, J. Jian, X. Jiang, P. Jungebluth, P. Kadhiresan, K. Kataoka, A. Khademhosseini, J. Kopeček, N. A. Kotov, H. F. Krug, D. S. Lee, C.-M. Lehr, K. W. Leong, X.-J. Liang, M. Ling Lim, L. M. Liz-Marzán, X. Ma, P. Macchiarini, H. Meng, H. Möhwald, P. Mulvaney,

A. E. Nel, S. Nie, P. Nordlander, T. Okano, J. Oliveira, T. H. Park, R. M. Penner, M. Prato, V. Puntes, V. M. Rotello, A. Samarakoon, R. E. Schaak, Y. Shen, S. Sjöqvist, A. G. Skirtach, M. G. Soliman, M. M. Stevens, H.-W. Sung, B. Z. Tang, R. Tietze, B. N. Udugama, J. S. VanEpps, T. Weil, P. S. Weiss, I. Willner, Y. Wu, L. Yang, Z. Yue, Q. Zhang, Q. Zhang, X.-E. Zhang, Y. Zhao, X. Zhou and W. J. Parak, *ACS Nano*, 2017, **11**, 2313.
15 S. Kunjachan, J. Ehling, G. Storm, F. Kiessling and T. Lammers, *Chem. Rev.*, 2015, **115**, 10907.
16 J. A. Barreto, W. O'Malley, M. Kubeil, B. Graham, H. Stephan and L. Spiccia, *Adv. Mater.*, 2011, **23**, H18.
17 M. Longmire, P. L. Choyke and H. Kobayashi, *Nanomedicine*, 2008, **3**, 703.
18 K. Zarschler, L. Rocks, N. Licciardello, L. Boselli, E. Polo, K. P. Garcia, L. De Cola, H. Stephan and K. A. Dawson, *Nanomed. Nanotechnol. Biol. Med.*, 2016, **12**, 1663.
19 L. A. Kunz-Schughart, A. Dubrovska, C. Peitzsch, A. Ewe, A. Aigner, S. Schellenburg, M. H. Muders, S. Hampel, G. Cirillo, F. Iemma, R. Tietze, C. Alexiou, H. Stephan, K. Zarschler, O. Vittorio, M. Kavallaris, W. J. Parak, L. Madler and S. Pokhrel, *Biomaterials*, 2017, **120**, 155.
20 X. Jiang, B. Du, Y. Huang and J. Zheng, *Nano Today*, 2018, **21**, 106.
21 J. Joo, X. Liu, V. R. Kotamraju, E. Ruoslahti, Y. Nam and M. J. Sailor, *ACS Nano*, 2015, **9**, 6233.
22 J.-H. Park, L. Gu, G. von Maltzahn, E. Ruoslahti, S. N. Bhatia and M. J. Sailor, *Nat. Mater.*, 2009, **8**, 331.
23 M. Rosso-Vasic, E. Spruijt, Z. Popovic, K. Overgaag, B. van Lagen, B. Grandidier, D. Vanmaekelbergh, D. Dominguez-Gutierrez, L. De Cola and H. Zuilhof, *J. Mater. Chem.*, 2009, **19**, 5926.
24 X. Cheng, S. B. Lowe, P. J. Reece and J. J. Gooding, *Chem. Soc. Rev.*, 2014, **43**, 2680.
25 J. H. Warner, A. Hoshino, K. Yamamoto and R. D. Tilley, *Angew. Chem., Int. Ed.*, 2005, **44**, 4550.
26 X. Cheng, S. B. Lowe, S. Ciampi, A. Magenau, K. Gaus, P. J. Reece and J. J. Gooding, *Langmuir*, 2014, **30**, 5209.
27 C. M. Hessel, M. R. Rasch, J. L. Hueso, B. W. Goodfellow, V. A. Akhavan, P. Puvanakrishnan, J. W. Tunnel and B. A. Korgel, *Small*, 2010, **6**, 2026.
28 Y. He, Y. Zhong, F. Peng, X. Wei, Y. Su, Y. Lu, S. Su, W. Gu, L. Liao and S.-T. Lee, *J. Am. Chem. Soc.*, 2011, **133**, 14192.
29 C. Song, Y. Zhong, X. Jiang, F. Peng, Y. Lu, X. Ji, Y. Su and Y. He, *Anal. Chem.*, 2015, **87**, 6718.
30 F. Erogbogbo, K.-T. Yong, I. Roy, G. Xu, P. N. Prasad and M. T. Swihart, *ACS Nano*, 2008, **2**, 873.
31 Y. He, Y. Su, X. Yang, Z. Kang, T. Xu, R. Zhang, C. Fan and S.-T. Lee, *J. Am. Chem. Soc.*, 2009, **131**, 4434.
32 Y. Zhong, F. Peng, X. Wei, Y. Zhou, J. Wang, X. Jiang, Y. Su, S. Su, S.-T. Lee and Y. He, *Angew. Chem., Int. Ed.*, 2012, **51**, 8485.
33 Y. Zhong, F. Peng, F. Bao, S. Wang, X. Ji, L. Yang, Y. Su, S.-T. Lee and Y. He, *J. Am. Chem. Soc.*, 2013, **135**, 8350.
34 L. Ruizendaal, S. P. Pujari, V. Gevaerts, J. M. Paulusse and H. Zuilhof, *Chem.–Asian J.*, 2011, **6**, 2776.
35 L. Wang, V. Reipa and J. Blasic, *Bioconjugate Chem.*, 2004, **15**, 409.

36 R. Intartaglia, A. Barchanski, K. Bagga, A. Genovese, G. Das, P. Wagener, E. Di Fabrizio, A. Diaspro, F. Brandi and S. Barcikowski, *Nanoscale*, 2012, **4**, 1271.
37 F. Erogbogbo and M. T. Swihart, *AIP Conf. Proc.*, 2010, **1275**(1), 35.
38 J. H. Ahire, M. Behray, C. A. Webster, Q. Wang, V. Sherwood, N. Saengkrit, U. Ruktanonchai, N. Woramongkolchai and Y. Chao, *Adv. Healthcare Mater.*, 2015, **4**, 1877.
39 C.-H. Lai, J. Hütter, C.-W. Hsu, H. Tanaka, S. Varela-Aramburu, L. De Cola, B. Lepenies and P. H. Seeberger, *Nano Lett.*, 2016, **16**, 807.
40 Z. F. Li and E. Ruckenstein, *Nano Lett.*, 2004, **4**, 1463.
41 J. H. Warner, A. Hoshino, K. Yamamoto and R. D. Tilley, *Angew. Chem., Int. Ed.*, 2005, **44**, 4550.
42 S. Ohta, P. Shen, S. Inasawa and Y. Yamaguchi, *J. Mater. Chem.*, 2012, **22**, 10631.
43 L. Ruizendaal, S. Bhattacharjee, K. Pournazari, M. Rosso-Vasic, L. H. J. de Haan, G. M. Alink, A. T. M. Marcelis and H. Zuilhof, *Nanotoxicology*, 2009, **3**, 339.
44 S. Bhattacharjee, I. M. Rietjens, M. P. Singh, T. M. Atkins, T. K. Purkait, Z. Xu, S. Regli, A. Shukaliak, R. J. Clark, B. S. Mitchell, G. M. Alink, A. T. Marcelis, M. J. Fink, J. G. Veinot, S. M. Kauzlarich and H. Zuilhof, *Nanoscale*, 2013, **5**, 4870.
45 S. Bhattacharjee, L. de Haan, N. Evers, X. Jiang, A. Marcelis, H. Zuilhof, I. Rietjens and G. Alink, *Part. Fibre Toxicol.*, 2010, 7, 1.
46 M. Dasog, Z. Yang, S. Regli, T. M. Atkins, A. Faramus, M. P. Singh, E. Muthuswamy, S. M. Kauzlarich, R. D. Tilley and J. G. C. Veinot, *ACS Nano*, 2013, **7**, 2676.
47 Q. Li, T. Y. Luo, M. Zhou, H. Abroshan, J. Huang, H. J. Kim, N. L. Rosi, Z. Shao and R. Jin, *ACS Nano*, 2016, **10**, 8385.
48 X. Chen, X. Zhang, L. Y. Xia, H. Y. Wang, Z. Chen and F. G. Wu, *Nano Lett.*, 2018, **18**, 1159.
49 X. Cheng, E. Hinde, D. M. Owen, S. B. Lowe, P. J. Reece, K. Gaus and J. J. Gooding, *Adv. Mater.*, 2015, **27**, 6144.
50 B. F. P. McVey, D. Konig, X. Cheng, P. B. O'Mara, P. Seal, X. Tan, H. A. Tahini, S. C. Smith, J. J. Gooding and R. D. Tilley, *Nanoscale*, 2018, **10**, 15600.
51 F. Erogbogbo, K. T. Yong, I. Roy, R. Hu, W. C. Law, W. Zhao, H. Ding, F. Wu, R. Kumar, M. T. Swihart and P. N. Prasad, *ACS Nano*, 2011, **5**, 413.
52 M. P. Singh, T. M. Atkins, E. Muthuswamy, S. Kamali, C. Tu, A. Y. Louie and S. M. Kauzlarich, *ACS Nano*, 2012, **6**, 5596.
53 C. Tu, X. Ma, P. Pantazis, S. M. Kauzlarich and A. Y. Louie, *J. Am. Chem. Soc.*, 2010, **132**, 2016.
54 S. Wu, Y. Zhong, Y. Zhou, B. Song, B. Chu, X. Ji, Y. Wu, Y. Su and Y. He, *J. Am. Chem. Soc.*, 2015, **137**, 14726.
55 M. Qiu, A. Singh, D. Wang, J. L. Qu, M. Swihart, H. Zhang and P. N. Prasad, *Nano Today*, 2019, **25**, 135.
56 C. Tu, X. Ma, A. House, S. M. Kauzlarich and A. Y. Louie, *ACS Med. Chem. Lett.*, 2011, **2**, 285.
57 N. Licciardello, S. Hunoldt, R. Bergmann, G. Singh, C. Mamat, A. Faramus, J. L. Z. Ddungu, S. Silvestrini, M. Maggini, L. De Cola and H. Stephan, *Nanoscale*, 2018, **10**, 9880.

58 J. B. Wacker, I. Lignos, V. K. Parashar and M. A. M. Gijs, *Lab Chip*, 2012, **12**, 3111.
59 P. A.-O. Debie, J. Van Quathem, I. Hansen, G. Bala, S. Massa, N. Devoogdt, C. Xavier and S. Hernot, *Mol. Pharm.*, 2017, **14**, 1145.
60 K. P. Conner, B. M. Rock, G. K. Kwon, J. P. Balthasar, L. Abuqayyas, L. C. Wienkers and D. A. Rock, *Drug Metab. Dispos.*, 2014, **42**, 1906.
61 K. Pant, D. Gröger, R. Bergmann, J. Pietzsch, J. Steinbach, B. Graham, L. Spiccia, F. Berthon, B. Czarny, L. Devel, V. Dive, H. Stephan and R. Haag, *Bioconjugate Chem.*, 2015, **26**, 906.
62 A. Eisfeld and J. S. Briggs, *Chem. Phys.*, 2006, **324**, 376.
63 N. Ryu, Y. Okazaki, E. Pouget, M. Takafuji, S. Nagaoka, H. Ihara and R. Oda, *Chem. Commun.*, 2017, **53**, 8870.
64 S. Zhang, P. Shao and M. Bai, *Bioconjugate Chem.*, 2013, **24**, 1907.
65 J. H. Ahire, Q. Wang, P. R. Coxon, G. Malhotra, R. Brydson, R. Chen and Y. Chao, *ACS Appl. Mater. Interfaces*, 2012, **4**, 3285.
66 M. Behray, C. A. Webster, S. Pereira, P. Ghosh, S. Krishnamurthy, W. T. Al-Jamal and Y. Chao, *ACS Appl. Mater. Interfaces*, 2016, **8**, 8908.
67 A. Z. Ruiz, D. Brühwiler, T. Ban and G. Calzaferri, *Monatsh. Chem.*, 2005, **136**, 77.
68 S. Thieme, M. Walther, H. J. Pietzsch, J. Henniger, S. Preusche, P. Mading and J. Steinbach, *Appl. Radiat. Isot.*, 2012, **70**, 602.

Silicon nanostructures for sensing and bioimaging: general discussion

Paola Ceroni, Yimin Chao, Carina Crucho, Luisa De Cola, Anna Fucikova, Ankit Goyal, Jinmyoung Joo, Ali Reza Kamali, Liubov Osminkina, Simone Silvestrini, Holger Stephan, Wei Sun and Ming Lee Tang

DOI: 10.1039/d0fd90006k

Holger Stephan opened a general discussion of the paper by Jinmyoung Joo: (1) Have you determined the size and zeta potential of the SiNPs conjugate with the peptide RPARPAR? (2) Have you considered the formation of a protein corona around the SiNPs-RPARPAR conjugates?

Jinmyoung Joo replied: Yes, we have measured both the size and zeta potential of the RPARPAR conjugated SiNPs using DLS. The size is increased to ~20 nm after RPARPAR conjugation, and the zeta potential is ~+17 mV.

This [second point] is a very important point to note. The protein corona may be formed on the SiNP-RPARPAR conjugate and the degree of protein corona formation should be different according to the RPARPAR coverage. This also significantly affects the receptor binding followed by intracellular uptake. We are undertaking further study on this aspect.

Anna Fucikova asked: I would like to ask why the authors did just a 24 h study and why the article does not present any cytotoxicity testing. It is known that silicon nanoparticles with different surface passivation, and especially with different levels of surface oxidation, have dramatically different toxicological profiles. In a 48 h study the effect of inflammation and secondary toxicity could also be observed.

Jinmyoung Joo answered: We have studied the cellular uptake for 24 h post-incubation with SiNPs because the PL intensity becomes negligible as shown in Fig. 3 and 4 (DOI: 10.1039/c9fd00124g). In addition, we have carried out cytotoxicity tests with the SiNPs for 24 h, and found no obvious toxicity effects caused by silicon degradation. It is assumed that the current concentration range might be safe in use, thus not included in the paper. However, long-term monitoring especially focusing on toxicological profiles arising by inflammation and secondary effects should be a good suggestion. We will study this issue further separately as this manuscript is focussed on the multivalent interactions of SiNPs to the cell. We appreciate this question.

Holger Stephan opened a general discussion of the paper by Liubov A. Osminkina: In the experimental part you mentioned that the hydrodynamic diameter was determined by DLS. Which solvents did you use and are there differences in size distribution for AD-Si QDs and SCD-Si QDs?

Liubov Osminkina responded: The initial microparticle size after manufacture was quantified by laser diffraction (not DLS) using a Malvern Mastersizer 2000. Particle dispersion in water was aided by adding a drop of 5wt% Igepal surfactant. After further particle size reductions the size distributions of AD and SCD batches were not quantified.

Ali Reza Kamali opened a general discussion of the paper by Yimin Chao: Scheme 1 exhibits functionalised Si nanoparticles. The material was produced by the anodic polarization of boron-doped porous silicon in HF solutions, followed by an extensive chemical processing. In the absence of characterization methods such as XRD, how can the stability of the Si core in the nanoparticles be confirmed?

Yimin Chao replied: There is typical XRD data in the supporting information to confirm the nature of the Si core of the nanoparticles. In addition, the XPS data shown in Figure 2 in our article also proves the existence of the Si core.

Ali Reza Kamali asked: Please can you comment on the environmental impact associated with the preparation of functionalised Si nanoparticles?

Yimin Chao answered: The final product of functionalized Si nanoparticles has almost no impact on the environment. However, the preparation method in this work involves corrosive HF and other chemicals that should be managed with care during the synthesis. Some other bottom-up methods developed in my lab are environmentally friendly.

Holger Stephan remarked: The isothiocyanate group is quite reactive to amino groups of peptides and proteins. Can you make a statement about which cell (membrane) proteins bind to the ITC SiNPs and how long it takes to form a thiourea group in cell culture?

Yimin Chao responded: Apparently previous studies suggested that free ITCs penetrate cells by diffusion. Once inside the cells, they are rapidly conjugated *via* their –N=C=S group with intracellular thiols, mainly GSH.[1–3] There are no data on cell membrane proteins binding to ITC so far. In the case of our ITC-functionalised silicon nanoparticles, we suppose that cellular uptake is mainly caused by endocytosis. Further studies will need to be performed to confirm which subtype of endocytosis occurs for our nanoparticles, whether it is clathrin-mediated, caveolae-dependent or another type.

1 Y. Zhang and E. C. Callaway, *Biochem. J.*, 2002, **364**, 301–307.
2 Y. Zhang, *Carcinogenesis*, 2012, **33**, 2–9.
3 Y. Zhang, *Carcinogenesis*, 2000, **21**, 1175–1182.

Carina Crucho queried: In Figure 1 in your paper, the peak at 2262 cm^{-1} in the FTIR spectrum of allyl bromide capped SiNPs is mislabeled. Continuing with the

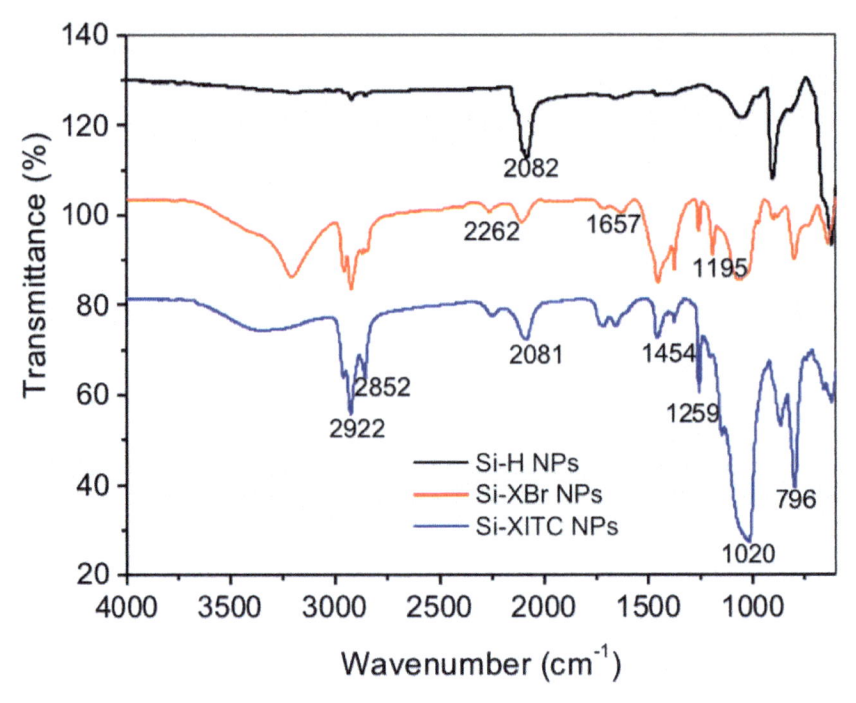

Fig. 1 FTIR spectra of H-terminated SiNPs, allyl bromide capped SiNPs, and ITC-capped SiNPs.

characterization, I wonder if the authors tried solution-state ^{1}H-NMR to monitor the surface functionalization, as for example in Han Zuilhof's article in this *Faraday Discussion*.[1]

1 DOI: 10.1039/c9fd00102f.

Yimin Chao replied: Thanks for pointing out this correction; the label 2262 cm^{-1} has been shifted to the feature to the left; see Fig. 1 in the discussion.

NMR is an analysis tool at the molecular level that is not always feasible for nanoparticles. We did try NMR during the synthesis, but no nice data was obtained for this sample. We did obtain some NMR results for our other nanoparticles, *e.g.* amine-capped Si nanoparticles published in 2012.[1]

1 J. H. Ahire, Q. Wang, P. R. Coxon, G. Malhotra, R. Brydson, R. Chen and Y. Chao, *ACS Appl. Mater. Interfaces*, 2012, **4**, 3285–3292.

Wei Sun opened a general discussion of the paper by Luisa De Cola: The authors conclude that the product is a mixture of surface-decorated SiO$_2$ nanoparticles and carbon QDs. Would the carbon QDs also contain a significant amount of Si? Prof. Yao He's article showed elemental mapping in which all particles seem to contain Si.[1] Personally I agree that the products should not be considered conventional Si(0) NPs, but they could still be called Si-containing NPs.

1 DOI: 10.1039/c9fd00088g.

Ming Lee Tang addressed Luisa De Cola and Holger Stephan: Prof. De Cola – I have read and cited your other papers on triplets. What a coincidence to discover you're working on Si nanoparticles too! I really enjoyed reading the careful characterization of the nanoparticles in your article "Shedding light on the aqueous synthesis of silicon nanoparticles by reduction of silanes with citrates",[1] where you concluded based on all the evidence that they were not Si(0). However, in your article "Ultrasmall silicon nanoparticles as a promising platform for multimodal imaging",[2] a very similar synthesis (with a different reference) was used to create the Si-based scaffold, but no mention was made of it being silica in nature. Is there a chemical difference in the Si nanomaterial in both these papers?

1 DOI: 10.1039/c9fd00127a.
2 DOI: 10.1039/c9fd00091g.

Holger Stephan responded: In both articles[1,2] the synthesis route of He and coworkers was applied, namely the reaction of APTES and trisodium citrate in a microwave[3,4] (sample name MW-CS in the paper "Shedding light on the aqueous synthesis of silicon nanoparticles by reduction of silanes with citrates"[1]). Thus the reaction products in both papers are identical. They are silica nanoparticles with 3-aminopropyl-decorated surfaces mixed with thermal degradation products of citrate (carbon quantum dots).

1 DOI: 10.1039/c9fd00127a.
2 DOI: 10.1039/c9fd00091g.
3 Y. He, Y. Zhong, F. Peng, X. Wei, Y. Su, Y. Lu, S. Su, W. Gu, L. Liao and S.-T. Lee, *J. Am. Chem. Soc.*, 2011, **133**, 14192–14195, DOI: 10.1021/ja2048804.
4 Y. Zhong, F. Peng, F. Bao, S. Wang, X. Ji, L. Wang, Y. Su, S.-T. Lee and Y. He, *J. Am. Chem. Soc.*, 2013, **135**, 8350–8356, DOI: 10.1021/ja4026227.

Simone Silvestrini replied: We stand by the answer given by Prof. Holger. We add that it was indeed during the experimental activities leading to the publication of "Ultrasmall silicon nanoparticles as a promising platform for multimodal imaging" (DOI: 10.1039/c9fd00091g) that we realized a more in-depth study on the chemical nature of the particles was needed. We were particularly curious about the chemical mechanism that would lead citrate to reduce alkoxysilanes to elemental silicon, an aspect that in our opinion remains unclear and overlooked in the interpretation given by other authors.

Paola Ceroni asked Luisa De Cola: I appreciated the accurate and critical discussion in your article. My question is: which technique would you suggest to use as a fast and easy test for a new reaction of silicon nanocrystals to be sure of the nature of the nanoparticles?

Simone Silvestrini answered: An important lesson we learnt during this study that we would like to stress once more before answering the question is that we should trust no single test when we go to assess the merits of a new process to produce silicon nanoparticles. As is often the case in materials science, we should strive to validate new processes using a broad array of techniques, possibly going some lengths to get overlapping data from different approaches that support each other.

In the case of silicon(0) nanoparticles in the size range we are dealing with, X-ray photoelectron spectroscopy (XPS) has the merit of quick discrimination of silica from silicon, as discussed in our report and in the referenced material (ref. 11 and 12, in particular[1,2]). For this reason, we suggest its use to verify the chemical identity of Si(0) nanoparticles prepared *via* novel routes. It should be noted, however, that XPS will not yield structural information – *i.e.* on the crystallinity (X-ray powder diffraction) and the size distribution (X-ray- or light-scattering techniques) of the nanoparticles.

FT-IR is another useful tool in this sense, and one that is too often overlooked. While data interpretation for unknown samples can be more complex than in the case of XPS, FT-IR represents a great opportunity for the routine characterization of reaction products of established synthetic routes. This is especially true once reference samples become available for the target material and possibly for its pollutants/alternative products, which are likely to depend on the reaction itself.

1 M. L. Mastronardi, F. Maier-Flaig, D. Faulkner, E. J. Henderson, C. Kübel, U. Lemmer and G. A. Ozin, *Nano Lett.*, 2012, **12**, 337–342.
2 K. Sato, T. Izumi, M. Iwase, Y. Show, H. Morisaki, T. Yaguchi and T. Kamino, *Appl. Surf. Sci.*, 2003, **216**, 376–381.

Carina Crucho opened a general discussion of the paper by Holger Stephan: I wonder if the authors considered using solution-state ^{1}H-NMR to monitor the amine surface functionalization. Quantification by solution-state ^{1}H-NMR it is a method widely used to quantify organic moieties (hydrophobic and hydrophilic) attached to silica nanoparticles.[1] In the case of amine groups, from NMR spectra in D_2O (as the protonated amines are soluble in water) it is possible to see the NMR peaks of the protons attached to the carbon next to the amine group or next to the silicon atom.

1 C. I. C. Crucho, C. Baleizão and J. P. S. Farinha, *Anal. Chem.*, 2017, **89**, 681–687, DOI: 10.1021/acs.analchem.6b03117.

Holger Stephan replied: Recent publications in *Analytical Chemistry*[1,2] describe suitable and reliable methods for the quantification of surface amino groups on slica nanoparticles (NPs) using solution NMR. These methods should in principle also be applicable for the determination of the amino groups of silicon NPs with 3-aminopropyl-decorated surfaces mixed with thermal degradation products of citrate.[3] However, we have not quantified the amino groups of silicon NPs by NMR. Here, a photometric assay (Kaiser Test) was employed to obtain information on the synthetically accessible surface amino groups. In a previous publication,[4] we used ^{1}H-NMR and ^{13}C-NMR spectroscopy to study the interactions of amino-functionalised silicon NPs with citric acid in D_2O. By integration of the citric acid signals and comparison with the signals assigned to the silicon NPs, we found a 1/6 ratio of citrate compared to the 3-aminopropyl residue.

1 C. I. C. Crucho, C. Baleizão and J. P. S. Farinha, *Anal. Chem.* 2017, **89**(1), 681–687, DOI: 10.1021/acs.analchem.6b03117.
2 F. Kunc, V. Balhara, A. Brinkmann, Y. Sun, D. M. Leek and L. J. Johnston, *Anal. Chem.*, 2018, **90**, 13322–13330, DOI: 10.1021/acs.analchem.8b02803.
3 DOI: 10.1039/c9fd00127a.
4 N. Licciardello, S. Hunoldt, R. Bergmann, G. Singh, C. Mamat, A. Faramus, J. L. Z. Ddungu, S. Silvestrini, M. Maggini, L. De Cola and H. Stephan, *Nanoscale*, 2018, **10**, 9880–9891, DOI: 10.1039/c8nr01063c.

Ankit Goyal asked: This study of the bio-application of luminescent Si NPs is interesting. You mentioned in the discussion that the reason for biocompatibility and rapid renal clearance is the minimum surface modifications performed on the Si NP surface to achieve the imaging goals without hampering the size and surface charges of the particles. Could you please state whether similar behavior was found in every mouse you studied? It is highly surprising that NPs (even Si) do not accumulate in any other tissue or hamper any other metabolic activities. Did you check WBC and platelet counts after exposing the mice to NPs? Did you confirm the non-accumulation of Si NPs in renal and liver tissues by TEM? There are several metabolic pathways in mice (or other animals) that could quench luminescence, and inactive Si NPs could still accumulate in the system, so these studies could be decisive. Please provide your views on this.

Holger Stephan answered: The most reliable method to obtain exact and even quantitative data on the bio-distribution is positron emission tomography (PET) with radio-labelled compounds. With this technique, matrix effects can be excluded. In our previous paper[1] we showed that ultrasmall silicon particles are excreted very quickly, whereas the citrate-stabilized particles have the best pharmacokinetic properties. There is of course considerable activity at early time points (<1 h), in particular in the kidneys and liver. The same behaviour is found for the dual labelled (nuclear and fluorescence) silicon particles (5 nu/nu mice). We could further show that the particles remain intact, and most importantly, PET and optical imaging give the same results. We have not analysed certain organs/tissues/red and white cells in detail yet. However, we agree that this is of great interest. In particular, *ex vivo* audiographic images and fluorescence microscopy can provide detailed insight.

1 N. Licciardello, S. Hunoldt, R. Bergmann, G. Singh, C. Mamat, A. Faramus, J. L. Z. Ddungu, S. Silvestrini, M. Maggini, L. De Cola and H. Stephan, *Nanoscale*, 2018, **10**, 9880–9891.

Conflicts of interest

There are no conflicts to declare.

Faraday Discussions

PAPER

Bridging energy bands to the crystalline and amorphous states of Si QDs†

Bruno Alessi, *[a] Manuel Macias-Montero,[b] Chiranjeevi Maddi,[a] Paul Maguire,[a] Vladimir Svrcek[c] and Davide Mariotti [a]

Received 5th October 2019, Accepted 10th January 2020
DOI: 10.1039/c9fd00103d

The relationship between the crystallization process and opto-electronic properties of silicon quantum dots (Si QDs) synthesized by atmospheric pressure plasmas (APPs) is studied in this work. The synthesis of Si QDs is carried out by flowing silane as a gas precursor in a plasma confined to a submillimeter space. Experimental conditions are adjusted to propitiate the crystallization of the Si QDs and produce QDs with both amorphous and crystalline character. In all cases, the Si QDs present a well-defined mean particle size in the range of 1.5–5.5 nm. Si QDs present optical bandgaps between 2.3 eV and 2.5 eV, which are affected by quantum confinement. Plasma parameters evaluated using optical emission spectroscopy are then used as inputs for a collisional plasma model, whose calculations yield the surface temperature of the Si QDs within the plasma, justifying the crystallization behavior under certain experimental conditions. We measure the ultraviolet-visible optical properties and electronic properties through various techniques, build an energy level diagram for the valence electrons region as a function of the crystallinity of the QDs, and finally discuss the integration of these as active layers of all-inorganic solar cells.

1. Introduction

The unique properties of silicon quantum dots (QDs) have attracted great attention in numerous fields of science such as photonics, photovoltaics, electronics and biomedicine[1–5] due to the unique interplay between quantum effects, surface states and direct/indirect transition dynamics and their biocompatibility.[6–8] Over the last decade, the focus has been on crystalline Si QDs,[9–14] and efforts made toward the study of amorphous Si QDs have been very limited[15–18] due to the difficulty of preserving the individual character of the QDs. Nonetheless, both

[a]Nanotechnology and Integrated Bio-Engineering Centre (NIBEC), Ulster University, Newtownabbey, BT37 0QB, UK. E-mail: alessi-b@ulster.ac.uk
[b]Laser Processing Group, Institute of Optics (CSIC), Serrano 121, 28006 Madrid, Spain
[c]Research Center for Photovoltaics, National Institute of Advanced Industrial Science and Technology (AIST), Central 2, Umezono 1-1-1, Tsukuba, 305-8568, Japan
† Electronic supplementary information (ESI) available. See DOI: 10.1039/c9fd00103d

phases present distinguished and remarkable features. Amorphous silicon thin films often offer better transport properties due to enhanced structural disorder preventing radiative recombination[19] and tunability of their optical absorption edge by controlling hydrogen content.[16] At the nanoscale, QDs can provide added functionalities not available in bulk silicon for amorphous and crystalline Si, in combination with other nanoscale properties (surface-to-volume ratio, surface chemistry, *etc.*). The synthesis of Si QDs by low-pressure plasma has been the focus of extensive research that has revealed the benefits of plasma processes for nanomaterials synthesis.[20–23] Non-thermal plasmas at atmospheric pressure offer beneficial and complementary features but have received limited attention.

Atmospheric pressure plasmas (APPs) present great versatility for the production and treatment of nanomaterials[24,25] as they allow flexible design and easy integration. Also, at this pressure, ion collisions with the nanoparticle surface are responsible for particle heating above the background gas temperature, allowing controlled crystallization[14] by carefully tuning the synthesis conditions. We have previously studied the synthesis and material properties of crystalline and amorphous silicon QDs, separately, by atmospheric pressure plasmas (APPs).[16,26,27] Our approach produced Si QDs with a well-defined particle size and observable quantum confinement effects.

Herein, we present an experimental and theoretical investigation of the Si QD phase transition in APPs, comparing the plasma conditions leading to or preventing crystallization.

We then perform various measurements on selected samples to assess the energy band diagrams and derive the relationships between structural features and opto-electronic properties as a function of the synthesis conditions. In this context we use different measurement techniques to build an energy level diagram of near-gap electron states, and critically compare methods and results. This approach is important for implementing nanomaterials in real-world applications.

Finally, we test the applicability of our Si QDs by integrating them as active layers in all-inorganic solar cells. While these devices still present very low efficiencies, here we demonstrate the viability of APP processes to be used in the manufacturing of next-generation photovoltaics.

2. Experimental details

The plasma reactor used for the synthesis of the Si QDs operates in a parallel electrode configuration at atmospheric pressure (760 Torr). A schematic diagram of the system is depicted in Fig. S1 in the ESI.† The plasma is generated inside a rectangular glass tube with a 0.5 mm gap and 0.3 mm wall thickness. Radio frequency (RF) power at 13.56 MHz and 120 W is applied through a matching unit to two rectangular copper electrodes with a cross section of 20 mm × 5 mm.

Argon and hydrogen are supplied as background gases, while silane (SiH_4) is used as a Si precursor with varying concentration between 50 ppm and 200 ppm. The flows of Ar and H_2 are set to 810–840 sccm and 150 sccm, respectively, in order to keep concentrations of approximately 99.7% and 0.3% at a fixed total flow of 1000 sccm. The plasma setup is accessorized with a two-axis stage and it is possible to directly deposit Si QDs on a substrate and form homogeneous films.[26] The plasma conditions are characterized using optical emission spectroscopy.

The equipment used to acquire the emission spectra is an Ocean Optics HR4000CG UV-NIR spectrometer (range 194–1122 nm) coupled with a 50 μm optic fiber. These measurements are carried out by locating one end of the optic fiber perpendicular to the plasma 10 mm away.

Silicon QDs are characterized using transmission electron microscopy (TEM) with a JEOL JEM-2100F microscope. The TEM analysis includes bright-field imaging to observe the morphology of the QDs and selected area electron diffraction (SAED) to characterize their crystallinity. For TEM, the QDs are collected directly in vials containing ethanol and then drop-casted onto an 3 nm ultrathin carbon film on Cu grid (Agar Scientific). Chemical analysis is performed using Fourier transform infrared spectroscopy (FTIR) and X-ray photoelectron spectroscopy (XPS) core level measurements. The FTIR instrument is a Nicolet iS5 from Thermo Scientific equipped with an attenuated total reflectance (ATR) iD5 accessory. XPS and ultraviolet photoelectron spectroscopy (UPS) measurements were performed using an ESCALAB 250 Xi microprobe spectrometer (Thermo Fisher Scientific, UK), equipped with an X-ray and UV source. XPS analysis was carried out with a focused XR6 monochromatic, micro-focused Al$_{K\alpha}$ ($h\nu$ = 1486.6 eV, <900 μm spot size) radiation source with a hemispherical energy analyzer. The binding energy was calibrated against the Pt$_{4f}$ peak taken to be located at 72.1 eV with a pass energy of 20 eV. XPS measurements were carried at a pressure of $1-5 \times 10^{-9}$ mbar. The valence band spectra were collected with a 20 eV pass energy. Optical absorption is obtained using a PerkinElmer 650S ultraviolet-visible (UV-vis) spectrometer equipped with a 150 mm integrating sphere. For UV-vis characterization, Si QDs were deposited on a quartz substrate, forming a homogeneous film. For valence electron analysis XPS in the valence region, UPS and a Kelvin probe were used. UPS spectra were collected with a UV source energy He(ɪ) ($h\nu$ = 21.22 eV) at a pressure of approximately 5.5×10^{-8} mbar, with 2 eV pass energy. A negative bias of 10 V was applied to the sample to shift the spectra from the spectrometer threshold. The energy resolution was around ∼100 meV. The Kelvin probe (KP Technologies APS04) is operated in air with a 2 mm gold alloy tip, after calibrating the tip work function against a sputtered Au thin film (W_{Au} = 4.69 ± 0.05 eV, W_{tip} = 4.4 ± 0.1 eV). Additionally, the Kelvin probe (KP) setup is equipped with a surface photovoltage module which measures the surface contact potential difference (CPD) induced by a monochromated white light source and an air photoemission module (APS), which uses a deuterium lamp source ($\Delta\lambda$ = 1 nm) to induce photoemission of electrons from the samples. For XPS, UPS and KP samples are directly deposited to form a film of QDs on ITO-coated glass (150 nm, 15 Ω sq^{-1}, VisionTek) in order to have good electrical contact with the stubs. Characterization of all the samples was carried out within 1 hour after synthesis to limit the effects of oxidation.[26]

3. Results and discussion

3.1 Structural and chemical characterization

The TEM results show that well-separated Si particles are produced for all values of precursor concentration introduced into the plasma (Fig. S1 in ESI†). However, high resolution TEM (HR-TEM) images reveal major differences between the particles depending on the precursor concentration. In particular, high silane concentrations (150–200 ppm) lead to the production of amorphous particles, while with low concentrations (50 ppm) crystalline particles are obtained (Fig. 1). As an example, Fig. 1a and b display the HR-TEM of a crystalline QD synthesized using a SiH$_4$ concentration of 50 ppm and an amorphous QD produced with 200 ppm of the precursor, respectively. In Fig. 1a, the particle exhibits fringes with spacing of 0.17 nm that correspond to the (311) plane of the silicon crystalline lattice. More detailed evidence of the crystalline or amorphous character of the Si

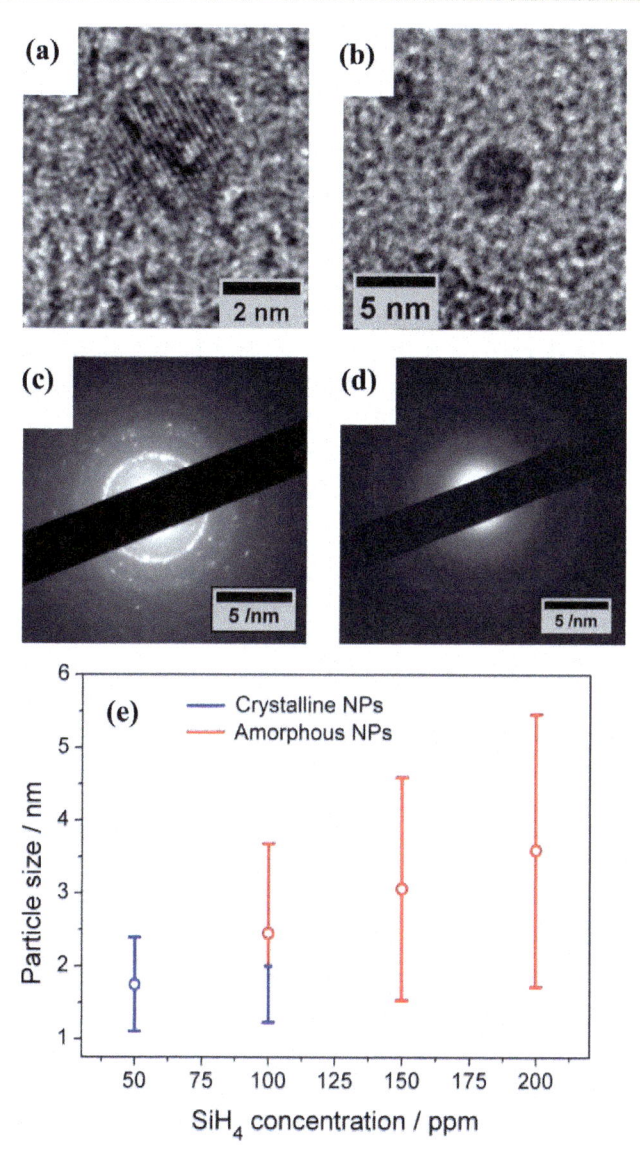

Fig. 1 Transmission electron microscopy (TEM) characterization of the Si quantum dots (QDs). High resolution TEM of (a) a crystalline and (b) an amorphous Si QD. (c) and (d) Selected area electron diffraction (SAED) pattern of (c) crystalline and (d) amorphous Si QDs. (e) Mean particle size (dots) and standard deviation (bars) of the Si QDs as a function of the precursor concentration used; blue and red lines respectively denote the crystalline or amorphous character of the QDs within the size distribution.

QDs is obtained using SAED. Fig. 1c and d show SAED patterns for the two extreme conditions of precursor concentration considered (50 ppm and 200 ppm). The crystallinity of the Si QDs produced using a low SiH$_4$ concentration results in the observation of sharp spots that together form well-defined rings in the SAED pattern (Fig. 1c). The spots detected in the diffractogram match well with the crystalline planes corresponding to the diamond lattice of silicon (see Fig. S2 in ESI†). On the other hand, a high concentration of precursor (150–200 ppm) results in faded diffuse rings in the SAED pattern that can be attributed to the amorphous character of the Si QDs (Fig. 1d). The conditions described above illustrate how our APP system is capable of producing highly crystalline QDs or purely amorphous QDs by controlling, in this case, the silicon precursor concentration. These results are in agreement with previous results published elsewhere.[14,16] Hence, it is possible to adjust from crystalline Si QD production with a low precursor concentration in the plasma (≤50 ppm) to amorphous Si QDs with a high concentration (≥150 ppm). For intermediate conditions the situation exhibits significant differences. TEM analysis of samples prepared using a SiH$_4$ concentration of 100 ppm showed that within the particle size distribution, only the smallest (<2 nm) particles exhibited crystalline character, while larger particles were amorphous. Thus, under these conditions both crystalline and amorphous particles can be generated simultaneously. To further understand the mechanism that makes possible the crystallization process of QDs inside the plasma region, a collision-corrected model (CCM) has been used and the results are described below. Regarding the particle size analysis, low magnification TEM images have been used, counting over 500 QDs for each of the conditions. The TEM micrographs used for the calculations and the particle size histograms are included in Fig. S3 in the ESI.† The overall results are presented in Fig. 1e, showing the mean value and standard deviation (obtained by fitting a log-normal distribution) of the QD size for various precursor concentrations, indicating crystalline or amorphous character in blue and red, respectively. In the graph it is possible to observe that increasing the concentration of precursor in the plasma leads to the production of QDs with larger size, with mean diameters varying from 1.7 nm to 3.6 nm. At the same time, the size dispersion also increases, starting with a value of 0.6 nm for 50 ppm of SiH$_4$ and reaching a value of 1.8 nm for 200 ppm of SiH$_4$.

The XPS technique was used to chemically characterize the Si QDs produced under different experimental conditions. The photoelectron spectra in the Si 2p region are shown in Fig. 2a, along with a deconvolution of the peaks in the different oxidation states of Si. We can observe that only the crystalline samples show a higher binding energy shoulder, which can readily be associated with limited oxidation. On the contrary, samples which have a least a fraction of amorphous particles do not show this feature.

Further details on the chemical composition of the Si QDs were obtained by FTIR analysis. Fig. 2b displays the FTIR spectra of Si QDs produced using the indicated SiH$_4$ concentration in the plasma. In the two selected regions of the infrared spectrum shown in Fig. 2b, it is possible to observe the vibrations that correspond to Si–O and Si–H$_x$ bonds. In particular, the absorption band at ~1075 cm^{-1} associated with the Si–O–Si stretching mode is shown. This peak has a very strong absorption cross section and therefore the low absorbance (lower than the Si–H$_x$ peaks) is evidence of a small level of oxidation even after exposure to the atmosphere. The peaks at 783 cm^{-1}, 862 cm^{-1}, 902 cm^{-1} and 2139 cm^{-1} are associated with the Si–H$_3$ bending, symmetric deformation, degenerate

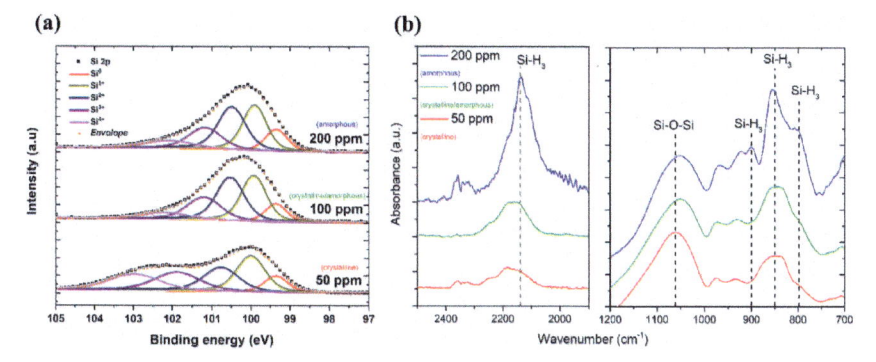

Fig. 2 (a) Deconvolution of XPS Si 2p signal into components relative to the different oxidation states of Si atoms, reflecting an increased number of Si–O bonds with different coordinations for the sample with crystalline nanoparticles. (b) Fourier transform infrared (FTIR) spectra of Si QDs for the different precursor concentrations indicated. The dashed lines indicate vibrational transitions at wavenumbers 783 cm^{-1}, 862 cm^{-1}, 902 cm^{-1} and 2139 cm^{-1} and are associated with the Si–H$_3$ bending, symmetric deformation, degenerate deformation and stretching modes, respectively, and a Si–O–Si absorption band at 1075 cm^{-1}.

deformation and stretching modes, respectively.[28,29] The intensity of these peaks becomes significant for Si QDs synthesized using a high precursor concentration (200 ppm) which correspond to amorphous samples, while it is negligible for lower concentrations when the particles are crystalline. This suggests that when the synthesis is carried out using a silane concentration above 100 ppm, *i.e.* whenever some amorphous material is present, the particles become partially hydrogenated, possibly due to hydrogen incorporation within the QDs.[16] In alternative, the amorphous particles may preferentially have silicon tri-hydride terminated surfaces as a result of the synthesis conditions.

We already reported the stability of Si nanocrystals through FTIR measurements in a previous study[26] over a period of 30 days and concluded that these nanocrystals, while being H-terminated, tend to oxidize from the interaction of inserted oxygen backbonds (O$_x$Si–H) and water vapour (*e.g.* from humidity in the air) condensing on the surface. However, this process is slow and self-limited, particularly when the QDs are deposited in films. Interestingly, the amorphous particles seem to be protected from oxidation at least in the first stages of exposure to the atmosphere (<1 h) and within the volume scoped by the XPS. This fact could be ascribed to the different kinetics of oxidation within amorphous particles[30,31] or the higher H concentration and the ability of hydrogen to passivate dangling bonds more evenly than in the crystalline case. The degree of hydrogenation therefore seems to be important both in the oxidation process as well as at some level determining the phase of the QDs.

The XPS instrument also allows the acquisition of reflection electron energy loss spectra using the flood gun as an electron source. This technique can be used to easily ascertain qualitatively the presence of incorporated hydrogen within a sample, *via* an energy loss feature which sits around 1.8 eV from the zero-loss peak. In our case we observe a distinguishable peak for the conditions in which we obtain amorphous particles (100 ppm and 200 ppm) of SiH$_4$ (Fig. S4, ESI†). We believe that under these conditions the level of hydrogenation is higher than in the other cases.

When the QDs are crystalline (50 ppm), the presence of hydrogen is limited at the surface and for this reason the hydrogen reflection signal is essentially absent.

3.2 Formation mechanisms leading to the synthesis of Si QDs

We used a model to calculate the temperature at the surface of the Si QDs (T_p) during the synthesis process within the APP (see Section S6 in ESI†). Fig. 3 presents the estimated values of T_p corresponding to our synthesis conditions (black, red, green and blue squares); on the same graph we also plot the experimental crystallization temperature (CT, grey points) of Si QDs from the literature.[32] The CT divides the graph into two regions: particles with a temperature below the CT are expected to be amorphous (blue region in Fig. 3) while particles with a temperature above the CT are expected to present a crystalline character (orange region in Fig. 3). Due to the intrinsic difficulty in measuring these values, it is not possible to define a sharp CT line to separate the two states.[32] Instead, a transition region represented by a white band can be defined (see Fig. 3). It is possible to observe that for low SiH_4 concentrations (50 ppm, black squares) the data points are all located entirely on the crystalline side of the graph. This result clearly agrees with the experimental evidence reported in Fig. 1, that is, the formation of purely crystalline Si QDs. For high precursor concentrations (>100 ppm) the opposite situation is observed. In this case, T_p is mainly located in the amorphous region with only the smallest, and less numerous, particles near the crystallization band. This is again in agreement with the experimental data in Fig. 1, where amorphous particles were observed under these conditions. The

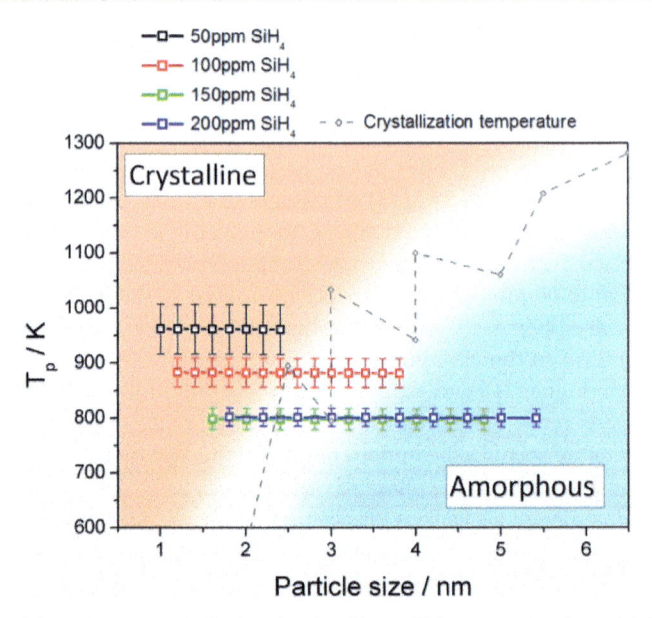

Fig. 3 Particle temperature calculated using the collision-corrected model (CCM) for different concentrations of precursor introduced in the plasma and experimental crystallization temperature from ref. 32. The temperature has been calculated for the particle size distribution obtained for each condition. The crystallization temperature divides the graph into crystalline (orange) and amorphous (blue) regions.

inability to produce a sufficiently high QD temperature for the crystallization of the Si QDs under these conditions can be partly attributed to the lower energetic plasma conditions (see Fig. S5†) but also to the larger particle size produced. At an intermediate SiH$_4$ flow (100 ppm), only the smallest particles are located on the crystalline side and the rest lie on the crystallization band, as indeed confirmed by our QD characterization (Fig. 1). These results provide theoretical justification for the experimental observations, since only particles smaller than 2 nm were found to be crystalline in this intermediate condition.

3.3 Valence band, Fermi level and bandgap measurements

The functional properties of semiconductors, in particular for energy applications, depend on the electronic structure and how the energy band parameters align with other application device components. We therefore perform here the experimental evaluation of energy band diagram (EBD) parameters such as valence/conduction band edges, Fermi level and bandgap for our samples. All measurements are therefore conducted on films of QDs deposited on solid substrates. Firstly, we focus on the valence band maximum (VBM) and Fermi level; in the next section we will complement these results with bandgap measurements to produce the EBD for both crystalline and amorphous QDs. To evaluate the electron energy levels near the valence band region and Fermi levels, we combined and compared results from different measurement techniques which offer different features, whose results are reported in Table 1.

UPS can be used to obtain the absolute value of the Fermi level of a semiconductor as well as the VBM. UPS measurements produce a cut-off energy ($E_{cut-off}$) and an on-set energy (E_{on-set}), which relate to the binding energy of electrons originating from the deepest levels of the material (ionized by the He–I source, $h\nu = 21.22$ eV) and from the valence band region, respectively (Fig. 4a). We should note that the UPS signal is referenced to the Fermi level of the semiconductor material under analysis. Therefore, the Fermi level can be extracted from the difference between $E_{cut-off}$ and the energy of the He–I photons, *i.e.* $E_{F-UPS} = E_{cut-off}$

Table 1 Fermi levels determined from Kelvin probe (E_{F-KP}) and UPS (E_{F-UPS}) measurements with corresponding uncertainties. For the Kelvin probe measurement, the uncertainty corresponds to std. deviation measurements within the scoped area and for the UPS measurement it is mostly due to the spectrometer energy resolution. VBM values determined from UPS (VBM$_{UPS}$) and APS (VBM$_{APS}$) with corresponding uncertainties are also reported. VBM$_{XPS}$ was calculated from XPS measurements and the Fermi levels produced by UPS. Uncertainties correspond to energy resolution for UPS and XPS, and for APS a rms sum of std. deviation relative to the fit. Energy bandgaps are obtained through Tauc plots of transmittance and reflectance measurements, assuming an indirect bandgap functional dependence between absorption coefficient and light energy. Uncertainties are a rms sum of uncertainties from fitting and uncertainty due to light source instabilities

Sample	Fermi level/eV		VBM/eV			E_g/eV
	E_{F-KP}	E_{F-UPS}	VBM$_{UPS}$	VBM$_{XPS}$	VBM$_{APS}$	$E_{g-UVVIS}$
50 ppm	-6.0 ± 0.1	-4.7 ± 0.1	-5.9 ± 0.1	-6.3 ± 0.6	-5.7 ± 0.2	2.47 ± 0.07
100 ppm	-4.0 ± 0.1	-4.3 ± 0.1	-6.2 ± 0.1	-6.4 ± 0.6	-6.1 ± 0.2	2.6 ± 0.1
200 ppm	-3.9 ± 0.1	-3.5 ± 0.1	-5.7 ± 0.1	-5.8 ± 0.6	-5.8 ± 0.3	2.3 ± 0.1

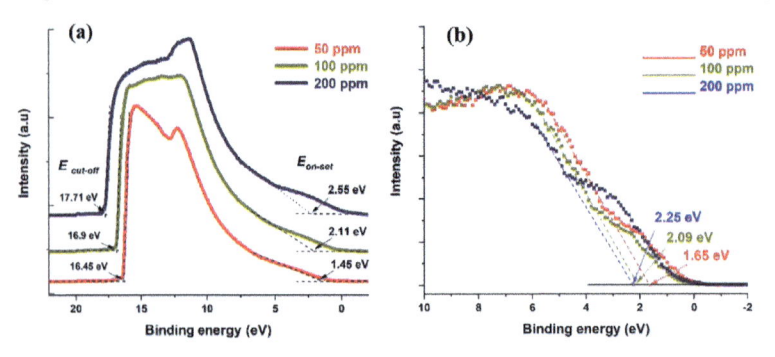

Fig. 4 (a) UPS spectra for the three different samples showing both cut-off and on-set values. (b) XPS valence band spectra showing the difference between VBM and Fermi level.

− 21.22 eV (see Table 1). The VBM is also calculated from the UPS measurements, where $VBM_{UPS} = E_{F\text{-}UPS} - E_{on\text{-}set}$ (Table 1).

The Fermi level values obtained with the Kelvin probe show the same trend as those obtained by UPS, even though the values are different, and in one case (50 ppm) notably higher.

Meanwhile, XPS produces the difference between the Fermi level and the VBM (Fig. 4b). In order to determine the values of the VBM reported in Table 1, we used the Fermi levels obtained by UPS (also in Table 1). Finally, the APS technique is similar in principle to the lower energy range of the UPS, but it is operated under atmospheric conditions and the source is a deuterium lamp, which is not able to reach a cut-off in the photoemitted electrons. From the APS signal it is possible to extract an absolute value of VBM (see ESI Section S9†).

All the values summarized in Table 1 are also reported in Fig. 5. The comparison shows that discrepancies between different measurements do exist, which in most cases are within measurement uncertainties. Even differences above 0.1 eV can be significant for applications, however we should note that comparisons of different measurement techniques are seldomly reported in the literature and, as such, our results highlight and underline the difficulties and limitations of current and available measurement methods.

With the exception of one of the measurements, the Fermi levels show similar values and exhibit the same trend, *i.e.* the Fermi energy becoming smaller with the particles going from crystalline to amorphous (Fig. 5a). The value of the Fermi level for the crystalline QDs (50 ppm) stands out and emphasizes the strong surface sensitivity of the KP technique (1–3 monolayers); while stray capacitance originating from inhomogeneities in the film can also impact the measurement, we tentatively ascribe this very large value of the Fermi level to the impact of even minor surface oxidation. The VBM values also show similar trends (Fig. 5b) with differences that can be justified by measurement uncertainties, with the exception of the VBM values of crystalline QDs (50 ppm) measured by XPS, which may be due to difficulties in extracting a good and reliable fit to the *x*-axis due to the limited resolution. A critical evaluation of the capabilities of each technique is therefore needed in order to determine which approach is most suitable and reliable in this specific case.

UPS measurements are implicitly limited by the general mechanism involved in photoemission as the light interacting with the sample induces a surface dipole, which complicates the evaluation of the Fermi level and VBM values, especially in

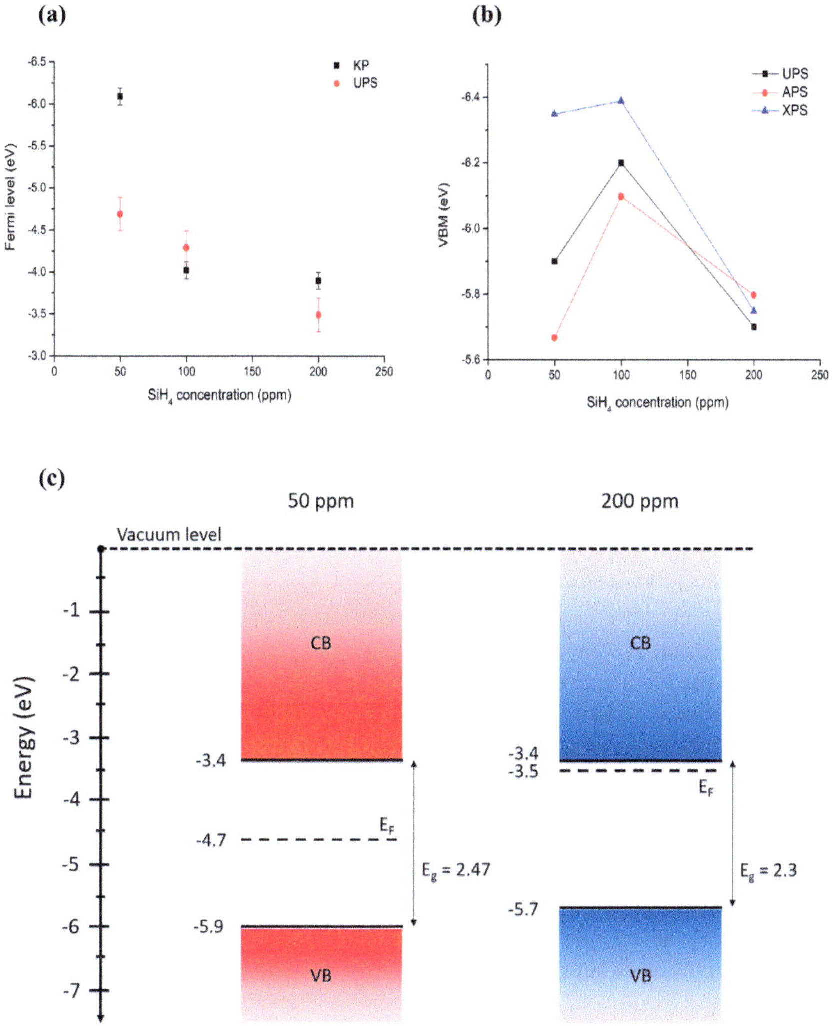

Fig. 5 (a) Comparison of Fermi level measured with Kelvin probe and UPS. (b) Comparison of VBM values measured (UPS, APS) and calculated (XPS) using different techniques. Error bars are omitted for clarity and can be found in Table 1. (c) Energy band diagrams obtained combining UPS values for Fermi level and VBM, and energy bandgaps through Tauc plots for the extreme cases of 50 ppm and 200 ppm of silane, which correspond to the smallest ($\langle r \rangle$ = 1.8 nm) all-crystalline particles and the biggest ($\langle r \rangle$ = 3.6 nm) all-amorphous ones.

conductive or highly doped samples.[33] In our case we assume that the samples do not develop a particularly strong surface dipole, since we do not expect the samples to be highly conductive. However, the benefits of UPS technique depend on the high photoemission yield, the narrow energy resolution achievable and the ability to scope only the occupied electron states of the very top surface of a solid (2–3 nm). Instead, XPS in the valence band region suffers from moderate energy resolution, lower yields and the absence of an energy cut-off, which does not permit values to be obtained relative to the vacuum level and generally scopes a "deeper" region beneath the surface (5–10 nm).[34] However, both techniques are operated in high vacuum, which can help to preserve the state of the surfaces.

KP and associated APS operating under atmospheric conditions are instead subject to surface dipoles and surface chemistry induced by environmental adsorbates which can easily shift the values up a few eV, raising or lowering the barriers felt by electrons escaping the material, or even induce completely new energy levels, especially in the valence electron region. These may be considered systematic errors in the values, which cannot be accounted for if there is not rigorous knowledge of the surface chemistry. While they are mostly useful for characterizing surfaces which will be exposed to the atmosphere (*e.g.* for corrosion studies), they also have the advantage of being cheap, easy and fast techniques. In addition, KP measurements are not affected by light-induced dipoles. However, APS operates under atmospheric conditions and photoelectrons experience different electrostatic environments as higher or lower potential barriers when escaping from the material surface.

These conditions depend on the atmospheric species and can change substantially with ambient temperature and humidity. In other words, while UPS and XPS relate to a potential energy of the ultra-high vacuum in the XPS chamber, the APS values relate to the electrostatic potential under atmospheric conditions at the time of measurement.[35] It is clear that various techniques can be used for the determination of VBM and Fermi levels, however care should be taken in the selection of the most appropriate technique.

UV-vis transmission and reflectance spectra were acquired for the Si QDs. In all cases, the Si QDs exhibit continuous and relatively featureless optical characteristics. Transmission and reflectance measurements can be used to determine the bandgap of the Si QDs. A full description of the bandgap determination can be found in the ESI (Section S7†), which was calculated using Tauc plots, with a 1/2 coefficient corresponding to an indirect bandgap.[36] A similar argument holds for amorphous silicon particles, for which the joint optical density of states is modeled by a square law.[37,38] For low precursor concentrations (50 ppm), QDs present a bandgap of roughly 2.5 eV (see Table 1), and as previously observed (Fig. 1e) these experimental conditions generate crystalline particles. This value is consistent with H-terminated Si QDs in the size range reported (1.8 nm). However, for SiH_4 concentrations greater than 100 ppm, the value at which we start to observe amorphous particles (Fig. 1e), the bandgap (more rigorously, the mobility gap) tends to a value of about 2.3–2.5 eV, which is consistent with the expected value for amorphous Si.[39] The bulk bandgap of amorphous silicon is reported in the wide range of 1.6–1.97 eV, under conditions that strongly depend on the content of hydrogen and the degree of structural disorder introducing or relieving stress components. The relative importance of the two mechanisms is still debated.[40–42] The higher values with respect to the bulk counterparts may be ascribed to quantum confinement effects, given the small particle size that we obtained. This fact can be either explained by an increased hydrogen content, which has been found in the REELS spectrum (Fig. S4 in ESI†). The 200 ppm silane samples, which resulted in the biggest amorphous-only particles, additionally show a long absorption tail to sub-bandgap energies that may be due to unsaturated bonds, ultimately acting as shallow dopant levels.

The surface photovoltage module in the KP can detect small amounts of photoinduced charge on a sample surface. The ability to build charge on the surface depends on both the light-induced bending of energy levels due to surface states, which results either in charge accumulation or depletion, and charge carrier mobilities.[43] We observe a detectable photovoltage only in the samples with amorphous particles (see Fig. S8 in ESI†). The energy threshold for the appearance of a photovoltage shows similar values to the optical gaps (2.3 eV to 2.5 eV) found *via* UV-vis absorption, and the sign of the shift indicates that the samples have n-type behavior. This implies that charge transport within the layer of amorphous particles is superior with respect to the films of crystalline nanoparticles.[43]

3.4 Energy level diagram

Based on our discussion in Section 3.2, we can expect XPS and APS to be somewhat less accurate for the determination of the VBM. The former is because of a lower count and resolution at lower energies, which tends to overestimate the difference between the VBM and Fermi level, given the inability to resolve weaker signals. The latter is susceptible to environmental conditions and adsorbates to a higher degree than other techniques. For the same reason, Fermi level through the Kelvin probe may be equally dependent on the calibration value, which is obtained by the APS method and may be affected by ambient temperature, humidity and adsorbates on the particle surface.

In conclusion, we believe that UPS values are more reliable for the determination of both the absolute Fermi level and the VBM, under the only assumption of a negligible light-induced surface dipole. Using energy bandgap values from UV-vis spectroscopy and UPS values, we have produced an energy band diagram of our single-phase samples (Fig. 5c).

The result is that, while the valence band maxima and bandgap do not vary significantly with the crystalline state of our nanoparticles, the value of the Fermi level tends to get closer to the conduction band edge with increased amounts of amorphous particles giving a stronger n-type character to the films. The UV-vis longer wavelength (sub-gap) absorption and the photovoltage response of amorphous samples support this picture.

4. Application as active layers in PV cells

Finally, in this section we provide some detail on the integration of Si QDs in photovoltaic (PV) devices. This is mainly to show that better knowledge of the EBD parameters can benefit application development, and also in part to demonstrate the capability of the APP process for direct integration of QDs in experimental devices. We have evaluated the performance of all-inorganic PV cells using the crystalline Si QDs synthesized with our method (50 ppm). The particles are directly deposited as a homogeneous film with the help of a two-axis stage placed 1 cm below the exit orifice of the capillary. Two device architectures were explored, both having Si QDs as the active component for the photogeneration of electron–hole pairs (see also ESI†).

One of the device architectures is illustrated in Fig. 6a, the other in ESI S10.† We used indium-doped tin oxide (ITO) strip-coated glass (VisionTek Systems Ltd., 15 Ω sq^{-1}, 150 nm ITO thickness) as our substrate and transparent conductive contact. A TiO$_2$ layer (40 nm thick) is formed using a sol–gel method and spin-coating following the protocol previously described[44] on top of the transparent conductive substrate. The TiO$_2$ coated ITO-glass is then used as a substrate to directly deposit the Si QDs (3

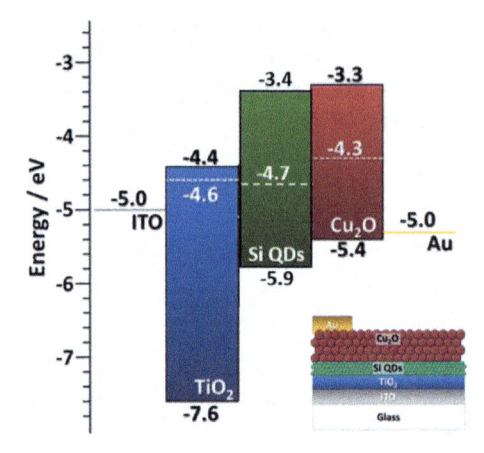

Fig. 6 Non-equilibrated band diagram for the PV device based on Si QDs as active layers, Cu_2O QDs as an electron blocking layer, a TiO_2 film hole blocking layer and corresponding contacts. Inset: diagram of the layer structure of the device.

μm thick film) exiting from the plasma reactor, using an X–Y stage to ameliorate the uniformity of the deposited layer. Spray-deposited Cu_2O is then used as an electron blocking layer and a sputter-deposited gold film is used as a top contact. The non-equilibrated band diagram of the cell is shown in Fig. 6. We should note that the selection of the Cu_2O transport layer was informed by our EBD measurements. The performance of the device was assessed with a solar simulator (Sub Femtoamp Keithley 6430) at standard AM1.5 irradiation. An open-circuit voltage and fill factor of 0.785 V and 87% were obtained, respectively. These are remarkable figures of merit that we can partly ascribe to exceptionally good energy level alignment of the Si QDs with the selected transport layer and contacts; however, the current density is very poor and affects the overall performance. Additionally, the measured value of series resistance is also indicative of relatively efficient electron transport in the cell (Table S10 in ESI†). Our current setup could not deliver thinner Si QD films as normally employed in this type of device; a reduction in the thickness could contribute to some improvements in the current density and overall device performance.

While the overall performance remains very low, the device parameters are very encouraging and demonstrate the usefulness of careful EBD parameter analysis. The utilization of both amorphous and crystalline Si QDs for PVs remains debatable and recent trends show that further manipulation through surface modification, alloying or other methodologies will be required.[45,46] In this context, the feasibility and integration of APP processes in the fabrication of next-generation devices is very promising and can give advantageous results.

5. Conclusions

We have demonstrated that Si QDs with tunable crystallinity can be grown in APPs. The study of a variety of experimental conditions has enabled us to produce Si QDs exhibiting crystalline or amorphous characteristics. The crystalline or amorphous character of the Si QDs was explained by efficient heating of the particles in non-thermal APPs. Analysis of the energy balance on the surface of the particles shows that the plasma parameters can be tuned to control the temperature of the particles when immersed in the plasma, and hence their crystallinity. We built an electron energy diagram of the valence electron regions for two selected samples, one with completely crystalline nanoparticles and one with completely amorphous nanoparticles and, after a critical assessment of measurements from different instruments, found that films formed from these free-standing particles tend to develop an n-type character for charge carriers in amorphous nanoparticles, despite similar values of valence band edges and optical bandgaps. We also showed the potential of APP processes for the fabrication of all-inorganic PV cells. Our analysis also highlights the need for improving analytical techniques and methodologies so that they could be used more extensively to dictate application-focused research directions.

Conflicts of interest

There are no conflicts to declare.

Acknowledgements

This work was supported by the EPSRC (EP/K022237/1, EP/M024938/1, EP/R008841/1) and the Leverhulme International Network (IN-2012-136).

References

1 F. Priolo, T. Gregorkiewicz, M. Galli and T. F. Krauss, *Nat. Nanotechnol.*, 2014, **9**, 19–32.
2 M. L. Mastronardi, E. J. Henderson, D. P. Puzzo and G. A. Ozin, *Adv. Mater.*, 2012, **24**, 5890–5898.
3 D. K. Kim, Y. Lai, B. T. Diroll, C. B. Murray and C. R. Kagan, *Nat. Commun.*, 2012, **3**, 1216.
4 J. Park, L. Gu, G. von Maltzahn, E. Ruoslahti, S. N. Bhatia and M. J. Sailor, *Nat. Mater.*, 2009, **8**, 331–336.
5 V. Švrček, D. Mariotti, Y. Shibata and M. Kondo, *J. Phys. D: Appl. Phys.*, 2010, **43**, 415402–415410.
6 D. Mariotti, S. Mitra and V. Švrček, *Nanoscale*, 2013, **5**, 1385–1398.
7 V. Švrček, M. Kondo, K. Kalia and D. Mariotti, *Chem. Phys. Lett.*, 2009, **478**, 224–229.
8 V. Švrček, D. Mariotti and M. Kondo, *Appl. Phys. Lett.*, 2010, **97**, 161502.
9 O. Wolf, M. Dasog, Z. Yang, I. Balberg, J. G. C. Veinot and O. Millo, *Nano Lett.*, 2013, **13**, 2516–2521.
10 M. Dasog, G. B. De los Reyes, L. V. Titova, F. A. Hegmann and J. G. C. Veinot, *ACS Nano*, 2014, **8**, 9636–9648.
11 K. Kusova, *et al.*, *ACS Nano*, 2010, **4**, 4495–4502.
12 V. Švrček, K. Dohnalova, D. Mariotti, M. T. Trinh, R. Limpens, S. Mitra, T. Gregorkiewicz, K. Matsubara and M. Kondo, *Adv. Funct. Mater.*, 2013, **23**, 6051–6058.
13 D. M. Sagar, J. M. Atkin, P. K. B. Palomaki, N. R. Neale, J. L. Blackburn, J. C. Johnson, A. J. Nozik, M. B. Raschke and M. C. Beard, *Nano Lett.*, 2015, **15**, 1511–1516.
14 S. Askari, I. Levchenko, K. Ostrikov, K. P. Maguire and D. Mariotti, *Appl. Phys. Lett.*, 2014, **104**, 163103.

15 S. Askari, M. Macias-Montero, T. Velusamy, P. Maguire, V. Svrcek and D. Mariotti, *J. Phys. D: Appl. Phys.*, 2015, **48**, 314002.
16 S. Askari, V. Svrcek, P. Maguire and D. Mariotti, *Adv. Mater.*, 2015, **27**, 8011–8016.
17 R. Anthony and U. Kortshagen, *Phys. Rev. B: Condens. Matter Mater. Phys.*, 2009, **80**, 115407.
18 Z. Shen, U. Kortshagen and S. A. Campbell, *J. Appl. Phys.*, 2004, **96**, 2204.
19 R. A. Street, *Adv. Phys.*, 1981, **30**, 593–676.
20 U. Kortshagen, *J. Phys. D: Appl. Phys.*, 2009, **42**, 113001.
21 U. Kortshagen, L. Mangolini and A. Bapat, *Nanotechnology and Occupational Health*, 2006, 39–52.
22 L. Mangolini, E. Thimsen and U. Kortshagen, *Nano Lett.*, 2005, **5**(4), 655–659.
23 R. Gresback, Z. Holman and U. Kortshagen, *Appl. Phys. Lett.*, 2007, **91**, 093119.
24 D. Mariotti and R. M. Sankaran, Perspectives on atmospheric-pressure plasmas for nanofabrication, *J. Phys. D: Appl. Phys.*, 2011, **44**, 174023.
25 A. J. Wagner, D. Mariotti, K. J. Yurchenko and T. K. Das, *Phys. Rev. E: Stat., Nonlinear, Soft Matter Phys.*, 2009, **80**, 065401.
26 M. Macias-Montero, S. Askari, S. Mitra, C. Rocks, C. Ni, V. Svrcek, P. A. Connor, P. Maguire, J. T. S. Irvine and D. Mariotti, *Nanoscale*, 2016, **8**, 6623–6628.
27 S. Askari, M. Macias-Montero, T. Velusamy, P. Maguire, V. Svrcek and D. Mariotti, *J. Phys. D: Appl. Phys.*, 2015, **48**, 314002.
28 F. S. Tautz and J. A. Schaefer, *J. Appl. Phys.*, 1998, **84**, 6636–6643.
29 Y. J. Chabal, *Phys. B*, 1991, **170**, 447–456.
30 Z. H. Lu, E. Sacher and A. Yelo, *Philos. Mag. B*, 1988, **58**, 385–388.
31 A. Szekeres and P. Danesh, *J. Non-Cryst. Solids*, 1995, **187**, 45–48.
32 A. N. Goldstein, *Appl. Phys. A*, 1996, **62**, 33–37.
33 M. M. Beerbom and B. Lagel, *J. Electron Spectrosc. Relat. Phenom.*, 2006, **152**, 12–17.
34 S. Hüfner, *Photoelectron Spectroscopy: Principles and Applications*, Springer Science & Business Media, 2013.
35 A. Kahn, *Mater. Horiz.*, 2016, **3**, 7–10.
36 J. Tauc, *Mater. Res. Bull.*, 1968, **3**, 37–46.
37 N. F. Mott and E. A. Davis, *Electronic Processes in Non-Crystalline Materials*, Clarendon-Press, Oxford, 1971.
38 M. L. Theye, *Phys. Scr.*, 1989, **T29**, 157–161.
39 N.-M. Park, T.-S. Kim and S.-J. Park, *Appl. Phys. Lett.*, 2001, **78**, 17.
40 Y. Abdulraheem, I. Gordon, T. Bearda, H. Meddeb and J. Poortmans, *AIP Adv.*, 2014, **4**, 057122.
41 A. M. Berntsen and W. F. van der Weg, *Phys. Rev. B: Condens. Matter Mater. Phys.*, 1993, **48**, 14656–14658.
42 G. D. Cody, T. Tiedje, B. Abeles, B. Brooks and Y. Goldstein, *Phys. Rev. Lett.*, 1981, **47**, 1480.
43 L. Kronik and Y. Shapira, *Surf. Sci. Rep.*, 1999, **37**, 1–206.
44 D. Carolan, C. Rocks, D. B. Padmanaban, P. Maguire, V. Svrcek and D. Mariotti, *Sustainable Energy Fuels*, 2017, **1**, 1611–1619.
45 V. Švrček, D. Mariotti, R. A. Blackley, W. Z. Zhou, T. Nagai, K. Matsubara and M. Kondo, *Nanoscale*, 2013, **5**, 6725–6730.
46 M. Bürkle, M. Lozac'h, C. McDonald, D. Mariotti, K. Matsubara and V. Švrček, *Adv. Funct. Mater.*, 2017, **27**, 1701898.

Faraday Discussions

PAPER

Silicon photosensitisation using molecular layers†

Lefteris Danos,[*a] Nathan R. Halcovitch,[a] Ben Wood,[a] Henry Banks,[a] Michael P. Coogan,[a] Nicholas Alderman,[b] Liping Fang,[c] Branislav Dzurnak[d] and Tom Markvart[de]

Received 29th September 2019, Accepted 11th November 2019
DOI: 10.1039/c9fd00095j

Silicon photosensitisation via energy transfer from molecular dye layers is a promising area of research for excitonic silicon photovoltaics. We present the synthesis and photophysical characterisation of vinyl and allyl terminated Si(111) surfaces decorated with perylene molecules. The functionalised silicon surfaces together with Langmuir–Blodgett (LB) films based on perylene derivatives were studied using a wide range of steady-state and time resolved spectroscopic techniques. Fluorescence lifetime quenching experiments performed on the perylene modified monolayers revealed energy transfer efficiencies to silicon of up to 90 per cent. We present a simple model to account for the near field interaction of a dipole emitter with the silicon surface and distinguish between the 'true' FRET region (<5 nm) and a different process, photon tunnelling, occurring for distances between 10–50 nm. The requirements for a future ultra-thin crystalline solar cell paradigm include efficient surface passivation and keeping a close distance between the emitter dipole and the surface. These are discussed in the context of existing limitations and questions raised about the finer details of the emitter–silicon interaction.

Introduction

Silicon photovoltaics (PV) dominate the world PV market with a global share of 90% and over 500 GW accumulated worldwide PV installations, and this dominance is likely to continue.[1,2] Being an indirect semiconductor, however, thick wafers are needed to ensure optimum sunlight absorption. This has an effect on

[a]Department of Chemistry, Energy Lancaster, Lancaster University, Lancaster, LA1 4YB, UK. E-mail: l.danos@lancaster.ac.uk
[b]Department of Chemistry, University of Ottawa, Ottawa, K1N 6N5, Canada
[c]Department of Electrical and Electronic Engineering, Southern University of Science and Technology, Shenzhen, Guangdong, 518055, China
[d]Centre for Advanced Photovoltaics, Czech Technical University, 166 27 Prague, Czech Republic
[e]Solar Energy Laboratory, School of Engineering, Faculty of Engineering and Physical Sciences, University of Southampton, Southampton, SO17 1BJ, UK
† Electronic supplementary information (ESI) available. See DOI: 10.1039/c9fd00095j

the cost of production, with wafer costs comprising up to 30% of the final PV module cost.[3] An attractive route to minimise the use of silicon is enhancing the excitation rate of electron–hole pairs by photosensitisation.

Photosensitisation of silicon has been postulated as a method to excite silicon *via* energy transfer from molecules close to the surface of the silicon.[4-7] This process results in the production of electron–hole pairs *via* non-radiative energy (rather than electron) transfer. Originally proposed by Dexter,[5] the process can be thought of as a near-field dipole–dipole interaction akin to Förster resonance energy transfer (FRET), between an excited molecule and the silicon substrate in close proximity (∼0–2 nm). The result of this interaction, is the direct generation of electron–hole pairs. Since the molecular excited state is localised, the process can by-pass the momentum selection rule at the root of the indirect band gap of silicon, turning silicon effectively into a direct-gap material and reducing the amount by up to 2 orders of magnitude. Indeed, by employing a similar approach to that found in light harvesting antenna in photosynthesis, we can envisage an ultra-thin nanostructured silicon solar cell photosensitised by light harvesting units without overall efficiency loss.[6,8] In this approach, the photovoltaic process is divided into two independent steps: an absorption/energy collection step carried out by the molecular light harvesting structure, and the charge generation step carried out by a thin silicon p/n junction. Each step can be optimised independently, resulting in a compact converter with minimum material requirements.

There have been experimental studies that looked at the interactions of excited states close to the surface of silicon to verify the presence of excited state energy transfer (FRET). The majority of these studies investigated the fluorescence quenching of the excited state as a function of distance to the silicon surface. A significant amount of quenching is observed as the excited state approaches closer to the silicon surface using evaporated dye layers,[9-11] quantum dots,[12-14] Langmuir–Blodgett (LB) monolayers,[15-18] and dye loaded zeolites.[19] The distance dependent data can be explained in terms of a simple damping oscillating dipole placed at a certain distance on the surface of the silicon, which is similar to fluorescence quenching near metal substrates.[20-23] We discuss the physics in some detail in the modelling section. We believe that only at distances less than about 5 nm can classical FRET be observed.[24] Furthermore, efficiencies in excess of the single junction Shockley–Queisser limit can be achieved, with recent advances in singlet fission[25-27] and triplet–triplet annihilation.[28-30]

The chemical modification of Si(111) surfaces has emerged as an area of intense research activity in the past three decades[7,31-33] that could provide solutions for different applications in optoelectronics.[34] The Si surface chemistry of interest produces oxide free surfaces with stable monolayers of molecules that are directly bonded on the Si surface (Si–C bond). Such surfaces can be generated from hydrogen terminated Si(111) surfaces routinely produced from wet chemistry.[35] A simple two-step chlorination/alkylation reaction[36] can give rise to an oxide free, well passivated surface that can be further functionalised with suitable chromophores while protecting the surface from oxidation.

In this work we used a chlorination/alkylation reaction to produce three types of oxide free silicon surfaces consisting of a methylated silicon surface (Si-Me), a vinyl silicon surface (Si-Vinyl) and an allyl silicon surface (Si-Allyl) and characterised them using XPS and FT-IR. We examined their passivation properties by measuring silicon fluorescence and lifetime decay to indicate the quality of the alkyl monolayer on the silicon surface. Further functionalisation of the Si-Vinyl and Si-Allyl surfaces *via* a Heck coupling reaction with a perylene derivative resulted in a controlled emitter to silicon surface distance in the 'true' FRET regime (<2 nm). The functionalised silicon surfaces were studied with time-resolved fluorescence confocal microscopy in order to estimate the lifetime quenching of the perylene dyes. The near field region was extended in the 2–4 nm distance range with the deposition of LB monolayers of suitable perylene derivatives. We observed fluorescence quenching of up to 90% from the anchored perylene molecules for <1 nm distances, which is indicative of efficient energy transfer to silicon. We modelled our results using a modification of an earlier model for dipole emitters near metallic surfaces.[21]

Modelling

Dipole emission near a silicon interface

An excited molecule can lose its energy in two ways when close to an interface: radiative energy transfer and non-radiative energy transfer. In their original paper,[21] Chance, Prock and Silbey (CPS) modelled the lifetime of an excited molecule near the interface between two media based on a forced damped oscillator and the complex dielectric constant of the media. The theory was able to fit and explain a series of experiments[20,37] based on the LB deposition technique[38] of how the lifetime of an emitter can be modified by the presence of gold, silver and copper surfaces. An excited fluorescent molecule can be modelled as an electric dipole, which can emit electromagnetic radiation not only in the far field but also in the near field. The CPS theory can be modified to model fluorescence near semiconductor surfaces instead of metals *via* the complex dielectric constant although it raises questions about the role of the imaginary component of the dielectric constant for a low absorbing indirect semiconductor such as silicon. In the following, we will outline the main findings of the CPS theory and how it can be modified to describe the effect of an indirect semiconductor surface on the excited state lifetime of an emitter.

The interaction between the molecular dipole and silicon is characterised by the dipole–silicon distance. When the distance between the dipole and silicon is larger than the wavelength of light, the fluorescence lifetime of the dipole oscillates periodically with the variation of the dipole–silicon distance. At this distance range, the main mechanism is light interference of the far field radiation of the source wave emitted by the dipole and the reflected wave from the silicon surface.

When the dipole–silicon distance is brought closer and is less than the wavelength of light, the near field radiation of the dipole starts to couple to the trapped modes in silicon, mediated by the evanescent waves propagating along the surface. We named this process photon tunnelling,[39] a process similar to quantum tunnelling through a potential barrier, which is induced by total internal reflection for our case. We estimate that for an emitter–silicon distance of 30 nm, up to 70% of emitted photons from a dye molecule are injected into the waveguide mode in silicon *via* tunnelling.[39] This tunnelling process is especially useful when silicon is in the form of a thin film (<1 μm), for which surface texturing is not feasible.

When the distance is just a few nanometres, an electron–hole pair could be directly excited in the silicon by the proximate molecular dipole. This process is similar to Förster's resonance energy transfer process between molecules.[40]

Calculation of the damping rates

For the system shown in Fig. 1, the CPS theory gives the fluorescence damping rate (inverse fluorescence lifetime) as follows:

$$\left(\frac{b}{b_0}\right)_{\text{VED}} = 1 + \frac{3q}{2}\text{Re}\left(\int_0^\infty du \frac{u^3}{\sqrt{1-u^2}} r_{12}^{\text{TM}} \exp\left(i2\sqrt{1-u^2}\,k_1 d\right)\right) \quad (1)$$

$$\left(\frac{b}{b_0}\right)_{\text{HED}} = 1 + \frac{3q}{4}\text{Re}\left(\int_0^\infty du \frac{u}{\sqrt{1-u^2}}\left[(u^2-1)r_{12}^{\text{TM}} + r_{12}^{\text{TE}}\right] \exp\left(i2\sqrt{1-u^2}\,k_1 d\right)\right) \quad (2)$$

$$\left(\frac{b}{b_0}\right)_{\text{ISO}} = \frac{1}{3}\left(\frac{b}{b_0}\right)_{\text{VED}} + \frac{2}{3}\left(\frac{b}{b_0}\right)_{\text{HED}} \quad (3)$$

where $u = k_p/k_1$ represents the normalised in-plane wavenumber, and k_1 is the wavenumber of light in the emitting matrix of the molecule. The three power dissipation regions of dipole emission could be calculated separately by setting the integral interval of eqn (1) or (2) in the related regions, as shown in Table 1.

This method was first applied to dipole emission near a metal surface by Ford & Weber.[41] By using this method, we arrived at the model curves plotted in Fig. 2a fitted to experimental fluorescence lifetime quenching results from a cyanine dye LB monolayer mixed with stearic acid.

The far field model was first given by Drexhage.[37] For photon tunnelling, we did a quantum mechanical calculation for the case that silicon is in the form of a 25 nm thick waveguide,[39] although evaluating the integral in the photon

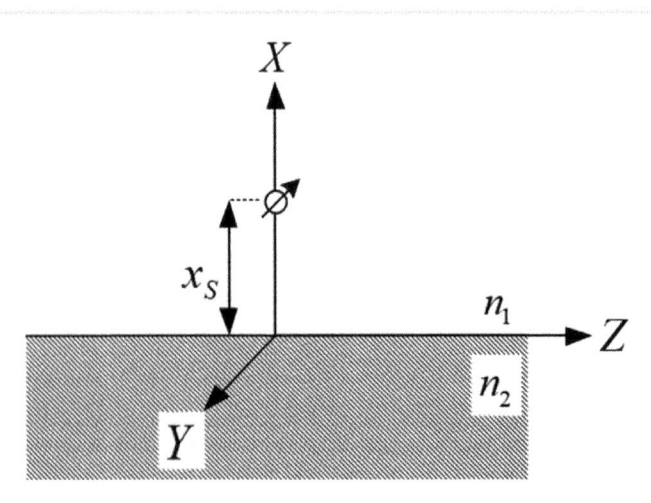

Fig. 1 Schematic of a dipole emission near a surface used in the calculation of the fluorescence damping rates.

Table 1 The integral interval for different regions

Far field	$k_p < k_1$	$u < 1$
Photon tunnelling	$k_1 < k_p < k_{Si}$	$1 < u < k_{Si}/k_1$
FRET	$k_p > k_{Si}$	$u > k_{Si}/k_1$

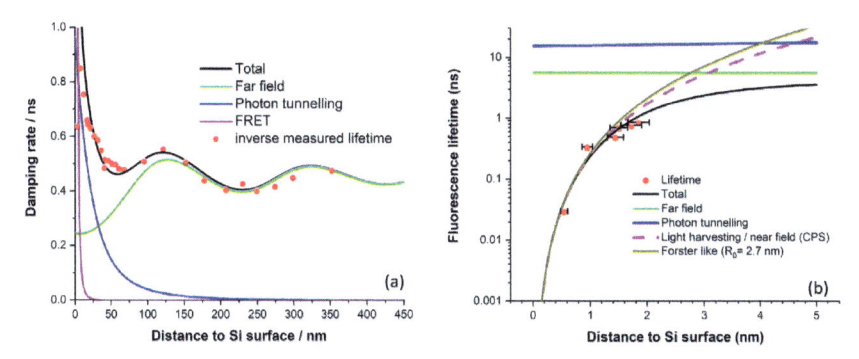

Fig. 2 (a) Fluorescence damping rate of a mixed dye LB monolayer vs. distance to bulk silicon (after ref. 39). (b) Normalised fluorescence lifetime of various chain length diol linked protoporphyrin silicon surfaces (after ref. 24).

tunnelling region yields similar results. For the FRET region, recent results using protoporphyrin IX molecules anchored to the silicon surface demonstrate efficient energy transfer with a fit to an inverse cubic law and an estimated Förster radius of 2.7 nm, Fig. 2b.[24]

Experimental

Synthesis of dibromo benzoic perylene diimide (Di-Br-Pe)

Using a modified literature procedure[42] a mixture of 0.5 g (1.27 mmol) of 3,4,9,10-perylene tetracarboxylic dianhydride, 0.433 g (3.18 mmol) of 4-bromoaniline, 20 g of imidazole and 0.1 g (0.456 mmol) of zinc acetate was heated at 100 °C for 2 h. The resulting mixture was further heated at 140 °C for 20 h under a nitrogen atmosphere (ESI, Fig. S1a†). The mixture was cooled to room temperature and acidified with 100 mL of 2 M hydrochloric acid with stirring for 20 minutes. The precipitate was collected by filtration and washed with copious amounts of water and methanol to remove impurities. The precipitate was finally dried under vacuum at 100 °C for 24 hours. Elemental analysis and the reaction scheme can be found in the ESI.†

Synthesis of bis(*n*-decylimido)perylene (PTCD-C10)

To perylene-3,4,9,10-tetracarboxylic dianhydride (100 mg, 0.26 mmol) in a boiling tube, decylamine (200 μL, 157 mg, 1 mmol) was added and thoroughly mixed (ESI, Fig. S1b†). The mixture was slowly heated with a heat gun until steam began to evolve. The mixture was heated, maintaining a steady evolution of gas, until the resulting red/brown residue became viscous. Following cooling, the boiling tube was broken open and the residue was extracted repeatedly into boiling chloroform. The combined chloroform extracts were filtered to remove broken glass and then concentrated under vacuum to around 10 mL and the product was collected by filtration to give a red powder (104 mg, 0.16 mmol, 59%). Elemental and NMR analysis can be found in the ESI.†

Synthesis of functionalised silicon surfaces

All solvents used in the silicon surface chemistry were anhydrous, stored in a glove box under nitrogen and used as received from Merck. The silicon substrates (10 × 10 mm squares, float zone, n-type, 25 Ω cm^{-1}, 500 μm thickness) were polished on both sides and cleaned in Decon 90, acetone, and isopropanol followed by exposure to UV light in a UV Ozone system on each side for 20 min at 60 °C. They were then placed in a solution of hydrogen peroxide in sulfuric acid (1 : 3, 45 minutes), followed by a thorough rinse in ultra-pure water. The surface of the silicon was etched in semiconductor grade ammonium fluoride (degassed, 15 minutes) to remove the native oxide layer and hydrogen-terminate the surface. After drying under flowing nitrogen, the sample was passed into a nitrogen-filled glovebox using a vacuum desiccator for further functionalisation.

The hydrogen-terminated silicon surfaces were immersed in a saturated solution of phosphorus pentachloride in chlorobenzene (110 °C, 2 hours) to which a few grains (<1 mg) of benzyl peroxide were added as a radical initiator. The sample was then washed in chlorobenzene followed by tetrahydrofuran (THF) and immersed in a 1.0–1.6 M Grignard solution of XMgCl (where X = methyl, vinyl and allyl groups) in THF (110 °C, 24 hours). After washing in THF and sonication in methanol, the samples were ready for further spectroscopic characterisation.

The vinyl ("Si-Vinyl") and allyl ("Si-Allyl") terminated Si(111) surfaces were washed in THF and acetonitrile, and immersed in a solution of dibromoperylene in 1 : 10 triethylamine : acetonitrile (Di-Br-Pe) (15 mL total) for a Heck coupling reaction.[43,44] To this, palladium acetate (<0.01 g) was added and the resulting solution was heated to 100 °C for 24 hours in a pressure vessel. The functionalised silicon surfaces were removed from the glovebox, and sonicated in acetonitrile (10 × 5 mL, 5 minutes per cycle) followed by methanol (2 × 5 mL, 5 minutes per cycle). The resulting perylene functionalised surfaces ("Si-Vinyl-Pe" and "Si-Allyl-Pe") were further washed with isopropanol and a lens tissue was dragged across the surface to ensure no physisorbed perylene moieties were left behind. After drying under nitrogen, the samples were characterised using spectroscopy.

Langmuir–Blodgett (LB) films

Substrates (quartz slides and silicon) were first washed in a warm solution of Decon 90 detergent, using abrasion to clean the surfaces. The slides were then sequentially sonicated in Decon 90, deionised water and ethanol. The slides were then dried and exposed to UV light in a UV Ozone system for 30 minutes, followed by exposure to piranha solution (3 parts H_2SO_4 to one part H_2O_2) for at least 30 min. The samples were exposed to propanol vapours to dry the surface and a final stream of nitrogen to remove any dust for further investigation. Once the slides had been cleaned, they were then exposed to hexamethydisilazane vapours (HMDS) overnight in a lidded glass staining jar to give good hydrophobic adhesion to the surface of the glass.

Stock solutions in chloroform of pure perylene dye (PTCD-C10)[45] and stearic acid (SA) were made up to concentrations of 1 mg per ml. Stock solutions of 10^{-4} M were prepared for mixtures of PTCD-C10 and SA with 1 : 1, 1 : 10, 1 : 25, 1 : 50 and 1 : 1000 molar ratios for the deposition of the mixed monolayers. Monolayer and multilayer LB films were fabricated using a Nima Trough (Nima Technology, UK) equipped with three dipping wells and barriers. Millipore ultrapure water was used with a resistivity > 18.2 Mohm and a pH of 6–7. Mixed chloroform solutions were deposited in 10–25 μl aliquots onto the surface of the water in the trough, depending on the amount of material needed, and were left for 15 minutes to allow the chloroform to evaporate. Three compressions were performed before any deposition in order to allow the film to anneal. All monolayers were deposited at a constant pressure of 26 mN m^{-1}. A schematic of each sample structure is shown in Fig. S3, in the ESI.†

Absorption steady state spectra

Absorption spectra were measured using a UV-Vis absorption spectrometer (Cary 60, Agilent) in the spectral range 300–800 nm. For all fluorescence measurements in solution (chloroform) we ensured the maximum absorption peak was always less than A < 0.1 in order to avoid any re-absorption effects.

Emission steady state spectra

Emission spectra were measured with a time-resolved fluorescence spectrometer (FluoTime 300, PicoQuant). The sample was excited with a 480 nm pulsed diode laser (LDH-P-C-485, PicoQuant) operated at 40 MHz. The emission spectrum was recorded at right angles to the excitation beam with a Peltier cooled photomultiplier (PMA-C 192-M, PicoQuant) in the spectral range of 560–850 nm and with a bandwidth of 5 nm. A 488 nm band edge filter (Semrock) was used to block unwanted laser scattering. Signals were digitised with a Time Harp 260 PCI card (PicoQuant) operated in steady state mode. Normalised absorption and emission spectra and decay curves from PTCD-C10 and Di-Br-Pe in solution are shown in Fig. S4, in the ESI.†

Silicon fluorescence emission and decays

Silicon emission spectra and decay curves were recorded at right angles to the excitation laser beam from different functionalised surfaces using a fluorescence spectrometer (PicoQuant, FT300) equipped with a thermoelectric cooled NIR-PMT unit (Hamamatsu, H10330A-45) with a spectral range of 950 nm to 1400 nm. The samples were excited with a 730 nm picosecond pulsed diode laser (P-C-730, PicoQuant) with 40 MHz repetition rate. Signals were digitised with a Time Harp 260 PCI card (PicoQuant). Decay curves were obtained at the maximum of the emission (1135 nm or 1200 nm) with bursts of multiple pulses in order to improve the signal sensitivity. This enabled the recovery of a high signal for the long lifetime samples. The time resolved decay curves were analysed using FLUOFIT software (PicoQuant) using a two-exponential model.

Time resolved emission spectra (TRES) and fluorescence decays

Time resolved emission spectra (TRES) and fluorescence decays from mixed perylene dye LB monolayers were measured using time correlated single photon counting (TCSPC) with a fluorescence spectrometer (PicoQuant, FT200 or FT300) equipped with a photomultiplier (PMA-185 or PMA-C 192-M, PicoQuant). The samples were excited with either 440 nm (LDH-P-C-440B, PicoQuant) or 480 nm (LDH-D-C-485, PicoQuant) picosecond pulsed diode lasers operated at a 40 MHz repletion rate and with a 200 ps full width half maximum (FWHM) instrument response function (IRF). The emission signals were digitised using a high-resolution TCSPC module (PicoHarp 300, PicoQuant) with 4 ps time width per channel. The emission from the LB monolayers on glass and silicon substrates were collected at right angles to the excitation laser beam and the emission arm was fitted with a 440 nm (Chroma) or 488 mm (Semrock) edge pass filter before the monochromator with a 8 nm or 5 nm spectral bandwidth.

Fluorescence lifetime imaging microscopy (FLIM)

Fluorescence lifetime measurements were performed on the organic monolayers on glass and silicon substrates with an inverse fluorescence lifetime microscope (MT200, PicoQuant). The body of the microscope consisted of a modified Olympus IX73 equipped with a 100× air lens objective with numerical aperture (NA) of 0.90 (MPlanFL N, Olympus). The MT200 system was configured with an objective scanning using a piezo XY stage (PI-721.CDQ) where the objective is moved instead of the sample. The samples were excited using a 485 nm pulsed diode laser (LDH-P-C-485, PicoQuant) operated at 10 or 20 MHz with an optical power between 0.1–0.25 µW. The emitted fluorescence was spectrally cleaned with a dichroic mirror and a transmission band edge filter (510 nm) or a transmission band pass filter (600–660 nm). A pinhole of 75 µm was employed to reject light that was out of focus. The fluorescence was detected using single photon counting with an avalanche diode (SPAD-100, PicoQuant) and digitised with a Time Harp 260 PCI card (PicoQuant). The IRF had a FWHM resolution of 250 ps and 24 ps per channel time increment. Image scans were performed over an area of 80 × 80 µm^2 with a varied pixel composition ranging from 256 × 256 pixels up to 640 × 640 pixels. The overall amount of photons per image was used as the measured decay curve. Lifetime image analysis was not suitable for the Si-Vinyl-Pe and Si-Allyl-Pe samples because of low photon counts detected per pixel but FLIM analysis was possible for the LB perylene mixed monolayer deposited on glass because of higher photon count rates. All decay curves were analysed using multi-exponential models *via* an iterative reconvolution process using SymPhoTime software (PicoQuant). Fit quality was assessed from the χ^2 parameter and weighted residuals. An example of a FLIM image and decay curve for a mixed (PTCD-C10 : SA) monolayer deposited on glass with SA in a 1 : 25 molar ratio is shown in Fig. S7, in the ESI.†

Variable angle spectroscopic ellipsometry (VASE)

The thickness of the organic monolayer was measured over a spectral range of 200 nm to 1000 nm at three different incidence angles (60°, 65°, and 70°) using a spectroscopic phase modulated ellipsometer (M-2000 V Automated Angle, J. A.

Woollam Co., Inc., USA). The data collected were analysed using a three-layer model (substrate/monolayer/air) using silicon for the substrate and a Cauchy model for the monolayer. Multiple readings were taken from each sample and averaged over all consistent measurements. Examples of model fits for a mixed perylene monolayer with stearic acid deposited on silicon are shown in Fig. S8.†

Infrared absorption spectra

All IR spectra were collected using a Shimadzu IRTracer-100 equipped with a deuterated lanthanum α alanine doped triglycine sulfate (DLATGS) detector or a mercury cadmium telluride (MCT) detector. Spectra were collected with the Shimadzu software package between 500 and 4000 cm^{-1} at a resolution of 4 cm^{-1}. The spectra were collected either in a transmitted geometry or using a variable angle specular reflectance accessory (Pike, VeeMax III) equipped with an infrared polariser (p-polarised) at an incidence angle of 60 degrees with the surface. Background scans of a native oxide, hydrogen terminated or chlorine terminated silicon surface were used to record the absorption spectra. Spectra were collected as the sum of 512 scans and an applied atmosphere background to remove peaks due to parasitic absorption of CO_2 and H_2O.

X-ray photoelectron spectroscopy (XPS)

XPS measurements were performed in triplicate on the samples using a Thermo Scientific K-Alpha spectrometer for a survey scan and with the Si 2p and C 1s regions investigated in detail. The Si-Cl, Si-Me, Si-Vinyl and Si-Vinyl-Pe samples were sent for XPS measurements at the Nexus national EPSRC XPS users service at Newcastle University.

The Si 2p (96–100 eV) and C 1s (280–297 eV) bonding regions were investigated in detail to reveal the bonding of the monolayer to the silicon surface. Spectra were fitted to a Shirley background and were subtracted. Deconvolution of the spectra was performed by fitting multiple Gaussian bands until the residual standard deviation reached a minimum using peak fitting software (CasaXPS). The position of the Si 2p(3/2) peak was found to be approximately 0.6 eV lower than that of the Si 2p(1/2) peak and double the integrated peak area. Three Gaussian functions were used to fit the peaks for the Si–C bond at 284.0 eV, C–C bond at 285.0 eV, and C–O or C=O bonds at 286.0 eV, although the peak centres slightly changed.[46] We report surface coverages of the vinyl terminated Si(111) surfaces relative to the methyl terminated Si(111) surfaces by comparing the areas of the Si–C bond peaks to the Si 2p peak for the same sample.[47]

Results and discussion

The different functionalised Si(111) surfaces prepared in this study are shown in Scheme 1. The methyl, vinyl and allyl Si(111) terminated surfaces were prepared using a modified chlorination/alkylation reaction that is known to produce well-passivated Si surfaces with good coverages.[48–50] The successful attachment of the molecular chains was confirmed using polarised infrared (IR) absorption (Fig. S9, in the ESI†) and X-ray photoelectron spectroscopy (XPS) measurements (Fig. S10, in the ESI†). A section of the IR spectra for the Si-Vinyl and Si-Allyl surfaces shows C–H stretches in agreement with previous studies (Fig. 3a and b).[50] XPS confirmed

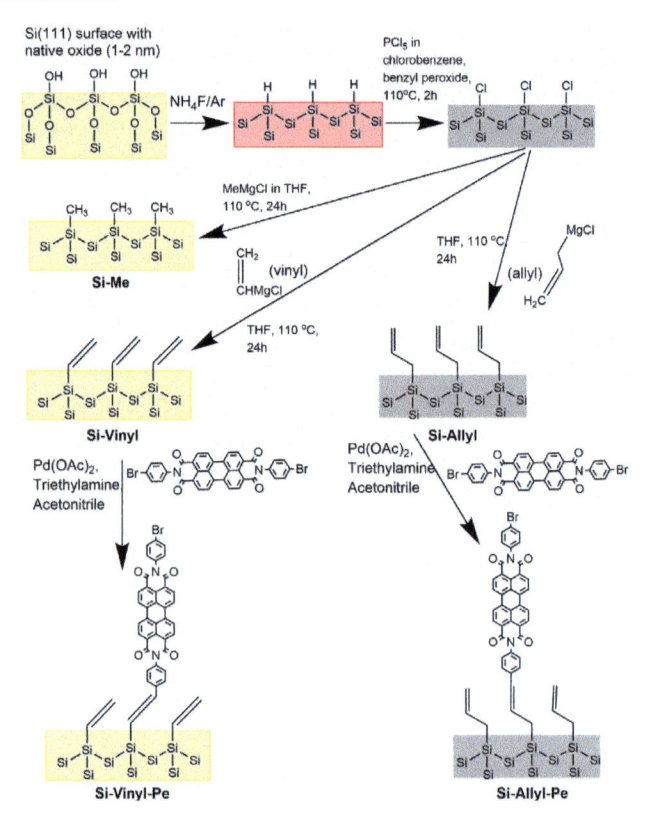

Scheme 1 Reaction scheme for the synthesis of the alkylated/perylene silicon surfaces.

that the vinyl group was attached on the silicon surface in the C 1s region and showed no silicon–oxygen peaks in the Si 2p region, confirming that an oxide free surface was prepared (Fig. 3c and d). Molecular coverage for the vinyl terminated surfaces was estimated to be 95% ± 5% by comparing the ratio of the XPS C–Si/Si 2p for the vinyl terminated surface to the methyl terminated surface.[49]

Passivation using alkyl layers

The alkylation of silicon surfaces has proven to provide excellent passivation properties using low-temperature surface modifications.[48,49] Surface photovoltage (SPV) measurements on methyl terminated Si(111) surfaces have revealed unusually large SPV signals that can only be explained by charge accumulation in the Si(111) interface during the alkylation procedure.[47–52] Similar surface chemistry treatments have resulted in a p/n junction near the surface of silicon.[53]

The passivation properties of the methyl, vinyl and allyl terminated silicon surfaces were investigated by measuring the silicon emission spectra and decay curves (Fig. 4). There is a visible difference in the spectral shape of the Si emission peak (λ_{max} = 1133 nm, and FWHM = 100 nm) between the methyl, vinyl and allyl Si(111) surfaces and chlorine terminated (λ_{max} = 1200 nm, and FWHM = 230 nm) Si(111) which is similar in shape with the Si emission peak from a clean Si(111)

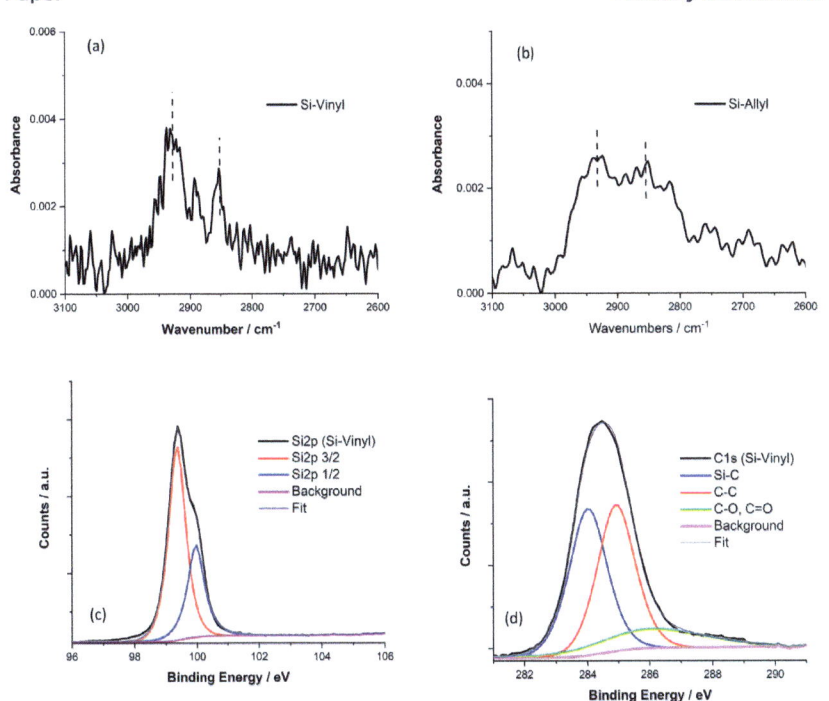

Fig. 3 (a) Polarised reflection IR spectrum of the C–H stretching region for the Si(111)-Vinyl surface. Peaks at 2920 cm^{-1} and 2985 cm^{-1} are indicated on the spectrum and correspond to the C–H asymmetric and symmetric stretching modes respectively. (b) Polarised reflection IR spectrum of the C–H stretching region for the Si(111)-Allyl surface. Peaks at 2920 cm^{-1} and 2850 cm^{-1} are indicated on the spectrum and correspond to the C–H asymmetric and symmetric stretching modes respectively of (CH$_2$). (c) High resolution XPS spectra of the Si 2p region for the Si(111)-Vinyl surface showing Si 2p$_{3/2}$ (red) and Si 2p$_{1/2}$ (blue). (d) High resolution XPS spectra of the C 1s region for the vinyl terminated Si(111) surface showing C–Si (blue), C–C (red), and C–O (green).

surface with the native oxide present (Fig. 4a). The observed difference in the emission peak shape is attributed to the improved surface passivation of the alkylated Si(111) surfaces and the elimination of surface traps. In contrast, the broad Si emission peak observed for the chlorine terminated Si surface (and for the native oxide) indicates poor surface passivation which is consistent with previous work.[47,50]

The passivation properties of the Si-Me, Si-Vinyl and Si-Allyl surfaces were further investigated by measuring the silicon fluorescence decay curves (Fig. 4b) and this revealed an effective recombination lifetime two orders of magnitude higher than those of the native oxide and chlorinated surfaces, indicating significant improved passivation. The very fast decay observed for the Si-Cl and Si (native oxide) samples is indicative of the presence of several surface traps.

Transient photoluminescence using single photon counting has been used to separate the surface and bulk recombination in silicon.[54] In this case, we applied a simple model and the measured effective recombination lifetime was converted to the surface recombination velocity (SRV) of electron–hole pairs assuming

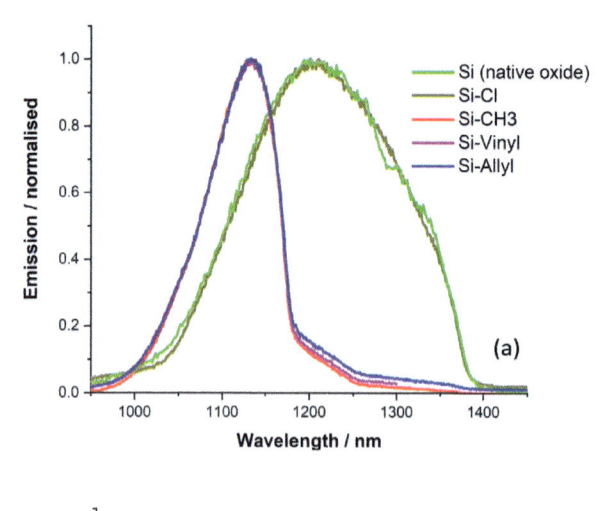

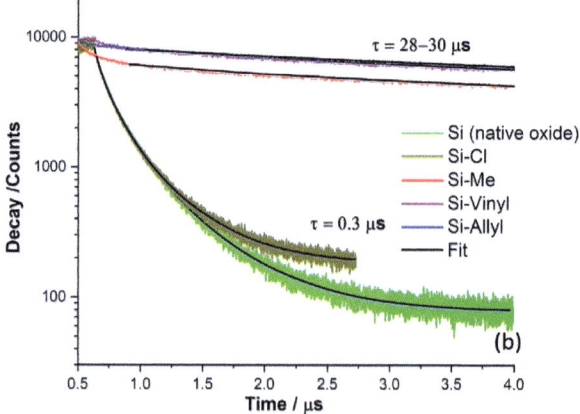

Fig. 4 (a) Emission spectra of various silicon treated surfaces; a clean silicon surface with native oxide present, Si (native oxide), a chlorinated silicon surface (Si-Cl), a methyl terminated silicon surface (Si-Me), a vinyl terminated silicon surface (Si-Vinyl) and an allyl terminated silicon surface (Si-Allyl). (b) Emission decay curves from the functionalised silicon surfaces at the maximum emission.

(guaranteed from the wafer) a high silicon bulk lifetime (>1000 μs).[48,49,51] For the double-sided polished surfaces, we estimated a surface recombination velocity of 830 cm s^{-1}. Similar and even better SRV values were obtained using covalently attached monolayers and showed as good passivation properties as those obtained using traditional passivation techniques. In this case, vinyl monolayers can be used for effective silicon surface passivation.

Attachment of perylene

The Heck coupling reaction was used, as this reaction has been shown to work for bromine and chlorine-halogenated chromophores, and is a low temperature process widely investigated in the literature.[43,44] Only trace amounts of the metal catalyst are required, making this a very attractive method for future scale-up (Scheme 1). The structures of the vinyl and allyl terminated surfaces resulting from the attachment of Di-Br-Pe are shown in Fig. S2 (ESI†).

The successful attachment of the perylene dye was confirmed by XPS and FT-IR. The integrated area ratio of the XPS N 1s bond peak relative to the Si–C/Si 2p XPS peak for the Si-Vinyl-Pe sample gave a 5% coverage accounting for the presence of two nitrogen atoms in the perylene derivative (Fig. S10e, ESI†). The carbon C 1s bond peak also indicated that C=O was present on the monolayer which can be attributed to the perylene molecules (Fig. S10d, ESI†). A large amount of Pd 3d was observed for the Si-Vinyl-Pe samples in the XPS spectrum. Furthermore, recombination lifetime measurements performed on a similar surface with a cyanine dye attached to a vinyl terminated surface coupled *via* the Heck reaction indicated a significant decrease in the recombination lifetime.[55] The recombination lifetime did not change after multiple washes, confirming that the metal contamination could not be removed by washing. A reason for this could be the incorporation of palladium on the silicon surface leading to heavy metal contamination and a dramatic reduction in the electron–hole recombination lifetimes. A repeated procedure using a longer alkene linker (1-decene) in order to move the palladium catalysis further from the surface produced similar reduced recombination lifetimes. The Heck reaction might be disadvantageous for the functionalisation of alkene terminated silicon surfaces for photosensitised solar cells and a catalyst free approach might be more suitable.[56]

LB monolayers of perylene

In previous work, we studied the energy transfer of an excited molecule to silicon using LB monolayers of mixed carbocyanine dyes with stearic acid.[16–18,39] Fluorescence lifetime quenching results were obtained for distances greater than 2 nm of the emitter to the surface of the silicon since the native oxide present in the silicon was not removed. The preparation of inert spacer structures can be achieved using stearic acid (SA) for distances up to 30 nm or thermal oxide for greater distances from the silicon surface. Accurate determination of the spacer thickness can be determined using spectroscopic ellipsometry.

In this study, we used a 10 carbon long chain modified perylene tetracarboxylic derivative (PTCD-C10) as a monolayer material to deposit LB films on glass and silicon substrates at different molar mixing ratios with stearic acid (SA) (Fig. S3, in the ESI†).

The fluorescence emission in dilute solutions from both perylene derivative molecules (PTCD-C10 and Di-Br-Pe) used in this study showed the expected three peak vibronic structure at 534 nm, 575 nm and 625 nm with a slight 2 nm blue shift observed for the emission peaks of Di-Br-Pe (Fig. S4a, in the ESI†). The emission spectra are essentially mirror images of their respective absorption spectra (Fig. S4b, in the ESI†). The solution fluorescence decays show the expected mono-exponential lifetime with excited state lifetimes τ(Di-Br-Pe) = 3.7 ns and τ(PTCD-C10) = 4.0 ns (Fig. S4c and d, in the ESI†).

In condensed phases such as the LB monolayers, concentration quenching occurs with the appearance of a red-shifted emission band due to excimer formation (Fig. S5a, in the ESI†).[57] The decay lifetimes were multi-exponential and the average lifetimes remained essentially the same irrespective of the mixing ratio with SA because the PTCD-C10 molecules pack close together forming aggregates on the water surface. Time resolved emission spectra (TRES) showed the absence of any monomers in the monolayer and emission occurred from aggregates and excimer formation in the monolayer. The excimer emission spectrum observed from different mixed perylene monolayers is broad with a peak between 650 and 670 nm depending on the concentration of perylene molecules in the monolayer (Fig. S6a–c, in the ESI†). Only at a very low PTCD-C10 concentration in the monolayer with SA (1 : 1000) does the PTCD-C10 lifetime approach the one observed for dilute solutions.

The measured emission and decay curves from the PTCD-C10 mixed monolayer (1 : 25) on glass were used as the true 'unquenched' data in order to normalise all measurements taken from the PTCD-C10 monolayers on silicon substrates (Fig. S7b, in the ESI†). PTCD-C10 monolayers were deposited on silicon in two different configurations (Pe-LB1 in a hydrophilic Si prepared surface and Pe-LB2 in a hydrophobic Si prepared surface), giving rise to emitter distances ranging from 2–4 nm (Fig. S3, in the ESI†). Spectroscopic ellipsometry was used to measure the distance to the surface of the silicon (Fig. S8, in the ESI†).

FLIM measurements and fit to the FRET region

Fluorescence lifetimes were measured using time-resolved fluorescence confocal microscopy. The overall decay intensities were analysed for each image with several image measurements taken from different regions and lifetimes were averaged. Because of the low photon count rate per pixel, a direct analysis of the lifetime images (FLIM) was not possible for Si-Vinyl-Pe and Si-Allyl-Pe samples. Selected fluorescence lifetime images are shown in Fig. 5. The functionalised Si(111) surfaces (Fig. 5a and b) show the perylene molecules distributed across the surface. Formation of aggregates on the surface is possible due to the presence of

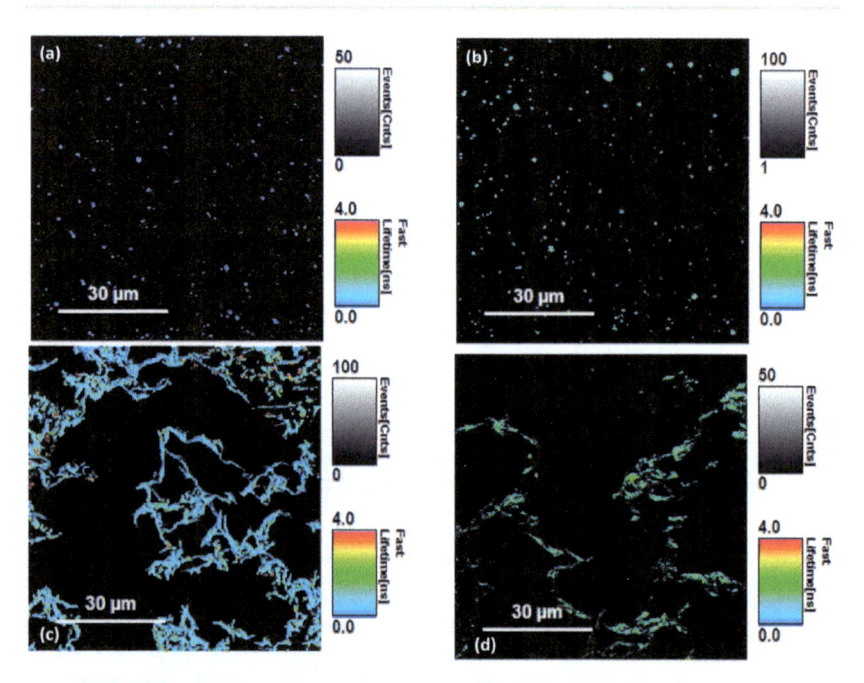

Fig. 5 Fluorescence lifetime imaging (FLIM) scans of (a) the Si-Vinyl-Pe surface, (b) the Si-Allyl-Pe surface, (c) the Pe-LB1 Si surface and (d) the Pe-LB2 Si surfaces.

the large pi-ring of the perylene molecule. The dark areas correspond to regions of the Si(111) surface that were not functionalised with the perylene molecules. The same pattern was observed for each perylene functionalised silicon surface studied. The LB monolayers deposited on silicon from the PTCD-C10 perylene derivatives mixed with SA showed different patterns (Fig. 5c and d). The strong intermolecular interactions of the perylene molecules on the surface of water in a Langmuir monolayer resulted in the molecules forming 'fibre like' structures on the deposited LB monolayers on the glass and silicon substrates. The dark areas observed in the FLIM images are due to the presence of stearic acid in the mixed PTCD-C10 : SA (1 : 25) monolayer. It is evident that the PTCD-C10 molecules pack closely together on the surface of water, forming aggregates irrespective of the SA mixing ratio. Only at high mixing ratios with SA greater than 1 : 1000 were we able to observe isolated PTCD-C10 molecules on the LB monolayer.

The overall decay curves obtained for each sample are shown in Fig. 6a together with decay fits. There is significant fluorescence quenching in the

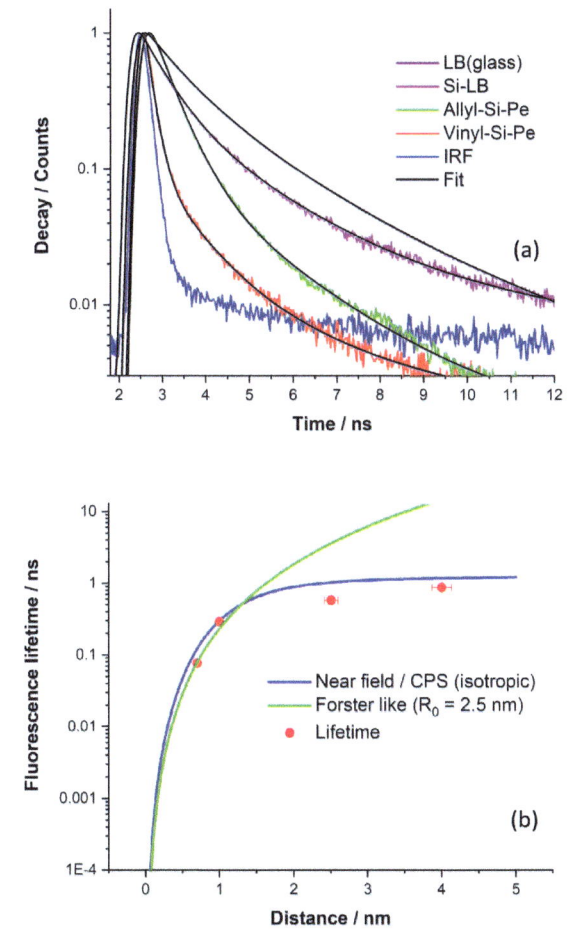

Fig. 6 (a) Normalised fluorescence decay curves for different perylene surfaces. (b) The experimental fluorescence lifetimes of different perylene functionalised surfaces fitted to Chance–Prock–Silbey (CPS, full blue line) and to an inverse cubic (Förster) fit.

lifetimes observed for the functionalised Si(111) surfaces (Si-Vinyl-Pe, and Si-Allyl-Pe) and the LB monolayer (PTCD-C10:SA) samples with respect to the 'unquenched' decay curves. In fact, over 90% quenching is observed for the fluorescence lifetime of Si-Vinyl-Pe, indicating efficient FRET from the perylene molecules to silicon. The fluorescence lifetimes measured for the samples prepared in this study are plotted in Fig. 6b as a function of the emitter–silicon distance in the near field regime (0.5–5 nm). We have modelled the observed fluorescence lifetime quenching with a modified CPS theory with input parameters: the refractive index of the silicon substrate at the maximum emission wavelength ($\lambda = 650$ nm), the refractive index of the capping layer, the fluorescence quantum yield, and the 'true' unquenched lifetime of the emitter. We have used an isotropic transition dipole orientation for the perylene molecules but a vertical and horizontal configuration gives similar model curves for such close distances to the surface of the silicon (Fig. S11, in the ESI†). We can also describe our results with a Förster-like energy transfer between the perylene molecules and the silicon with an inverse cubic dependence giving rise to an estimate Förster radius, $R_0 = 2.5$ nm.

The CPS model adequately describes the observed lifetime data but there is room for improvement. In particular, the fluorescence quantum yield input parameter for the emitter was set to be low ($q \sim 0.1$) in order to achieve a good fit. This might be true because of the presence of aggregates in the PTCD-C10 monolayer and perhaps for the perylene attachment on the silicon as revealed by the FLIM measurements. The other input parameter is the true unquenched lifetime, which for the PTCD-C10 LB monolayer, was estimated from the fluorescence lifetime on glass ($\tau = 1.3$ ns). But for the Si-Vinyl-Pe and Si-Allyl-Pe samples, we expect this to be higher for isolated molecules anchored on the surface of glass, approaching the one observed in solution ($\tau = 3.7$ ns). Despite the limitations of the model, the observed fluorescence quenching is described well by the fitted model in the near field 'true' FRET region which takes place for emitter–silicon distances up to 5 nm. The estimation of the Förster radius for the FRET observed between the perylene molecules and the silicon gives similar values with a previous study.[24] Further work is needed to understand this type of interaction in the near field regime (<5 nm) with not only singlet emitters but also triplet emitters and we will present our recent results in an upcoming publication. This puts constraints on observing 'true' FRET from excited states close to the surface of the silicon since silicon substrates have a native oxide of approximate thickness of 1–2 nm. There are questions that need to be tackled, in particular, the importance of dielectric screening at small dipole–dipole separations and the type of electron–hole transitions (direct or indirect) for this type of interaction.

Conclusions

Silicon photosensitisation *via* FRET from functionalised molecular layers offers an exciting area of research into reducing the thickness of crystalline silicon solar cells by up to two orders of magnitude and enabling the ability to go beyond the single junction efficiency limit. We have presented a modified model based on the CPS theory that describes the interaction of a dipole emitter as a function of distance from the surface of silicon. Interaction at distances between 10–100 nm can be described as a photon tunnelling process that excites the trapped modes of silicon from the evanescent field of the chromophore.[39] In keeping with recent work,[24] we confirm that only for distances less than about 3 nm we observe 'true' FRET phenomena. Vinyl and allyl Si(111) terminated surfaces have been shown to offer good electrical passivation and protection from oxidation while providing the means for further functionalisation *via* a palladium catalysed reaction. LB films can be deposited on the passivated silicon surfaces, acting as light harvesting layers and further boosting the absorption in the silicon but control of chromophore aggregation is required to maximise energy transfer efficiency. Fluorescence lifetime measurements with time-resolved confocal microscopy revealed the distribution of the molecules on the Si(111) surface and we observed energy transfer efficiencies of up to 90% for distances less than 1 nm. A simple model is presented to fit the active FRET area (<5 nm) with an estimated Förster radius $R_0 = 2.5$ nm. There are still questions about the role of the dielectric constant of silicon on this type of interaction, the type of electron–hole transition and the importance of dielectric screening for such small dipole–dipole interactions. Notwithstanding, surface chemistry offers an exciting opportunity for the passivation/functionalization of the silicon interface with molecular layers for applications in electronic devices.

Conflicts of interest

The authors declare no conflicts of interest.

Acknowledgements

H. B. acknowledges financial support from the Department of Chemistry at Lancaster University for a PhD studentship. B. W. acknowledges the financial support provided by the Leverhulme Trust (DS - 2017-036) for a Materials Social Futures (MSF) PhD studentship. L. F. acknowledges financial support from the National Science Foundation of China (No. 61604138). The Centre for Advanced Photovoltaics is supported by the Czech Ministry of Education, Youth and Sport. CZ.02.1.01/0.0/0.0/15_003/0000464.

Notes and references

1 C. Battaglia, A. Cuevas and S. De Wolf, *Energy Environ. Sci.*, 2016, **9**, 1552.
2 J. Jean, P. R. Brown, R. L. Jaffe, T. Buonassisi and V. Bulović, *Energy Environ. Sci.*, 2015, **8**, 1200.
3 D. M. Powell, M. T. Winkler, H. J. Choi, C. B. Simmons, D. B. Needleman and T. Buonassisi, *Energy Environ. Sci.*, 2012, **5**, 5874.
4 T. Markvart, *Prog. Quantum Electron.*, 2000, **24**, 107–186.
5 D. L. Dexter, *J. Lumin.*, 1979, **18–19**, 779.
6 N. Alderman, L. Danos, L. Fang, T. Parel and T. Markvart, *2014 IEEE 40th Photovoltaic Spec. Conf.*, 2014, **2014**, 17.
7 W. Peng, S. M. Rupich, N. Shafiq, Y. N. Gartstein, A. V. Malko and Y. J. Chabal, *Chem. Rev.*, 2015, **115**, 12764.
8 T. Markvart, L. Danos, L. Fang, T. Parel and N. Soleimani, *RSC Adv.*, 2012, **2**, 3173.
9 T. Hayashi, T. G. Castner and R. W. Boyd, *Chem. Phys. Lett.*, 1983, **94**, 461.

10 A. P. Alivisatos, M. F. Arndt, S. Efrima, D. H. Waldeck and C. B. Harris, *J. Chem. Phys.*, 1987, **86**, 6540.
11 G. B. Piland, J. J. Burdett, T.-Y. Hung, P.-H. Chen, C.-F. Lin, T.-L. Chiu, J.-H. Lee and C. J. Bardeen, *Chem. Phys. Lett.*, 2014, **601**, 33.
12 H. M. Nguyen, O. Seitz, D. Aureau, A. Sra, N. Nijem, Y. N. Gartstein, Y. J. Chabal and A. V. Malko, *Appl. Phys. Lett.*, 2011, **98**, 161904.
13 H. M. Nguyen, O. Seitz, W. Peng, Y. N. Gartstein, Y. J. Chabal and A. V. Malko, *ACS Nano*, 2012, **6**, 5574.
14 M. T. Nimmo, L. M. Caillard, W. De Benedetti, H. M. Nguyen, O. Seitz, Y. N. Gartstein, Y. J. Chabal and A. V. Malko, *ACS Nano*, 2013, 7, 3236.
15 M. I. Sluch, A. G. Vitukhnovsky and M. C. Petty, *Phys. Lett. A*, 1995, **200**, 61.
16 L. Danos, R. Greef and T. Markvart, *Thin Solid Films*, 2008, **516**, 7251.
17 L. Danos and T. Markvart, *Chem. Phys. Lett.*, 2010, **490**, 194.
18 L. Fang, N. Alderman, L. Danos and T. Markvart, *Mater. Res. Innovations*, 2014, **18**, 494.
19 S. Huber and G. Calzaferri, *ChemPhysChem*, 2004, **5**, 239–242.
20 H. Kuhn, *J. Chem. Phys.*, 1970, **53**, 101.
21 R. R. Chance, A. Prock and R. Silbey, in *Advances in Chemical Physics*, 1978, vol. XXXVII, pp. 1–64.
22 M. Stavola, D. L. Dexter and R. S. Knox, *Phys. Rev. B: Condens. Matter Mater. Phys.*, 1985, **31**, 2277.
23 W. L. Barnes, *J. Mod. Opt.*, 1998, **45**, 661.
24 N. Alderman, L. Danos, L. Fang, M. C. Grossel and T. Markvart, *Chem. Commun.*, 2017, **53**, 12120.
25 J. Lee, P. Jadhav, P. D. Reusswig, S. R. Yost, N. J. Thompson, D. N. Congreve, E. Hontz, T. Van Voorhis and M. A. Baldo, *Acc. Chem. Res.*, 2013, **46**, 1300.
26 M. B. Smith and J. Michl, *Chem. Rev.*, 2010, **110**, 6891.
27 M. Einzinger, T. Wu, J. F. Kompalla, H. L. Smith, C. F. Perkinson, L. Nienhaus, S. Wieghold, D. N. Congreve, A. Kahn, M. G. Bawendi and M. A. Baldo, *Nature*, 2019, **571**, 90.
28 Y. C. Simon and C. Weder, *J. Mater. Chem.*, 2012, **22**, 20817.
29 J. De Wild, A. Meijerink, J. K. Rath, W. G. J. H. M. Van Sark and R. E. I. Schropp, *Energy Environ. Sci.*, 2011, **4**, 4835.
30 Y. Y. Cheng, T. Khoury, R. G. C. R. Clady, M. J. Y. Tayebjee, N. J. Ekins-Daukes, M. J. Crossley and T. W. Schmidt, *Phys. Chem. Chem. Phys.*, 2010, **12**, 66.
31 J. M. Buriak, *Chem. Rev.*, 2002, **102**, 1271.
32 B. Fabre, *Chem. Rev.*, 2016, **116**, 4808.
33 A. B. Sieval, R. Linke, H. Zuilhof and E. J. R. Sudhölter, *Adv. Mater.*, 2000, **12**, 1457.
34 V. M. Agranovich, Y. N. Gartstein and M. Litinskaya, *Chem. Rev.*, 2011, **111**, 5179.
35 G. S. Higashi, Y. J. Chabal, G. W. Trucks and K. Raghavachari, *Appl. Phys. Lett.*, 1990, **56**, 656.
36 A. Bansal, X. Li, I. Lauermann, N. S. Lewis, S. I. Yi and W. H. Weinberg, *J. Am. Chem. Soc.*, 1996, **118**, 7225.
37 K. H. Drexhage, *J. Lumin.*, 1970, **1–2**, 693.
38 H. Kuhn, D. Mobius and H. Bucher, in *Techniques of Chemistry*, eds. A. Weisberger and B. Rossiter, Wiley, New York, 1978, pp. 577–702.

39 L. Fang, K. S. Kiang, N. P. Alderman, L. Danos and T. Markvart, *Opt. Express*, 2015, **23**, A1528.
40 T. Förster, *Discuss. Faraday Soc.*, 1959, **27**, 7.
41 G. W. Ford and W. H. Weber, *Phys. Rep.*, 1984, **113**, 195.
42 M. E. Bhosale and K. Krishnamoorthy, *Chem. Mater.*, 2015, **27**, 2121.
43 W. Cabri and I. Candiani, *Acc. Chem. Res.*, 1995, **28**, 2–7.
44 H. Mizuno and J. M. Buriak, *ACS Appl. Mater. Interfaces*, 2010, **2**, 2301.
45 P. A. Antunes, C. J. L. Constantino, R. F. Aroca and J. Duff, *Langmuir*, 2001, **17**, 2958.
46 S. R. Puniredd, O. Assad and H. Haick, *J. Am. Chem. Soc.*, 2008, **130**, 13727.
47 N. Alderman, L. Danos, M. C. Grossel and T. Markvart, *RSC Adv.*, 2013, **3**, 20125.
48 A. B. Sieval, C. L. Huisman, A. Schönecker, F. M. Schuurmans, A. S. H. van der Heide, A. Goossens, W. C. Sinke, H. Zuilhof and E. J. R. Sudhölter, *J. Phys. Chem. B*, 2003, **107**, 6846.
49 E. J. Nemanick, P. T. Hurley, B. S. Brunschwig and N. S. Lewis, *J. Phys. Chem. B*, 2006, **110**, 14800.
50 K. E. Plass, X. Liu, B. S. Brunschwig and N. S. Lewis, *Chem. Mater.*, 2008, **20**, 2228.
51 N. Alderman, L. Danos, M. C. Grossel and T. Markvart, *RSC Adv.*, 2012, **2**, 7669.
52 N. Alderman, M. Adib Ibrahim, L. Danos, M. C. Grossel and T. Markvart, *Appl. Phys. Lett.*, 2013, **103**, 81603.
53 M. G. Walter, X. Liu, L. E. O'Leary, B. S. Brunschwig and N. S. Lewis, *J. Phys. Chem. C*, 2013, **117**, 14485.
54 F. D. Heinz, W. Warta and M. C. Schubert, *Appl. Phys. Lett.*, 2017, **110**, 042105.
55 N. Alderman, PhD thesis, University of Southampton, 2013.
56 S. R. Puniredd, O. Assad, T. Stelzner, S. Christiansen and H. Haick, *Langmuir*, 2011, **27**, 4764.
57 S. A. Hussain, P. K. Paul and D. Bhattacharjee, *J. Phys. Chem. Solids*, 2006, **67**, 2542.

Faraday Discussions

PAPER

The next big thing for silicon nanostructures – CO$_2$ photocatalysis

Wei Sun,[†ab] Xiaoliang Yan,[†bc] Chenxi Qian,[†bd] Paul N. Duchesne,[b] Sai Govind Hari Kumar[b] and Geoffrey A. Ozin[*b]

Received 7th October 2019, Accepted 26th November 2019
DOI: 10.1039/c9fd00104b

Silicene is a relatively new member of the growing family of two-dimensional single-element materials. Both top-down and bottom-up approaches provide access to silicene, the former *via* vapor deposition on a substrate and the latter *via* exfoliation of the layered CaSi$_2$ precursor. Most top-down research has been concerned with understanding the various electronic, optical, magnetic, mechanical, electrical, thermal transport and gas-adsorption properties of silicene. By contrast, the focus on bottom-up silicene has primarily been on its synthesis, structure and chemical properties as they relate to its function and utility. Herein, emphasis is placed on the bottom-up strategy because of its scalability and the ease of subsequent silicene modification, with both qualities being important prerequisites for heterogeneous catalysis applications. In this context, synthetic freestanding silicene exists as single sheets or multilayer assemblies, depending on the CaSi$_2$ exfoliation synthesis conditions. The structure of a sheet comprises three connected chair-configuration silicon 6-rings. This connectivity creates buckled sheets in which the hybridization around the unsaturated silicon atoms is sp^2–sp^3. By adjusting the CaSi$_2$ exfoliation synthesis conditions, either layered silane (Si$_6$H$_6$) or siloxene (Si$_6$H$_3$(OH)$_3$) nanosheets can be obtained. In our studies, we have explored the nucleation and growth of different transition metal nanoparticles on and within the layer spaces of these nanosheets, and explored their thermochemical and photochemical reactivity in CO$_2$ hydrogenation reactions. An overview of these findings, related works and a new-and-optimized catalyst are provided in this article.

[a]*State Key Laboratory of Silicon Materials, School of Materials Science and Engineering, Zhejiang University, Hangzhou, Zhejiang 310027, P. R. China*
[b]*Materials Chemistry and Nanochemistry Research Group, Solar Fuels Cluster, Department of Chemistry, University of Toronto, 80 St. George Street, Toronto, ON M5S 3H6, Canada. E-mail: g.ozin@utoronto.ca*
[c]*College of Chemistry and Chemical Engineering, Taiyuan University of Technology, Taiyuan, Shanxi 030024, P. R. China*
[d]*California Institute of Technology, 1200 E California Blvd, Pasadena, CA 91125, USA*
† These authors contributed equally.

Introduction

When one thinks of the wide-ranging technologies based on the bulk and nanoscale forms of the second most earth-abundant element (silicon) in electronics, optics, photonics, energy storage, actuation, sensing, imaging and even medicine, notably absent from the list is any type of heterogeneous catalysis.

In this article we ask: could catalysis be the next "big thing" for nano-silicon? Special focus is placed on solar powered, gas-phase heterogeneous hydrogenation chemistry and the catalytic conversion of carbon dioxide into various chemicals and fuels (Fig. 1).

CO$_2$ reduction to CO on silicon nanostructures

Our earliest demonstration that nanostructured silicon was capable of reducing CO$_2$ to CO was inspired by the ability of molecular silanes to function as reducing agents. This activity is exemplified by the reduction of CO$_2$ to carbon by silane (SiH$_4$) in a fast, highly exothermic reaction (SiH$_4$ + 2CO$_2$ → SiO$_2$ + 2C + 2H$_2$O), as well as the capacity of silanes to catalytically reduce CO$_2$ to methanol or methane.[1-3]

Recent studies have shown that surface hydrides on nanoscale silicon are capable of reducing CO$_2$, under both solution- and gas-phase conditions, with gaseous products in the latter case having been confirmed by ^{13}C-labeling experiments.[4,5] The gas-phase reduction of CO$_2$ to CO using 3 to 5 nm silicon hydride nanocrystals was enabled by light, and exhibited conversion rates of up to hundreds of micromoles per gram of catalyst per hour. The large surface area and broadband light harvesting capability of these nanocrystals, together with the reducing power of their surface SiH sites, played key roles in the stoichiometric conversion of CO$_2$ to CO (Fig. 2a). In an attempt to transform this material from a stoichiometric reagent into a true catalyst, the silicon hydride nanocrystals were

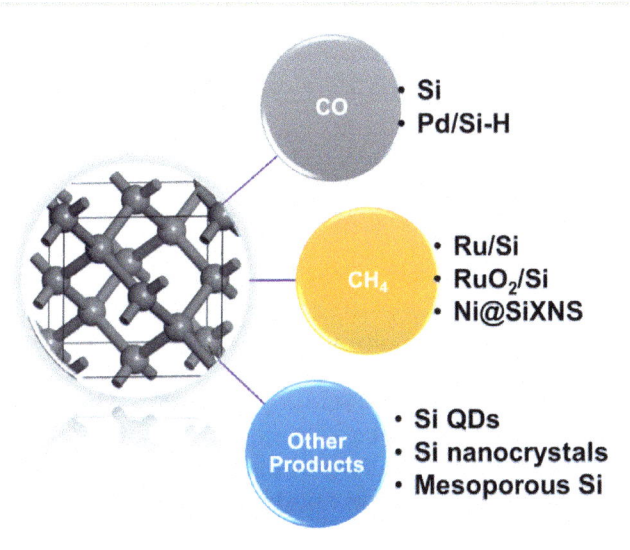

Fig. 1 Silicon nanostructures for the catalytic conversion of CO$_2$ to value-added products.

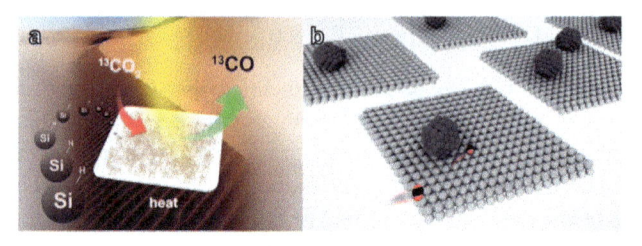

Fig. 2 Illustration of CO_2 hydrogenation strategies using (a) stoichiometric silicon hydride nanocrystals and (b) catalytic nanosheets.

doped with boron and phosphorus, but to no avail; thus remained the challenge of making them catalytic.[6]

This challenge inspired a change of course to explore the reactivity of silicon hydride nanosheets rather than nanocrystals, a breakthrough that enabled the first example of catalytic CO_2 hydrogenation by silicon hydride nanostructures (Fig. 2b). To expand upon this discovery, silicon hydride nanosheets synthesized via $CaSi_2$ exfoliation and decorated with palladium nanoclusters, denoted Pd@SiNS, were found to effectively photocatalyze the reverse water gas shift reaction.[7] These Pd@SiNS exhibited a ^{13}CO production rate of 10 µmol g_{cat}^{-1} h^{-1} that was maintained during repeated catalytic cycles. Insight into the surface chemistry responsible for this photocatalytic cycling was obtained using in situ diffuse reflectance Fourier transform infrared spectroscopy (DRIFTS) coupled with $^{13}CO_2$ labelling studies and supported by density functional theory calculations. As a critical step in facilitating the catalytic CO_2 reduction cycle, Pd nanoclusters were found to dissociate H_2 and, through hydrogen spillover, replenish surface hydrides. In comparison, the palladium-free sample, denoted SiNS, showed a much lower and rapidly decreasing ^{13}CO production rate, consistent with the silicon hydride nanocrystal studies mentioned above.[5] Without the palladium nanoclusters, the surface Si–H hydrides do react stoichiometrically with CO_2 to form SiOH, but have no mechanism for being replenished.[5]

CO_2 methanation on silicon nanosheets

The reduction of CO_2 by H_2 to CH_4 is known as methanation. It was recently discovered that Ru or Ni nanocluster-decorated silicon nanowires (SiNW), silicon inverse opals (i-Si-o) and siloxene nanosheets (SiXNS) function as photocatalysts for CO_2 methanation[8–11] (Fig. 3). These silicon architectures are broadband absorbers of solar radiation that, through the photothermal effect, generate high local temperatures and facilitate the hydrogenation of CO_2.

Expounding upon these results, Ru/SiNW exhibits greater than 97% absorbance over the solar spectral region (i.e., between ca. 200 and 1000 nm) and exhibits superior catalytic activity compared to control samples of Ru on planar silicon or silica.[8] Similarly, Ru on a silicon inverse opal (Ru/i-Si-o) displayed a quantum efficiency three times higher than that of Ru on a silica opal (Ru/SiO_2).[10] Moreover, RuO_2 on an inverse opal (RuO_2/i-Si-o) produced an impressive light-assisted CH_4 formation rate of 4.4 mmol g_{cat}^{-1} h^{-1} under ambient conditions.[9]

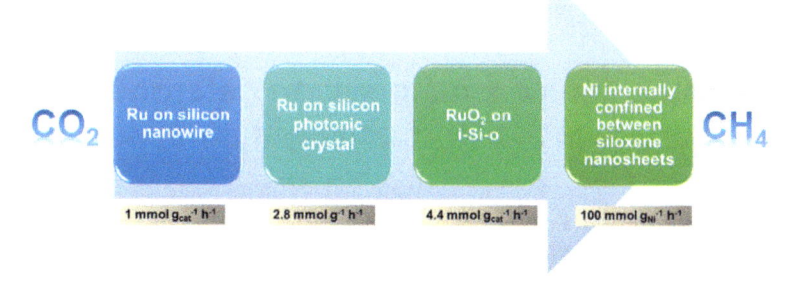

Fig. 3 Progression of conversion rates for Ru/SiNW, Ru/i-Si-o, RuO₂/i-Si-o, and Ni/SiXNS photomethanation catalysts.

Inspired by these early successes, earth-abundant and low-cost Ni nanoclusters on siloxene nanosheets (SiXNS) were explored as catalysts for CO_2 photomethanation. It was discovered that the solvent used in the synthesis of Ni/SiXNS (*i.e.*, ethanol or water) determined whether Ni nanoclusters were formed within or on the external surface of the nanosheets.[11] In water, hydrolytic condensation of Si–H induced crosslinking of the silicon hydride nanosheets that, in the presence of the nickel nitrate precursor, caused the Ni nanoclusters to preferably form on the exterior surface; in ethanol, however, this cross-linking did not occur and the Ni nanoclusters preferably formed between the nanosheets. These two forms of Ni/SiXNS showed distinct catalytic behaviors and reaction pathways. The Ni on the external surface of SiXNS had a low CH_4 selectivity of 54.3% and a light-assisted CH_4 production rate of 41.0 mmol g_{Ni}^{-1} h^{-1}. In contrast, internally confined Ni displayed high CH_4 selectivity (above 90%) and a CH_4 production rate of 52.9 mmol g_{Ni}^{-1} h^{-1} that increased to 100 mmol g_{Ni}^{-1} h^{-1} after 12 h of continuous testing. *In situ* DRIFTS studies further demonstrated that externally confined Ni favored a dissociative reaction pathway, in which chemisorbed *CO_2 dissociates into *CO and *O, then from *CO to *C, which is then hydrogenated to form CH_4. Evidence suggested that this methanation pathway was favored by terminally bonded *CO, whereas bridge-bonded *CO favored the reverse water gas shift pathway. In contrast, the internally confined Ni exhibited an associative CO_2 reaction pathway involving a formate intermediate instead of bridging *CO.

Beyond reverse water gas shift and methanation

In addition to CO and CH_4, other products have been observed from CO_2 reduction on nanostructured forms of silicon. For example, silicon quantum dots of 1 to 2 nm in diameter photocatalytically reduced CO_2 in a Na_2CO_3 solution to formaldehyde and formic acid.[12] In another study, pristine silicon nanocrystals synthesized *via* ball milling also produced formaldehyde under similar conditions, whereas their alkyl-capped analogues were inactive.[13] Under a pressure of 10 bar and 150 °C in toluene, hydride-terminated silicon nanocrystals produced surface acetal, a precursor to formaldehyde.[4] It was also found that hydride-terminated mesoporous Si could trap and reduce CO_2 to produce methanol.[14,15]

In all cases, surface oxidation significantly limited the reactivity and stability of nanostructured silicon materials.

Rational design of a nano silicon heterogeneous CO₂ hydrogenation photocatalyst

Thus far, research into enabling catalytic activity in nanostructured forms of silicon have provided a few key insights:

(1) Employing a H_2 spillover co-catalyst in combination with nanostructured silicon hydrides can impart catalytic activity to a nominally stoichiometric CO_2 hydrogenation reaction;

(2) Internal (*versus* external) confinement of co-catalysts in silicon nanosheets influences both the selectivity and stability of the catalyst's CO_2 hydrogenation performance; and

(3) Silicon hydride nanostructures capable of strongly absorbing broadband optical radiation can enable photothermal CO_2 catalysis, initiating reaction pathways that are both photochemical (*via* electron–hole separation) and thermochemical (*via* electron–hole recombination).

To explore these ideas analytically, the well-defined Pd–SiNS composite system can serve as a valuable archetype. Specifically, in this work a uniform 2 wt% coating of Pd nanoclusters on SiNS was prepared *via* the spontaneous reduction of a Pd precursor by the Si–H groups on the SiNS surface as previously described.[7] A thermal post-treatment step was used to encapsulate and stabilize the Pd nanoclusters between cross-linked nanosheets, as described previously for Ni/SiXNS[11] (Fig. 4a). The reverse water gas shift reaction of the resulting material was then tested in a flow reactor at 200 °C, using light and a 3 : 1 mixture of H_2/CO_2. Unsurprisingly, the untreated sample exhibited a relatively low CO production rate of 10 μmol g_{cat}^{-1} h^{-1}, consistent with earlier observations.[7] Under identical conditions, however, the post-treated sample had a significantly enhanced CO production rate of around 80 μmol g_{cat}^{-1} h^{-1}, which gradually increased over 40 h of continuous testing to reach 150 μmol g_{cat}^{-1} h^{-1} without reaching a visible plateau (Fig. 4b).

The gradually increasing CO production rate may correspond with the "slow release" concept first described for the Ni/SiXNS system, whereby surface-protected Pd nanoclusters are gradually exposed to the H_2 reactant as it

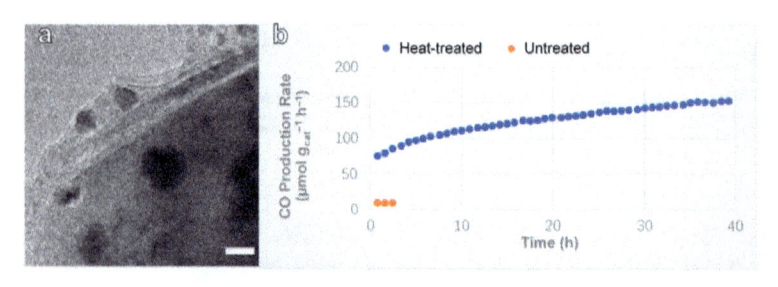

Fig. 4 (a) TEM image showing the stacking of Si nanosheets and the higher-contrast contrast Pd nanoclusters (scale bar = 20 nm). (b) Flow reactor test results showing the CO production rate for the heat-treated and untreated Pd/SiNS samples over a 40 h period.

diffuses between the nanosheets, while the sintering and deactivation of Pd nanoclusters is largely impeded.[16] This apparent enhancement of the reaction rate is particularly surprising given that, below 300 °C, the reactivity of traditional Pd/SiO$_2$ catalysts is negligible.[17]

In situ DRIFTS provided useful insight into this phenomenon (Fig. 5). Following an initial reduction step in H$_2$, the differential spectrum revealed an apparent loss of Si–O–Si and Si–O signals, as well as the emergence of OH, indicating that close contact between Pd and Si had successfully enabled reductive spillover of any oxidized Si sites, consistent with our previous observation.[7] When the CO$_2$ hydrogenation reaction was driven by both heat and light, diagnostic vibrational modes of gaseous CO appeared, which signaled reactivity promoted by the photothermal effect.[18] The lost oxidized Si species as negative peaks were also more pronounced and several adsorbed C-containing species became observable, including *HCO, *CH, and different forms of *CO, which are typically seen in reverse water gas shift reactions. No linearly adsorbed CO at 2054 cm^{-1} was observable, in contrast with the strong signals of gaseous CO and multiply-coordinated CO. Upon purging with He the gaseous CO totally disappeared, and only multiply-coordinated CO and a trace amount of *HCO remained. This behavior is distinct from the linearly adsorbed CO on Pd/SiO$_2$ catalysts[19] and likely originates from the encapsulated nature of the Pd nanoparticles in heat-treated SiNS.[20] It is also likely that the photothermal effect could raise the local temperature of the encapsulated Pd nanoclusters, thereby enhancing catalytic performance.[18]

The reverse water gas shift catalytic cycle on Pd/SiNS is believed to occur at the interface between the Pd and SiNS as summarized in Fig. 6. In brief, CO$_2$ reacts with surface Si–H to form Si–OH and Si–O–Si groups according to the equations:

$$Si–H + CO_2 \rightarrow Si–OH + CO$$

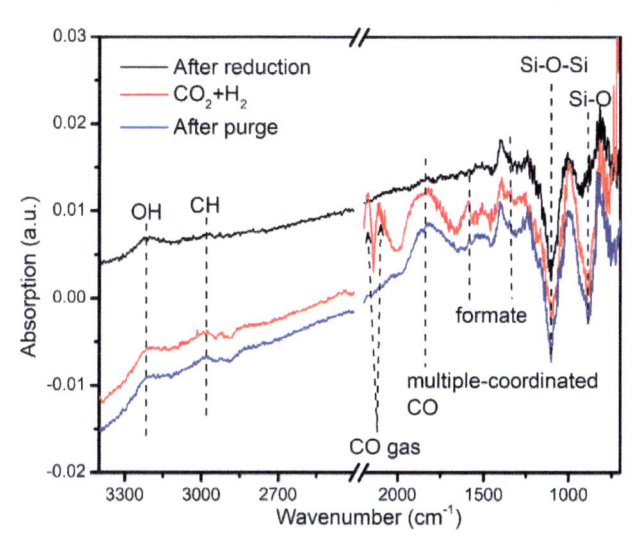

Fig. 5 *In situ* DRIFTS spectra of heat-treated 2 wt% Pd–SiNS, recorded in differential mode.

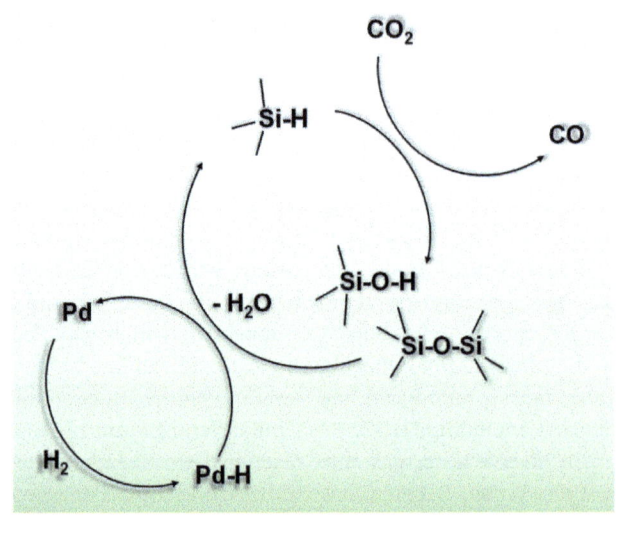

Fig. 6 Catalytic cycle for the RWGS reaction on Pd/SiNS. Originally published in ref. 7 and modified with permission.

$$Si–H + CO_2 \rightarrow 1/2Si–O–Si + 1/2H_2O + CO$$

The formation of bridging siloxane Si–O–Si induces strain into SiNS, which increases the reactivity of Si–OH and Si–O–Si towards H-atoms formed by H_2 spillover on the Pd nanoclusters according to the reactions:

$$Si–OH + H_2 \rightarrow Si–H + H_2O$$

$$Si–O–Si + 2H_2 \rightarrow 2Si–H + H_2O$$

This forms H_2O and reinstates Si–H, thereby completing the catalytic cycle, Fig. 6.

Conclusions

This brief overview of heterogeneous CO_2 photocatalysis using nanostructured forms of hydride-terminated silicon could signal the next "big thing" for this earth-abundant, low-cost, environmentally-friendly element. The design of these silicon nanostructures has allowed the implementation of several strategies that are vital for sustainable and effective catalysis: combining them with metal, spatially confining nanoparticles and employing the photothermal effect.

The ability to transform silicon hydride nanostructures from stoichiometric CO_2 hydrogenation agents into catalysts driven by the light and heat of the sun, and to simultaneously influence the dominant reaction pathway through structure design represents an advance capable of elevating silicon catalysis from a scientific curiosity to a relevant technology. By employing all the above strategies, we have demonstrated the use of a silicene-derived nano-silicon structure that accomplished highly active and stable conversion of CO_2 into CO with light illumination. In the quest to industrialize both the capture of CO_2 and its subsequent conversion into value-added chemicals and fuels, this represents a momentous step towards a sustainable future.

Experimental

Information regarding the flow reactor, reagents, and the *in situ* DRIFTS setup have been previously described.[7] The Pd–SiNS were first precipitated from their acetone dispersion using gravity and then dried under vacuum to form a solid powder. Subsequent calcination overnight at 300 °C in air ensured that the Pd nanoparticles could be encapsulated by crosslinked, oxidized Si layers. For the flow reactor test, H_2 and CO_2 were flowed in a 3 : 1 ratio with a total volumetric flow rate of 1 sccm. During the test, the reactor was irradiated using a 300 Xe lamp with a light intensity of 40 suns, and the temperature was kept constant at 200 °C using an automated temperature controller. For *in situ* DRIFTS data collection, the sample was purged with He (20 sccm) at room temperature for 10 min and then again at 300 °C for another 50 min. The background spectrum was taken at this point, and so the spectra collected afterward were differential spectra. The sample was then treated in H_2 (9 sccm) at 300 °C and simultaneously illuminated by a 120 W Xe lamp for 1 h. The reactant gases H_2 and CO_2 (3 : 1 ratio) were subsequently introduced into the reactor under the same conditions for 1 h. Finally, the sample was purged with He for 20 min. TEM images of the annealed Pd–SiNS sample were obtained using a Hitachi HF-3300 environmental transmission electron microscope at an operating voltage of 300 kV.

Conflicts of interest

There are no conflicts to declare.

Acknowledgements

G. A. O. acknowledges the financial support of the Ontario Ministry of Research and Innovation (MRI), the Ministry of Economic Development, Employment and Infrastructure (MEDI), the Ministry of the Environment and Climate Change's (MOECC) Best in Science (BIS) Award, the Ontario Center of Excellence Solutions 2030 Challenge Fund, the Ministry of Research Innovation and Science (MRIS) Low Carbon Innovation Fund (LCIF), Imperial Oil, the University of Toronto's Connaught Innovation Fund (CIF), the Connaught Global Challenge (CGC) Fund, and the Natural Sciences and Engineering Research Council of Canada (NSERC). W. S. acknowledges the financial support of the ZJU100 Young Professor Program from Zhejiang University and the National Natural Science Foundation of China (Grant No. 51902287). X. L. Y. acknowledges the National Natural Science Foundation of China (Grant No. 21878203), Shanxi International Cooperation Project (Grant No. 201703D421037), and Natural Science Foundation of Shanxi Province (Grant No. 201801D121061). C. Q. and P. N. D. acknowledge the support of the Natural Sciences and Engineering Research Council of Canada (NSERC) for postdoctoral scholarships. The authors wish to thank Dr Abdinoor A. Jelle for his help in performing TEM characterization.

References

1. T. Matsuo and H. Kawaguchi, *J. Am. Chem. Soc.*, 2006, **128**, 12362–12363.
2. A. Berkefeld, W. E. Piers and M. Parvez, *J. Am. Chem. Soc.*, 2010, **132**, 10660–10661.
3. S. N. Riduan, J. Y. Ying and Y. Zhang, *ChemCatChem*, 2013, **5**, 1490–1496.
4. M. Dasog, G. B. De los Reyes, L. V. Titova, F. A. Hegmann and J. G. C. Veinot, *ACS Nano*, 2014, **8**, 9636–9648.
5. W. Sun, C. Qian, L. He, K. K. Ghuman, A. P. Y. Wong, J. Jia, A. A. Jelle, P. G. O'Brien, L. M. Reyes, T. E. Wood, A. S. Helmy, C. A. Mims, C. V. Singh and G. A. Ozin, *Nat. Commun.*, 2016, **7**, 12553.
6. A. P. Y. Wong, W. Sun, C. Qian, A. A. Jelle, J. Jia, Z. Zheng, Y. Dong and G. A. Ozin, *Adv. Sustainable Syst.*, 2017, **1**, 1700118.
7. C. Qian, W. Sun, D. L. H. Hung, C. Qiu, M. Makaremi, S. G. Hari Kumar, L. Wan, M. Ghoussoub, T. E. Wood, M. Xia, A. A. Tountas, Y. F. Li, L. Wang, Y. Dong, I. Gourevich, C. V. Singh and G. A. Ozin, *Nat. Catal.*, 2019, **2**, 46–54.
8. P. G. O'Brien, A. Sandhel, T. E. Wood, A. A. Jelle, L. B. Hoch, D. D. Perovic, C. A. Mims and G. A. Ozin, *Adv. Sci.*, 2014, **1**, 1400001.
9. A. A. Jelle, K. K. Ghuman, P. G. O'Brien, M. Hmadeh, A. Sandhel, D. D. Perovic, C. V. Singh, C. A. Mims and G. A. Ozin, *Adv. Energy Mater.*, 2018, **8**, 1870041.
10. P. G. O'Brien, K. K. Ghuman, A. A. Jelle, A. Sandhel, T. E. Wood, J. Y. Y. Loh, J. Jia, D. Perovic, C. V. Singh, N. P. Kherani, C. A. Mims and G. A. Ozin, *Energy Environ. Sci.*, 2018, **11**, 3443–3451.
11. X. Yan, W. Sun, L. Fan, P. N. Duchesne, W. Wang, C. Kübel, D. Wang, S. G. H. Kumar, Y. F. Li, A. Tavasoli, T. E. Wood, D. L. H. Hung, L. Wan, L. Wang, R. Song, J. Guo, I. Gourevich, A. A. Jelle, J. Lu, R. Li, B. D. Hatton and G. A. Ozin, *Nat. Commun.*, 2019, **10**, 2608.
12. Z. Kang, C. H. A. Tsang, N.-B. Wong, Z. Zhang and S.-T. Lee, *J. Am. Chem. Soc.*, 2007, **129**, 12090–12091.
13. F. Peng, J. Wang, G. Ge, T. He, L. Cao, Y. He, H. Ma and S. Sun, *Mater. Lett.*, 2013, **92**, 65–67.
14. M. Dasog, S. Kraus, R. Sinelnikov, J. G. C. Veinot and B. Rieger, *Chem. Commun.*, 2017, **53**, 3114–3117.
15. S. A. Martell, Y. Lai, E. Traver, J. MacInnis, D. D. Richards, S. MacQuarrie and M. Dasog, *ACS Appl. Nano Mater.*, 2019, **2**, 5713–5719.
16. E. D. Goodman, A. C. Johnston-Peck, E. M. Dietze, C. J. Wrasman, A. S. Hoffman, F. Abild-Pedersen, S. R. Bare, P. N. Plessow and M. Cargnello, *Nat. Catal.*, 2019, **2**, 748–755.
17. J. Ye, Q. Ge and C.-j. Liu, *Chem. Eng. Sci.*, 2015, **135**, 193–201.
18. J. Jia, H. Wang, Z. Lu, P. G. O'Brien, M. Ghoussoub, P. Duchesne, Z. Zheng, P. Li, Q. Qiao, L. Wang, A. Gu, A. A. Jelle, Y. Dong, Q. Wang, K. K. Ghuman, T. Wood, C. Qian, Y. Shao, C. Qiu, M. Ye, Y. Zhu, Z.-H. Lu, P. Zhang, A. S. Helmy, C. V. Singh, N. P. Kherani, D. D. Perovic and G. A. Ozin, *Adv. Sci.*, 2017, **4**, 1700252.
19. X. Jiang, X. Wang, X. Nie, N. Koizumi, X. Guo and C. Song, *Catal. Today*, 2018, **316**, 62–70.
20. L. Wang, L. Wang, X. Meng and F. S. Xiao, *Adv. Mater.*, 2019, **31**, 1901905.

Faraday Discussions

DISCUSSIONS

Silicon nanostructures for energy conversion and devices: general discussion

Pavel Galář, Ankit Goyal, Ali Reza Kamali, Davide Mariotti and Wei Sun

DOI: 10.1039/D0FD90007A

Ali Reza Kamali opened discussion of the paper by Davide Mariotti: In the Experimental section, it is mentioned that "The plasma setup is accessorized with a two-axis stage and it is possible to directly deposit Si QDs on a substrate and form homogeneous films". What was the substrate used in these experiments?

Davide Mariotti replied: Quartz for UV-Vis, ITO-coated glass for XPS, UPS and KP, and ITO-coated glass + TiO_2 film in the cells. These were described in the Experimental section of the paper.

Ali Reza Kamali asked: It is mentioned that the laser decomposition of silane at high concentrations led to the production of amorphous particles, while low silane concentrations could lead to the preparation of crystalline particles, and this is attributed to the temperature. Please can you provide an explanation of how the temperature was calculated/measured during the synthesis?

Davide Mariotti responded: Our synthesis process does not involve lasers. The measurement of relevant temperatures is described in the supplementary information, see sections S5 and S6. The gas temperature is determined by optical emission spectroscopy and fitting the emission of the OH band to synthetic spectra. The nanoparticle temperature is then derived from a theoretical model that uses as input various experimental parameters. The full description is provided in sections S5 and S6.

Ali Reza Kamali enquired: Fig. 1a shows an amorphous nanoparticle. How was it confirmed that the nanoparticle is silicon?

Davide Mariotti answered: Fig. 1a shows a crystalline nanoparticle and in this case the lattice spacing was verified to correspond to silicon, this was discussed in the paper just above the figure.

Fig. 1b reports an amorphous nanoparticle and in this case there was no chemical information we could extract from the image of this specific particle. However we have carried out EDX analysis of a highly populated area of particles *vs.* an empty one. Furthermore, the macroscopic samples were analysed by XPS and FTIR which all reported consistent results.

Ankit Goyal asked: The study presented here is a nice controlled experiment though I couldn't find the reason how the higher precursor gas concentration would reduce crystallization? Could you please comment on it? Crystallization by the thermal process is a function of temperature and time so please tell whether you have varied the time too or if not then why?

Davide Mariotti replied: The discussion of the crystallization is presented in Fig. 3 of the manuscript and the corresponding section 3.2. The higher precursor concentration essentially changes the plasma conditions, which in turn change the temperature of the QDs in the plasma. This is depicted in Fig. 3 where at 200 ppm silane, the QDs temperature reaches only 800 K, which is below their crystallization temperature as depicted in the figure by the grey dashed line/white region. At low silane concentrations (50 ppm), the plasma conditions allow for the QDs to reach higher temperature (>900 K) as depicted in Fig. 3, which is higher than the crystallization temperature. In part, also the size of the QDs also contributes to place these within the respective temperature regimes as shown in Fig. 3.

The time that the QDs are exposed to any given temperature is determined by the QDs residence time in the plasma, which is determined by the total flow of the process (1000 sccm). This has not been changed and therefore we expect all the QDs to have the same residence time within the plasma.

We have not changed the total flow, so that we could study the effect of silane concentration without changing other parameters. Changing the total flow would again change the plasma conditions and therefore add another parameter to evaluate.

Pavel Galář added: Based on the FTIR results, the authors claim that there are higher concentrations of hydrogen and oxygen in the Si QDs in amorphous samples in contrast to the crystalline one. What should be then the surface termination of the small crystalline Si QDs? Taking into account the expected nucleation process (chapter 2.5 in ref. 1), the fact that the samples are prepared in argon plasma and the chamber is in a nitrogen protection atmosphere (Fig. S1), one would expect compact hydrogen related termination of the small crystalline QDs.

1 U. R. Kortshagen, R. M. Sankaran, R. N. Pereira, S. L. Girshick, J. J. Wu and E. S. Aydil, *Chem. Rev.*, 2016, **116**, 11061–11127, DOI: 10.1021/acs.chemrev.6b00039.

Davide Mariotti responded: Yes we agree that we expect hydrogen-terminations for the crystalline QDs and this is consistent with our results, although, as discussed, some degree of oxidation is observed.

We infer that hydrogen is present in higher concentrations in the amorphous QDs compared to the crystalline ones. This is supported by our FTIR and REELS analysis as well as corroborated by the shift towards the conduction band of the Fermi level in the amorphous QDs. We do not make any specific comment in this respect with regards to oxygen.

We should note that chapter 2.5 of the *Chemical Reviews* paper quoted is mainly written on the basis of low-pressure plasma processes.[1] For atmospheric pressure plasmas, nucleation can be somewhat different, however there is still very limited understanding and experimental evidence that relate to nucleation and growth of nanoparticles in atmospheric pressure plasmas.

1 U. R. Kortshagen, R. M. Sankaran, R. N. Pereira, S. L. Girshick, J. J. Wu and E. S. Aydil, *Chem. Rev.*, 2016, **116**, 11061–11127, DOI: 10.1021/acs.chemrev.6b00039.

Ali Reza Kamali opened discussion of the paper by Wei Sun: The article highlights the application of nanostructured hydride-terminated silicon in heterogeneous CO_2 photocatalysis, as a possible big application for silicon nanostructures, which is an interesting topic from a science point of view. However, considering the cost and environmental impacts related to the preparation of such nanostructured Si, how do the authors evaluate the opportunity of this approach for large-scale conversion of CO_2?

Wei Sun replied: Thank you Prof. Kamali for your attention and question. Cost and environmental impact are indeed important factors for potential large-scale applications. We notice that you are the right person to consult with on this point. Traditional production of silicon materials usually needs very high temperatures above the material's melting point, which is very energy-consuming and which emits a vast amount of the greenhouse gas CO_2. However, recent advances like yours (ultra-fast shock-wave combustion synthesis) has brought down the temperature and processing time, as well as achieving production of silicon nanostructures from cheap sources like beach sand or biomass. These have brought more promise to not only the catalysis application, but also other energy-related fields, *e.g.* Li-ion batteries, in which you have much expertise.

We also value the idea to bring down the temperature and negative environmental impacts by using alternative precursors for the synthesis. For instance, we employed commercial SiO (around $1 per gram) as the precursor to replace the traditional $HSiO_{1.5}$ solution in methyl isobutyl ketone (MIBK) for the thermal disproportionation reaction to produce silicon nanocrystals. As for silicon nanosheets, we used $CaSi_2$ as the precursor which reacts with HCl (instead of HF) to obtain H-terminated nanostructures.

I also want to bring to your notice that in this same *Faraday Discussions* volume Prof. Mita Dasog has provided useful information on fuel production from mesoporous silicon (DOI: 10.1039/c9fd00098d). In summary, the production of silicon and nanosilicon has already been carried out at large-scale with fairly low cost compared with many other nanomaterials. Although it must be worthwhile and scientifically important to develop even more economic and feasible synthetic solutions, the earth-abundant and non-toxic Si should shine bright like a diamond (it does have diamond structure, and can reduce CO_2, which can also be converted to diamond as you demonstrated) for catalysis, just like it did in Li-ion batteries and photovoltaic applications.

Conflicts of interest

There are no conflicts to declare.